Climate change and agriculture

It is widely recognised that agriculture is a significant contributor to global warming and climate change. Agriculture needs to reduce its environmental impact and adapt to current climate change whilst still feeding a growing population, i.e. become more 'climate-smart'. Burleigh Dodds Science Publishing is playing its part in achieving this by bringing together key research on making the production of the world's most important crops and livestock products more sustainable.

Based on extensive research, our publications specifically target the challenge of climate-smart agriculture. In this way we are using 'smart publishing' to help achieve climate-smart agriculture.

Burleigh Dodds Science Publishing is an independent and innovative publisher delivering high quality customer-focused agricultural science content in both print and online formats for the academic and research communities. Our aim is to build a foundation of knowledge on which researchers can build to meet the challenge of climate-smart agriculture.

For more information about Burleigh Dodds Science Publishing simply call us on +44 (0) 1223 839365, email info@bdspublishing.com or alternatively please visit our website at www.bdspublishing.com.

Related titles:

Achieving carbon-negative bioenergy systems from plant materials
Print (ISBN 978-1-78676-252-8); Online (ISBN 978-1-78676-254-2, 978-1-78676-255-9)

Achieving sustainable management of boreal and temperate forests
Print (ISBN 978-1-78676-292-4); Online (ISBN 978-1-78676-294-8, 978-1-78676-295-5)

Assessing the environmental impact of agriculture
Print (ISBN 978-1-78676-228-3); Online (ISBN 978-1-78676-228-3, 978-1-78676-231-3)

Achieving sustainable urban agriculture
Print (ISBN 978-1-78676-316-7); Online (ISBN 978-1-78676-318-1, 978-1-78676-319-8)

Chapters are available individually from our online bookshop:
https://shop.bdspublishing.com

climate•SMART•publishing
IN AGRICULTURAL SCIENCE

BURLEIGH DODDS SERIES IN AGRICULTURAL SCIENCE
NUMBER 78

Climate change and agriculture

Edited by Dr Delphine Deryng, NewClimate Institute/Integrative Research Institute on Transformations of Human-Environment Systems (IRI THESys), Humboldt-Universität zu Berlin, Germany

burleigh dodds
SCIENCE PUBLISHING

Published by Burleigh Dodds Science Publishing Limited
82 High Street, Sawston, Cambridge CB22 3HJ, UK
www.bdspublishing.com

Burleigh Dodds Science Publishing, 1518 Walnut Street, Suite 900, Philadelphia, PA 19102-3406, USA

First published 2020 by Burleigh Dodds Science Publishing Limited

Library of Congress Control Number: 2020934455

British Library Cataloguing in Publication Data
A catalogue record for this book is available from the British Library

ISBN 978-1-78676-320-4 (Print)
ISBN 978-1-78676-323-5 (PDF)
ISBN 978-1-78676-322-8 (ePub)
ISSN 2059-6936 (print)
ISSN 2059-6944 (online)

DOI 10.19103/AS.2020.0064

Typeset by Deanta Global Publishing Services, Dublin, Ireland

Contents

Series list

Title	Series number
Achieving sustainable cultivation of maize - Vol 1 From improved varieties to local applications Edited by: Dr Dave Watson, CGIAR Maize Research Program Manager, CIMMYT, Mexico	001
Achieving sustainable cultivation of maize - Vol 2 Cultivation techniques, pest and disease control Edited by: Dr Dave Watson, CGIAR Maize Research Program Manager, CIMMYT, Mexico	002
Achieving sustainable cultivation of rice - Vol 1 Breeding for higher yield and quality Edited by: Prof. Takuji Sasaki, Tokyo University of Agriculture, Japan	003
Achieving sustainable cultivation of rice - Vol 2 Cultivation, pest and disease management Edited by: Prof. Takuji Sasaki, Tokyo University of Agriculture, Japan	004
Achieving sustainable cultivation of wheat - Vol 1 Breeding, quality traits, pests and diseases Edited by: Prof. Peter Langridge, The University of Adelaide, Australia	005
Achieving sustainable cultivation of wheat - Vol 2 Cultivation techniques Edited by: Prof. Peter Langridge, The University of Adelaide, Australia	006
Achieving sustainable cultivation of tomatoes Edited by: Dr Autar Mattoo, USDA-ARS, USA & Prof. Avtar Handa, Purdue University, USA	007
Achieving sustainable production of milk - Vol 1 Milk composition, genetics and breeding Edited by: Dr Nico van Belzen, International Dairy Federation (IDF), Belgium	008
Achieving sustainable production of milk - Vol 2 Safety, quality and sustainability Edited by: Dr Nico van Belzen, International Dairy Federation (IDF), Belgium	009
Achieving sustainable production of milk - Vol 3 Dairy herd management and welfare Edited by: Prof. John Webster, University of Bristol, UK	010
Ensuring safety and quality in the production of beef - Vol 1 Safety Edited by: Prof. Gary Acuff, Texas A&M University, USA & Prof. James Dickson, Iowa State University, USA	011
Ensuring safety and quality in the production of beef - Vol 2 Quality Edited by: Prof. Michael Dikeman, Kansas State University, USA	012
Achieving sustainable production of poultry meat - Vol 1 Safety, quality and sustainability Edited by: Prof. Steven C. Ricke, University of Arkansas, USA	013
Achieving sustainable production of poultry meat - Vol 2 Breeding and nutrition Edited by: Prof. Todd Applegate, University of Georgia, USA	014

Introduction

It has been suggested that agriculture may account for up to 24% of the greenhouse gas emissions (GHGs) contributing to climate change. Agricultural production accounts for over 80% of food system emissions, and nearly 60% of global non-CO_2 GHGs. At the same time climate change is threatening to disrupt agricultural production. This collection reviews key research addressing this challenge. Part 1 of the book reviews current research on the impacts of climate change on agriculture, such as the effects of increased temperatures, as well as the ways these impacts can be modelled. Part 2 assesses what we know about the contribution of agriculture to climate change, including the impacts of both crop and livestock production as well as land use. Part 3 surveys mitigation strategies to achieve a more 'climate-smart' agriculture such as the role of integrated crop-livestock and agroforestry systems.

Part 1 Impacts of climate change on agriculture

Atmospheric changes, including high carbon dioxide (CO_2) concentrations, air temperature and tropospheric ozone concentrations , are challenging our crop cultivation systems, with potential negative implications for global food production. Chapter 1 investigates how these atmospheric changes directly affect different physiological processes, and further modify agro-ecosystem functioning through complex interactions with other environmental factors, including nutrient or water availability. The chapter reviews current research on the impact of elevated CO_2 concentrations, higher temperatures and high ozone concentrations on crop growth and yields. This research shows, for example, that elevated CO_2 levels may have a fertilizing and positive impact on crop yield and water use in some cases, whereas heat stress and increased ozone concentrations may reduce crop yields.

As Chapter 2 indicates, the high atmospheric concentrations of GHGs such as CO_2, nitrous oxide (N_2O) and methane (CH_4) have had significant negative impacts on soil processes. The impacts range from a decline in carbon sequestration and soil health to increased soil temperatures, decomposition of soil organic matter, release and leaching of nutrients, increased microbial activity, salinization, alkalization and moisture stress. The chapter highlights current knowledge on the effects of climate change on soil properties and processes, and a set of climate-smart technologies for climate change mitigation. Climate-smart technologies include conservation agriculture, precision agriculture, integrated nutrient management, residue retention, soil and water conservation, agroforestry, controlled grazing and stocking rates, crop rotations, cover crops, biochar and improved plant varieties.

Crop models are powerful tools to explore climate change impacts on crop productivity and the effect of different agricultural management practices. Chapter 3 reviews how crop models take into consideration the effects of biotic factors such as climate variables, soils and crop genetics, and how they are used in climate change impact assessments. The chapter emphasizes how the wide use of crop models raises challenges that need to be addressed in future research, highlighting the need for model improvements to consider extreme weather events and stresses related to low input agricultural systems.

Part 2 The contribution of agriculture to climate change

Despite gains in productivity often accompanied by reduced emissions, the livestock sector remains the largest anthropogenic emitter of CH_4 and has significantly reduced global carbon storage and photosynthetic capacities as well as releasing nitrogen and phosphorus to air, water and/or soil. Chapter 4 explores some of the many facets of livestock's contributions to climate change and the difficulties involved in quantifying them. It provides a closer look into the contribution of livestock CH_4 emissions to changing atmospheric CH_4 concentrations over the last few decades. The chapter discusses the production of livestock CH_4, global atmospheric concentrations and the global CH_4 cycle, ways of quantifying enteric fermentation and emissions from manure. It includes a case study to show measurement issues in practice.

While gains in agricultural productivity have enabled rising levels of food security, they have also made the food system a major contributor to global climate change. Chapter 5 reviews the contribution to climate change of various GHGs from global crop cultivation. Topics include fertilizer management, the effect of land use changes on peatlands, crop residue management, cropland soil organic matter changes and bioenergy production. In each case the chapter reviews the current state of knowledge and data availability as well as the outstanding uncertainties in current estimates. The chapter also considers cropland GHG emissions in the context of food security and other sustainable development goals. Finally, the chapter provides an overview of how future scenarios of agricultural development can be used to project local and regional GHG emissions within integrated assessment frameworks.

Building on the previous chapter, Chapter 6 discusses the role of agricultural expansion and the effects of land cover and land-use change in contributing to climate change. It begins by reviewing the impacts of land-use change on climate, specifically focusing on carbon emission, surface energy fluxes - such as reflection of solar radiation and evapotranspiration and hydrological impacts - and the emission of reactive gases from vegetation. It assesses current techniques to estimate the impacts of land-use change on climate. It also reviews the role of the land sector in climate change mitigation, highlighting how the reduction of deforestation, the increase of reforestation, restoration and afforestation programmes and the growth of bioenergy crops have the potential help to mitigate and adapt to climate change. Finally, the chapter reviews future land-use trajectories and their use in assessing future potential climate change as a result of different levels of GHGs emissions. It concludes by providing future research trends and resources for further information.

Chapter 7 explores how the accurate assessment of agricultural GHGs emissions for accounting and mitigation options is still a key concern. While an extensive body of data is available, limitations of the different measuring approaches have often been ignored. Despite some constraints, chamber-based approaches have dominated annual assessments of GHGs to provide information on spatial and temporal variations. The Eddy Covariance (EC) technique has become the approach of choice and the basis of international monitoring networks. However, as the chapter points, the method has limitations such as footprint constraints, poor replicability, complexity and high cost. In the case of measuring enteric CH_4 emissions from livestock systems, respiratory chamber, tracer or GreenfeedTM techniques also have limitations for confined animals and requiring significant technical expertise can be used. An overview of other livestock-associated approaches is given in this chapter, arguing for further investment in alternative methods for routine on-farm measurements and the identification of mitigation options.

Part 3 Adaptation and mitigation strategies

Climate-smart cropping options aim to simultaneously improve crop productivity, adapt to climate change and reduce GHGs emissions. Echoing themes in Chapter 2, Chapter 8 provides a concise overview of a wide range of climate-smart cropping options and investigates in detail the potential of conservation agriculture and soil fertility management practices to contribute to the three pillars of climate-smart agriculture (CSA), taking in the issues of possible trade-offs and constraints to adoption. The chapter shows that gaps in understanding and assessing impacts may lead to an overestimation of the potential of CSA. Two contrasting case studies from intensive and low-input agriculture illustrate that reliable contextualized and quantitative information can be obtained with participatory, integrated and cross-scale assessments. Such information is key to support decision and policy making towards holistic solutions that can enable a successful transformation to CSA.

Chapter 9 illustrates through a contrasting set of examples how current crop-livestock systems contribute to the food and nutrition security of smallholder livelihoods. It highlights how farmers use these systems to cope with a variable climate. The chapter provides an overview of recent research on climate change and mixed crop-livestock systems. It also assesses ways of quantifying how crop-livestock systems contribute to food supply, dietary diversity and income of smallholder farmers. The chapter then reviews analyses of short-term climate variability coping strategies that farmers currently apply in mixed crop-livestock systems, highlighting most effective adaptation options to climate change in mixed farming systems. It concludes by consolidating these findings to provide a review of the resilience of mixed crop-livestock systems in the face of existing and future climate variability.

With appropriate management, agroforestry contributes to the provision of food, fiber and wood products and helps maintain ecosystem services such as nutrient cycling and biodiversity. Chapter 10 reviews recent research on the role of trees in integrating climate mitigation and adaptation goals, including the development of novel bioenergy solutions based on intercropping trees and shrubs in croplands. Many perennial staple crops are native to the African continent, and expanded production of these species could make a notable contribution to supply the future food and nutrition needs of vulnerable communities. The chapter discusses how agroforestry can be seen as an effective ecosystem-based adaptation strategy and an efficient carbon sink. The chapter discusses examples such as the semi-arid Sahel region in Africa, where community-driven tree regeneration has helped increase carbon sequestration in agricultural lands.

Part 1

The impacts of climate change on agriculture

Chapter 1

The effects on crop cultivation of increased CO_2, temperature and ozone levels due to climate change

Eline Vanuytrecht, Flemish Institute for Technological Research (VITO) and KU Leuven Department of Earth & Environmental Sciences, Belgium

1 Introduction

Making our world food secure is a challenge today but nothing less in the future. A 60–100% increase in food supply is required towards 2050 to accommodate nutritional needs of the expected 9 billion people and their changing consumption pattern (Alexandratos and Bruinsma, 2012; FAO et al., 2018; Godfray et al., 2010; Ingram and Porter, 2015). Moreover, agricultural productivity has to increase sustainably, knowing its resources like water, land and fertilizers are scarce and are increasingly to be shared with other sectors (i.e. industry and households) (Keating et al., 2014). The challenge is intensified by atmospheric changes and air pollution, which are already present today and will further affect crop yield in the future (Bindi et al., 2015; Elbehri, 2015; Gornall et al., 2010; Ingram and Porter, 2015; Field et al., 2014; Tai et al., 2014; Wheeler and Braun, 2013). Weather variability is responsible for over 30–50% of yield variability of major crops (Frieler et al., 2017; Ray et al., 2015; Zampieri et al., 2017) and further atmospheric changes can have a large impact.

In this chapter, I present observed and projected future effects on crop cultivation for food production caused by three evident changes of the

http://dx.doi.org/10.19103/AS.2020.0064.01

atmosphere, that is atmospheric CO_2 concentration ([CO_2]), air temperature and tropospheric ozone concentration ([O_3]). They went through a notable development since the Industrial Revolution, and all three are projected to further increase. Each atmospheric phenomenon is here covered by:

1 a presentation of the current status and future projections of its physical conditions,
2 a concise description of observed physiological and agronomic consequences in experiments, and
3 an impact assessment at the global scale.

Thereafter, interactive effects of several (atmospheric) changes are discussed and one local-scale case study is presented. The chapter finishes with an outlook on potential future contributions of research to sustainable crop production under altered atmospheric conditions. Ultimately, recommended reading is presented for those who are hungry for more information.

2 The effects of elevated CO_2 concentration on crop cultivation

2.1 The physical phenomenon

Atmospheric [CO_2] has steadily increased since the Industrial Revolution from 280 ppm to above 400 ppm today, and is expected to continue to rise for decades by approximately 2 ppm per year, even with stringent emission reductions (Stocker et al., 2013). Projected end-of-the-century [CO_2] depends on the assumed representative concentration pathway (RCP) that denotes the approximate radiative forcing (W/m^2) in 2100 relative to 1750 (van Vuuren et al., 2011). By the year 2100, [CO_2] is projected to reach 421 ppm after an earlier peak following the low RCP 2.6; a stable 538 to 670 ppm following the stabilization RCPs 4.5 and 6.0, respectively; or an intermediate 936 ppm following the high RCP 8.5 (Stocker et al., 2013). Since plants assimilate CO_2 to build biomass, these changes affect crop production.

2.2 Physiological and agronomic consequences

Major effects of elevated atmospheric [CO_2] ([CO_2]e) on plants include improved photosynthetic efficiency and reduced stomatal conductance (Ainsworth and Long, 2005; Drake et al., 1997), resulting in improved crop water productivity, that is the ratio of crop yield over evapotranspiration (Vanuytrecht et al., 2012). In C3 plants, [CO_2]e increases the carboxylation rate of photosynthesis and improves its efficiency by suppressing photorespiration (Ainsworth and Rogers,

2007). Within the C3 category, tubers and nitrogen-fixers respond more vigorously to CO_2 fertilization (e.g. McGrath and Lobell, 2013a; Vanuytrecht et al., 2012). C4 plants on the other hand are saturated at ambient [CO_2] and avoid photorespiration, hence little direct CO_2 stimulation is expected for those plants (Leakey, 2009). Even though Rubsico activity and content decrease at [CO_2]e (Ainsworth et al., 2002; Ainsworth and Rogers, 2007; Wang et al., 2013), improved photosynthesis translates generally in boosted biomass production of both vegetative and reproductive plant parts (Ainsworth, 2008; Ainsworth et al., 2002; Jablonski et al., 2002; Vanuytrecht et al., 2012; Wang et al., 2013). Leaf stomata respond to (inter-cellular) [CO_2]e by decreasing conductance (Ainsworth, 2008; Ainsworth and Rogers, 2007; Wang et al., 2013). Indirectly, this effect additionally benefits crop production, also for C4s, and more so in dry than in wet conditions (Bishop et al., 2014; van der Kooi et al., 2016).

Other effects include reduced overall plant water use (Kimball, 2016; Shimono et al., 2013; Vanuytrecht et al., 2012), increased leaf area but reduced specific leaf area hence thicker leaves (Ainsworth et al., 2002; Burkart et al., 2011; Kimball et al., 2002), changed carbon allocation (Vanuytrecht et al., 2012; Wang et al., 2014b) and altered elemental composition. Concentrations of nitrogen, proteins and nutrients decrease in general (Broberg et al., 2017; Dong et al., 2018; Myers et al., 2014; Taub et al., 2008; Walker et al., 2017; Wang et al., 2013; Zhu et al., 2018), while starch and sugar concentrations increase (Dong et al., 2018). Reduced nutrient concentrations result from an interplay of mechanisms (including dilution by additional carbon uptake, reduced stomata-mediated nutrient uptake and within-plant transport), which is yet to be fully understood (Broberg et al., 2017; McGrath and Lobell, 2013b; Taub and Wang, 2008). Notwithstanding the stimulatory effect of [CO_2]e on quantitative biomass and yield production, altered nutrient concentrations have a serious impact on food nutrition and security. The responses at individual plant level induce changes in plant communities and ecosystems (Ziska and Bunce, 2006).

Crop responses to [CO_2]e are being studied in fully controlled growth chambers (e.g. Leisner et al., 2018; Reddy and Zhao, 2005), as well as in the field in open-top chambers (OTC, e.g. Saha et al., 2015; Wang et al., 2018) or free air CO_2 enrichment (FACE) experiments (e.g. Bourgault et al., 2018; Gray et al., 2016; Manderscheid et al., 2018; Zhu et al., 2018). The latter mimic natural conditions more closely by ensuring free root grow, minimizing boundary effects and allowing to assess ecosystem effects (Hendrey et al., 1993). Meta-analyses summarizing the wealth of observational data contribute to our general understanding of crop responses to [CO_2]e. For a variety of (C3) crops, meta-analysis of FACE observations confirmed biomass and yield gains by +18–19% for a typical increase in [CO_2] to 550 ppm. Evapotranspiration decreased for both C3 and C4 by −5–10% on average, resulting from the reduction in stomatal conductance but increase in leaf area (Kimball, 2016; Vanuytrecht et al., 2012). Studies for soybean (Ainsworth et al., 2002), rice (Ainsworth, 2008) and wheat revealed +23–24% yield increase for an average increase in [CO_2] to 620–700 ppm. Associated increases in pod number but unaltered seed mass for soybean (Ainsworth et al., 2002) were reported, as well as increases in grain mass, grain and panicle/ear number for rice and wheat (Ainsworth, 2008; Wang et al., 2013). Interestingly, Ainsworth (2008) and Wang et al. (2013) found roughly half the CO_2 stimulation in free-air experiments compared to more enclosed (or pot-rooted) systems. Environmental conditions including drought and nitrogen limitation affected the response magnitude with magnified versus suppressed responses, respectively (Wang et al., 2013).

When focusing on plant nitrogen dynamics at [CO_2]e (~ 666 ppm), Lam et al. (2012) found increased nitrogen uptake (+17%), but reduced yield nitrogen concentration (−8 to 10%) except in legumes. An equivalent effect was found for protein concentration of barley, rice and wheat grains (−10 to 15%) and potato tubers (−14%), and a smaller but nevertheless significant effect for soybean (−1.4%) (Taub et al., 2008). Also the concentration of other essential nutrients tended to change with, for example, strong negative effects for important elements including Fe, S, Zn and Mg in wheat (Broberg et al., 2017). Table 1 summarizes general impacts of [CO_2]e on yield quantity, yield quality and water use. Note that responses depend on the magnitude of change in [CO_2] and on the interaction with other stresses determined by the environmental setting.

2.3 Global assessment of CO_2 induced impact

The global impact of [CO_2]e, and climatic changes in general, on crop cultivation has now been studied for over 20 years, by biophysical process-based crop or ecosystem models or by statistical relations. Those assessments can be made with an ensemble of impact models driven by future climate projections from multiple climate models (Challinor et al., 2013, 2014a, 2018; Rosenzweig et al., 2013; Wallach et al., 2018). An in-depth discussion of modelling methodologies for climate impact assessment on crop production can be found in Chapter 6 of this book. In this first chapter, I mainly present the large-scale projected impacts of the atmospheric changes by these studies.

Many global assessments evaluate the CO_2 effect in combination with associated climatic changes to present a realistic future scenario (e.g. Asseng et al., 2013; Challinor et al., 2014b; Rosenzweig et al., 2014). For research purposes, though, isolating the CO_2 effect in modelling exercises can nevertheless be informative. Global yield benefits of CO_2 (hence, when isolated from the integrated whole of climatic changes) were simulated for 2080 for maize (+13%), rice (+24%), wheat (+26%) and particularly soybean (+35%) under RCP 8.5 by a model ensemble of six global gridded crop models driven

Table 1 Summary of crop responses to atmospheric changes as observed in field-scale experiments (FACE, OTC and open air in general)

Atmospheric change	Impact level		Response (relative to ambient conditions)
Increased [CO_2]			
	Yield		
	quantity	▲▲	for C3 due to photosynthesis stimulus
		yet: △	for nitrogen deficient conditions
		△△△	for drought conditions
		▲	for C4 in drought conditions due to water savings
	quality	▼	in nitrogen, protein and nutrient concentration
	Water use	▼	in canopy-scale transpiration
		yet: ▷	for drought conditions
		▽▽▽	in stomatal conductance
Increased temperature (T)			
	Yield		
	quantity	▼▼	for T rise close to crop's optimal temperature range due to a distorted balance between photosynthesis and respiration
		yet: △	for moderate T increases in cooler climates
		△	for alternative cultivars with longer growing season
		△	for new crops in previously uncultivated regions
		▼▼▼	for extreme heat
	quality	▼	due to adverse effects on yield traits (e.g. milling quality or grain chalkiness) and due to visual damage (especially horticultural crops)
	Water use	▲	in evapotranspiration due to higher evaporative demand
		yet: ▽	in stomatal closure for drought conditions
Increased [O_3]			
	Yield		
	quantity	▼▼	due to molecular, biochemical and metabolic responses
	quality	▲	in nitrogen, protein and nutrient concentration
		▼	due to adverse effects on yield traits (e.g. starch concentration or grain chalkiness) and visual injury
	Water use	▼▼	in stomatal conductance as initial defence strategy
		yet: ▷ - △	due to subsequent lower stomatal sensitivity

▼ decrease; ▲ increase; ▶ status-quo.
number of ◀◀◀ indicates strength of the response: ◀ moderate; ◀◀ strong; ◀◀◀ very strong.
This table summarizes general responses. In reality, responses highly depend on the magnitude of change (in [CO_2], [O_3] or T) and on the interaction with other stresses determined by the environmental setting.

by data from five global climate models (GCMs) (Deryng et al., 2016). Note that these projected CO_2 fertilization effects could not counterbalance the negative impacts of associated climatic changes for maize and rice but contributed to a break-even for soybean and a slight improvement in wheat yield. CO_2-induced yield gains and associated reductions in crop evapotranspiration further increased global crop water productivity by 2080 by +26–44% on average if CO_2 alone was considered (and still by +10–27% with the consideration of associated climatic changes). Notably, arid regions benefitted particularly (with overall gains up to +48% for wheat in rainfed conditions) (Deryng et al., 2016).

3 The effects of increased temperature on crop cultivation

3.1 The physical phenomenon

Elevated atmospheric [CO_2] and other greenhouse gases create a greenhouse effect, whose radiative forcing leads to global warming. Since the Industrial Revolution, average surface temperatures have risen by 0.85°C, and are expected to further rise at a pace of roughly 0.2°C per decade over the coming 30 years. Global air temperatures will increase by 0.3–1.7°C according to the most optimistic projections (RCP 2.6) and by 2.6–4.8°C according to the business-as-usual scenario (RCP 8.5) by the end of the century (Stocker et al., 2013). Manifested changes are irreversible in the short term due to the slow reaction of the earth's system. Crop growing season temperatures are projected to rise more than annual averages. With a mean warming of 2°C, maximum temperature increases of 3°C and more will not be rare (Schleussner et al., 2015). Also, more extremes are expected, including intensification of heat waves in terms of duration, frequency and temperature (Teixeira et al., 2013). The duration of warm spells are projected to increase particularly in the (sub)tropics (Schleussner et al., 2015). Projected future temperatures are often above critical levels for many currently cultivated crop species, especially at night (Field et al., 2014). Since temperature governs many processes in plants, warming can seriously impact on crop production.

3.2 Physiological and agronomic consequences

Rates of many physiological processes governed by temperature slow down when temperatures fall outside crop-specific optimal temperature ranges (Porter and Gawith, 1999; Sánchez et al., 2014). The effects are highly non-linear, with excessive negative impacts when crossing critical thresholds (Kawasaki and Uchida, 2016; Schlenker and Roberts, 2009). Apart from being crop- or even cultivar-specific, temperature responses vary also with growth stage (Kawasaki and Uchida, 2016). Often, changes in minimum and maximum temperatures during critical growing stages explain observed yield losses better than average growing season temperatures (Duncan et al., 2015).

High temperatures affect crops in various ways, both directly and indirectly, with major effects including altered photosynthesis and (photo)respiration (e.g. Eyshi Rezaei et al., 2015), accelerated growth and development rates (e.g.

Wolkovich et al., 2012), increased transpiration (at least if not in combination with drought; Crawford et al., 2012; Wall et al., 2011), changed carbon allocation and root growth (Awal et al., 2003; Eyshi Rezaei et al., 2015; Gray and Brady, 2016; Nagel et al., 2009), and visible injury (e.g. Bisbis et al., 2018), in general leading to biomass and yield losses (Ainsworth and Ort, 2010; Akter and Rafiqul Islam, 2017; Asseng et al., 2011; Wahid et al., 2007). Ultimately, net carbon gain in plants is governed by the balance between photosynthetic assimilation and respirational loss. While photosynthesis improves linearly between a crop-specific minimum and maximum temperature, it falls steeply beyond a maximum threshold. High temperature affects Rubisco activity (thus reducing the carbon fixation capacity) as well as the plant cells' light harvesting capacity and membrane stability (Eyshi Rezaei et al., 2015; Sage et al., 2008; Sage and Kubien, 2007). Respiration increases logistically in a temperature range of 0–40°C (Atkin and Tjoelker, 2003; Taiz and Zeiger, 2014). The involved metabolics and bio-chemics of the photosynthetic-respiratory balance are nicely summarized by Eyshi Rezaei et al. (2015).

Further, increased mean growing season temperature accelerates crop development, resulting in shorter growth stages and overall a shorter growth cycle with accelerated senescence. This implicates a shorter period for photo-assimilation and thus less production, with additional negative implications for several yield components (Asseng et al., 2015; Cleland et al., 2007; Craufurd and Wheeler, 2009; Menzel et al., 2006). Reproductive phases (including anthesis and grain filling) are most vulnerable to acceleration (Barlow et al., 2015; Wolkovich et al., 2012). Also, chilling and vernalization requirements, that is days of low temperatures to induce cold hardiness, may not be met when average temperatures increase and distort plant development (Bisbis et al., 2018).

Extreme heat stress during reproductive development can cause severe effects on both crop morphology and functioning, including flower abortion, distorted allocation, reduced head development and grain number, and sterility (Barnabás et al., 2008; Deryng et al., 2014; Ferris et al., 1998; Lobell et al., 2012; Semenov and Shewry, 2011; Tahir and Nakata, 2005; Zhang et al., 2018). During heat waves, extreme temperatures and radiation lead also to visible sunburn (Bisbis et al., 2018). For yield quality as well, extreme high temperature can have a severe impact (Kawasaki and Uchida, 2016). In addition to those high maximum temperatures, high night (minimum) temperatures have been proven to have different, but at least equally negative, impacts (Chen et al., 2017).

In contrast to the above-mentioned negative effects, high temperatures can indirectly promote extension of the growing season and reduce cold stress (Zhang et al., 2018). In mid- to high-latitudes (in the northern hemisphere), increased temperatures create opportunities to cultivate crops in areas unsuitable for crop production before, grow cultivars with a longer cycle or even have multiple harvests in one season (Peltonen-Sainio et al., 2018), if soil water and fertility status allow it and heat days are limited (Asseng et al., 2015; Mueller et al., 2015; Ostberg et al., 2018; Ramankutty et al., 2002). By the end of the century, about 76% of the boreal region might reach suitable temperatures for crop cultivation (compared to the actual 32%), as a result of a northward expansion of the cultivation area by over 1,000 km (King et al., 2018). Although average temperatures may rise and expand the potential growing season, they may still involve a higher risk of late cold stress, especially given the expected higher occurrence of late cold events under stable 'blocking' atmospheric conditions in the future (Cattiaux et al., 2010).

Experiments testing crop-specific responses to high temperature are numerous and include studies in growth chambers (e.g. Hatfield and Prueger, 2015) and greenhouse (e.g. Lizaso et al., 2018), as well as field experiments with OTCs, semi-enclosed or covered systems (e.g. Chen et al., 2017), portable heat chambers (e.g. Talukder et al., 2014), free-air enhancements (T-FACE or FATE/I, e.g. Ruiz-Vera et al., 2018) and translocation studies (e.g. Tito et al., 2018). Existing reviews and meta-analyses have usefully compiled data from different studies and summarized findings. For rice, Xiong et al. (2017) composed temperature dose-response relations based on observations from 95 experiments at high temperature. Grain yield was reduced by −39.6% on average due to increases of 1–15°C and 2–12°C in day temperature and night (minimum) temperature, respectively. Reduced photosynthesis and increased respiration at high day and night temperature led respectively to decreased seed set and to decreased biomass production, spikelet number and grain weight. Also yield quality was negatively affected by decreased head rice percentage and increased rice chalkiness (Xiong et al., 2017). For wheat, reduced grain number and grain weight were identified as major physiological responses to heat stress and causal for reduced final yield (Akter and Rafiqul Islam, 2017; Barlow et al., 2015; Eyshi Rezaei et al., 2015). Maize yield was particularly affected by heat stress right before and during anthesis, with reduced grain number largely affecting final yield (Eyshi Rezaei et al., 2015; Sánchez et al., 2014).

How temperature finally affects crop cultivation cannot be generalized but depends much on the environmental conditions of an agricultural system. Table 1 summarizes general impacts of increased temperature on yield quantity, yield quality and water use. As with $[CO_2]$e, responses in the real world depend largely on the magnitude of temperature change and on the interaction with other stresses determined by the environmental setting. This was for example shown in a set of nighttime warming field experiment with rice in China (Chen et al., 2017). While shorter rice cultivars cultivated in warmer seasons showed yield loss with warming less than 1°C due to shortening of the growth phases, increased respiration and lower grain-filling rates, longer rice cultivars cultivated in cooler seasons or locations showed increased yields because mitigated low temperature stress dominated higher respiration rates

(Chen et al., 2017). A similar observation was made for wheat in China in a study compiling results from more than 40 warming experiments (Zhao et al., 2016).

3.3 Global assessment of temperature induced impact

By regressing historical yield observations to past weather data on a global scale, a number of studies diagnosed a higher incidence of extreme heat episodes and associated yield losses in recent years (comparing with 1960–1980) (Cammarano et al., 2016; Lesk et al., 2016; Moore and Lobell, 2015; Zampieri et al., 2017; Zhang et al., 2018). Cereal production decreased significantly by −9–10% due to extreme heat and drought on a global scale, but with more damage in developing countries (Lesk et al., 2016). For wheat, heat was found to have a more significant negative impact than drought, especially in temperate regions (Zampieri et al., 2017). For maize, high temperatures decreased yields in important production regions, that is by 0.6% per decade in India and by 1.1% in China, the United States and overall globally. In Mexico on the other hand, where maize production is constrained by low temperatures, historical temperature increases improved maize yield by 0.5% per decade (Zhang et al., 2018).

Future heat-on-yield impacts on regional to global scale have been modelled by process-based (e.g. Asseng et al., 2015; Balkovič et al., 2018; Liu et al., 2016; Rose et al., 2016; Schleussner et al., 2015, 2018) as well as statistical methods (e.g. Davenport et al., 2018; Liu et al., 2016; Lobell and Asseng, 2017), which confirm overall reduced crop productivity when warming of 1°C or above is considered, with similar or occasionally slightly positive effects at a smaller scale. Positive yield effects for major crops were mostly associated with temperature increases in mountainous and high-latitude regions (Balkovič et al., 2018; Ostberg et al., 2018). As a directive, each 1°C increase in global mean temperature was estimated to reduce, on average, global yields by about −7.5% for maize, −6% for wheat and −3% for rice and soybean (Zhao et al., 2017; Fig. 1). Apart from yield losses, also increased yield variability is expected to challenge global future food production (Asseng et al., 2015). Furthermore, significantly more severe impacts on crop production were simulated with an increase of 2°C as compared to 1.5°C. And above 2°C warming, effects cannot be simply extrapolated as temperature effects are highly non-linear (Asseng et al., 2015; Schlenker and Roberts, 2009; Zhao et al., 2017).

Tropical agricultural production systems, located in developing countries where yields are already low - the so-called hotspots for climate change (Müller et al., 2014) - are more heavily hit than their mid- to high-latitude counterparts (Asseng et al., 2015; Ostberg et al., 2018; Rosenzweig et al., 2014; Schleussner et al., 2015). Schleussner et al. (2015) projected slightly more optimistic global yield changes than Zhao et al. (2017), but still more

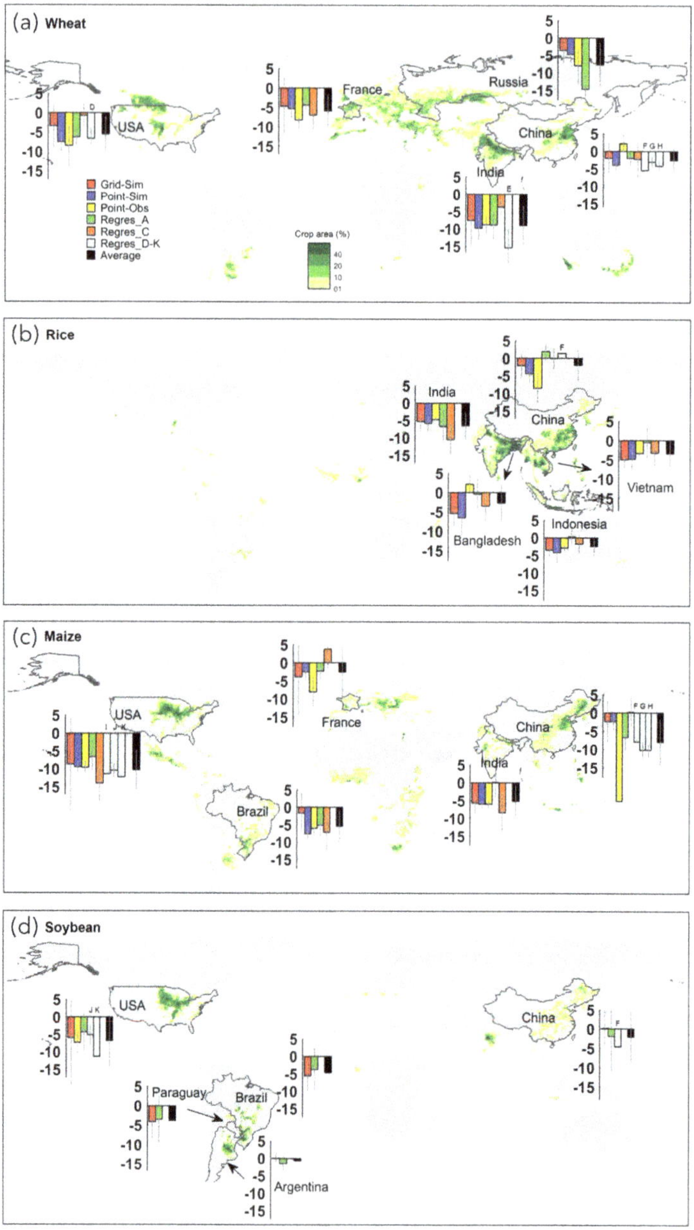

Figure 1 Temperature effect (%) on yield for a 1°C increase in global temperature for the five major producing countries for (a) wheat, (b) rice, (c) maize, (d) soybean as simulated with grid-based simulations (red bar), point-based simulation models (blue bar), point-based field observations (yellow bar), country level regression analyses (green, orange and grey bars) methods and the overall average (black bar). Source: reprinted with permission from Zhao et al. (2017, p. 9328).

severe impacts in the tropics with yield changes from −9 to −16% for wheat, −3 to −6% for maize, +6 to +7% for soybean and +6 for rice from +1.5°C to +2°C temperature increases, respectively. Of course, all model estimates depend on assumptions about crops, crop responses and cropping systems, on included equations, climate projections and simplification of real processes (e.g. Angulo et al., 2013; Challinor et al., 2009; Knox et al., 2012; Müller et al., 2011). How this may affect yield loss projections and other impacts is discussed elsewhere in this book.

4 The effects of high ozone concentration on crop cultivation

4.1 The physical phenomenon

Our troposphere, that is the lower 10 km of our atmosphere, always contains low concentrations of ozone ([O_3]) from natural sources. Yet, air pollution has quadrupled background [O_3] from 10 ppb before the Industrial Revolution (Volz and Kley, 1988) to daytime summer concentrations exceeding 40 ppb in many parts of the northern hemisphere (Cooper et al., 2014; Mills et al., 2018a). Tropospheric ozone originates predominantly from photochemical reactions of its precursors, that is nitrogen oxides (NO_x), volatile organic compounds (VOC) including methane (CH_4) and carbon monoxide (CO) (Stevenson et al., 2006), mostly emitted by human activities. Its background concentrations are still on the rise and expected to further increase depending on regional emissions and climatic and atmospheric policy to reduce air pollution and the emission of precursors (Stocker et al., 2013). Peak [O_3] of over 60 ppb during hot and dry weather, though, has been gradually decreasing recently in highly polluted areas in Europe and North America due to lower emission of ozone precursors. Intensified industrialization and emission of NO_x in India, China and other parts of Asia on the other hand, have increased the background [O_3] by 20% and 13%, respectively, between 1990 and 2013 (Brauer et al., 2016), and are further projected to increase. Monitoring is hard though because observation networks are sparse in the southern hemisphere (Cooper et al., 2014; Sofen et al., 2016; Wang et al., 2017a). Ozone concentrations are highly spatially variable, varying with meteorological conditions and topography. Yet, ozone can easily be dispersed from highly polluted areas to adjacent remote regions. Many of the world's most productive growing regions are already exposed to high [O_3], which are often exceeding critical ozone levels for vegetation (Ainsworth, 2017; Mills et al., 2018a). Being a major toxic air pollutant with a strong impact on human health and plant production, ozone is already responsible for significant crop yield losses (Ainsworth et al., 2012; Avnery et al., 2011a; Fuhrer, 2009; Mills and Harmens, 2011).

4.2 Physiological and agronomic consequences

Ozone penetrates plant leaves by diffusion through the stomata, whose functioning is affected initially by stomatal closure as defence strategy and ultimately by reduced stomatal sensitivity (Ainsworth, 2017). In the plant cells, ozone degrades to reactive oxygen species and basically causes oxidative stress, interferes with the photosynthesis process and ultimately reduces primary production. The effect of excessive ozone uptake is multifold, including effects on primary crop physiology and metabolism, growth and carbon allocation, and (reproductive) development (e.g. Ainsworth, 2017; Ainsworth et al., 2012).

Primarily, ozone decreases Rubisco activity (e.g. Lehnherr et al., 1987) causing reduced carbon assimilation and early senescence (Black et al., 2000; Fuhrer, 2009; Wang and Frei, 2011). Further, carbon translocation processes are affected, which results in lower harvest index, altered leaf composition (Fuhrer, 2009; Fuhrer and Booker, 2003) and altered root-shoot ratio (Feng et al., 2008). The reproductive process is negatively affected by, among others, reduced pollen germination and flower and seed production (Black et al., 2000; Leisner and Ainsworth, 2012). The molecular, biochemical and metabolic processes after exposure to ozone are described in other articles (see for example Ainsworth, 2017; Ainsworth et al., 2012; Castagna and Ranieri, 2009; Dizengremel et al., 2009; Fiscus et al., 2005; Ludwikow and Sadowski, 2008; Wilkinson and Davies, 2010).

Ozone effects manifest themselves also in visible leaf injury (e.g. Bisbis et al., 2018; Feng et al., 2014), and in yield quality changes in both negative and positive direction (Booker et al., 2009; Wang and Frei, 2011). Negative effects include increased grain chalkiness in rice (Wang et al., 2012b, 2014a) and decreased starch in potato (Vorne et al., 2002). Positive effects include improved baking quality for wheat (Broberg et al., 2015) and increased protein and nutrient concentrations in general (Broberg et al., 2015; Wang et al., 2012b, 2014a) resulting from a concentration effect due a decreased biomass production, as well as nutrient reallocation after early senescence and faster grain filling (Frei, 2015). At larger scale, there are ozone-induced changes in species composition and ecosystem functioning (e.g. Fuhrer et al., 2016), soil water budget (e.g. Bernacchi et al., 2011) or soil carbon sequestration and nitrogen cycling (e.g. Andersen, 2003; Kou et al., 2014) with consequences for greenhouse gas emission and nitrate leaching (e.g. Decock and Six, 2012; Tang et al., 2015).

Plant sensitivities to ozone pollution are commonly expressed as species-specific critical levels above which damage occurs. Those levels are determined by dose-response relations based on experimental data. Today, it is accepted that not the atmospheric $[O_3]$, yet the ozone flux into the plant best determines the damage to plants (Ainsworth et al., 2012; Mills et al., 2011; Moura et al.,

2018; Sanz et al., 2016). Still, exposure-based metrics including AOT40 (i.e. the accumulated hourly [O_3] at the plant level above 40 ppb during a specified period; CLRTAP, 2017) and W126 (i.e. a weighed metric of among others hourly mean [O_3] above 100 ppb over 12 daytime hours; Lefohn et al., 2018) are commonly used as these methods do not require complex simulation of stomatal conductance (Ashmore, 2005). A flux-based metric that wins ground, is the phytotoxic ozone dose (POD_y), which defines the accumulated plant uptake of ozone above a threshold *Y* during a specified period (CLRTAP, 2017; Mills et al., 2011; Sanz et al., 2016). This kind of relationships has for example been established for wheat (Danielsson et al., 2003; Pleijel et al., 2004) and potato (Pleijel et al., 2004).

OTCs are popular to study the effects of ozone on crops, and also FACE experiments are increasingly developed (Tang et al., 2011) to study physiological responses of soybean (e.g. Betzelberger et al., 2010; Morgan et al., 2006, 2004), rice (e.g. Pang et al., 2009; Shi et al., 2009), wheat (e.g. Zhu et al., 2011) and sugarcane (Moura et al., 2018). Notably, much research is focused on C3 plants (Mills et al., 2018a), while less is known about C4 plants including maize and sorghum (Yendrek et al., 2017). Common in all FACE experiments was the decrease in leaf-level photosynthesis rate and the drop in leaf area, resulting in significant yield reduction (e.g. Dermody et al., 2008; Feng et al., 2011; Pang et al., 2009). For sugarcane, Moura et al. (2018) reported a 20% decrease in aboveground biomass production, with a more than proportional reduction in the stalks, which are relevant for sugar/ethanol production. The sensitivity of sugarcane to ozone damage was ascribed to its relatively high stomatal conductance and its preference for high growth temperatures which concur with high levels of ozone formation (Moura et al., 2018).

Numerous studies have summarized ozone impact on crops in qualitative and quantitative reviews (Ainsworth, 2008; Booker et al., 2009; Feng et al., 2008; Feng and Kobayashi, 2009; Fiscus et al., 2005; Fuhrer, 2009; Morgan et al., 2003; Pleijel et al., 2018). Feng and Kobayashi (2009) for example quantitatively summarized 406 experimental observations of ozone-induced yield effects, and found mean yield reductions of −21% for wheat (whereas Feng et al. (2008) found a slightly higher mean yield reduction of −29%), −28% for rice (whereas Ainsworth (2008) found only half this reduction), −22% for soybean (an order of magnitude confirmed by Morgan et al. (2003)), −12% for potato, −14% for barley and −42% for beans, for [O_3] equal to 60 ppb. For current peak [O_3] of 40 ppb, yields already decreased by −10% for wheat, −17% for rice, −8% for soybean, −5% for potato, −9% for barley and −19% for beans as compared to ozone-free air (Feng and Kobayashi, 2009). Regression techniques and standardization of observations among crops have led to an ozone sensitivity ranking with wheat, soybean, pulses, cotton, onion and lettuce being most ozone-sensitive, whereas rice, maize, barley, oats, potato, sugar beet, oilseed rape and grape are among the more tolerant crops (Mills et al., 2007; Mills and Harmens, 2011). Table 1 summarizes also the general impacts of high [O_3] on yield quantity, yield quality and water use. Also for [O_3], real responses depend on the magnitude of change and on the interaction with other stresses determined by the environmental setting.

4.3 Global assessment of O_2 induced impact

With ozone dose-response functions and impact metrics (e.g. Oue et al., 2011), current mean global yield losses were estimated to be −11.2% for soybean, −9.6% for wheat, −3.9% for maize and −3% for rice (Avnery et al., 2011a; Mills et al., 2018b; Van Dingenen et al., 2009). The highest relative yield losses for wheat (Mills et al., 2018b), maize and rice were registered for the Middle East and South Asia, while soybean losses were the highest in Europe, Canada and China (Mills et al., 2018b). Chuwah et al. (2015) confirm global reductions in rice and maize yield of −4%, with peaks of −10% in the Middle East, China and India (Fig. 2). Associated global economic losses were estimated to be between US$11 and 26 billion (Mills and Harmens, 2011), and assumed to be as high as those of recent climate change impacts on agricultural production (Avnery et al., 2011a; Mills et al., 2018b; Van Dingenen et al., 2009). At a regional scale, similar losses were simulated for Europe (Mills and Harmens, 2011), the United States (Yue and Unger, 2014), China (Feng et al., 2015) and Southeast Asia (Tang et al., 2013).

Pessimistic scenarios without stringent air pollution regulation for 2030 projected wheat yield losses ranging between −5 and −26% (an additional 1.5–10% loss as compared to the year 2000 losses), soybean losses of −15–19% (an additional 1–11% from 2000) and maize losses of −4–9% (an additional 2–3% from 2000), resulting in $17–35 billion losses (an additional $6–17 billion from 2000). The highest yield losses would be in China (Feng et al., 2015), South Asia and the Middle East, except for soybean with also high losses in Europe (Avnery et al., 2011b). Chuwah et al. (2015) warn for even higher additional maize and rice yield losses of −20% towards 2050. Still, because the ozone monitoring network is not dense in developing countries, models may underestimate the extent of the problem in Asia or Africa. Moreover, responses of typical crops that thrive in tropical conditions are not often studied, so the impact of ozone may be further underestimated for those regions (Emberson et al., 2001).

5 Interaction effects of atmospheric changes on crop cultivation

Even more complex than the effects of a single atmospheric change on different aspects of crop cultivation, is the interactive impact of multiple altered

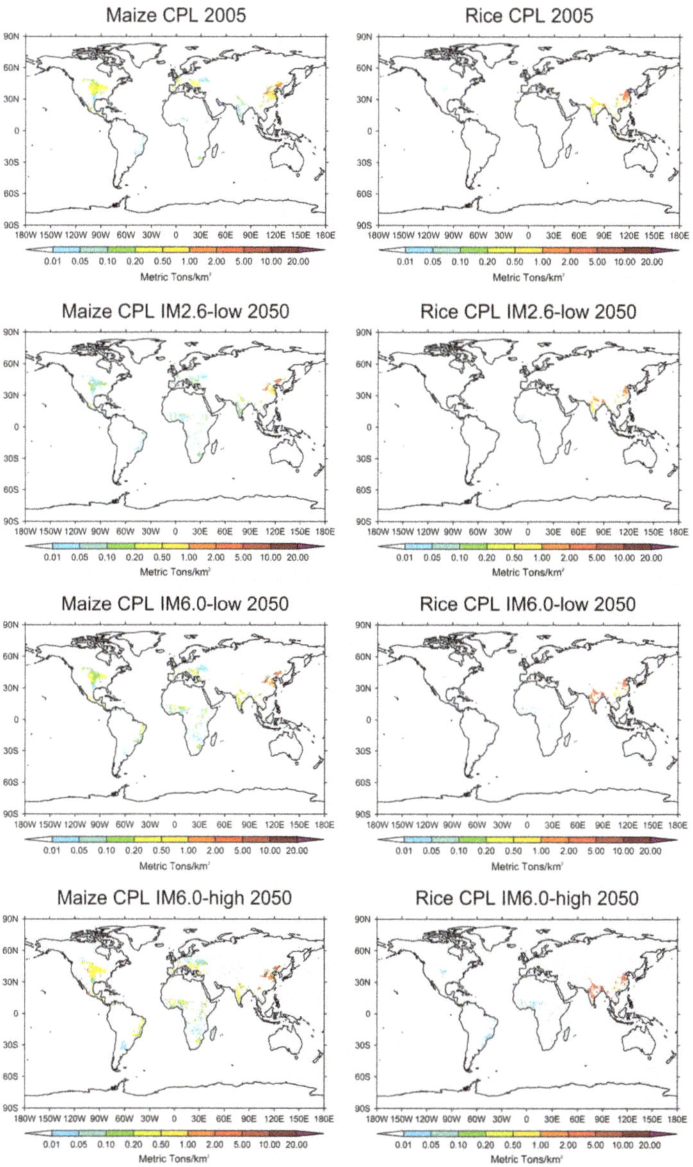

Figure 2 Ozone $[O_3]$ effect (ton/km² loss) resulting in crop production loss (CPL) for 2005 (upper row) and 2050 (lower rows) with $[O_3]$ associated with different emission and land use scenarios including IM6.0-high (RCP6.0 type scenario, that is weak climate policy with high air pollutant emissions), IM6.0-low (RCP6.0 type scenario, that is weak climate policy with low air pollutant emissions) and IM2.6-low (RCP2.6 type scenario, that is stringent climate policy with low air pollutant emissions) for maize (left panels) and rice (right panels) as simulated with the IMAGE model (Bouwman et al., 2006). Source: reprinted with permission from Chuwah et al. (2015, p. 19).

environmental stressors. The severity and timing of different atmospheric changes along with crop sensitivity determine the final outcome of their balance. Environmental constraints like nitrogen and soil water availability, disease and pest prevalence furthermore interact with atmospheric changes. Those constraints have the potential to change the magnitude and even direction of primary impacts of atmospheric changes. Experiments that combine several atmospheric, climatic and/or environmental changes have been set up to adequately investigate interaction impacts because effects are not simply additive. Two-factor experiments are gaining importance (e.g. Craigon et al., 2002; Ruiz-Vera et al., 2015), but multifactorial experiments, combining, for example warming, CO_2, ozone and water stress, are still limited (e.g. Mikkelsen et al., 2015, 2008).

5.1 Physiological and agronomic consequences

Most common are studies that combine elevated temperature and [CO_2]. For soybean and maize grown in the field, high temperatures (+2–3°C) have been shown to potentially negate beneficial CO_2 effects yield (Ruiz-Vera et al., 2015, 2013). Also for rice, positive yield responses to [CO_2]e were decreasing with increasing temperature (Hasegawa et al., 2013). Combining results from different species and experiments showed that the interactive effects of temperature and [CO_2]e vary among temperature regimes, crops and the response variables of interest (Bishop et al., 2014; Wang et al., 2012a). Overall, growth effects were not only explained by effects on net photosynthesis, probably due to differential contribution of two opposed effects at higher temperatures, that is the greater CO_2 benefit from decreasing photorespiration versus the larger negative temperature impact on reproductive processes (Bishop et al., 2014; Wang et al., 2012a). A positive interaction between [CO_2]e and drought, often induced by high temperature, is adding complexity (Bishop et al., 2014; Leakey et al., 2009). While C4 crops like maize showed beneficial CO_2-induced mitigation of drought stress (Erbs et al., 2015; Manderscheid et al., 2014), wheat yield responses were shown to depend on the severity and timing (before and after flowering) of both temperature and water stress (Fitzgerald et al., 2016).

Less common experiments combining [CO_2]e and [O_3]e showed counteracting effects of the two components on yield quantity and quality (Wang et al., 2014a; Wang and Frei, 2011). Mostly, [CO_2]e protected crops partly against ozone-induced damage thanks to CO_2-induced stomatal closure, whereas ozone reduced the positive effects of CO_2 fertilization (e.g. Booker et al., 2004; Cardoso-Vilhena, 2004; Pleijel et al., 2000; Wang et al., 2014a). An analogous mediating effect on ozone damage was found for drought, which also triggered stomatal closure hence reduced ozone uptake (Bernacchi et al., 2011; Feng et al., 2008). Morgan et al. (2003) on the other hand found no such alleviating effect for soybean, possibly because ozone also reduced stomatal responsiveness in drought conditions (Fuhrer, 2003; Mills et al., 2009; Wilkinson and Davies, 2010).

Altered atmospheric conditions also mean altered living conditions for pests and diseases. Moreover, crops weakened by (atmospheric) stresses may be more vulnerable to pests and diseases.

5.2 Global and local yield impact

Biophysical models easily combine two or more atmospheric/environmental factors, assuming that interactions are correctly simulated by combining the model algorithms for isolated effects (Asseng et al., 2013; Challinor et al., 2014a; Deryng et al., 2016; Rosenzweig et al., 2014). Only those studies that combine several co-existing environmental changes give insight in realistic impacts that can be expected in the (near) future, and can show regional disparities because of locally differing environmental conditions (McGrath and Lobell, 2013a), as opposed to studies with one changing factor at a time. The latter remain relevant, nevertheless, for improving our fundamental understanding (Schleussner et al., 2018).

Ozone was shown to aggravate negative impacts of rising temperature with co-varying temperature and [O_3] towards 2050 (Tai et al., 2014). Under a pessimistic RCP-8.5 scenario (without accounting for the CO_2 effect on crops though), global production losses of up to 11% and nearly doubling of the undernourishment rates were projected by statistical relations. Under RCP 4.5, though, controlled ozone pollution potentially reduced the negative effect of temperature. Crop species' sensitivity to either heat or ozone governed overall impacts (Tai et al., 2014). Despite plenty of evidence for the dramatic effect of high [O_3] on crop yield, ozone is often not considered in holistic studies on future crop production. Reasons that are mentioned include a lack of data, and a lack of recognition of or confidence in observed ozone responses from outside the ozone research community (Mills et al., 2018b).

By 2080 under a pessimistic scenario of peaking [CO_2] and associated changes in temperature and rainfall (RCP 8.5), the median impact on global production was found negative for maize and rice (−8.5 and −3%, respectively) but break-even and positive for soybean and wheat yield (0 and 3%, respectively) (Deryng et al., 2016). There was a clear positive impact of elevated [CO_2], that mitigated the negative impact of increased air temperature. Also overall crop water use (evapotranspiration) reduced from −3 to −17%, which

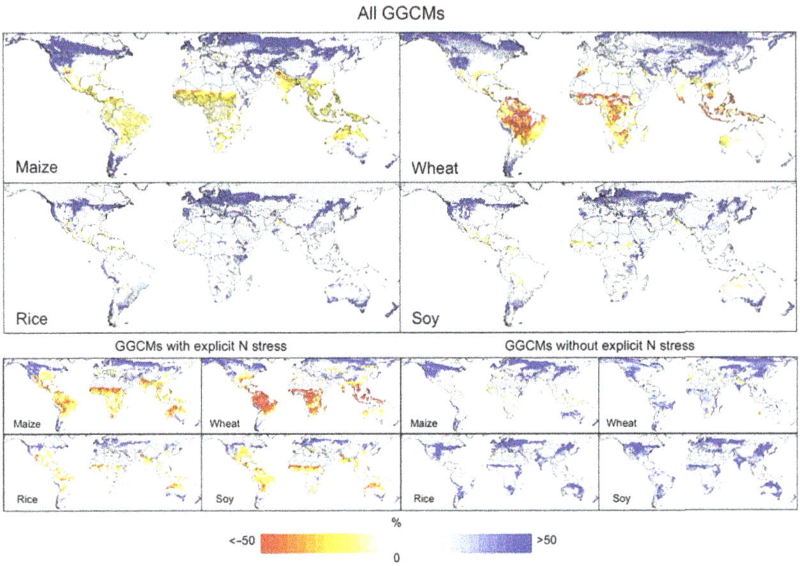

Figure 3 Median effect of a combination of increased [CO_2], temperature and precipitation (%) on yield for 2080 assuming RCP 8.5 for rainfed maize wheat, rice and soybean as simulated by seven crop models driven by five global climate models. Results are splitted for conditions assuming nitrogen stress being present (bottom four panels left) and absent (bottom four panels right). Source: reprinted with permission from Rosenzweig et al. (2014, p. 3271).

together with the productivity enhancement lead to improvement in crop water productivity under future climates, which is also vital to assess challenges to future food security (Deryng et al., 2016). Rosenzweig et al. (2014) confirmed a rather strong overall negative effect, as that the benefits of [CO_2]e could not counteract impacts of associated changes in temperature and precipitation. Specifically negative effects were found for those scenarios and locations with higher temperature, and if potential nitrogen limitation was considered (Fig. 3). On a regional scale, mean simulated climate change impact on crop yield across Africa and South Asia was estimated to be −8% by the 2050s with decreases of −17% (wheat), −5% (maize), −15% (sorghum) and −10% (millet) in Africa and of −16% (maize) and −11% (sorghum) in South Asia but no changes for rice (Knox et al., 2012).

By way of example, I spotlight one local study on climate impact on the future production potential of Malawi's staple food maize and related food security (Stevens and Madani, 2016). In the near future (2020s), the beneficial CO_2 effect on crop water use combined with projections of slight temperature increases and moderate rainfall reductions (as for this near-future time line similarly forced by RCP 4.5 and RCP 8.5) would increase maize yield from

4.6% to 5.4% in Malawi's capital district, Lilongwe. This would allow meeting the (growing) population's maize requirements. Further in the future, opposing effects of increased temperature, $[CO_2]$e and high rainfall variability resulted in maize yield changes between −1 to +1% (2050s) and −3 to 0% (2080s). Lower future maize production was estimated to put around 12% of the population at food insecurity risk in the 2080s (Stevens and Madani, 2016). Expectations may have to be revised if the impact of growing soil nutrient deficiency (caused by increased crop growth hence nutrient demand under CO_2 fertilization) or pest pressure would increase, or if technological developments would also be considered.

6 Conclusion and future trends

The previous sections have made clear that atmospheric changes already have a big impact on crop cultivation, and will do nothing less in the future (Lobell and Tebaldi, 2014). Adapting our agricultural systems is the only option to secure global food production (Ainsworth and Ort, 2010; Lobell et al., 2008). Several adaptation options are discussed in Part 3 of this book.

More benefit can result from adopting mitigation strategies simultaneously with adaptation plans (Smith and Olesen, 2010). Indeed, agriculture is an important emission source of greenhouse gases and ozone precursors (see Part 2 of this book). Limiting the mean global warming to 1.5°C as compared to 2°C by overall mitigation would seriously reduce the risk for crop yield reduction and other climate change-related impacts (Schleussner et al., 2015). A shift towards more sustainable cropping practices can contribute to a lower footprint of crop cultivation (Smith et al., 2013). Ways for reducing the footprint (e.g. by rationalizing nitrogen fertilizer use, matching irrigation and fertility management and increasing soil carbon storage) are also presented in Part 3 of this book.

There is, undoubtedly, need to further improve our knowledge and fill existing knowledge gaps, for example, about the impact of nitrogen availability on crop responses to atmospheric changes, or about the impact of atmospheric changes on yield quality. Collecting more experimental evidence is therefore key. Future research in controlled as well as field conditions will contribute to better comprehend physiological responses to extremes, responses of neglected crops, typical impacts in (sub)tropical areas, adaptation options, responses to combined changes and so on.

Major food-insecure regions where millions of mal-nourished people live, primarily in Africa and Asia, specifically deserve the attention of research and innovation (Challinor et al., 2007; Lobell et al., 2008; Müller et al., 2014). Today, much of current's research prioritizes the four major crops in their primary production areas including rice in South(east) Asia, China and Japan,

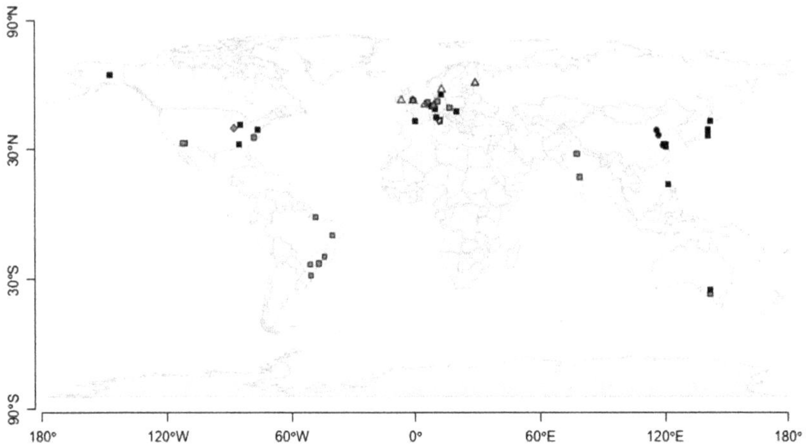

Figure 4 Geographical distribution of crop field experiments with manipulation of atmospheric [CO_2], [O_3] or air temperature (in FACE, OTC or open air with infrared heaters/reflective curtains). Different symbols represent different atmospheric changes (□ = [CO_2] in FACE or OTC; ○ = temperature with infrared or curtains; ▽ = [O_3] in FACE; ◇ = [CO_2] x temperature and [CO_2] × [O_3]; △ = [CO_2] × [O_3]). Black-filled symbols represent experiments with major crops (wheat, rice, maize or soybean). Open symbols represent experiments with other crops (including apple, banana, barley, beans, brassica spp., cotton, coffee, grape, mango, orange, pea, peach, potato, sorghum, sugarcane, sugar beet, tobacco). Grey-filled symbols represent experiments with both major as well as other crops. Source: map drawing based on tabulated data by the CLIMMANI Networking Group (https://climmani.org/).

wheat in China, the United States and India, and maize and soybean in the United States, which is justifiable because together they supply almost half of the world population's food calories (FAOSTAT, 2018). Still, typical local crops often constitute the staple food of millions of people, including millet, groundnut and rapeseed in southern Asia, and sorghum in the Sahel, and are facing highly negative impacts (Lobell et al., 2008). Figure 4 visualizes the hiatus in geographical distribution and the dominance of the four major crops in agricultural field experiments with manipulation of atmospheric [CO_2], [O_3] or air temperature (in FACE, OTC or open air with infrared heaters/reflective curtains) around the world. Studies of under-represented crops would contribute to make informed decisions about crop choice and adaptation strategies. The high response of cassava to [CO_2]e and its potential for future production, for example was highlighted by Rosenthal et al. (2012) and Rosenthal and Ort (2012). Besides global and regional impact assessments, impacts on smallholder farmers are needed (e.g. Hochman et al., 2017; Tingem et al., 2009) but unfortunately not less constrained by limited data availability (Jones et al., 2017).

Breeding crops that are tolerant to atmospheric threats offers opportunities (e.g. Ceccarelli et al., 2010; Mir et al., 2012; Singh et al., 2017), but the process is long (Challinor et al., 2016) and not seldom cultivars that are doing well for one aspect are more sensitive to another aspect (Osborne et al., 2016). Models can assist in identification of desired tolerance traits, a first step in the breeding process (e.g. Semenov and Stratonovitch, 2013; Singh et al., 2017). Breeding programmes are promoting for example drought-tolerant rice cultivars in Asian rainfed agriculture (Pray et al., 2011).

Models can be useful tools to assess the impacts of various scenarios of future climatic/atmospheric changes and support adaptation and innovation. Limited data accessibility for developing, evaluating and applying the models is as stringent as closing the gaps in our understanding of physiological theory and concepts (Jones et al., 2017; Maiorano et al., 2017; Wang et al., 2017b). Together with desired improvements of existing models and modelling techniques, data availability for modelling is also discussed more extensively in Chapter 6 of this book. Existing initiatives for sharing experimental data like the Global Open Data for Agriculture and Nutrition (GODAN; http://www.godan.info/) or the Open Data Journal for Agricultural Research (https://odjar.org) respond already to this need.

7 Where to look for further information

A good handbook that gives insight in topical data, methods and technology to assess the climate change impact on agricultural systems is Rosenzweig and Hillel (2015).

Important research projects that quantify climate impacts aiming simultaneously at improving insights and connecting researchers worldwide, include:

- The Agricultural Modelling Intercomparison and Improvement Project, AgMIP (https://www.agmip.org/),
- The InterSectoral Impact Model Improvement Project, ISIMIP (https://www.isimip.org/), and
- The CGIAR research program on Climate Change, Agriculture and Food Security, CCFAS (https://ccafs.cgiar.org/research/projects#.W-UtiadKiM8).

Centres of expertise include:

- The Potsdam Institute on Climate impact research, PIK (https://www.pik-potsdam.de/), and
- The Oregon Climate Change Research Institute, OCCRI (http://www.occri.net/).

8 References

Ainsworth, E. A. 2008. Rice production in a changing climate: a meta-analysis of responses to elevated carbon dioxide and elevated ozone concentration. *Glob. Change Biol.* 14, 1642–50.

Ainsworth, E. A. 2017. Understanding and improving global crop response to ozone pollution. *Plant J.* 90, 886–97. https://doi.org/10.1111/tpj.13298.

Ainsworth, E. A. and Long, S. P. 2005. What have we learned from 15 years of free-air CO_2 enrichment (FACE)? A meta-analytic review of the responses of photosynthesis, canopy properties and plant production to rising CO_2. *New Phytol.* 165, 351–71.

Ainsworth, E. A. and Ort, D. R. 2010. How do we improve crop production in a warming world? *Plant Physiol.* 154, 526–30. https://doi.org/10.1104/pp.110.161349.

Ainsworth, E. A. and Rogers, A. 2007. The response of photosynthesis and stomatal conductance to rising [CO_2]: mechanisms and environmental interactions. *Plant Cell Environ.* 30, 258–70. https://doi.org/10.1111/j.1365-3040.2007.01641.x.

Ainsworth, E. A., Davey, P., Bernacchi, C., Dermody, O., Heaton, E. and Moore, D. 2002. A meta-analysis of elevated [CO_2] effects on soybean (*Glycine max*) physiology, growth and yield. *Glob. Change Biol.* 8, 695–709.

Ainsworth, E. A., Yendrek, C. R., Sitch, S., Collins, W. J. and Emberson, L. D. 2012. The effects of tropospheric ozone on net primary productivity and implications for climate change. *Annu. Rev. Plant Biol.* 63, 637–61. https://doi.org/10.1146/annurev-arplant-042110-103829.

Akter, N. and Rafiqul Islam, M. 2017. Heat stress effects and management in wheat. A review. *Agron. Sustain. Dev.* 37, 37. https://doi.org/10.1007/s13593-017-0443-9.

Alexandratos, N. and Bruinsma, J. 2012. *World Agriculture Towards 2030/2050: The 2012 Revision* (ESA Working paper No. 12- 03). FAO, Rome.

Andersen, C. P. 2003. Source-sink balance and carbon allocation below ground in plants exposed to ozone. New Phytol. 157, 213–28. https://doi.org/10.1046/j.1469-8137.2003.00674.x.

Angulo, C., Rötter, R., Lock, R., Enders, A., Fronzek, S. and Ewert, F. 2013. Implication of crop model calibration strategies for assessing regional impacts of climate change in Europe. Agric. For. Meteorol. 170, 32–46. https://doi.org/10.1016/j.agrformet.2012.11.017.

Ashmore, M. R. 2005. Assessing the future global impacts of ozone on vegetation. Plant Cell Environ. 28, 949–64. https://doi.org/10.1111/j.1365-3040.2005.01341.x.

Asseng, S., Foster, I. and Turner, N. C. 2011. The impact of temperature variability on wheat yields. Glob. Change Biol. 17, 997–1012. https://doi.org/10.1111/j.1365-2486.2010.02262.x.

Asseng, S., Ewert, F., Rosenzweig, C., Jones, J. W., Hatfield, J. L., Ruane, A. C., Boote, K. J., Thorburn, P. J., Rotter, R. P., Cammarano, D., Brisson, N., Basso, B., Martre, P., Aggarwal, P. K., Angulo, C., Bertuzzi, P., Biernath, C., Challinor, A. J., Doltra, J., Gayler, S., Goldberg, R., Grant, R., Heng, L., Hooker, J., Hunt, L. A., Ingwersen, J., Izaurralde, R. C., Kersebaum, K. C., Muller, C., Naresh Kumar, S., Nendel, C., O'Leary, G., Olesen, J. E., Osborne, T. M., Palosuo, T., Priesack, E., Ripoche, D., Semenov, M. A., Shcherbak, I., Steduto, P., Stockle, C., Stratonovitch, P., Streck, T., Supit, I., Tao, F., Travasso, M., Waha, K., Wallach, D., White, J. W., Williams, J. R. and Wolf, J. 2013. Uncertainty in simulating wheat yields under climate change. *Nat. Clim. Change* 3, 827–32. https://doi.org/10.1038/nclimate1916.

Asseng, S., Ewert, F., Martre, P., Rötter, R. P., Lobell, D. B., Cammarano, D., Kimball, B. A., Ottman, M. J., Wall, G. W., White, J. W., Reynolds, M. P., Alderman, P. D., Prasad, P. V. V., Aggarwal, P. K., Anothai, J., Basso, B., Biernath, C., Challinor, A. J., De Sanctis, G., Doltra, J., Fereres, E., Garcia-Vila, M., Gayler, S., Hoogenboom, G., Hunt, L. A., Izaurralde, R. C., Jabloun, M., Jones, C. D., Kersebaum, K. C., Koehler, A.-K., Müller, C., Naresh Kumar, S., Nendel, C., O'Leary, G., Olesen, J. E., Palosuo, T., Priesack, E., Eyshi Rezaei, E., Ruane, A. C., Semenov, M. A., Shcherbak, I., Stöckle, C., Stratonovitch, P., Streck, T., Supit, I., Tao, F., Thorburn, P. J., Waha, K., Wang, E., Wallach, D., Wolf, J., Zhao, Z. and Zhu, Y. 2015. Rising temperatures reduce global wheat production. *Nat. Clim. Change* 5, 143-7. https://doi.org/10.1038/nclimate2470.

Atkin, O. K. and Tjoelker, M. G. 2003. Thermal acclimation and the dynamic response of plant respiration to temperature. *Trends Plant Sci.* 8, 343-51. https://doi.org/10.1016/S1360-1385(03)00136-5.

Avnery, S., Mauzerall, D. L., Liu, J. and Horowitz, L. W. 2011a. Global crop yield reductions due to surface ozone exposure: 1. Year 2000 crop production losses and economic damage. *Atmos. Environ.* 45, 2284-96. https://doi.org/10.1016/j.atmosenv.2010.11.045.

Avnery, S., Mauzerall, D. L., Liu, J. and Horowitz, L. W. 2011b. Global crop yield reductions due to surface ozone exposure: 2. Year 2030 potential crop production losses and economic damage under two scenarios of O_3 pollution. *Atmos. Environ.* 45, 2297-309. https://doi.org/10.1016/j.atmosenv.2011.01.002.

Awal, M. A., Ikeda, T. and Itoh, R. 2003. The effect of soil temperature on source-sink economy in peanut (*Arachis hypogaea*). *Environ. Exp. Bot.* 50, 41-50. https://doi.org/10.1016/S0098-8472(02)00111-9.

Balkovič, J., Skalský, R., Folberth, C., Khabarov, N., Schmid, E., Madaras, M., Obersteiner, M. and van der Velde, M. 2018. Impacts and uncertainties of +2°C of climate change and soil degradation on European crop calorie supply. *Earths Future* 6, 373-95. https://doi.org/10.1002/2017EF000629.

Barlow, K. M., Christy, B. P., O'Leary, G. J., Riffkin, P. A. and Nuttall, J. G. 2015. Simulating the impact of extreme heat and frost events on wheat crop production: a review. *Field Crops Res.* 171, 109-19. https://doi.org/10.1016/j.fcr.2014.11.010.

Barnabás, B., Jäger, K. and Fehér, A. 2008. The effect of drought and heat stress on reproductive processes in cereals. *Plant Cell Environ.* 31, 11-38. https://doi.org/10.1111/j.1365-3040.2007.01727.x.

Bernacchi, C. J., Leakey, A. D. B., Kimball, B. A. and Ort, D. R. 2011. Growth of soybean at future tropospheric ozone concentrations decreases canopy evapotranspiration and soil water depletion. *Environ. Pollut.* 159, 1464-72. https://doi.org/10.1016/j.envpol.2011.03.011.

Betzelberger, A. M., Gillespie, K. M., Mcgrath, J. M., Koester, R. P., Nelson, R. L. and Ainsworth, E. A. 2010. Effects of chronic elevated ozone concentration on antioxidant capacity, photosynthesis and seed yield of 10 soybean cultivars. *Plant Cell Environ.* 33, 1569-81. https://doi.org/10.1111/j.1365-3040.2010.02165.x.

Bindi, M., Palosuo, T., Trnka, M. and Semenov, M. 2015. Modelling climate change impacts on crop production for food security. *Clim. Res.* 65, 3-5. https://doi.org/10.3354/cr01342.

Bisbis, M. B., Gruda, N. and Blanke, M. 2018. Potential impacts of climate change on vegetable production and product quality - A review. *J. Clean. Prod.* 170, 1602-20. https://doi.org/10.1016/j.jclepro.2017.09.224.

Bishop, K. A., Leakey, A. D. B. and Ainsworth, E. A. 2014. How seasonal temperature or water inputs affect the relative response of C3 crops to elevated [CO_2]: a global analysis of open top chamber and free air CO_2 enrichment studies. *Food Energy Secur.* 3, 33–45. https://doi.org/10.1002/fes3.44.

Black, V. J., Black, C. R., Roberts, J. A. and Stewart, C. A. 2000. Impact of ozone on the reproductive development of plants. *New Phytol.* 147, 421–47. https://doi.org/10.1046/j.1469-8137.2000.00721.x.

Booker, F. L., Fiscus, E. L. and Miller, J. E. 2004. Combined effects of elevated atmospheric carbon dioxide and ozone on soybean whole-plant water use. *Environ. Manage.* 33, S355–62. https://doi.org/10.1007/s00267-003-9144-z.

Booker, F., Muntifering, R., McGrath, M., Burkey, K., Decoteau, D., Fiscus, E., Manning, W., Krupa, S., Chappelka, A. and Grantz, D. 2009. The ozone component of global change: potential effects on agricultural and horticultural plant yield, product quality and interactions with invasive species. *J. Integr. Plant Biol.* 51, 337–51. https://doi.org/10.1111/j.1744-7909.2008.00805.x.

Bourgault, M., Löw, M., Tausz-Posch, S., Nuttall, J. G., Delahunty, A. J., Brand, J., Panozzo, J. F., McDonald, L., O'Leary, G. J., Armstrong, R. D., Fitzgerald, G. J. and Tausz, M. 2018. Effect of a heat wave on lentil grown under free-air CO_2 enrichment (FACE) in a semi-arid environment. *Crop Sci.* 58, 803–12. https://doi.org/10.2135/cropsci2017.09.0565.

Bouwman, A. F., Kram, T. and Klein Goldewijk, K. 2006. *Integrated Modelling of Global Environmental Change. An Overview of IMAGE 2.4.* Netherlands Environmental Assessment Agency (MNP), Bilthoven, The Netherlands.

Brauer, M., Freedman, G., Frostad, J., van Donkelaar, A., Martin, R. V., Dentener, F., Dingenen, R. van, Estep, K., Amini, H., Apte, J. S., Balakrishnan, K., Barregard, L., Broday, D., Feigin, V., Ghosh, S., Hopke, P. K., Knibbs, L. D., Kokubo, Y., Liu, Y., Ma, S., Morawska, L., Sangrador, J. L. T., Shaddick, G., Anderson, H. R., Vos, T., Forouzanfar, M. H., Burnett, R. T. and Cohen, A. 2016. Ambient air pollution exposure estimation for the global burden of disease 2013. *Environ. Sci. Technol.* 50, 79–88. https://doi.org/10.1021/acs.est.5b03709.

Broberg, M. C., Feng, Z., Xin, Y. and Pleijel, H. 2015. Ozone effects on wheat grain quality – a summary. *Environ. Pollut.* 197, 203–13. https://doi.org/10.1016/j.envpol.2014.12.009.

Broberg, M. C., Högy, P. and Pleijel, H. 2017. CO_2-induced changes in wheat grain composition: meta-analysis and response functions. *Agronomy* 7, 32–49.

Burkart, S., Manderscheid, R., Wittich, K.-P., Löpmeier, F. J. and Weigel, H.-J. 2011. Elevated CO_2 effects on canopy and soil water flux parameters measured using a large chamber in crops grown with free-air CO_2 enrichment. *Plant Biol.* 13, 258–69. https://doi.org/10.1111/j.1438-8677.2010.00360.x.

Cammarano, D., Zierden, D., Stefanova, L., Asseng, S., O'Brien, J. J. and Jones, J. W. 2016. Using historical climate observations to understand future climate change crop yield impacts in the Southeastern US. *Clim. Change* 134, 311–26. https://doi.org/10.1007/s10584-015-1497-9.

Cardoso-Vilhena, J. 2004. Mechanisms underlying the amelioration of O_3-induced damage by elevated atmospheric concentrations of CO_2. *J. Exp. Bot.* 55, 771–81. https://doi.org/10.1093/jxb/erh080.

Castagna, A. and Ranieri, A. 2009. Detoxification and repair process of ozone injury: from O_3 uptake to gene expression adjustment. *Environ. Pollut.* 157, 1461–9. https://doi.org/10.1016/j.envpol.2008.09.029.

Cattiaux, J., Vautard, R., Cassou, C., Yiou, P., Masson-Delmotte, V. and Codron, F. 2010. Winter 2010 in Europe: a cold extreme in a warming climate. *Geophys. Res. Lett.* 37. https://doi.org/10.1029/2010GL044613.

Ceccarelli, S., Grando, S., Maatougui, M., Michael, M., Slash, M., Haghparast, R., Rahmanian, M., Taheri, A., Al-Yassin, A., Benbelkacem, A., Labdi, M., Mimoun, H. and Nachit, M. 2010. Plant breeding and climate changes. *J. Agric. Sci.* 148, 627–37. https://doi.org/10.1017/S0021859610000651.

Challinor, A., Wheeler, T., Garforth, C., Craufurd, P. and Kassam, A. 2007. Assessing the vulnerability of food crop systems in Africa to climate change. *Clim. Change* 83, 381–99. https://doi.org/10.1007/s10584-007-9249-0.

Challinor, A. J., Ewert, F., Arnold, S., Simelton, E. and Fraser, E. 2009. Crops and climate change: progress, trends, and challenges in simulating impacts and informing adaptation. *J. Exp. Bot.* 60, 2775–89. https://doi.org/10.1093/jxb/erp062.

Challinor, A. J., Smith, M. S. and Thornton, P. 2013. Use of agro-climate ensembles for quantifying uncertainty and informing adaptation. *Agric. For. Meteorol.* 170, 2–7. https://doi.org/10.1016/j.agrformet.2012.09.007.

Challinor, A. J., Martre, P., Asseng, S., Thornton, P. and Ewert, F. 2014a. Making the most of climate impacts ensembles. *Nat. Clim. Change* 4, 77–80.

Challinor, A. J., Watson, J., Lobell, D. B., Howden, S. M., Smith, D. R. and Chhetri, N. 2014b. A meta-analysis of crop yield under climate change and adaptation. *Nat. Clim. Change* 4, 287–91. https://doi.org/10.1038/nclimate2153.

Challinor, A. J., Koehler, A.-K., Ramirez-Villegas, J., Whitfield, S. and Das, B. 2016. Current warming will reduce yields unless maize breeding and seed systems adapt immediately. *Nat. Clim. Change* 6, 954–8. https://doi.org/10.1038/nclimate3061.

Challinor, A. J., Müller, C., Asseng, S., Deva, C., Nicklin, K. J., Wallach, D., Vanuytrecht, E., Whitfield, S., Ramirez-Villegas, J. and Koehler, A.-K. 2018. Improving the use of crop models for risk assessment and climate change adaptation. *Agric. Syst.* 159, 296–306. https://doi.org/10.1016/j.agsy.2017.07.010.

Chen, J., Chen, C., Tian, Y., Zhang, X., Dong, W., Zhang, B., Zhang, J., Zheng, C., Deng, A., Song, Z., Peng, C. and Zhang, W. 2017. Differences in the impacts of nighttime warming on crop growth of rice-based cropping systems under field conditions. *Eur. J. Agron.* 82, 80–92. https://doi.org/10.1016/j.eja.2016.10.006.

Chuwah, C., van Noije, T., van Vuuren, D. P., Stehfest, E. and Hazeleger, W. 2015. Global impacts of surface ozone changes on crop yields and land use. *Atmos. Environ.* 106, 11–23. https://doi.org/10.1016/j.atmosenv.2015.01.062.

Cleland, E. E., Chuine, I., Menzel, A., Mooney, H. A. and Schwartz, M. D. 2007. Shifting plant phenology in response to global change. *Trends Ecol. Evol.* 22, 357–65. https://doi.org/10.1016/j.tree.2007.04.003.

CLRTAP. 2017. Manual on methodologies and criteria for modelling and mapping critical loads and levels and air pollution effects, risks and trends. Chapter III: mapping critical levels for vegetation. UNECE Convention on Long-range Transboundary Air Pollution.

Cooper, O. R., Parrish, D. D., Ziemke, J., Balashov, N. V., Cupeiro, M., Galbally, I. E., Gilge, S., Horowitz, L., Jensen, N. R., Lamarque, J.-F., Naik, V., Oltmans, S. J., Schwab, J., Shindell, D. T., Thompson, A. M., Thouret, V., Wang, Y. and Zbinden, R. M. 2014. Global distribution and trends of tropospheric ozone: an observation-based review. *Elem. Sci. Anth.* 2. https://doi.org/10.12952/journal.elementa.000029.

Craigon, J., Fangmeier, A., Jones, M., Donnelly, A., Bindi, M., De Temmerman, L., Persson, K. and Ojanpera, K. 2002. Growth and marketable-yield responses of potato to increased CO_2 and ozone. *Eur. J. Agron.* 17, 273–89. https://doi.org/10.1016/S1161-0301(02)00066-7.

Craufurd, P. Q. and Wheeler, T. R. 2009. Climate change and the flowering time of annual crops. *J. Exp. Bot.* 60, 2529–39.

Crawford, A. J., McLachlan, D. H., Hetherington, A. M. and Franklin, K. A. 2012. High temperature exposure increases plant cooling capacity. *Curr. Biol.* 22, R396-7. https://doi.org/10.1016/j.cub.2012.03.044.

Danielsson, H., Karlsson, G. P., Karlsson, P. E. and Håkan Pleijel, H. 2003. Ozone uptake modelling and flux-response relationships–an assessment of ozone-induced yield loss in spring wheat. *Atmospheric Environment* 37, 475–85. doi:10.1016/S1352-2310(02)00924-X.

Davenport, F., Funk, C. and Galu, G. 2018. How will East African maize yields respond to climate change and can agricultural development mitigate this response? *Clim. Change*. https://doi.org/10.1007/s10584-018-2149-7.

Decock, C. and Six, J. 2012. Effects of elevated CO_2 and O_3 on N-cycling and N_2O emissions: a short-term laboratory assessment. *Plant Soil* 351, 277–92. https://doi.org/10.1007/s11104-011-0961-1.

Dermody, O., Long, S. P., McConnaughay, K. and Delucia, E. H. 2008. How do elevated CO_2 and O_3 affect the interception and utilization of radiation by a soybean canopy? *Glob. Change Biol.* 14, 556–64.

Deryng, D., Conway, D., Ramankutty, N., Price, J. and Warren, R. 2014. Global crop yield response to extreme heat stress under multiple climate change futures. *Environ. Res. Lett.* 9, 034011. https://doi.org/10.1088/1748-9326/9/3/034011.

Deryng, D., Elliott, J., Folberth, C., Müller, C., Pugh, T. A. M., Boote, K. J., Conway, D., Ruane, A. C., Gerten, D., Jones, J. W., Khabarov, N., Olin, S., Schaphoff, S., Schmid, E., Yang, H. and Rosenzweig, C. 2016. Regional disparities in the beneficial effects of rising CO_2 concentrations on crop water productivity. *Nat. Clim. Change* 6, 786-90. https://doi.org/10.1038/nclimate2995.

Dizengremel, P., Thiec, D. L., Hasenfratz-Sauder, M.-P., Vaultier, M.-N., Bagard, M. and Jolivet, Y. 2009. Metabolic-dependent changes in plant cell redox power after ozone exposure. *Plant Biol.* 11, 35–42. https://doi.org/10.1111/j.1438-8677.2009.00261.x.

Dong, J., Gruda, N., Lam, S. K., Li, X. and Duan, Z. 2018. Effects of elevated CO_2 on nutritional quality of vegetables: a review. *Front. Plant Sci.* 9. https://doi.org/10.3389/fpls.2018.00924.

Drake, B. G., Gonzalezmeler, M. A. and Long, S. P. 1997. More efficient plants: a consequence of rising atmospheric CO_2? *Annu. Rev. Plant Physiol. Plant Mol. Biol.* 48, 609–39.

Duncan, J. M. A., Dash, J. and Atkinson, P. M. 2015. Elucidating the impact of temperature variability and extremes on cereal croplands through remote sensing. *Glob. Change Biol.* 21, 1541–51. https://doi.org/10.1111/gcb.12660.

Elbehri, A. (Ed.), 2015. *Climate Change and Food Systems: Global Assessments and Implications for Food Security and Trade*. FAO, Food and Agriculture Organization of the United Nations, Rome.

Emberson, L. D., Ashmore, M. R., Murray, F., Kuylenstierna, J. C. I., Percy, K. E., Izuta, T., Zheng, Y., Shimizu, H., Sheu, B. H., Liu, C. P., Agrawal, M., Wahid, A., Abdel-Latif, N. M., van Tienhoven, M., de Bauer, L. I. and Domingos, M. 2001. Impacts of air pollutants on vegetation in developing countries. *Water. Air. Soil Pollut.* 130, 107–18. https://doi.org/10.1023/A:1012251503358.

Erbs, M., Manderscheid, R., Hüther, L., Schenderlein, A., Wieser, H., Dänicke, S. and Weigel, H.-J. 2015. Free-air CO_2 enrichment modifies maize quality only under drought stress. Agron. Sustain. Dev. 35, 203–12. https://doi.org/10.1007/s13593-014-0226-5.

Eyshi Rezaei, E., Webber, H., Gaiser, T., Naab, J. and Ewert, F. 2015. Heat stress in cereals: mechanisms and modelling. *Eur. J. Agron.* 64, 98–113. https://doi.org/10.1016/j.eja.2014.10.003.

FAO, IFAD, UNICEF, WFP and WHO. 2018. *The State of Food Security and Nutrition in the World 2018*. FAO, Rome.

FAOSTAT. 2018. Database on agriculture [WWW Document]. URL http://faostat.fao.org/.

Feng, Z. and Kobayashi, K. 2009. Assessing the impacts of current and future concentrations of surface ozone on crop yield with meta-analysis. *Atmos. Environ.* 43, 1510–19. https://doi.org/10.1016/j.atmosenv.2008.11.033.

Feng, Z., Kobayashi, K. and Ainsworth, E. A. 2008. Impact of elevated ozone concentration on growth, physiology, and yield of wheat (*Triticum aestivum* L.): a meta-analysis. *Glob. Change Biol.* 14, 2696–708. https://doi.org/10.1111/j.1365-2486.2008.01673.x.

Feng, Z., Pang, J., Kobayashi, K., Zhu, J. and Ort, D. R. 2011. Differential responses in two varieties of winter wheat to elevated ozone concentration under fully open-air field conditions. *Glob. Change Biol.* 17, 580–91. https://doi.org/10.1111/j.1365-2486.2010.02184.x.

Feng, Z., Sun, J., Wan, W., Hu, E. and Calatayud, V. 2014. Evidence of widespread ozone-induced visible injury on plants in Beijing, China. *Environ. Pollut.* 193, 296–301. https://doi.org/10.1016/j.envpol.2014.06.004.

Feng, Z., Hu, E., Wang, X., Jiang, L. and Liu, X. 2015. Ground-level O_3 pollution and its impacts on food crops in China: a review. *Environ. Pollut.* 199, 42–48. https://doi.org/10.1016/j.envpol.2015.01.016.

Ferris, R., Ellis, R. H., Wheeler, T. R. and Hadley, P. 1998. Effect of high temperature stress at anthesis on grain yield and biomass of field-grown crops of wheat. *Ann. Bot.* 82, 631–9.

Field, C. B., Barros, V. R., Dokken, D. J., Mach, K. J., Mastrandrea, M. D., Bilir, T. E., Chatterjee, M., Ebi, K. L., Estrada, Y. O., Genova, R. C., Girma, B., Kissel, E. S., Levy, A. N., MacCracken, S., Mastrandrea, P. R., and White, L. L. 2014. *Climate Change 2014: Impacts, Adaptations, and Vulnerability. Part A: Global and Sectoral Aspects. Contribution of Working Group II to the Fifth Assessment Report (AR5) of the Intergovernmental Panel on Climate Change*. Cambridge University Press, Cambridge, UK and New York, NY.

Fiscus, E. L., Booker, F. L. and Burkey, K. O. 2005. Crop responses to ozone: uptake, modes of action, carbon assimilation and partitioning. *Plant Cell Environ.* 28, 997–1011. https://doi.org/10.1111/j.1365-3040.2005.01349.x.

Fitzgerald, G. J., Tausz, M., O'Leary, G., Mollah, M. R., Tausz-Posch, S., Seneweera, S., Mock, I., Löw, M., Partington, D. L., McNeil, D. and Norton, R. M. 2016. Elevated atmospheric [CO_2] can dramatically increase wheat yields in semi-arid environments and buffer against heat waves. *Glob. Change Biol.* 22, 2269–84. https://doi.org/10.1111/gcb.13263.

Frei, M. 2015. Breeding of ozone resistant rice: relevance, approaches and challenges. *Environ. Pollut.* 197, 144–55. https://doi.org/10.1016/j.envpol.2014.12.011.

Frieler, K., Schauberger, B., Arneth, A., Balkovič, J., Chryssanthacopoulos, J., Deryng, D., Elliott, J., Folberth, C., Khabarov, N., Müller, C., Olin, S., Pugh, T. A. M., Schaphoff, S., Schewe, J., Schmid, E., Warszawski, L. and Levermann, A. 2017. Understanding the weather signal in national crop-yield variability. *Earths Future* 5, 605–16. https://doi.org/10.1002/2016EF000525.

Fuhrer, J. 2003. Agroecosystem responses to combinations of elevated CO_2, ozone, and global climate change. *Agric. Ecosyst. Environ.* 97, 1–20. https://doi.org/10.1016/S0167-8809(03)00125-7.

Fuhrer, J. 2009. Ozone risk for crops and pastures in present and future climates. *Naturwissenschaften* 96, 173–94. https://doi.org/10.1007/s00114-008-0468-7.

Fuhrer, J. and Booker, F. 2003. Ecological issues related to ozone: agricultural issues. *Environ. Int.* 29, 141–54. https://doi.org/10.1016/S0160-4120(02)00157-5.

Fuhrer, J., Martin, M. V., Mills, G., Heald, C. L., Harmens, H., Hayes, F., Sharps, K., Bender, J. and Ashmore, M. R. 2016. Current and future ozone risks to global terrestrial biodiversity and ecosystem processes. *Ecol. Evol.* 6, 8785-99. https://doi.org/10.1002/ece3.2568.

Godfray, H. C. J., Beddington, J. R., Crute, I. R., Haddad, L., Lawrence, D., Muir, J. F., Pretty, J., Robinson, S., Thomas, S. M. and Toulmin, C. 2010. Food security: The challenge of feeding 9 billion people. *Science* 327, 812–18. https://doi.org/10.1126/science.1185383.

Gornall, J., Betts, R., Burke, E., Clark, R., Camp, J., Willett, K. and Wiltshire, A. 2010. Implications of climate change for agricultural productivity in the early twenty-first century. *Philos. Trans. R. Soc. Lond. B Biol. Sci.* 365, 2973–89. https://doi.org/10.1098/rstb.2010.0158.

Gray, S. B. and Brady, S. M. 2016. Plant developmental responses to climate change. *Dev. Biol., Plant Dev.* 419, 64–77. https://doi.org/10.1016/j.ydbio.2016.07.023.

Gray, S. B., Dermody, O., Klein, S. P., Locke, A. M., McGrath, J. M., Paul, R. E., Rosenthal, D. M., Ruiz-Vera, U. M., Siebers, M. H., Strellner, R., Ainsworth, E. A., Bernacchi, C. J., Long, S. P., Ort, D. R. and Leakey, A. D. B. 2016. Intensifying drought eliminates the expected benefits of elevated carbon dioxide for soybean. *Nat. Plants* 2, 16132. https://doi.org/10.1038/nplants.2016.132.

Hasegawa, T., Sakai, H., Tokida, T., Nakamura, H., Zhu, C., Usui, Y., Yoshimoto, M., Fukuoka, M., Wakatsuki, H., Katayanagi, N., Matsunami, T., Kaneta, Y., Sato, T., Takakai, F., Sameshima, R., Okada, M., Mae, T. and Makino, A. 2013. Rice cultivar responses to elevated CO_2 at two free-air CO_2 enrichment (FACE) sites in Japan. *Funct. Plant Biol.* 40, 148–59.

Hatfield, J. L. and Prueger, J. H. 2015. Temperature extremes: effect on plant growth and development. *Weather Clim. Extrem.* 10, 4–10. https://doi.org/10.1016/j.wace.2015.08.001.

Hendrey, G. R., Lewin, K. F. and Nagy, J. 1993. Free air carbon dioxide enrichment: development, progress, results. *Vegetatio* 104–5, 17–31. https://doi.org/10.1007/BF00048142.

Hochman, Z., Horan, H., Reddy, D. R., Sreenivas, G., Tallapragada, C., Adusumilli, R., Gaydon, D. S., Laing, A., Kokic, P., Singh, K. K. and Roth, C. H. 2017. Smallholder farmers managing climate risk in India: 2. Is it climate-smart? *Agric. Syst.* 151, 61-72. https://doi.org/10.1016/j.agsy.2016.11.007.

Ingram, J. S. I. and Porter, J. R. 2015. Plant science and the food security agenda. *Nat. Plants* 1, 15173. https://doi.org/10.1038/NPLANTS.2015.173.

Jablonski, L. M., Wang, X. Z. and Curtis, P. S. 2002. Plant Reproduction Under Elevated Co_2 Conditions: a Meta-Analysis of Reports on 79 Crop and Wild Species. *New Phytol.* 156, 9–26.

Jones, J. W., Antle, J. M., Basso, B., Boote, K. J., Conant, R. T., Foster, I., Godfray, H. C. J., Herrero, M., Howitt, R. E., Janssen, S., Keating, B. A., Munoz-Carpena, R., Porter, C. H., Rosenzweig, C. and Wheeler, T. R. 2017. Toward a new generation of agricultural system data, models, and knowledge products: state of agricultural systems science. *Agric. Syst.* https://doi.org/10.1016/j.agsy.2016.09.021.

Kawasaki, K. and Uchida, S. 2016. Quality matters more than quantity: asymmetric temperature effects on crop yield and Quality grade. *Am. J. Agric. Econ.* 98, 1195–209. https://doi.org/10.1093/ajae/aaw036.

Keating, B. A., Herrero, M., Carberry, P. S., Gardner, J. and Cole, M. B. 2014. Food wedges: framing the global food demand and supply challenge towards 2050. *Glob. Food Secur.* 3, 125–32. https://doi.org/10.1016/j.gfs.2014.08.004.

Kimball, B. A. 2016. Crop responses to elevated CO_2 and interactions with H_2O, N, and temperature. *Curr. Opin. Plant Biol.* 31, 36–43. https://doi.org/10.1016/j.pbi.2016.03.006.

Kimball, B. A., Kobayashi, K. and Bindi, M. 2002. Responses of agricultural crops to free-air CO_2 enrichment. *Adv. Agron.* 77, 293–368.

King, M., Altdorff, D., Li, P., Galagedara, L., Holden, J. and Unc, A. 2018. Northward shift of the agricultural climate zone under 21st-century global climate change. *Sci. Rep.* 8. https://doi.org/10.1038/s41598-018-26321-8.

Knox, J., Hess, T., Daccache, A. and Wheeler, T. 2012. Climate change impacts on crop productivity in Africa and South Asia. *Environ. Res. Lett.* 7, 034032. https://doi.org/10.1088/1748-9326/7/3/034032.

Kou, T., Wang, L., Zhu, J., Xie, Z. and Wang, Y. 2014. Ozone pollution influences soil carbon and nitrogen sequestration and aggregate composition in paddy soils. *Plant Soil* 380, 305–13. https://doi.org/10.1007/s11104-014-2096-7.

Lam, S. K., Chen, D., Norton, R., Armstrong, R. and Mosier, A. R. 2012. Nitrogen dynamics in grain crop and legume pasture systems under elevated atmospheric carbon dioxide concentration: a meta-analysis. *Glob. Change Biol.* 18, 2853–9. https://doi.org/10.1111/j.1365-2486.2012.02758.x.

Leakey, A. D. B. 2009. Rising atmospheric carbon dioxide concentration and the future of C4 crops for food and fuel. *Proc. R. Soc. B-Biol. Sci.* 276, 2333–43.

Leakey, A. D. B., Ainsworth, E. A., Bernacchi, C. J., Rogers, A., Long, S. P. and Ort, D. R. 2009. Elevated CO_2 effects on plant carbon, nitrogen, and water relations: six important lessons from FACE. *J. Exp. Bot.* 60, 2859–76.

Lefohn, A. S., Malley, C. S., Smith, L., Wells, B., Hazucha, M., Simon, H., Naik, V., Mills, G., Schultz, M. G., Paoletti, E., Marco, A. D., Xu, X., Zhang, L., Wang, T., Neufeld, H. S., Musselman, R. C., Tarasick, D., Brauer, M., Feng, Z., Tang, H., Kobayashi, K., Sicard, P., Solberg, S. and Gerosa, G. 2018. Tropospheric ozone assessment report: global ozone metrics for climate change, human health, and crop/ecosystem research. *Elem. Sci. Anth.* 6. https://doi.org/10.1525/elementa.279.

Lehnherr, B., Grandjean, A., Mächler, F. and Fuhrer, J. 1987. The effect of ozone in ambient air on ribulosebisphosphate carboxylase/oxygenase activity decreases photosynthesis and grain yield in wheat. *J. Plant Physiol.* 130, 189–200. https://doi.org/10.1016/S0176-1617(87)80223-7.

Leisner, C. P. and Ainsworth, E. A. 2012. Quantifying the effects of ozone on plant reproductive growth and development. *Glob. Change Biol.* 18, 606–16. https://doi.org/10.1111/j.1365-2486.2011.02535.x.

Leisner, C. P., Wood, J. C., Vaillancourt, B., Tang, Y., Douches, D. S., Robin Buell, C. and Winkler, J. A. 2018. Impact of choice of future climate change projection on growth chamber experimental outcomes: a preliminary study in potato. *Int. J. Biometeorol.* 62, 669–79. https://doi.org/10.1007/s00484-017-1475-1.

Lesk, C., Rowhani, P. and Ramankutty, N. 2016. Influence of extreme weather disasters on global crop production. *Nature* 529, 84–87. https://doi.org/10.1038/nature16467.

Liu, B., Asseng, S., Müller, C., Ewert, F., Elliott, J., Lobell, D. B., Martre, P., Ruane, A. C., Wallach, D., Jones, J. W., Rosenzweig, C., Aggarwal, P. K., Alderman, P. D., Anothai, J., Basso, B., Biernath, C., Cammarano, D., Challinor, A., Deryng, D., Sanctis, G. D., Doltra, J., Fereres, E., Folberth, C., Garcia-Vila, M., Gayler, S., Hoogenboom, G., Hunt, L. A., Izaurralde, R. C., Jabloun, M., Jones, C. D., Kersebaum, K. C., Kimball, B. A., Koehler, A.-K., Kumar, S. N., Nendel, C., O'Leary, G. J., Olesen, J. E., Ottman, M. J., Palosuo, T., Prasad, P. V. V., Priesack, E., Pugh, T. A. M., Reynolds, M., Rezaei, E. E., Rötter, R. P., Schmid, E., Semenov, M. A., Shcherbak, I., Stehfest, E., Stöckle, C. O., Stratonovitch, P., Streck, T., Supit, I., Tao, F., Thorburn, P., Waha, K., Wall, G. W., Wang, E., White, J. W., Wolf, J., Zhao, Z. and Zhu, Y. 2016. Similar estimates of temperature impacts on global wheat yield by three independent methods. *Nat. Clim. Change* 6, 1130–6. https://doi.org/10.1038/nclimate3115.

Lizaso, J. I., Ruiz-Ramos, M., Rodríguez, L., Gabaldon-Leal, C., Oliveira, J. A., Lorite, I. J., Sánchez, D., García, E. and Rodríguez, A. 2018. Impact of high temperatures in maize: phenology and yield components. *Field Crops Res.* 216, 129–40. https://doi.org/10.1016/j.fcr.2017.11.013.

Lobell, D. B. and Asseng, S. 2017. Comparing estimates of climate change impacts from process-based and statistical crop models. *Environ. Res. Lett.* 12, 015001. https://doi.org/10.1088/1748-9326/aa518a.

Lobell, D. B. and Tebaldi, C. 2014. Getting caught with our plants down: the risks of a global crop yield slowdown from climate trends in the next two decades. *Environ. Res. Lett.* 9, 074003. https://doi.org/10.1088/1748-9326/9/7/074003.

Lobell, D. B., Burke, M. B., Tebaldi, C., Mastrandrea, M. D., Falcon, W. P. and Naylor, R. L. 2008. Prioritizing climate change adaptation needs for food security in 2030. *Science* 319, 607–10.

Lobell, D. B., Sibley, A. and Ortiz-Monasterio, J. I. 2012. Extreme heat effects on wheat senescence in India. *Nat. Clim. Change* 2, 186–9. https://doi.org/10.1038/nclimate1356

Ludwikow, A. and Sadowski, J. 2008. Gene networks in plant ozone stress response and tolerance. *J. Integr. Plant Biol.* 50, 1256–67. https://doi.org/10.1111/j.1744-7909.2008.00738.x.

Maiorano, A., Martre, P., Asseng, S., Ewert, F., Müller, C., Rötter, R. P., Ruane, A. C., Semenov, M. A., Wallach, D., Wang, E., Alderman, P. D., Kassie, B. T., Biernath, C., Basso, B., Cammarano, D., Challinor, A. J., Doltra, J., Dumont, B., Rezaei, E. E., Gayler, S., Kersebaum, K. C., Kimball, B. A., Koehler, A.-K., Liu, B., O'Leary, G. J., Olesen, J. E., Ottman, M. J., Priesack, E., Reynolds, M., Stratonovitch, P., Streck, T., Thorburn, P. J., Waha, K., Wall, G. W., White, J. W., Zhao, Z. and Zhu, Y. 2017. Crop model improvement reduces the uncertainty of the response to temperature of multi-model ensembles. *Field Crops Res.* 202, 5–20. https://doi.org/10.1016/j.fcr.2016.05.001.

Manderscheid, R., Erbs, M. and Weigel, H.-J. 2014. Interactive effects of free-air CO_2 enrichment and drought stress on maize growth. *Eur. J. Agron.* 52, 11-21. https://doi.org/10.1016/j.eja.2011.12.007.

Manderscheid, R., Dier, M., Erbs, M., Sickora, J. and Weigel, H.-J. 2018. Nitrogen supply - A determinant in water use efficiency of winter wheat grown under free air CO_2 enrichment. *Agric. Water Manag.* 210, 70-77. https://doi.org/10.1016/j.agwat.2018.07.034.

McGrath, J. M. and Lobell, D. B. 2013a. Regional disparities in the CO_2 fertilization effect and implications for crop yields. *Environ. Res. Lett.* 8, 014054. https://doi.org/10.1088/1748-9326/8/1/014054.

McGrath, J. M. and Lobell, D. B. 2013b. Reduction of transpiration and altered nutrient allocation contribute to nutrient decline of crops grown in elevated CO_2 concentrations. *Plant Cell Environ.* 36, 697-705. https://doi.org/10.1111/pce.12007.

Menzel, A., Sparks, T. H., Estrella, N., Koch, E., Aasa, A., Ahas, R., Alm-Kübler, K., Bissolli, P., Braslavská, O., Briede, A., Chmielewski, F. M., Crepinsek, Z., Curnel, Y., Dahl, Å., Defila, C., Donnelly, A., Filella, Y., Jatczak, K., Måge, F., Mestre, A., Nordli, Ø., Peñuelas, J., Pirinen, P., Remišová, V., Scheifinger, H., Striz, M., Susnik, A., Vliet, A. J. H. V., Wielgolaski, F.-E., Zach, S. and Zust, A. 2006. European phenological response to climate change matches the warming pattern. *Glob. Change Biol.* 12, 1969-76. https://doi.org/10.1111/j.1365-2486.2006.01193.x.

Mikkelsen, T. N., Beier, C., Jonasson, S., Holmstrup, M., Schmidt, I. K., Ambus, P., Pilegaard, K., Michelsen, A., Albert, K., Andresen, L. C., Arndal, M. F., Bruun, N., Christensen, S., Danbæk, S., Gundersen, P., Jørgensen, P., Linden, L. G., Kongstad, J., Maraldo, K., Priemé, A., Riis-Nielsen, T., Ro-Poulsen, H., Stevnbak, K., Selsted, M. B., Sørensen, P., Larsen, K. S., Carter, M. S., Ibrom, A., Martinussen, T., Miglietta, F. and Sverdrup, H. 2008. Experimental design of multifactor climate change experiments with elevated CO_2, warming and drought: the CLIMAITE project. *Funct. Ecol.* 22, 185-95. https://doi.org/10.1111/j.1365-2435.2007.01362.x.

Mikkelsen, B. L., Jørgensen, R. B. and Lyngkjær, M. F. 2015. Complex interplay of future climate levels of CO_2, ozone and temperature on susceptibility to fungal diseases in barley. *Plant Pathol.* 64, 319-27. https://doi.org/10.1111/ppa.12272.

Mills, G. and Harmens, H. 2011. *Ozone Pollution: A Hidden Threat to Food Security*. NERC/Centre for Ecology & Hydrology, Bangor, UK.

Mills, G., Buse, A., Gimeno, B., Bermejo, V., Holland, M., Emberson, L. and Pleijel, H. 2007. A synthesis of AOT40-based response functions and critical levels of ozone for agricultural and horticultural crops. Atmos. Environ. 41, 2630-43. https://doi.org/10.1016/j.atmosenv.2006.11.016.

Mills, G., Hayes, F., Wilkinson, S. and Davies, W. J. 2009. Chronic exposure to increasing background ozone impairs stomatal functioning in grassland species. *Glob. Change Biol.* 15, 1522-33. https://doi.org/10.1111/j.1365-2486.2008.01798.x.

Mills, G., Pleijel, H., Braun, S., Büker, P., Bermejo, V., Calvo, E., Danielsson, H., Emberson, L., Fernández, I. G., Grünhage, L., Harmens, H., Hayes, F., Karlsson, P.-E. and Simpson, D. 2011. New stomatal flux-based critical levels for ozone effects on vegetation. *Atmos. Environ.* 45, 5064-8. https://doi.org/10.1016/j.atmosenv.2011.06.009.

Mills, G., Pleijel, H., Malley, C. S., Sinha, B., Cooper, O. R., Schultz, M. G., Neufeld, H. S., Simpson, D., Sharps, K., Feng, Z., Gerosa, G., Harmens, H., Kobayashi, K., Saxena,

P., Paoletti, E., Sinha, V. and Xu, X. 2018a. Tropospheric ozone assessment report: present-day tropospheric ozone distribution and trends relevant to vegetation. *Elem. Sci. Anth.* 6, 47. https://doi.org/10.1525/elementa.302.

Mills, G., Sharps, K., Simpson, D., Pleijel, H., Broberg, M., Uddling, J., Jaramillo, F., Davies, W. J., Dentener, F., Berg, M. V. den, Agrawal, M., Agrawal, S. B., Ainsworth, E. A., Büker, P., Emberson, L., Feng, Z., Harmens, H., Hayes, F., Kobayashi, K., Paoletti, E. and Dingenen, R. V. 2018b. Ozone pollution will compromise efforts to increase global wheat production. *Glob. Change Biol.* 24, 3560–74. https://doi.org/10.1111/gcb.14157.

Mir, R. R., Zaman-Allah, M., Sreenivasulu, N., Trethowan, R. and Varshney, R. K. 2012. Integrated genomics, physiology and breeding approaches for improving drought tolerance in crops. *Theor. Appl. Genet.* 125, 625–45. https://doi.org/10.1007/s00122-012-1904-9.

Moore, F. C. and Lobell, D. B. 2015. The fingerprint of climate trends on European crop yields. *Proc. Natl. Acad. Sci.* 112, 2670–5. https://doi.org/10.1073/pnas.1409606112.

Morgan, P. B., Ainsworth, E. A. and Long, S. P. 2003. How does elevated ozone impact soybean? A meta-analysis of photosynthesis, growth and yield. *Plant Cell Environ.* 26, 1317–28.

Morgan, P. B., Bernacchi, C. J., Ort, D. R. and Long, S. P. 2004. An in vivo analysis of the effect of season-long open-air elevation of ozone to anticipated 2050 levels on photosynthesis in soybean. *Plant Physiol.* 135, 2348–57. https://doi.org/10.1104/pp.104.043968.

Morgan, P. B., Mies, T. A., Bollero, G. A., Nelson, R. L. and Long, S. P. 2006. Season-long elevation of ozone concentration to projected 2050 levels under fully open-air conditions substantially decreases the growth and production of soybean. *New Phytol.* 170, 333–43. https://doi.org/10.1111/j.1469-8137.2006.01679.x.

Moura, B. B., Hoshika, Y., Ribeiro, R. V. and Paoletti, E. 2018. Exposure- and flux-based assessment of ozone risk to sugarcane plants. *Atmos. Environ.* 176, 252–60. https://doi.org/10.1016/j.atmosenv.2017.12.039.

Mueller, B., Hauser, M., Iles, C., Rimi, R. H., Zwiers, F. W. and Wan, H. 2015. Lengthening of the growing season in wheat and maize producing regions. Weather Clim. *Extrem.* 9, 47–56. https://doi.org/10.1016/j.wace.2015.04.001.

Müller, C., Cramer, W., Hare, W. L. and Lotze-Campen, H. 2011. Climate change risks for African agriculture. *Proc. Natl. Acad. Sci.* 108, 4313–15. https://doi.org/10.1073/pnas.1015078108.

Müller, C., Waha, K., Bondeau, A. and Heinke, J. 2014. Hotspots of climate change impacts in sub-Saharan Africa and implications for adaptation and development. *Glob. Change Biol.* 20, 2505–17. https://doi.org/10.1111/gcb.12586.

Myers, S. S., Zanobetti, A., Kloog, I., Huybers, P., Leakey, A. D. B., Bloom, A. J., Carlisle, E., Dietterich, L. H., Fitzgerald, G., Hasegawa, T., Holbrook, N. M., Nelson, R. L., Ottman, M. J., Raboy, V., Sakai, H., Sartor, K. A., Schwartz, J., Seneweera, S., Tausz, M. and Usui, Y. 2014. Increasing CO_2 threatens human nutrition. *Nature* 510, 139. https://doi.org/10.1038/nature13179.

Nagel, K. A., Kastenholz, B., Jahnke, S., van Dusschoten, D., Aach, T., Mühlich, M., Truhn, D., Scharr, H., Terjung, S., Walter, A. and Schurr, U. 2009. Temperature responses of roots: impact on growth, root system architecture and implications for phenotyping. *Funct. Plant Biol.* 36, 947–59. https://doi.org/10.1071/FP09184.

Osborne, S. A., Mills, G., Hayes, F., Ainsworth, E. A., Büker, P. and Emberson, L. 2016. Has the sensitivity of soybean cultivars to ozone pollution increased with time? An analysis of published dose-response data. *Glob. Change Biol.* 22, 3097–111. https://doi.org/10.1111/gcb.13318.

Ostberg, S., Schewe, J., Childers, K. and Frieler, K. 2018. Changes in crop yields and their variability at different levels of global warming. *Earth Syst. Dyn.* 9, 479–96. https://doi.org/10.5194/esd-9-479-2018.

Oue, H., Kobayashi, K., Zhu, J., Guo, W. and Zhu, X. 2011. Improvements of the ozone dose response functions for predicting the yield loss of wheat due to the elevated ozone. *J. Agric. Meteorol.* 67, 21–32. https://doi.org/10.2480/agrmet.67.1.2.

Pang, J., Kobayashi, K. and Zhu, J. 2009. Yield and photosynthetic characteristics of flag leaves in Chinese rice (*Oryza sativa* L.) varieties subjected to free-air release of ozone. *Agric. Ecosyst. Environ.* 132, 203–11. https://doi.org/10.1016/j.agee.2009.03.012.

Peltonen-Sainio, P., Palosuo, T., Ruosteenoja, K., Jauhiainen, L. and Ojanen, H. 2018. Warming autumns at high latitudes of Europe: an opportunity to lose or gain in cereal production? *Reg. Environ. Change* 18, 1453–65. https://doi.org/10.1007/s10113-017-1275-5.

Pleijel, H., Gelang, J., Sild, E., Danielsson, H., Younis, S., Karlsson, P.-E., Wallin, G., Skärby, L. andSelldén, G. 2000. Effects of elevated carbon dioxide, ozone and water availability on spring wheat growth and yield. *Physiol. Plant.* 108, 61–70. https://doi.org/10.1034/j.1399-3054.2000.108001061.x.

Pleijel, H., Danielsson, H., Ojanperä, K., Temmerman, L. D., Högy, P., Badiani, M. and Karlsson, P. E. 2004. Relationships between ozone exposure and yield loss in European wheat and potato—a comparison of concentration- and flux-based exposure indices. *Atmos. Environ.* 38, 2259–69. doi:10.1016/j.atmosenv.2003.09.076.

Pleijel, H., Broberg, M. C., Uddling, J. and Mills, G. 2018. Current surface ozone concentrations significantly decrease wheat growth, yield and quality. *Sci. Total Environ.* 613–14, 687–92. https://doi.org/10.1016/j.scitotenv.2017.09.111.

Porter, J. R. and Gawith, M. 1999. Temperatures and the growth and development of wheat: a review. *Eur. J. Agron.* 10, 23–36. https://doi.org/10.1016/S1161-0301(98)00047-1.

Pray, C., Nagarajan, L., Li, L., Huang, J., Hu, R., Selvaraj, K. N., Napasintuwong, O. and Babu, R. C. 2011. Potential impact of biotechnology on adaption of agriculture to climate change: the case of drought tolerant rice breeding in Asia. *Sustainability* 3, 1723–41. https://doi.org/10.3390/su3101723.

Ramankutty, N., Foley, J. A., Norman, J. and McSweeney, K. 2002. The global distribution of cultivable lands: current patterns and sensitivity to possible climate change. *Glob. Ecol. Biogeogr.* 11, 377–92. https://doi.org/10.1046/j.1466-822x.2002.00294.x.

Ray, D. K., Gerber, J. S., MacDonald, G. K. and West, P. C. 2015. Climate variation explains a third of global crop yield variability. *Nat. Commun.* 6, 5989. https://doi.org/10.1038/ncomms6989.

Reddy, K. R. and Zhao, D. 2005. Interactive effects of elevated CO_2 and potassium deficiency on photosynthesis, growth, and biomass partitioning of cotton. *Field Crops Res.* 94, 201–13. https://doi.org/10.1016/j.fcr.2005.01.004.

Rose, G., Osborne, T., Greatrex, H. and Wheeler, T. 2016. Impact of progressive global warming on the global-scale yield of maize and soybean. *Clim. Change* 134, 417–28. https://doi.org/10.1007/s10584-016-1601-9.

Rosenthal, D. M. and Ort, D. R. 2012. Examining Cassava's potential to enhance food security under climate change. *Trop. Plant Biol.* 5, 30–38. https://doi.org/10.1007/s12042-011-9086-1.

Rosenthal, D. M., Slattery, R. A., Miller, R. E., Grennan, A. K., Cavagnaro, T. R., Fauquet, C. M., Gleadow, R. M. and Ort, D. R. 2012. Cassava about-FACE: greater than expected yield stimulation of cassava (*Manihot esculenta*) by future CO_2 levels. *Glob. Change Biol.* 18, 2661–75. https://doi.org/10.1111/j.1365-2486.2012.02726.x.

Rosenzweig, C. and Hillel, D. 2015. *Handbook of Climate Change and Agroecosystems: The Agricultural Model Intercomparison and Improvement Project (AgMIP) Integrated Crop and Economic Assessments – Joint Publication with American Society of Agronomy, Crop Science Society of America, and Soil Science Society of America(In 2 Parts), ICP Series on Climate Change Impacts, Adaptation, and Mitigation*. Imperial College Press.

Rosenzweig, C., Jones, J. W., Hatfield, J. L., Ruane, A. C., Boote, K. J., Thorburn, P., Antle, J. M., Nelson, G. C., Porter, C., Janssen, S., Asseng, S., Basso, B., Ewert, F., Wallach, D., Baigorria, G. and Winter, J. M. 2013. The agricultural model intercomparison and improvement project (AgMIP): protocols and pilot studies. *Agric. For. Meteorol.* 170, 166–82. https://doi.org/10.1016/j.agrformet.2012.09.011.

Rosenzweig, C., Elliott, J., Deryng, D., Ruane, A. C., Müller, C., Arneth, A., Boote, K. J., Folberth, C., Glotter, M., Khabarov, N., Neumann, K., Piontek, F., Pugh, T. A. M., Schmid, E., Stehfest, E., Yang, H. and Jones, J. W. 2014. Assessing agricultural risks of climate change in the 21st century in a global gridded crop model intercomparison. *Proc. Natl. Acad. Sci.* 111, 3268–73. https://doi.org/10.1073/pnas.1222463110.

Ruiz-Vera, U. M., Siebers, M., Gray, S. B., Drag, D. W., Rosenthal, D. M., Kimball, B. A., Ort, D. R. and Bernacchi, C. 2013. Global warming can negate the expected CO_2 stimulation in photosynthesis and productivity for soybean grown in the Midwest United States. *Plant Physiol.*, 112.211938. https://doi.org/10.1104/pp.112.211938.

Ruiz-Vera, U. M., Siebers, M. H., Drag, D. W., Ort, D. R. and Bernacchi, C. J. 2015. Canopy warming caused photosynthetic acclimation and reduced seed yield in maize grown at ambient and elevated [CO_2]. *Glob. Change Biol.* 21, 4237–49. https://doi.org/10.1111/gcb.13013.

Ruiz-Vera, U. M., Siebers, M. H., Jaiswal, D., Ort, D. R. and Bernacchi, C. J. 2018. Canopy warming accelerates development in soybean and maize, offsetting the delay in soybean reproductive development by elevated CO_2 concentrations. *Plant Cell Environ.* 41(12), 2806–20. https://doi.org/10.1111/pce.13410.

Sage, R. F. and Kubien, D. S. 2007. The temperature response of C-3 and C-4 photosynthesis. *Plant Cell Environ.* 30, 1086–106. https://doi.org/10.1111/j.1365-3040.2007.01682.x.

Sage, R. F., Way, D. A. and Kubien, D. S. 2008. Rubisco, Rubisco activase, and global climate change. *J. Exp. Bot.* 59, 1581–95. https://doi.org/10.1093/jxb/ern053.

Saha, S., Sehgal, V. K., Chakraborty, D. and Pal, M. 2015. Atmospheric carbon dioxide enrichment induced modifications in canopy radiation utilization, growth and yield of chickpea [*Cicer arietinum* L.)]. *Agric. For. Meteorol.* 202, 102–11. https://doi.org/10.1016/j.agrformet.2014.12.004.

Sánchez, B., Rasmussen, A. and Porter, J. R. 2014. Temperatures and the growth and development of maize and rice: a review. *Glob. Change Biol.* 20, 408–17. https://doi.org/10.1111/gcb.12389.

Sanz, J., González-Fernández, I., Elvira, S., Muntifering, R., Alonso, R. and Bermejo-Bermejo, V. 2016. Setting ozone critical levels for annual Mediterranean pasture species: combined analysis of open-top chamber experiments. *Sci. Total Environ.* 571, 670–9. https://doi.org/10.1016/j.scitotenv.2016.07.035.

Schlenker, W. and Roberts, M. J. 2009. Nonlinear temperature effects indicate severe damages to U.S. crop yields under climate change. *Proc. Natl. Acad. Sci.* 106, 15594–8. https://doi.org/10.1073/pnas.0906865106.

Schleussner, C.-F., Lissner, T. K., Fischer, E. M., Wohland, J., Perrette, M., Golly, A., Rogelj, J., Childers, K., Schewe, J., Frieler, K., Mengel, M., Hare, W. and Schaeffer, M. 2015. Differential climate impacts for policy-relevant limits to global warming: the case of 1.5°C and 2°C. *Earth Syst. Dyn. Discuss.* 6, 2447–505. https://doi.org/10.5194/esdd-6-2447-2015.

Schleussner, C.-F., Deryng, D., Müller, C., Elliott, J., Saeed, F., Folberth, C., Liu, W., Wang, X., Pugh, T. A. M., Thiery, W., Seneviratne, S. I. and Rogelj, J. 2018. Crop productivity changes in 1.5°C and 2°C worlds under climate sensitivity uncertainty. *Environ. Res. Lett.* 13, 064007. https://doi.org/10.1088/1748-9326/aab63b.

Semenov, M. A. and Shewry, P. R. 2011. Modelling predicts that heat stress, not drought, will increase vulnerability of wheat in Europe. *Sci. Rep.* 1, 1–5. https://doi.org/10.1038/srep00066.

Semenov, M. A. and Stratonovitch, P. 2013. Designing high-yielding wheat ideotypes for a changing climate. *Food Energy Secur.* 2, 185–96. https://doi.org/10.1002/fes3.34.

Shi, G., Yang, L., Wang, Yunxia, Kobayashi, K., Zhu, J., Tang, H., Pan, S., Chen, T., Liu, G. and Wang, Y. 2009. Impact of elevated ozone concentration on yield of four Chinese rice cultivars under fully open-air field conditions. *Agric. Ecosyst. Environ.* 131, 178–84. https://doi.org/10.1016/j.agee.2009.01.009.

Shimono, H., Nakamura, H., Hasegawa, T. and Okada, M. 2013. Lower responsiveness of canopy evapotranspiration rate than of leaf stomatal conductance to open-air CO_2 elevation in rice. *Glob. Change Biol.* 19, 2444–53. https://doi.org/10.1111/gcb.12214.

Singh, P., Boote, K. J., Kadiyala, M. D. M., Nedumaran, S., Gupta, S. K., Srinivas, K. and Bantilan, M. C. S. 2017. An assessment of yield gains under climate change due to genetic modification of pearl millet. *Sci. Total Environ.* 601-2, 1226–37. https://doi.org/10.1016/j.scitotenv.2017.06.002.

Smith, P. and Olesen, J. E. 2010. Synergies between the mitigation of, and adaptation to, climate change in agriculture. *J. Agric. Sci.* 148, 543–52. https://doi.org/10.1017/S0021859610000341.

Smith, P., Haberl, H., Popp, A., Erb, K., Lauk, C., Harper, R., Tubiello, F. N., Pinto, A. de S., Jafari, M., Sohi, S., Masera, O., Böttcher, H., Berndes, G., Bustamante, M., Ahammad, H., Clark, H., Dong, H., Elsiddig, E. A., Mbow, C., Ravindranath, N. H., Rice, C. W., Abad, C. R., Romanovskaya, A., Sperling, F., Herrero, M., House, J. I. and Rose, S. 2013. How much land-based greenhouse gas mitigation can be achieved without compromising food security and environmental goals? *Glob. Change Biol.* 19, 2285–302. https://doi.org/10.1111/gcb.12160.

Sofen, E. D., Bowdalo, D. and Evans, M. J. 2016. How to most effectively expand the global surface ozone observing network. *Atmospheric Chem. Phys.* 16, 1445–57. https://doi.org/10.5194/acp-16-1445-2016.

Stevens, T. and Madani, K. 2016. Future climate impacts on maize farming and food security in Malawi. *Sci. Rep.* 6. https://doi.org/10.1038/srep36241.

Stevenson, D. S., Dentener, F. J., Schultz, M. G., Ellingsen, K., Noije, T. P. C. van, Wild, O., Zeng, G., Amann, M., Atherton, C. S., Bell, N., Bergmann, D. J., Bey, I., Butler, T., Cofala, J., Collins, W. J., Derwent, R. G., Doherty, R. M., Drevet, J., Eskes, H. J., Fiore, A. M., Gauss, M., Hauglustaine, D. A., Horowitz, L. W., Isaksen, I. S. A., Krol, M. C., Lamarque, J.-F., Lawrence, M. G., Montanaro, V., Müller, J.-F., Pitari, G., Prather, M. J., Pyle, J. A., Rast, S., Rodriguez, J. M., Sanderson, M. G., Savage, N. H., Shindell, D. T., Strahan, S. E., Sudo, K. and Szopa, S. 2006. Multimodel ensemble simulations of present-day and near-future tropospheric ozone. *J. Geophys. Res. Atmospheres* 111. https://doi.org/10.1029/2005JD006338.

Stocker, T. F., Qin, D., Plattner, G. -K., Tignor, M., Allen, S. K., Boschung, J., Nauels, A., Xia, Y., Bex, V., and Midgley, P. M. 2013. *Climate Change 2013: The Physical Science Basis. Contribution of Working Group I to the Fifth Assessment Report of the Intergovernmental Panel on Climate Change*. Cambridge University Press, Cambridge, UK and New York, NY.

Tahir, I. S. A. and Nakata, N. 2005. Remobilization of nitrogen and carbohydrate from stems of bread wheat in response to heat stress during grain filling. *J. Agron. Crop Sci.* 191, 106–15. https://doi.org/10.1111/j.1439-037X.2004.00127.x.

Tai, A. P. K., Martin, M. V. and Heald, C. L. 2014. Threat to future global food security from climate change and ozone air pollution. *Nat. Clim. Change* 4, 817–21. https://doi.org/10.1038/nclimate2317.

Taiz, L. and Zeiger, E. 2014. *Plant Physiology*, 6th ed. SAGE Publications Inc, Los Angeles, London, New Delhi, Singapur, Washington DC.

Talukder, A. S. M. H. M., McDonald, G. K. and Gill, G. S. 2014. Effect of short-term heat stress prior to flowering and early grain set on the grain yield of wheat. *Field Crops Res.* 160, 54–63. https://doi.org/10.1016/j.fcr.2014.01.013.

Tang, H., Liu, G., Han, Y., Zhu, J. and Kobayashi, K. 2011. A system for free-air ozone concentration elevation with rice and wheat: control performance and ozone exposure regime. *Atmos. Environ.* 45, 6276–82. https://doi.org/10.1016/j.atmosenv.2011.08.059.

Tang, H., Takigawa, M., Liu, G., Zhu, J. and Kobayashi, K. 2013. A projection of ozone-induced wheat production loss in China and India for the years 2000 and 2020 with exposure-based and flux-based approaches. *Glob. Change Biol.* 19, 2739–52. https://doi.org/10.1111/gcb.12252.

Tang, H., Liu, G., Zhu, J. and Kobayashi, K. 2015. Effects of elevated ozone concentration on CH_4 and N_2O emission from paddy soil under fully open-air field conditions. *Glob. Change Biol.* 21, 1727–36. https://doi.org/10.1111/gcb.12810.

Taub, D. R. and Wang, X. 2008. Why are nitrogen concentrations in plant tissues lower under elevated CO_2? A critical examination of the hypotheses. *J. Integr. Plant Biol.* 50, 1365–74. https://doi.org/10.1111/j.1744-7909.2008.00754.x.

Taub, D. R., Miller, B. and Allen, H. 2008. Effects of elevated CO_2 on the protein concentration of food crops: a meta-analysis. *Glob. Change Biol.* 14, 565–75. https://doi.org/10.1111/j.1365-2486.2007.01511.x.

Teixeira, E. I., Fischer, G., van Velthuizen, H., Walter, C. and Ewert, F. 2013. Global hot-spots of heat stress on agricultural crops due to climate change. *Agric. For. Meteorol.* 170, 206–15. https://doi.org/10.1016/j.agrformet.2011.09.002.

Tingem, M., Rivington, M. and Bellocchi, G. 2009. Adaptation assessments for crop production in response to climate change in Cameroon. *Agron. Sustain. Dev.* 29, 247-56. https://doi.org/10.1051/agro:2008053.

Tito, R., Vasconcelos, H. L. and Feeley, K. J. 2018. Global climate change increases risk of crop yield losses and food insecurity in the tropical Andes. *Glob. Change Biol.* 24, e592–e602. https://doi.org/10.1111/gcb.13959.

van der Kooi, C. J., Reich, M., Löw, M., De Kok, L. J. and Tausz, M. 2016. Growth and yield stimulation under elevated CO_2 and drought: a meta-analysis on crops. *Environ. Exp. Bot.* 122, 150–7. https://doi.org/10.1016/j.envexpbot.2015.10.004.

van Dingenen, R., Dentener, F. J., Raes, F., Krol, M. C., Emberson, L. and Cofala, J. 2009. The global impact of ozone on agricultural crop yields under current and future air quality legislation. *Atmos. Environ.* 43, 604–18. https://doi.org/10.1016/j.atmosenv.2008.10.033.

Vanuytrecht, E., Raes, D., Geerts, S. and Willems, P. 2012. Quantifying field-scale effects of elevated carbon dioxide concentration on crops. *Clim. Res.* 54, 35–47. https://doi.org/10.3354/cr01096.

van Vuuren, D. P., Edmonds, J., Kainuma, M., Riahi, K., Thomson, A., Hibbard, K., Hurtt, G. C., Kram, T., Krey, V., Lamarque, J.-F., Masui, T., Meinshausen, M., Nakicenovic, N., Smith, S. J. and Rose, S. K. 2011. The representative concentration pathways: an overview. *Clim. Change* 109, 5. https://doi.org/10.1007/s10584-011-0148-z.

Volz, A. and Kley, D. 1988. Evaluation of the Montsouris series of ozone measurements made in the nineteenth century. *Nature* 332, 240–2. https://doi.org/10.1038/332240a0.

Vorne, V., Ojanperä, K., De Temmerman, L., Bindi, M., Högy, P., Jones, M. B., Lawson, T. and Persson, K. 2002. Effects of elevated carbon dioxide and ozone on potato tuber quality in the European multiple-site experiment [`]CHIP-project'. *Eur. J. Agron.* 17, 369–81. https://doi.org/10.1016/S1161-0301(02)00072-2.

Wahid, A., Gelani, S., Ashraf, M. and Foolad, M. 2007. Heat tolerance in plants: an overview. *Environ. Exp. Bot.* 61, 199–223. https://doi.org/10.1016/j.envexpbot.2007.05.011.

Walker, C., Armstrong, R., Panozzo, J., Partington, D. and Fitzgerald, G. 2017. Can nitrogen fertiliser maintain wheat (*Triticum aestivum*) grain protein concentration in an elevated CO_2 environment? *Soil Res.* 55, 518–23. https://doi.org/10.1071/SR17049.

Wall, G. W., Kimball, B. A., White, J. W. and Ottman, M. J. 2011. Gas exchange and water relations of spring wheat under full-season infrared warming. *Glob. Change Biol.* 17, 2113–33. https://doi.org/10.1111/j.1365-2486.2011.02399.x.

Wallach, D., Martre, P., Liu, B., Asseng, S., Ewert, F., Thorburn, P. J., van Ittersum, M., Aggarwal, P. K., Ahmed, M., Basso, B., Biernath, C., Cammarano, D., Challinor, A. J., De Sanctis, G., Dumont, B., Eyshi Rezaei, E., Fereres, E., Fitzgerald, G. J., Gao, Y., Garcia-Vila, M., Gayler, S., Girousse, C., Hoogenboom, G., Horan, H., Izaurralde, R. C., Jones, C. D., Kassie, B. T., Kersebaum, K. C., Klein, C., Koehler, A.-K., Maiorano, A., Minoli, S., Müller, C., Naresh Kumar, S., Nendel, C., O'Leary, G. J., Palosuo, T., Priesack, E., Ripoche, D., Rötter, R. P., Semenov, M. A., Stöckle, C., Stratonovitch, P., Streck, T., Supit, I., Tao, F., Wolf, J. and Zhang, Z. 2018. Multimodel ensembles improve predictions of crop-environment-management interactions. *Glob. Change Biol.* 24, 5072–83. https://doi.org/10.1111/gcb.14411.

Wang, Y. and Frei, M. 2011. Stressed food – The impact of abiotic environmental stresses on crop quality. *Agric. Ecosyst. Environ.* 141, 271–86. https://doi.org/10.1016/j.agee.2011.03.017.

Wang, D., Heckathorn, S. A., Wang, X. and Philpott, S. M. 2012a. A meta-analysis of plant physiological and growth responses to temperature and elevated CO_2. *Oecologia* 169, 1–13. https://doi.org/10.1007/s00442-011-2172-0.

Wang, Y., Yang, L., Han, Y., Zhu, J., Kobayashi, K., Tang, H. and Wang, Y. 2012b. The impact of elevated tropospheric ozone on grain quality of hybrid rice: a free-air gas concentration enrichment (FACE) experiment. *Field Crops Res.* 129, 81–89. https://doi.org/10.1016/j.fcr.2012.01.019.

Wang, L., Feng, Z. and Schjoerring, J. K. 2013. Effects of elevated atmospheric CO_2 on physiology and yield of wheat (*Triticum aestivum* L.): a meta-analytic test of current hypotheses. *Agric. Ecosyst. Environ.* 178, 57–63. https://doi.org/10.1016/j.agee.2013.06.013.

Wang, Y., Song, Q., Frei, M., Shao, Z. and Yang, L. 2014a. Effects of elevated ozone, carbon dioxide, and the combination of both on the grain quality of Chinese hybrid rice. *Environ. Pollut.* 189, 9–17. https://doi.org/10.1016/j.envpol.2014.02.016.

Wang, X., Taub, D. R. and Jablonski, L. M. 2014b. Reproductive allocation in plants as affected by elevated carbon dioxide and other environmental changes: a synthesis using meta-analysis and graphical vector analysis. *Oecologia* 177, 1075–87. https://doi.org/10.1007/s00442-014-3191-4.

Wang, T., Xue, L., Brimblecombe, P., Lam, Y. F., Li, L. and Zhang, L. 2017a. Ozone pollution in China: a review of concentrations, meteorological influences, chemical precursors, and effects. *Sci. Total Environ.* 575, 1582-96. https://doi.org/10.1016/j.scitotenv.2016.10.081.

Wang, E., Martre, P., Zhao, Z., Ewert, F., Maiorano, A., Rötter, R. P., Kimball, B. A., Ottman, M. J., Wall, G. W., White, J. W., Reynolds, M. P., Alderman, P. D., Aggarwal, P. K., Anothai, J., Basso, B., Biernath, C., Cammarano, D., Challinor, A. J., Sanctis, G. D., Doltra, J., Dumont, B., Fereres, E., Garcia-Vila, M., Gayler, S., Hoogenboom, G., Hunt, L. A., Izaurralde, R. C., Jabloun, M., Jones, C. D., Kersebaum, K. C., Koehler, A.-K., Liu, L., Müller, C., Kumar, S. N., Nendel, C., O'Leary, G., Olesen, J. E., Palosuo, T., Priesack, E., Rezaei, E. E., Ripoche, D., Ruane, A. C., Semenov, M. A., Shcherbak, I., Stöckle, C., Stratonovitch, P., Streck, T., Supit, I., Tao, F., Thorburn, P., Waha, K., Wallach, D., Wang, Z., Wolf, J., Zhu, Y. and Asseng, S. 2017b. The uncertainty of crop yield projections is reduced by improved temperature response functions. *Nat. Plants* 3, 17102. https://doi.org/10.1038/nplants.2017.102.

Wang, B., Li, J., Wan, Y., Li, Y., Qin, X., Gao, Q., Waqas, M. A., Wilkes, A., Cai, W., You, S. and Zhou, S. 2018. Responses of yield, CH_4 and N_2O emissions to elevated atmospheric temperature and CO_2 concentration in a double rice cropping system. *Eur. J. Agron.* 96, 60–69. https://doi.org/10.1016/j.eja.2018.01.014.

Wheeler, T. and Braun, J. von, 2013. Climate change impacts on global food security. *Science* 341, 508–13. https://doi.org/10.1126/science.1239402.

Wilkinson, S. and Davies, W. J. 2010. Drought, ozone, ABA and ethylene: new insights from cell to plant to community. *Plant Cell Environ.* 33, 510–25. https://doi.org/10.1111/j.1365-3040.2009.02052.x.

Wolkovich, E. M., Cook, B. I., Allen, J. M., Crimmins, T. M., Betancourt, J. L., Travers, S. E., Pau, S., Regetz, J., Davies, T. J., Kraft, N. J. B., Ault, T. R., Bolmgren, K., Mazer, S. J., McCabe, G. J., McGill, B. J., Parmesan, C., Salamin, N., Schwartz, M. D. and Cleland, E. E. 2012. Warming experiments underpredict plant phenological responses to climate change. *Nature* 485, 494–7. https://doi.org/10.1038/nature11014.

Xiong, D., Ling, X., Huang, J. and Peng, S. 2017. Meta-analysis and dose-response analysis of high temperature effects on rice yield and quality. *Environ. Exp. Bot.* https://doi.org/10.1016/j.envexpbot.2017.06.007.

Yendrek, C. R., Erice, G., Montes, C. M., Tomaz, T., Sorgini, C. A., Brown, P. J., McIntyre, L. M., Leakey, A. D. B. and Ainsworth, E. A. 2017. Elevated ozone reduces photosynthetic carbon gain by accelerating leaf senescence of inbred and hybrid maize in a genotype-specific manner. *Plant Cell Environ.* 40, 3088–100. https://doi.org/10.1111/pce.13075.

Yue, X. and Unger, N. 2014. Ozone vegetation damage effects on gross primary productivity in the United States. *Atmospheric Chem. Phys.* 14, 9137–53. https://doi.org/10.5194/acp-14-9137-2014.

Zampieri, M., Ceglar, A., Dentener, F. and Toreti, A. 2017. Wheat yield loss attributable to heat waves, drought and water excess at the global, national and subnational scales. *Environ. Res. Lett.* 12, 064008. https://doi.org/10.1088/1748-9326/aa723b.

Zhang, L., Zhang, Z., Chen, Y., Wei, X. and Song, X. 2018. Exposure, vulnerability, and adaptation of major maize-growing areas to extreme temperature. *Nat. Hazards* 91, 1257–72. https://doi.org/10.1007/s11069-018-3181-7.

Zhao, C., Piao, S., Huang, Y., Wang, X., Ciais, P., Huang, M., Zeng, Z. and Peng, S. 2016. Field warming experiments shed light on the wheat yield response to temperature in China. *Nat. Commun.* 7, 13530. https://doi.org/10.1038/ncomms13530.

Zhao, C., Liu, B., Piao, S., Wang, X., Lobell, D. B., Huang, Y., Huang, M., Yao, Y., Bassu, S., Ciais, P., Durand, J.-L., Elliott, J., Ewert, F., Janssens, I. A., Li, T., Lin, E., Liu, Q., Martre, P., Müller, C., Peng, S., Peñuelas, J., Ruane, A. C., Wallach, D., Wang, T., Wu, D., Liu, Z., Zhu, Y., Zhu, Z. and Asseng, S. 2017. Temperature increase reduces global yields of major crops in four independent estimates. *Proc. Natl. Acad. Sci.* 114, 9326-31. https://doi.org/10.1073/pnas.1701762114.

Zhu, X., Feng, Z., Sun, T., Liu, X., Tang, H., Zhu, J., Guo, W. and Kobayashi, K. 2011. Effects of elevated ozone concentration on yield of four Chinese cultivars of winter wheat under fully open-air field conditions. *Glob. Change Biol.* 17, 2697–706. https://doi.org/10.1111/j.1365-2486.2011.02400.x.

Zhu, C., Kobayashi, K., Loladze, I., Zhu, J., Jiang, Q., Xu, X., Liu, G., Seneweera, S., Ebi, K. L., Drewnowski, A., Fukagawa, N. K. and Ziska, L. H. 2018. Carbon dioxide (CO_2) levels this century will alter the protein, micronutrients, and vitamin content of rice grains with potential health consequences for the poorest rice-dependent countries. *Sci. Adv.* 4, eaaq1012. https://doi.org/10.1126/sciadv.aaq1012.

Ziska, L. H. and Bunce, J. A. 2006. Plant responses to rising atmospheric carbon dioxide, in: Morison, J. I. L., Morecroft, M. D. (Eds), *Plant Growth and Climate Change*. Blackwell Publishing Ltd, Oxford, pp. 17–47.

Chapter 2

Effects of climate change on agricultural soils

Kennedy Were, Kenya Agricultural and Livestock Research Organisation, Kenya; and Bal Ram Singh, Norwegian University of Life Sciences, Norway

1 Introduction

The atmospheric concentrations of greenhouse gases (GHGs) have increased substantially over the years due to a combination of natural (e.g. changes in the sun's intensity, ocean current circulation and volcanic eruptions) and human factors (e.g. burning of fossil fuels, intensive agriculture and deforestation). However, with the high population growth and advances in technology, anthropogenic activities are contributing more to GHG emissions and changes in the composition of the Earth's atmosphere than the natural processes. In 2017, the atmospheric CO_2 concentrations had risen to about 405.5 ppm, N_2O concentrations had reached a new high of 329.9 ppb and CH_4 concentration was about 1859 ppb (WMO, 2018). The elevated atmospheric concentrations of these long-lived GHGs have in turn increased the amount of heat being trapped in the climate system (i.e. greenhouse effect) leading to global warming, with the attendant changes in the precipitation patterns. An analysis of the globally averaged combined land and ocean surface temperature data had indicated a warming of 0.85°C (0.65–1.06°C) between 1880 and 2012 (Fig. 1), and the temperatures are expected to rise between 1.8°C and 4.0°C by 2100, although with varying intensities of the warming across seasons and regions (IPCC, 2014). For example, greater warming is expected across all seasons in Africa ranging from 0.2°C per decade to more than 0.5°C per decade (Mutegi et al., 2018), with more intensity in Central and Southern Africa and the semi-arid tropical margins of the Sahara (Cairns et al., 2013). Similarly, rainfall patterns will change at all locations, with varying magnitudes, rates and

http://dx.doi.org/10.19103/AS.2020.0064.05

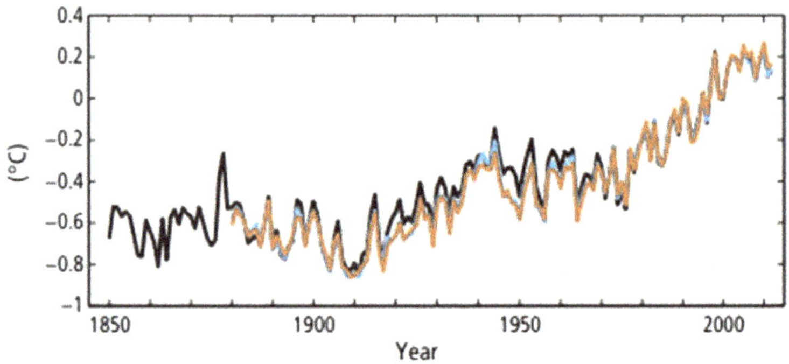

Figure 1 Globally averaged combined land and ocean surface temperature anomaly. Source: adapted from IPCC (2014).

directions, although, the predictions are subject to considerable uncertainties. In the western margins of Southern Africa, the annual precipitation is likely to decrease by up to 30% under a 4°C warming scenario (AGRA, 2014), while in East Africa, rainfall is predicted to increase by 5–20% in the December–January–February (DJF) season and decrease by 5–10% in the June–July–August (JJA) season by 2050 (Christensen et al., 2007).

Such changes in temperature, rainfall and GHGs for extended periods will have significant impacts on agroecosystems, including the soils. Soils will especially be affected because of the intricate linkage with the climate system through the carbon (C), nitrogen (N) and hydrological cycles (Brevik, 2013). The C and N are important components of soil organic matter (SOM), and higher temperatures might accelerate the decomposition of SOM leading to emissions of CO_2, N_2O, and CH_4 into the atmosphere. The direction and magnitude of the impacts will, however, depend on the amount of change in atmospheric gases, temperature, and precipitation amounts and pattern cycles (Brevik, 2013). Soils form the base on which human beings and all terrestrial ecosystems function, delivering an array of precious goods and services, such as food, fibre and water. As part of the global system, soils influence species distribution, productivity, water and biogeochemical cycling, and determine the agricultural potential of land (Rounsevell et al., 1999). However, soils are a non-renewable resource and any adverse changes on their properties and processes may be unrecoverable within the human lifespan, with serious implications for agriculture. Therefore, climate change trends and the response of soils to the changes have become central to the scientific discourse. Such knowledge is crucial for the formulation of soil management strategies that are effective and climate-smart.

In this chapter, we review and highlight the current knowledge on the effects of climate change on soil properties and processes, as well as on the set of climate-smart technologies, innovations and practices that are available for soil management in the face of climate change. In doing so, we draw on the materials reported by the IPCC and previous reviews.

2 Effects of climate change on agricultural soil properties and processes

Without rapid cuts in GHGs, climate change will have increasingly destructive and irreversible impacts on the soil's physical, chemical and biological properties and processes (Table 1). This is because it will alter the spatial and temporal patterns of atmospheric CO_2 concentrations, temperature, rainfall, solar radiation and wind, affecting soil moisture and temperature conditions. The changes in soil moisture and temperature regimes will ultimately affect the amount of SOM, which is an important determinant of soil health and functions. Soil organic matter influences qualities, such as soil structure, aggregate stability, nutrient availability, water retention, infiltration rates, oxygen holding capacity, microbial activities and cation exchange capacity (CEC) of coarse-textured and low-activity clay soils.

In terms of soil biology, expectations are that higher temperatures will enhance the decomposition of SOM, resulting in an increased release of CO_2 in the atmosphere and depletion of the soil organic C and N reserves, creating a positive feedback that could push temperatures even higher. Considering this, the group of soils that are most vulnerable to climate change is the peat soils, which are dominated by organic matter at all depths. The effects of climate change on SOM remains an active area of research because, globally, the

Table 1 Summary of the effects of climate change on soil properties and processes

Direct and indirect drivers of climate change	Effect on soil properties and processes
a) Increase in temperature	• Increased soil temperatures • Decomposition of SOM and reduction in SOC reserves • Release of nutrients and increased plant growth • Increase in soil microbial activity and growth • Decline in soil health/ quality indicators, for example, loss of soil structure, aggregate stability and water holding capacity
b) Increase in CO_2 concentrations	• Increase in soil C sequestration, but it might be counteracted by the increasing levels of ozone and soil C:N ratio
c) Increase in rainfall	• Leaching of nutrients • Increased erosion susceptibility • Increase chemical weathering of minerals and rocks • Increased soil acidification
d) Decrease in rainfall	• Salinization and alkalization • Reduced soil moisture and transportation of nutrients to root surface

amount of C in the soils (1500 Pg C to 1 m depth; 1 Pg = 10^{15} g) is twice the amount stored in the atmospheric C pool (750 Pg C); so, any slight changes might have far-reaching consequences on C cycle, climate and soil properties. However, as mentioned by Karmakar et al. (2016), the potential negative effects occasioned by the decomposition of SOM might be counteracted by enhanced plant nutrient releases and cycling leading to increased net primary productivity (NPP) and organic matter supplies to soils. It is well-documented that increased productivity is mostly accompanied by more litter (or crop residues), greater total root mass and root exudation, increased mycorrhizal colonization and activity of other micro-organisms, including symbiotic and root-zone N fixers, which have a positive effect on N supply to plants. Similarly, enhanced photosynthetic rates and plant growth under elevated CO_2 (i.e. CO_2 fertilization) could also have counteractive effects. For instance, increases in the amount of C entering the soil might stimulate microbial growth, biomass accumulation and activity, which are often C-limited. Given the important roles played by soil micro-organisms (e.g. fungi and bacteria) in SOM degradation, nutrient cycling, plant nutrition and soil aggregate formation, shifts in microbial communities might have an impact on soil functioning.

On the contrary, some recent studies have also shown that elevated CO_2 levels and fertilization will not necessarily lead to increased soil C sequestration, but may instead result in more C turnover. For instance, in his review article, Brevik (2013) pointed out that the increasing levels of ozone as the climate changes may offset the CO_2 fertilization effect leading to reduced plant growth under elevated CO_2. In addition, the CO_2 enrichment will increase the C:N ratio in soil. Hence, the decomposing organisms (e.g. bacteria) in the soil will be in need of more N, which can reduce N mineralization. If N mineralization is reduced, plant-available N levels in the soil also reduce, negatively affecting plant productivity.

Apart from the biological properties, climate change is also expected to affect various soil chemical parameters and processes. For example, high-intensity rainfall will increase the leaching of water-soluble plant nutrients in well-drained soils with high infiltration rates leading to acidification and loss of salts and nutrient cations, depending on the buffering pools in the soils (Karmakar et al., 2016). Salinization and alkalization, on the other hand, are likely in areas that will experience reduced rainfall and higher evaporation rates (aridity), especially depressional sites and where the groundwater table is high. Further, drivers of climate change, such as elevated temperatures and variable precipitation, will affect the availability and acquisition of plant nutrients. That is, higher temperature and rainfall will increase the soil temperatures, decomposition of SOM and release of nutrients, while lower rainfall will reduce the soil moisture and solution, affecting the transportation of nutrients to the root surface. Consequently, this will reduce crop yields, productivity and SOC levels in the agroecosystems.

Moreover, changes in temperature, precipitation patterns and atmospheric CO_2 concentrations will also have ramifications on soil physical properties and processes, which are responsible for gains, losses, transfers and movement of water, air, organic matter, soluble salts, carbonates, silicate clay minerals, sesquioxides and silica in the soil (Patil and Lamnganbi, 2018). For example, as already stated, higher temperatures will subject soils to increased decomposition rates and decline of SOM levels culminating in a decrease in soil structure and aggregate stability, and an increase in bulk density. This will have implications for C and N cycles, infiltration capacity, movement and storage of water, air and nutrients, rooting depth, microbial activities, and susceptibility to erosion and compaction. Furthermore, soil water available for plant growth may respond rapidly to variable and high-intensity rainfall, as well as to droughts. The plant-available water and aeration capacity is governed and linked to a range of soil physical properties, including porosity and pore size distribution. Thus, if climate change affects soil porosity through changes in SOM levels and bulk density, then it is not only the plant available water and aeration capacities that will be affected, but also root growth and soil enzyme activities. Lastly, the world's soils will become more susceptible to erosion by wind and water through changes in the intensity of precipitation and wind.

Other soil processes that are likely to be affected are soil formation and development. Climatic elements, particularly temperature and rainfall, determine the various stages of weathering of rocks and minerals, and the resultant chemical and mineralogical composition of soils. Hence, the same primary minerals and rocks will give rise to different secondary minerals and soil profiles when the conditions for weathering are altered. Climate change is expected to alter soil moisture and temperature conditions through direct climatic effects (e.g. temperature effects on evaporation), climate-induced changes in vegetation, plant growth rates, rates of soil water extraction by plants and the effect of enhanced CO_2 levels on plant transpiration (Pareek, 2017).

From the above highlights, it is apparent that soil properties and processes are intertwined and influence each other; hence, unravelling how climate change will affect them is rather complex. Models can provide a means of unravelling this complexity, as well as allow future scenarios to be considered and long-term predictions made. It is also clear that although soils are largely depicted as victims, they can equally be culprits of climate change when they become net sources of atmospheric C. Soil is an integral part of the global C and N cycles, and the management strategies adopted can either increase or reduce emissions of GHGs, such as CO_2, N_2O and CH_4, with repercussions on climate change variables. Proven climate-smart technologies, innovations and management practices (CS-TIMPs) are essential to adapt soils and mitigate the adverse climate change effects, build resilience in farming systems, reduce

GHG emissions and advance food security in a sustainable manner. An array of CS-TIMPs (e.g. conservation tillage and no-till practices) exists for judicious soil management, some of which are described in the following sub-section:

3 Climate-smart soil management options

Soils must be managed in a judicious, smart and sustainable manner in order to mitigate climate change effects and advance ecological stability, food security, viable rural livelihoods and agricultural resilience. This entails managing soils with a view to mitigate CO_2 emissions, enhance C sinks by removing CO_2 from the atmosphere for later storage as SOM and create resilient systems, among other socio-economic and environmental benefits. CS-TIMPs for soil management are manifold and well-documented (Were et al., 2016, 2019; Table 2), for example, conservation agriculture (CA), precision agriculture (PA),

Table 2 Summary of the effects of climate change on agricultural soils and the potential climate-smart soil management options

Effect of climate change on soil properties and processes	Potential climate-smart soil management options
a) Increased soil temperatures, reduced soil moisture	• Mulching including live mulch • Cover cropping, crop residue retention, conservation agriculture
b) Decomposition of SOM and reduction in SOC reserves; Decline in soil health and quality indicators, for example, loss of soil structure, aggregate stability and water-holding capacity; Increase in soil microbial activity and growth	• Conservation agriculture, precision agriculture, integrated nutrient management, residue retention, mulching, agroforestry, controlled grazing, use of manure, cover crops, biochar, improved plant varieties, crop rotations, micro-dosing of fertilizer, introducing improved grass species, reducing the frequency and extent of fires, rehabilitation of degraded rangelands, improving activity and diversity of soil biological resources, adjusting the timing of grazing, managing the stocking rates, restoration of drained wetlands and degraded soils, minimal tillage, selection of high biomass-producing crops, re-vegetation, cut-off drains, contour ploughing, integration of green manure and leguminous cover crops
c) Leaching and release of nutrients	• Integrated nutrient management • Precision agriculture and judicious fertilization with right fertilizer with right dose at right time and right place (4 Rs)
d) Increased erosion susceptibility	• Soil and water management practices (e.g. terracing, agroforestry, residue retention and mulching) and conservation agriculture
e) Salinization and alkalization	• Flushing, organic and green manuring, legume integration, crop rotation, growing suitably tolerant crops, mulching, crop residue retention, mineral fertilization • Use of chemical and biological amendments

integrated nutrient management (INM), residue retention, soil biological management, soil and water conservation (e.g. terracing and mulching), water harvesting, agroforestry, controlled grazing at optimal stocking rates and the use of organic amendments (e.g. manure), cover crops, biochar, improved plant varieties and crop rotations. Lal (2015) argued that the most desirable CS-TIMPs should aim at (i) minimizing losses of water and nutrients out of the ecosystem; (ii) creating positive ecosystem C, nutrient and water budgets; (iii) enhancing biodiversity; (iv) strengthening plant-soil feedback; and (v) minimizing soil disturbances and risk of erosion.

Conservation agriculture as one of the CS-TIMPs, aims at accumulating SOC and creating a healthy soil ecosystem by not tilling the soil prior to planting and establishing crops on the residues left on the ground after harvesting. It is underpinned by minimum soil disturbance, maintenance of at least 30% permanent organic mulch on the soil surface and a diversified cropping system through intercropping or crop rotation (Aune and Coulibaly, 2015; FAO, 2013). If executed correctly on suitable soil types, CA has multiple co-benefits, including (i) increased infiltration capacity, soil water content, SOM levels, soil biological activity and flexibility in planting and harvesting; (ii) reduced weeds, soil erosion, soil surface temperatures, soil compaction, emission of soil C to the atmosphere and fossil fuel consumption by avoiding ploughing; and (iii) improved soil tilth and fertility.

Apart from CA, integrated nutrient management (INM) is also a vital soil management practice in the face of climate change. It integrates the best of organic and conventional technologies in a way that is environmentally appropriate, productive and sustainable (Agriculture for Impact, 2014). That is, INM maximizes the use of organic resources (e.g. manure, compost and sludge) and improved germplasm, minimizes nutrient losses, allows timely and judicious use of inorganic fertilizers based on need and economic viability, and maintains and enhances beneficial soil organisms and biological processes (Oladele and Braimoh, 2014; Sanginga and Woomer, 2009). What's more, it recognizes the importance of knowledge on how to adapt the technologies and practices to local conditions in order to maximize nutrient and water-use efficiency, as well as to improve crop productivity. A typical example of INM includes intercropping and rotating improved maize with a legume (e.g. beans), and applying both organic and inorganic fertilizers in conjunction with soil and water conservation technologies, such as fallowing, strip cropping and contour ploughing (Zingore et al., 2015; Lal, 2009). By boosting the soil's physical, chemical and biological properties (e.g. enhanced supply of essential nutrients, reduced water and nutrient losses, as well as improved porosity and infiltration) INM strategies stimulate biomass production, which determines the amount of C stored in the plant biomass and the inputs of detritus to the soil.

Moreover, improving the efficiency of nutrients and water use through CS-TIMPs, such as micro-irrigation (e.g. drip irrigation), micro-dosing of mineral fertilizer and precision farming can be effective in building climate-smart soils. These technologies and practices ensure that the right amounts and type of fertilizer are placed close to the growing plant (or planting pocket) at the right time and that fewer drops of water are used per crop. This can boost NPP and the amount of organic materials returned to the soil C pool, create a positive effect on soil quality, promote a more efficient resource use and alleviate problems associated with parasitic weeds, such as *Striga hermonthica* (Aune and Coulibaly, 2015). Micro-dosing can be complemented with other technologies that can aid in establishing the exact nutrient and water requirements, as well as in mapping the specific sites in need.

Similarly, CS-TIMPs that enhance SOC sequestration through the uptake and transfer of atmospheric CO_2 into the soil reservoir by plants, plant residues and other organic solids are instrumental in managing soil productivity for sustained food security despite the changing climate. This is because SOC is a major determinant of soil quality and functions. For instance, it fosters the storage of water and nutrients available to the plant, the activity and diversity of soil biota, soil structure and tilth, and susceptibility of soil to erosion. When the concentration of SOC falls below a certain threshold (i.e. 1–1.5% depending on the soil type, climate and land use), fundamental soil properties are adversely affected inhibiting plant growth. Thus, CS-TIMPs that replenish and maintain the SOC pool above the critical level are essential for reducing soil degradation risks and reversing degradation trends. In addition to CA, micro-dosing and INM, which have been discussed above, the SOC pool can also be maintained through grazing and pasture management. By introducing improved grass species with higher productivity and C allocation to deeper roots, reducing the frequency and extent of fires through effective fire suppression means, burning when less amounts of CO_2, CH_4 and N_2O are emitted, reducing the fuel load through appropriate vegetation management, rehabilitation of degraded rangelands through exclosures, improving activity and diversity of soil biological resources (e.g. earthworms and termites), irrigation, fertilization, adjusting the timing of grazing and managing the stocking rates, optimal soil C sequestration and re-carbonization can be realized (UNFCCC, 2008; Smith et al., 2007). High stocking rates degrade the rangelands by reducing plant cover due to overgrazing and trampling effect and adversely affect the soil's physical qualities (e.g. water infiltration), increase soil erosion risks and reduce biomass productivity.

Furthermore, restorative land uses that favour conservation and improvement of soil, water and air quality have the potential for soil C sequestration in agroecosystems. For instance, adopting agroforestry systems, reverting marginal croplands to native systems or switching from annual crops to perennial vegetation can culminate in the accrual of SOC owing to higher aboveground biomass C, reduced soil disturbance and minimal removal of residue and harvested products. Apart from these, restoration of drained wetlands (or peatlands) and degraded soils can also enhance soil C sequestration and mitigate CO_2 emissions. Wetlands are characterized by peaty soils, which accumulate C because of the slow rate of decomposition under anaerobic conditions even though such conditions can also instigate CH_4 emissions if not properly managed. For the degraded soils, C gains can be achieved through practices, such as crop residue recycling, application of nutrient amendments and organic materials (e.g. manures), minimal tillage, selection of high biomass-producing crops, re-vegetation, and soil and water conservation measures (e.g. terracing, cut-off drains, mulching, tied ridges, agroforestry and contour ploughing). Soil and water conservation structures and measures can be effective in maintaining pedological processes, reducing the loss of soil and water, strengthening the cycling of C and other vital elements (e.g. N, P and S) and, ultimately, building climate-resilient soils in agroecosystems.

Finally, integration of green manure and leguminous cover crops (e.g. cow peas, groundnuts and velvet beans) in a rotation cycle is also a promising technology for C sequestration in agricultural soils. Legume-based cropping systems reduce C and N losses from the soil by improving soil aggregate stability and reducing erosion, improve soil quality by increasing SOC through their biomass and enhance the quality of residue input and soil biodiversity (Lal, 2004). Agroecosystems with high biodiversity are also known to absorb and sequester more C.

4 Conclusion and future trends

Climate change represents a serious threat to global food security and livelihoods because of the envisaged effects on agricultural soils. Changes in CO_2 concentrations, temperature and rainfall patterns owing to human activities and natural processes can have a great impact on soil properties and processes, as well as on the plants that grow on them. In this chapter, we have reviewed some of the effects of climate change on soil biological, chemical and physical properties and processes. In this chapter, we have also made it clear that soils are victims of climate change; but, because of their intricate linkage with the climate system, they can equally be culprits of climate change. Besides the effects, we have also explored some of the CS-TIMPs available to manage and build resilient soils in the face of climate-change. The CS-TIMPs are diverse; hence, we propose that the best fit package for scaling should be selected based not only on their technical feasibility and social acceptability, but also on their capacity to deliver multiple environmental benefits (e.g. carbon sequestration and biodiversity benefits) and to maximize returns on farmers investment (i.e. economic viability).

The capacity of these CS-TIMPs to deliver multiple environmental benefits, including climate change mitigation by reducing GHG emissions and sequestering C both in soils and biomass cannot be impugned. However, understanding the overall impact of soil management activities on C stocks and GHG balance can be rather complex, which partly explains the scarcity of research and information in this domain. Future research should endeavour to use novel and existing models and tools (e.g. the Carbon Benefits Project modelling system) to unravel this complexity, as well as to estimate information where comprehensive field measurement campaigns are not possible owing to constraints of cost and time. Such modelling efforts will also allow future scenarios and worlds to be considered; hence, will be useful in planning and long-term predictions.

5 References

AGRA. 2014. Africa agriculture report: climate change and smallholder agriculture in sub-Saharan Africa, No. 2. Nairobi, Kenya. Available at: http://www.agra.org/our-results/agra-statusreports/ (accessed on 13 March 2016).

Agriculture for Impact. 2014. No ordinary matter: conserving, restoring and enhancing Africa's soils. A Montpellier Panel report.

Aune, J. B. and Coulibaly, A. 2015. Microdosing of mineral fertilizer and conservation of agriculture in sustainable agricultural intensification in sub-Saharan Africa. In: Lal, R., Singh, B. R., Mwaseba, D. L., Kraybill, D., Hansen, D. O. and Eik, L. O. (Eds), *Sustainable Intensification to Advance Food Security and Enhance Climate Resilience in Africa*. Springer, Berlin, Germany, pp. 223–34.

Brevik, E. C. 2013. The potential impact of climate change on soil properties and processes and corresponding influence on food security. *Agriculture* 3(3), 398–417. doi:10.3390/agriculture3030398.

Cairns, J. E., Hellin, J., Sonder, K., Araus, J. L., MacRobert, J. F., Thierfelder, C. and Prasanna, B. M. 2013. Adapting maize production to climate change in sub-Saharan Africa. *Food Security* 5(3), 345–60. doi:10.1007/s12571-013-0256-x.

Christensen, J. H., Hewitson, B., Busuioc, A., Chen, A., Gao, X., Held, I., Jones, R., Kolli, R. K., Kwon, W.-T., Laprise, R., Magaña Rueda, V., Mearns, L., Menéndez, C.G., Räisänen, J., Rinke, A., Sarr, A. and Whetton, P. 2007. Regional climate projections. In: Solomon, S., Qin, D., Manning, M., Chen, Z., Marquis, M., Averyt, K. B., Tignor, M. and Miller, H. L. (Eds), *Climate Change 2007: The Physical Science Basis. Contribution of Working Group I to the Fourth Assessment Report of the Intergovernmental Panel on Climate Change*. Cambridge University Press, Cambridge, UK and New York, NY.

FAO. 2013. Climate-smart agriculture sourcebook. Available at: http://www.fao.org/docrep/018/i3325e/i3325e.pdf (accessed on 13 March 2016).

IPCC. 2014. Climate change 2014: impacts, adaptation, and vulnerability. Report from the Intergovernmental Panel on Climate Change. Available at: http://www.ipcc.ch/report/ar5/.

Karmakar, R., Das, I., Dutta, D. and Rakshit, A. 2016. Potential effects of climate change on soil properties: a review. *Science International* 4(2), 51–73. doi:10.17311/sciintl.2016.51.73.

Lal, R. 2004. Soil carbon sequestration to mitigate climate change. *Geoderma* 123(1-2), 1-22. doi:10.1016/j.geoderma.2004.01.032.

Lal, R. 2009. Soil degradation as a reason for inadequate human nutrition. *Food Security* 1(1), 45–57. doi:10.1007/s12571-009-0009-z.

Lal, R. 2015. Restoring soil quality to mitigate soil degradation. *Sustainability* 7(5), 5875-95. doi:10.3390/su7055875.

Mutegi, J., Ameru, J., Harawa, R., Kiwia, A. and Njue, A. 2018. Soil health and climate change: implications for food security in sub-Saharan Africa. *International Journal of Development and Sustainability* 7(1), 21-33.

Oladele, O. and Braimoh, A. 2014. Potential of agricultural land management activities for increased soil carbon sequestration in Africa- a review. *Applied Ecology and Environmental Research* 12(3), 741-51. doi:10.15666/aeer/1203_741751.

Pareek, N. 2017. Climate change impact on soils: adaptation and mitigation. *MOJ Ecology and Environmental Sciences* 2(3), 136–9. doi:10.15406/mojes.2017.02.00026.

Patil, A. and Lamnganbi, M. 2018. Impact of climate change on soil health: a review. *International Journal of Chemical Studies* 6(3), 2399–404.

Rounsevell, M. D. A., Evans, S. P. and Bullock, P. 1999. Climate change and agricultural soils: impacts and adaptations. *Climatic Change* 43(4), 683–709. doi:10.1023/A:1005597216804.

Sanginga, N. and Woomer, P. L. (Eds). 2009. *Integrated Soil Fertility Management in Africa: Principles, Practices and Development Process*. Tropical Soil Biology and Fertility Institute of the Centre for Tropical Agriculture, Nairobi, Kenya, 263p.

Smith, P., Martino, D., Cai, Z., Gwary, D., Janzen, H., Kumar, P., McCarl, B., Ogle, S., O'Mara, F., Rice, C., Scholes, B. and Sirotenko, O. 2007. Agriculture. In: Metz, B., Davidson, O. R., Bosch, P. R., Dave, R. and Meyer, L. A. (Eds), *Climate change 2007: Mitigation. Contribution of Working Group III to the Fourth Assessment Report of the Intergovernmental Panel on Climate Change*. Cambridge University Press, Cambridge, UK and New York, NY.

UNFCCC. 2008. Challenges and opportunities for mitigation in the agricultural sector. Technical Paper GE-08-64351.

Were, K. O., Gelaw, A. M. and Singh, B. R. 2016. Smart strategies for enhanced agricultural resilience and food security under a changing climate in sub-Saharan Africa. In: Lal, R., Singh, B. R, Mwaseba, D. L., Kraybill, D., Hansen, D. O. and Eik, L. O. (Eds), *Climate Change and Multi-Dimensional Sustainability in African Agriculture*. Springer International Publishing, Switzerland, pp. 431–53.

Were, K. O., Singh, B. R. and Ayaga, G. 2019. Degradation and climate-smart options for restoring the East African soils. In: Lal, R. and Stewart, B. A. (Eds), *Advances in Soil Science - Soil Degradation and Restoration in Africa*. CRC Press, Taylor & Francis Group, Boca Raton, FL and New York, NY, pp. 255–70.

WMO. 2018. WMO Greenhouse Gas Bulletin (GHG Bulletin) - No. 14: the state of greenhouse gases in the atmosphere based on global observations through 2017. Available at: https://library.wmo.int/doc_num.php?explnum_id=5455 (accessed on 27 September 2019).

Zingore, S., Mutegi, J., Agesa, B., Tamene, L. and Kihara, J. 2015. Soil degradation in sub-Saharan Africa and crop production options for soil rehabilitation. *Better Crops* 99(1), 24–6.

Chapter 3

Modeling the effects of climate change on agriculture: a focus on cropping systems

M. Adam, CIRAD, Burkina Faso; K. J. Boote, University of Florida-Gainesville, USA; G. N. Falconnier, CIRAD, France; C. H. Porter, University of Florida-Gainesville, USA; E. Eyshi Rezaei, University of Göttingen, Germany; and H. Webber, University of Bonn and Leibniz Centre for Agricultural Research (ZALF), Germany

1 Introduction

A number of agroecosystem models have been developed to simulate the integrated effects of climate, soils, crop management practices, and crop genetics on agricultural production. These models use code written in various computer-programming languages to simulate the crop growth, crop phenology, and final yield of crops from sowing to maturity as a function of weather inputs, soil properties, crop management inputs, inputs of crop cultivar traits, in addition to the crop-specific information embedded in the model structure (Adam et al., 2011). As such, these models can be powerful tools to explore agricultural impacts and adaptation to climate change. As a result, they are used extensively to simulate the effects of climate change on agriculture in climate change impact assessments at global (e.g. Rosenzweig et al., 2014; Schleussner et al., 2018), continental (e.g. Webber et al., 2018a), regional (e.g. Faye et al., 2018), and local (e.g. Freduah et al., 2019) scales. However, models currently lack explicit processes to account for combinations of extreme events and biotic stresses, and improvements are urgently needed. Also, given the strong focus on near-potential, high-input agriculture and main major crops

http://dx.doi.org/10.19103/AS.2020.0064.07

(maize, rice, wheat) in past research, a renewed focus on low-input agricultural systems, and the ability of current crop models to deal with them is also required.

Recently, two major international initiatives, the Agricultural Model Intercomparison and Improvement Project (AgMIP, Rosenzweig et al., 2012) and the Modelling European Agriculture with Climate Change for Food Security (MASCUR, Ewert et al. 2015), emerged to foster (crop) modelers' exchange of expertise, with a particular focus on integrated assessment and climate change projection for cropping systems, with many publications resulting from this work. This chapter should be seen as a guide for the reader to further explore this and related literature, and to understand challenges around the use of crop models for climate change studies.

We first introduce commonly used modeling approaches in assessing the effect of climate change on agricultural production. The rest of the chapter is focused on how crop models simulate the effects of climatic factors and abiotic stresses related to or impacted by climate variables (including potential extreme stresses) on crop growth and development. We present the recent impact assessment studies carried out for climate change, mainly through crop model intercomparisons leading to identification of possible model improvements. It is also important to note that we did not deal with biotic stresses and crop quality, though we recognize the urgent need to address them in crop models for climate change studies. For each of the sections of this chapter, we first emphasize the importance of the data needed and then describe how crop models deal with most of the abiotic stresses influenced by climate. Finally, we conclude with perspectives for the use and improvement of crop models in the framework of climate change impact assessment studies (extreme events, minor crops, biotic stresses).

2 Agroecosystem models: tools to simulate the effect of climate, soils, crop management practices, and crop genetics on cropping systems

2.1 Approaches to assess the effect of climate change on agricultural production

A wide spectrum of modeling approaches have been developed to estimate the effect of climate change on agricultural production. These models differ in the spatial scales considered and the diversity of agricultural dimensions (e.g. crop, livestock, household decision-making) that they integrate. In this paragraph, we acknowledged the diversity of the modeling tools used to address impact of climate change on agriculture. However, the focus of this chapter will be on the effect of climate change on the cropping systems rather than the whole agricultural production (crop, livestock, agroforestry).

Farm models combine crop and livestock components together with household decision-making and economics into an integrated assessment framework. Several approaches exist, ranging from econometric production models and risk behavior model to general equilibrium models and integrated bio-economic models (see Jones et al. (2017) for a description of the different approaches). These models allow assessing how food security and economic performance would be affected by household decision-making and farm management under climate change. For instance, Claessens et al. (2012) used the Tradeoff Analysis for Multi-Dimensional Impact Assessment (TOA-MD) model to analyze the effects of different adaptation strategies to climate change on poverty rates, and the percentage loss relative to the base system agricultural income. They showed that introducing an improved maize variety in the cropping system of Machakos (Kenya) might be sufficient to offset the negative effects of climate change. However, in Vihiga (another agro-ecological zone of Kenya), there was a need to adopt improved feed quantity and quality in combination with improved livestock breeds to perform better under climate change and improve farmers' livelihoods. However, it is important to note that data needed for the calibration of the different components of these models can be missing and can constrain their application (Claessens et al., 2012).

For the livestock component, there are models that simulate livestock performances (e.g. milk and meat production) and methane emissions at animal level (Tedeschi et al., 2014). Other types of models take into account herd dynamics and compute variation of animal numbers over time. Animal performance and herd dynamics can in turn be combined into integrated livestock models that also consider farmer decision on feeding and grazing management. A good example of use of such model in the literature is Descheemaeker et al. (2018) that assessed the effect of climate change on the livestock component of mixed farming systems of Zimbabwe. They showed the importance of the dry season feed gap that affected livestock production (especially for more densely stocked farms), leading to the vulnerability of the livestock production linked with the year-to-year climate variability that increased with all climate change scenarios tested.

For the cropping component, a number of approaches are found in the literature to assess the effect of climate change on crops: process-based crop models, statistical models, and global vegetation models can be used to explore the impact of an altered climate on crop production (Challinor et al., 2014). Process-based crop models incorporate a mechanistic description of key dynamic processes affecting crop yields and are described extensively in the following sections. We also emphasize that the many yield-limiting and yield-reducing factors, particularly for low-intensity cropping systems, their management and interaction with climate are generally not well captured by process-based crop models (Webber et al., 2014.)

Statistical models are based on statistical relationships linking historical weather records with yield observations (Rowhani et al., 2011). Statistical models have the advantage of low computational cost to produce robust estimates compared with process-based models (Lobell and Asseng, 2017). They however lack the ability to, for example, account for the impact of elevated CO_2 on crop production (Lobell and Asseng, 2017) and to unravel the physiological processes responsible for a statistical correlation between climate and yield, which can be problematic when dealing with the design of adaptive technologies (Roberts et al., 2017).

Modeling the impacts of global climate change can also be performed with global-scale agro-ecosystem models such as LPJmL (Bondeau et al., 2007) and Pegasus (Deryng et al., 2011). These models were primarily developed to simulate carbon and nitrogen dynamics, surface energy balance, and soil water balance at a global scale, and may include dynamic global vegetation models that simulate the full global carbon and water cycles (Rosenzweig et al., 2014). Agro-ecological zone modeling methodology, for example GAEZ-IMAGE (Leemans and Solomon, 1993), are climate zone type models developed to assess agricultural resources and potential at regional and global scales and have been embedded into integrated assessment models for global environmental change. Global vegetation models (e.g. Deryng et al., 2011) integrate both natural and agricultural ecosystems in one single coherent modeling framework and can incorporate explicit simulation of carbon and water cycles and land-atmosphere exchange flows. These models are applied at the global scale and offer the strength to address questions going beyond pure crop yield simulations, for example assessing the impact of agriculture on global terrestrial carbon balance (Bondeau et al., 2007).

Although these models represent crop growth processes, calibration of model outputs against field observation is however not possible with global vegetation models. These types of models often fail to reproduce yield variability in low-input regions, as they assume homogenous soil properties and management (e.g. irrigation, cultivar choice, timing and amount of fertilizer) across large areas due to the lack of available information (Müller et al., 2017). While much of the content of this chapter is generally applicable to global vegetation models, the focus is on the use of process-based crop models to assess the effect of climate change on cropping systems; therefore, the term 'model' refers to the latter. This choice was based on the main strength of process-based crop models to mechanistically represent the effect of key climate variables on crop growth.

2.2 Climate and other data needs for crop modeling

Models and data are inextricably linked. Models allow us to make sense of data, while data are necessary to develop, improve, evaluate, and run models.

Numerous crop simulation models were developed for specific purposes, crops, or regions and although most crop models use similar concepts and data inputs (Boote et al., 2013), they differ widely on input data formats (Porter et al., 2014). The majority of crop models are point-based and require data specific to location, time, and scenario. In general, input data for models fall into three categories: environment (climate, soil), management, and genetics.

- Environmental data include information on the location, weather, and soils for the field or area being simulated. Location data include latitude and longitude, elevation, and slope of the field. Weather data for crop models generally include daily rainfall, maximum and minimum temperature, and solar radiation. Soil data quantify the water-holding characteristics of the soil and chemical properties, as well as initial conditions at the start of simulation. Some models required hourly climate data for more precise processes that are later integrated to the daily step.
- Management data include details of all field operations including planting, sowing density, fertilizer, irrigation, and tillage.
- Genetic data provide the growth and life cycle development characteristics of the crop and cultivar. The genetic data are generally model-specific and not transferrable or directly translatable among models.

Calibration of models, especially soils and genetic parameters, using observed data from the target regions and crop is needed to improve the accuracy of the model. Data for calibration requires, at least, crop growth, development, and composition observations; soil water and nutrient dynamics data. Model calibration generally involves a step-wise calibration of soil parameters, then phenological genetic parameters, finally growth genetic parameters (He et al., 2017).

The International Benchmark Sites Network for Agrotechnology Transfer project (IBSNAT, Uehara and Tsuji, 1993) described minimum data required to run a model (level 1), evaluate a model (level 2), and develop new model capability (level 3). The minimum data concept was developed to identify data which are relatively easily collected in field experiments and which are sufficient to allow reasonable simulations of crop growth and development. These are summarized in Box 1. Level 1 data include information describing location, weather, soils, management operations, and genetics. Level 2 adds to level 1 with field measurements of crop and soil variables, which allows for evaluation of model reliability. Level 3 requires more detailed data from focused research on the model component being developed. Many models have additional, more detailed data (e.g., relative humidity, or stable organic C), but should be able to run in a basic mode with the minimum data. As model complexity increases, the level of data required to run the model also increases. Figure 1 illustrates how data complexity increase with model complexity.

Box 1 Minimum data for crop models

1. *Level 1–Data to run models*
   a. Location–latitude, longitude, slope, elevation
   b. Weather: daily maximum and minimum temperature, precipitation, solar radiation
   c. Soil properties
      i. Surface properties: color, permeability, drainage
      ii. Soil profile properties: water-holding characteristics, nitrogen, organic carbon, bulk density
      iii. Initial states of soil water, nitrogen, and carbon
   d. Crop management
      i. Crop, cultivar, planting date, row and plant spacing
      ii. Irrigation dates and amount
      iii. Fertilizer dates and amount and type
      iv. Other management operations (tillage, chemical applications)
   e. Crop and cultivar genetic information
2. Level 2–Data to evaluate models
   a. Level 1 data
   b. Crop measurements
      i. Yield and yield components–biomass, seed number, seed size, etc., at harvest
      ii. Phenology–dates of flowering, physiological maturity, harvest maturity, first seed, and so forth
      iii. Growth analysis–biomass components–leaf, stem, seed/grains, etc., at regular time intervals
   c. Soil measurements at different depths and over time
      i. Soil moisture
      ii. Soil N, C
3. Level 3–Data to develop new model capability
   a. Level 1 and 2 data
   b. Research reports / publications
   c. Detailed experiments including response to temperature, water, nitrogen, and other factors
   d. Specific experiments to address knowledge gaps

Source: adapted from Hunt and Boote (1998).

2.3 Modeling CO_2 effect, temperature, heat, and frost stress on crop growth and development

All crop models simulate crop development and crop life cycle, including rate of leaf appearance and/or leaf expansion, rate of progress to flowering

Increasing Model Complexity

Increasing Data Complexity

Data components	Model components	Potential production	Water limited production	Water and N limited production	Water, N, P limited production
Weather data	Solar radiation	✓	✓	✓	✓
	Max and min temperature	✓	✓	✓	✓
	Precipitation		✓	✓	✓
Genetics	Cultivar characteristics	✓	✓	✓	✓
Soil characteristics	Physical properties		✓	✓	✓
	Chemical properties			✓	✓
Management data	Planting data	✓	✓	✓	✓
	Irrigation management		✓	✓	✓
	Tillage management		✓	✓	✓
	N fertilizer management			✓	✓
	Organic matter management			✓	✓
	P fertilizer management				✓

Figure 1 Increased model complexity requires an increase in data complexity.

(anthesis) and reproductive growth, and then to maturity. There is usually no direct CO_2 effect on rate of crop life cycle progression. All the models follow the concepts of cardinal temperatures such as the base temperature at which development is zero and an optimum temperature at which development is most rapid (leaf appearance or time to anthesis). Models typically use growing degree-day accumulation approaches for this, but also consider photoperiod (short-day or long-day type/effects) to delay anthesis and maturation (Dingkuhn et al., 2008). Models for some species such as wheat include vernalization, in which there must be a minimum number of chill units accumulated prior to the crop being ready to sense daylength and then progress on toward flowering.

The models differ in the complexity of how they simulate primary productivity. Most site-specific crop models use relatively simple approaches based on the radiation-use efficiency (RUE) approach, in which RUE is multiplied by the intercepted daily solar radiation (as a function of leaf area index (LAI)), going directly to crop dry matter without considering respiration losses. The CO_2 effect on RUE is included by scalar functions computed from past experiments (Free-Air-CO_2-Enrichment (FACE) and other). For the temperature effects on primary productivity, all crop models must be parameterized from literature, where leaf photosynthesis or dry-matter accumulation rates are measured at different temperatures, as there are no equations that result in 'emergent' temperature responses. The more mechanistic models also simulate growth respiration: conversion to dry matter is not temperature-dependent but is based on approximate composition of plant organs in terms of protein, lipid, lignin, and so forth which have different synthesis costs as reported by Penning deVries et al. (1974) and the models also have maintenance respiration (latter rises with temperature).

A smaller number of crop models are more mechanistic and simulate leaf-level photosynthesis, some with the full mechanistic Farquhar-von Caemmer Rubisco kinetics (Farquhar and von Caemmerer, 1982) in which the response to CO_2 and temperature is an emergent outcome. Others use semi-complex leaf photosynthesis where the CO_2 responses of quantum efficiency and light-saturated leaf rate are set by values taken from the literature and/or derived from the Farquhar approach (Pickering et al., 1995). In these approaches, the photosynthesis is defined by three key parameters: (i) the Rubisco activity (Vcmax), (ii) the photosynthetic electron transport rate (Jmax) translating the ability to regenerate RuBP substrate for Rubisco, and (iii) the Triose Phosphate Utilization (TPU) translating the ability to regenerate RuBP by Pi through the conversion of triose phosphate into starch and sucrose. Busch and Sage (2017) showed that Rubisco limits carbon assimilation at high temperatures, while TPU limits at lower temperatures and at higher CO_2 concentrations. Further, recent work (Lombardozzi et al., 2018, Fabre et al. 2019) also showed that at high CO_2 concentration with plant of low sink capacity, TPU might downregulate photosynthesis. Such processes need to be included in crop models.

For biomass allocation among organs, most crop models follow concepts of differential partitioning of assimilate to leaves, stems, and roots during the vegetative phase, followed by subsequent allocation of assimilate to reproductive organs. Some models follow an externally prescribed allocation patterns as a function of development units, while others follow internally coded allocation following allometric principles, and yet others use combinations of approaches in which there is specific addition of grains and other organs that have defined sink strength. In the latter approach, the crops typically add a number of sinks (grains) until the number of grains carried is equivalent to the source assimilate supplying capacity. The effect of CO_2 and temperature on partitioning and sinks is primarily a result of effects on photoassimilation, although temperature does affect the rate of addition of sinks and the rate of sink growth.

In general terms, rate of crop development accelerates with rising temperature, thus reducing the crop life cycle and the grain-filling duration. Grain yield is reduced with rising temperature as a result of less resource capture associated with the shorter life cycle. There may also be additional effects of rising temperature to reduce or increase photosynthesis and grain growth rate, depending on the current temperatures for the region relative to the species optimum temperatures for processes.

In complement to this general overview of crop productivity responses to climate change factors of CO_2 and temperature, there are some recent crop models' improvements with respect to heat and frost stresses (more extreme events). Indeed, the latest activities in the AgMIP project have led many modelers to extend model temperature response to not only consider development rates and effects on photosynthesis, respiration, and RUE (Wang et al., 2017), but to also include heat stress effects (White et al., 2011a). Various models at differing levels of complexity now consider heat stress effects on grain-set and abortion, either directly or indirectly via effects on partitioning (Boote et al., 2018; Challinor et al., 2005; Gabaldón-Leal et al., 2016; Lobell et al., 2015; Moriondo et al., 2011; Raes et al., 2009) or advancing senescence (Asseng et al., 2011). Heat stress is defined here broadly as a short period of a few days occurring near anthesis with temperatures above crop-specific thresholds (Barlow et al., 2015; Eyshi Rezaei et al., 2015a) that damage reproductive structures and processes in the plant (Wheeler et al., 2000). Deryng et al. (2014) also considered crop yield response to heat stress in a global climate change impact study with the model Pegasus, showing the importance of maize losses due to extreme heat stress at anthesis.

Further, as farmers adapt longer season varieties and phenology of winter crops is accelerated with warming (Eyshi Rezaei et al., 2015b, 2017; Zheng et al., 2015), the risk of frost damage is projected to increase for some regions currently constrained by cold temperatures (Rigby and Porporato, 2008). Frost already limits maize and fruit production in large parts of Northern Europe (Olesen et al., 2011). Frost damage can cause large losses in grain yield, depending on the crop, development stage, and severity of the frost (Barlow et al., 2015). Frost damage is primarily related to proteins and enzymes losing functionality upon freezing (Nagao et al., 2005), with cell membrane being most affected (Murray et al., 1989). As a result, photosynthesis is impaired (Ensminger et al., 2006), as well as assimilate translocation within the plant (Cromey et al., 1998). Largest yield losses occur when frost occurs just after heading (Cromey et al., 1998; Frederiks et al., 2012), with flag leaves (Fuller et al., 2007), although acclimation (Rihan and Fuller, 2013) provides some protection against yield losses. Some models (e.g. EPIC, CERES-Wheat,

APSIM, InfoCrop, STICS) do consider frost damage and various approaches consider effects on different processes or structures (e.g. leaf area expansion, senescence, assimilate accumulation, grain number reduction, crop death and density; Asseng et al., 2011; Brisson et al., 2003; Stöckle et al., 2003). The CropSyst (Stöckle et al., 2003) model includes effects of freezing temperature impairing soil water movement.

2.4 Modeling water stresses (drought and flood) via soil water and evaporative demand

Drought is a main factor limiting yields and driving year-to-year yield variation in many environments (Webber et al., 2018a). Drought stress is commonly estimated by considering the ratio of soil water supply to crop water demand on a daily basis, employing various approaches to simulate supply and demand terms (Boote et al., 2013). The water balance of nearly all models requires computing the evapotranspiration (ET) demand based on various ET equations, and the total ET demand is partitioned to soil evaporation and crop transpiration as a function of LAI. Crop response to drought may occur on a very immediate timescale as plants sense dry soil and involves stomatal regulation and reduced leaf area expansion (Tardieu et al., 2018); however, most crop models are not set up to function in this manner. These responses are observed as cumulative effects on root growth and leaf area that ultimately lead to reduced radiation interception, biomass accumulation, and lower grain yields (Tardieu et al., 2018). It is difficult to generalize the impact of drought on grain yield as the dynamics of drought development, as affected by both crop traits and environment, act on various intermediate processes in crop growth which in turn affect subsequent water use (Richards et al., 2002; Tardieu, 2012). Most widely applied crop models consider the effects of drought in their simulations (Boote et al., 2013; White et al., 2011a).

The soil water balance in most crop models requires that all soil layers (to specified depths for the different soils) be characterized for their water-holding capacity (drained upper limit = field capacity, and lower-limit = point of no further extraction). The soil layers are also characterized by saturated upper limit above which the soil air space is totally filled with water, which is also the point at which roots begin to die because the soil oxygen is rapidly depleted (flooding). Water extraction from the soils is modeled differently among the various models, some considering root length density distribution with depth and uptake per unit root length, while other models provide extraction rates that vary with soil depth (no explicit roots). Some more summary-models for the soil water dynamics completely ignore rooting, water extraction profiles, and define a given amount of soil water (mm) available.

The ability to supply the crop transpiration demand will depend on the extent to which soil water uptake can meet the transpiration demand. If the soil water uptake is insufficient, water stress is simulated and that reduces photoassimilation as well as reducing leaf area expansion and other processes. Crop transpiration increases as vapor pressure deficit and temperature increase. The crop models have various approaches by which the transpiration is reduced with rising CO_2. In a typical approach, the stomatal conductance is computed for both elevated CO_2 and ambient CO_2, and those stomatal conductances are placed in a set of Penman-Montieth type equations to compute a reduction factor on daily transpiration (Boote et al., 2010). The APSIM crop models simulate dry-matter accumulation first, then compute transpiration using a transpiration efficiency that depends on vapor pressure deficit and CO_2 (Wang et al., 2005). Intercomparisons among 23 maize models (Bassu et al., 2014) have shown there is considerable variation among the models in their simulation of transpiration and soil evaporation. Model evaluations against observed ET flux data confirm this variability among maize models, although the ensemble projection is best (Kimball et al., 2019). Furthermore, the degree of CO_2 reduction on transpiration has not been adequately tested under field conditions and is not possible from flux methods at FACE sites because of the small fetch involved. There is considerable uncertainty for degree of CO_2 reduction of transpiration among the models, and also in the data available, in part because CO_2 causes increased LAI for some crops which offset transpiration reduction, and the higher foliage temperature under rising CO_2 offsets CO_2 reduction on transpiration.

Most models include expected effects of drought by reducing leaf expansion; changing the assimilate allocation between above and belowground biomass; alter translocation of stem carbohydrates to reproductive organs; and/ or limit either photosynthesis or RUE (Boote et al., 2013). A particular challenge with modeling drought stress is the need to correctly simulate many processes: atmospheric evaporative demand, evaporation and transpiration partitioning, soil evaporation, water balance, water uptake by the roots, reduction in transpiration or photosynthesis on sensing dry soil, and the various feedbacks between these responses and aboveground growth (Jamieson et al., 1998). Correct simulation of crop water supply and demand is further complicated by the need to correctly specify soil properties which can be highly variable even in a single field and the fact that models are rarely evaluated on soil water use or root growth (Ewert et al., 2015). The reliance on aboveground biomass and yield for model calibration under drought stress has inevitably resulted in models getting the right yield under drought but with incorrect simulated water use, as suggested in a maize multi-model comparison (Bassu et al., 2014).

Finally, climate change is projected to increase the incidence of high-intensity rainfalls in many parts of the world (IPCC, 2014; Christensen and

Christensen, 2003). For crops, heavy rainfalls can increase the chance of waterlogging and lead to lodging-induced yield losses, creating more conducive conditions for crop diseases (Kristensen et al., 2011). Waterlogging is already widespread (Ren et al., 2014) with very large negative impacts on yields (Kuai et al., 2015). Flooding cuts off crops' oxygen supply (Ahmed et al., 2013; Armstrong et al., 1983), affecting most crop growth processes (Araki et al., 2012; Herzog et al., 2016; Irfan et al., 2010; Malik et al., 2002; Wiengweera and Greenway, 2004). Root functions (Malik et al., 2002) can be considered as the most sensitive to flooding with resulting damage on aboveground growth being most severely impacted when flooding occurs near flowering (Ahmed et al., 2013). Impacts of flooding on crop growth is considered by a number of crop models (e.g. AquaCrop, DRAINMOD, APSIM, DSSAT, SWAGMAN), usually by applying reduction functions to leaf and/or root growth rates, water use, and grain yield (Shaw and Meyer, 2015; Shaw et al., 2013) as a function of the phenological development stage. However, the ability to simulate soil saturation depends critically on setting soil saturated conductivity (Ksat) in slow-percolating soil layers, and that is what limits predictability.

2.5 Modeling nitrogen stress as a result of soil nitrogen mineralization from soil organic carbon pools, influenced by climate factors

Correctly simulating the N status of the crops is also critical because crop response to CO_2 is reduced dramatically under N-deficient conditions. For instance, rice yield response to CO_2 for unfertilized rice was null (Ziska et al., 1996) and even at 40 kg N ha^{-1}, was less than half of that of N-fertilized plots (Nakagawa et al., 1994). However, the availability of N from the soil is simulated quite differently among the different crop models for multiple reasons. Most models (should) require inorganic N (nitrate and ammonium) available for different soil profile depths at initiation of simulation, along with requiring amounts of fertilizer N at specified dates. Unfortunately, initial nitrate and ammonium are almost never available, except in soil fertility experiments. Likewise, residue carryover deposited from prior crops (that contributes mineralized N to next year's crop) is often unknown. As a result, those initial conditions are often guessed which causes uncertainty. Further, fertilizer compounds applied do vary, and fertilizers containing urea or ammonium are subject to volatilization losses if not incorporated or not rained in or irrigated into the soil within 1–2 days of application. This creates additional uncertainty. Most crop models input and simulate mineralization of soil organic carbon (values are generally available from soil survey mapping for at least the topsoil or even second layer, but they are rarely deep enough). So, this is where model and input similarity ends, and uncertainty among models begins.

The different models have different methods of mineralizing the native soil organic C. For instance, DSSAT (Jones et al., 2003) has two approaches available: the Godwin-Papran method (Godwin and Singh, 1998), which has an active and a slow pool, and the CENTURY method (Gijsman et al., 2002) which has three pools: active, intermediate, and slow-stable. APSIM crop models use yet another approach that considers two pools (active and inactive; van Oosterom et al., 2001). There are yet other methods for simulating N mineralization (Corbeels et al., 2005). Furthermore, the crop models have different methods by which last year's crop residues left in the field are decomposed and by which animal manures (if applied) are decomposed. The rate of availability of inorganic N for crop N uptake depends on all of the following processes: initial inorganic N, rate of mineralization from native soil organic C, rate of decomposition of crop residues (and manure if applied), and fertilizer applied. Soil water content also influences N mineralization rate and N uptake. Thus, soil N availability is simulated in very different ways.

To add to the uncertainties, the crop models themselves have different ways of simulating the rate of N uptake from the soil. Some use N uptake per unit root length density, while others have N uptake with the bulk water uptake by transpiration, and there are other methods. The crop's demand for N varies considerably over the crop growth cycle, being small when plant size is small, but increasing over time. Typically, the crop N demand on a given day depends on the crop's daily dry-matter accumulation multiplied by the target N concentration for that tissue, or for the whole crop depending on the model. There are different ways of modeling the target N concentration over time. If root N uptake is insufficient, then the N concentration of the vegetative tissue declines initially, and if it declines below a certain 'critical' low limit, then dry-matter accumulation by the crop is reduced. That is one method for reducing growth as a function of N deficit. Alternatively, those crop models that simulate leaf photosynthesis use the declining N concentration of the foliage to reduce the assimilation rate (thus a different way to reduce growth under N deficit). Another possible difference in crop models is how they accumulate N during vegetative growth, and subsequently mobilize it from vegetative tissue to the grain during the grain growth phase, and the extent to which the grain N concentration (a quality aspect) is reduced under N stress or under CO_2 enrichment.

3 Use of crop models for climate change impact assessment

3.1 Data requirement for climate change impact assessment studies

Most crop models are point-based in that they simulate for conditions at a single latitude and longitude, and were developed to simulate field-scale conditions. These models are often calibrated and validated with data from agronomic field experiments. Examples include DSSAT (Jones et al., 2003), APSIM (Holzworth et al., 2014), EPIC (Izaurralde et al., 2006), among many others. Climate assessments may be done using these models on individual sites, or on multiple sites within a region (e.g. Adiku et al., 2015).

Gridded modeling frameworks (e.g. Elliott et al., 2014; Robertson, 2017) delineate regional, country-scale, or global land masses into regular grid cells. Site-based crop models or global agroecosystem models are used to simulate one or more homogeneous sets of conditions at each grid cell. For gridded systems, data are the same as for the point-based simulations, but data are provided in GIS raster or vector data layers which include all of the management, weather, soils, and genetic data for each grid cell (or pointers to those data). Data layers are also used to assign areal coverage of cropped land in each cell so that simulated yields can be aggregated into total production for the area of interest. As Gbegbelegbe et al. (2017) pointed out, uncertainties in pixel information for climate, soil, and crop management input can cause great uncertainties in crop models outputs. It remains important to be able to properly define at global and regional scales of the diversity of conditions (cultivar, soil, management practices). This remains a great challenge, but recent studies (Gbegbelegbe et al., 2017; Faye et al., 2018) tried to address such complexity.

Well-parameterized models can be used to simulate crop and environmental outcomes for hypothetical scenarios by modifying baseline model inputs in order to mimic the scenario of interest. Hypothetical scenarios are useful for optimizing resource-use efficiency, maximizing profit, minimizing environmental degradation, as well as our focus here, modeling the impacts of climate change and the benefits of climate adaptation and mitigation. Modeling future climate scenarios requires that projections of future daily weather be generated based on the output from climate models driven by various scenarios of greenhouse gas and aerosol emissions, or assumption on incremental increase or decrease of key climatic variables (e.g. temperature, rainfall). Because these climate-related hypothetical scenarios are assumed to occur at some future time slice, such as mid-century or end of century, they may include assumptions about technology trends and climate change adaptation packages. For some analyses, climate change mitigation, including greenhouse gas emissions and carbon sequestration, is an important modeling outcome.

Data from climate models must be processed prior to use in crop modeling applications to remove the biases associated with mean climate, climate variability, and extreme events (Ruane et al., 2015). Downscaling of global climate models (GCM) outputs is used to increase resolution and to provide better statistics for climate variability and extremes. One such product for downscaled and cleaned weather data for future climate scenarios is MarkSim

(Jones and Thornton, 2013). In many cases, daily weather data for modeling future climate scenarios are generated using historical, observed weather data. Monthly mean differences (projected by the climate models) are applied to historical observations of minimum and maximum temperatures, while rainfall differences are applied as a multiplicative factor. Means plus variances from the climate models can also be used to modify observed data. These methods preserve the historical weather patterns, but superimpose the selected future climatology.

Because of the critical need for data inputs to simulate climate impacts, there is a growing movement to create open access to data collected by governments, international organizations, research institutions, and businesses. Governments and funding organizations are increasingly mandating that project data be made publicly accessible. The FAIR Data Principles (Wilkinson et al., 2016) go beyond open access and promote data which are Findable, Accessible, Interoperable, and Reusable. These principles were designed as a concise and measurable set of guidelines for those wishing to enhance the reusability of their data holdings with specific emphasis on enhancing the ability of machines to automatically find and use the data. In the crop modeling domain, much effort has been put into harmonizing data for use in crop models. The AgMIP (Rosenzweig et al., 2012) promotes the use of ensemble modeling for model intercomparisons and for climate impact assessments. Data harmonization methods were developed to facilitate the use of common datasets with multiple models (Porter et al., 2014) as depicted in Fig. 2. The AgMIP data interoperability system was developed collaboratively with the crop model development teams participating in AgMIP between 2010 and 2015. The main components of this interoperability system are: (1) a data format standard; (2) a standard vocabulary based on the ICASA Data Dictionary (White et al., 2013); (3) and libraries of data translation tools, including model-specific translators for various crop models. The AgMIP harmonized data format consists of semi-structured key-value pairs stored as javascript object notation (json) objects. Each key in the key-value pairs is identified by a variable in the ICASA Data Dictionary which is associated with specific definitions and units.

3.2 Climate change effect in high-input systems

The evaluation of climate change factors on wheat production by Asseng et al. (2013) reveals that there is greater uncertainty for response at elevated temperature in the models than for response to CO_2, leading them to evaluate and improve their models against elevated temperature data on wheat in the hot serial cereal experiments (Webber et al., 2018b). Conversely, Bassu et al. (2014) found that the maize models were more uncertain for their CO_2 response than for their temperature response. They explained this higher uncertainty

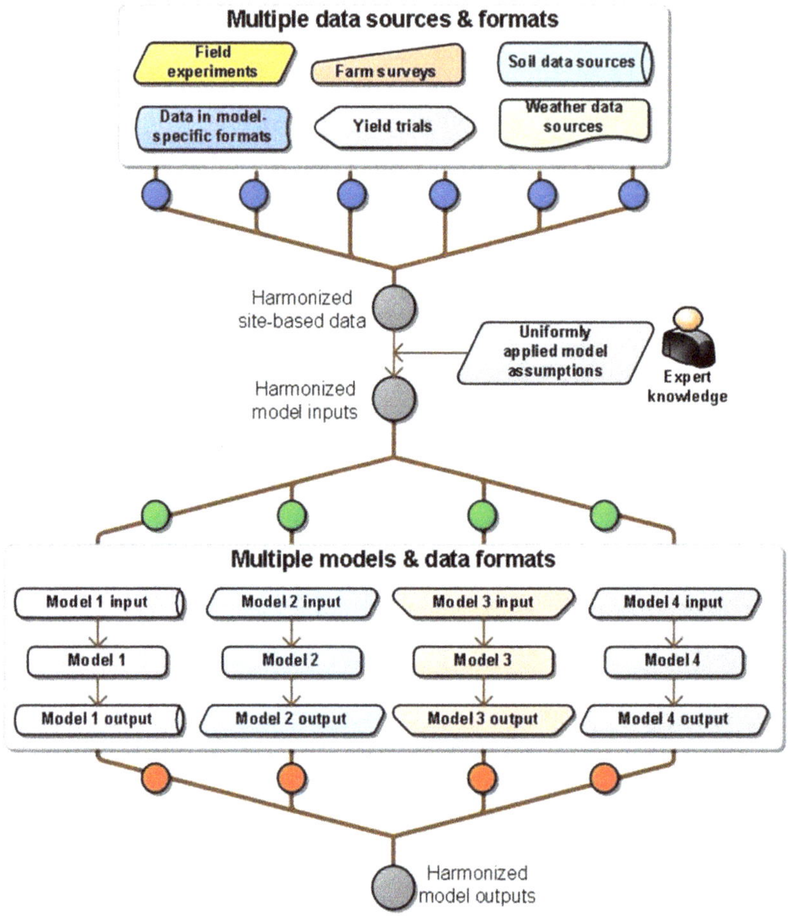

Figure 2 Schematic diagram of AgMIP data harmonization system. Each colored dot is associated with a translator which will convert raw data to harmonized AgMIP data format (blue), from AgMIP format to model-specific formats (green), or from model output formats to harmonized AgMIP model output format (orange). Harmonized data products are denoted by the larger gray dots. Source: adapted from Porter et al. (2014).

on CO_2 response on maize because of the prior controversy on the degree of CO_2 responsiveness of C-4 crops, and partly because most of the maize models use similar degree-day approaches for accelerating life cycle with rising temperature. This led Durand et al. (2018) to evaluate 23 maize models against CO_2-FACE data from Braunschweig, Germany, where they found that model responses to CO_2 were (correctly) very small under irrigated conditions, but that model responses under water-limited conditions were less than half that observed by Manderscheid et al. (2014). The issue here is the extent to which elevated CO_2 reduces transpiration (most models were insufficiently reducing transpiration), although the water conservation aspects under elevated CO_2 were most critical just at the time of setting grain number (and models that did not set explicit grain numbers performed less well). Similarly, Li et al. (2015) evaluated sensitivities of 16 rice models to both elevated CO_2 and rising temperature and found that model uncertainties increased as the conditions deviated from ambient, showing greater uncertainty at elevated CO_2 and at elevated temperature. Next, Hasegawa et al. (2017) tested multi-model response against observed data from CO_2-FACE experiments in Japan and China. There was considerable variation among the models in CO_2 responsiveness, although the ensemble of the models was quite close to the observed response. An interesting feature learned in that study was that the models overestimated CO_2-response for low-N treatments, confirming that models cannot yet fully handle interactions of CO_2 by N.

All these studies focused on major crops (i.e. wheat, maize, and rice), and could highlight the importance of simulating properly the effect of temperature and CO_2 response. However, given the strong water and nutrient limitations in low-input systems, models also need to rigorously account for soil water and nutrient dynamics (Jones et al., 2012). Mineralization of soil organic carbon and crop residues are important determinants of nutrient supply during the growing season, and soil moisture dynamic affect supply of nutrients and the ability of the plant to uptake these nutrients. Furthermore, soil organic carbon levels are likely to decrease with future climate scenarios (Basso et al., 2018); the higher temperatures decreasing biomass and hence the amount of crop residues returned to the soil. Eventually high-frequency rainfall will lead to more N leaching (Esse et al., 2001) and high rainfall is expected to increase in parts of the world hosting low-input systems, for example in the Sahelian region (Taylor et al., 2017). Accounting for interactions between soil processes, plant aboveground growth, and the changes in CO_2, temperature, and rainfall induced by climate change will therefore be crucial.

3.3 Climate change effect in low-input systems

Several soil-crop models have been evaluated in sub-Saharan conditions with low inorganic fertilizer and irrigation inputs for maize, sorghum, and millet: in Ghana (Dzotsi et al., 2010; Fosu-Mensah et al., 2012; MacCarthy et al., 2010), in Mali (Akinseye et al., 2017; Traore et al., 2017; Adam et al., 2018), in Nigeria (Adnan et al., 2017; Jagtap et al., 1993), in Benin (Amouzou et al., 2018), in Niger (Fatondji et al., 2012), recently in the West African Soudan savanna in general (Faye et al., 2018), and in Zambia (Corbeels et al., 2016). The tested soil/crop models differ strongly in their structure and the processes implanted to simulate soil nutrient and water dynamics (Webber et al., 2014). The abovementioned studies in low-input conditions generally pointed to satisfactory simulation of plant growth and grain yield. Interestingly, none of these studies investigated the accuracy of simulated grain yield, plant aboveground growth, where soil water and soil nutrient limitations were considered altogether. Therefore, an evaluation of the realism and consistency of how models describe both soil and plant processes, leading to grain yield, is still lacking in low-input smallholder conditions. Soil-crop models and their current parameterization for low-input conditions therefore run the risk of being right for the wrong reason (Bellocchi et al., 2010), model parameterization compensating for structural inconsistency (Martre et al., 2015). Assuming that such compensation will be effective or accurate with changing CO_2, water, and rainfall conditions under climate change is questionable. Overall, studies combining detailed measurements of crop growth variables as well as soil water and nutrient dynamics are costly and still scarce in smallholders low-input conditions. Strikingly, none of the abovementioned studies of model applications in low-input conditions (with the exception of Amouzou et al. (2018)) incorporated such a comprehensive dataset. Also, among the 132 listed studies in the work of Brilli et al. (2017) on comparing different soil-model structures, only three studies were in low-input conditions. The research agenda for modeling the effects of climate change under low-input conditions should therefore aim at implementing such detailed soil-crop monitoring in contrasting low-input sites. There is a critical need for soil-crop model intercomparison studies that address the specificities of smallholder low-input condition facing climate change.

In response to this need, Rosenzweig and Hillel (2015), with help of national researchers from different sub-Saharan and South Asia regions, initiated a CO_2, temperature, water/rainfall, and nitrogen (CTWN) analysis of the crop models in low-input systems, by choosing data representing typical fields taken from village survey data in these regions (i.e., sparse agronomic information, estimated yields, and economic information). Most of these field sites were unfertilized degraded soils (low-input systems). The soils were poorly parameterized from rather general soil maps, and there was inadequate information on soil organic C and its quality. Further, there was a lack of information on initial conditions of inorganic N, crop residues, manure applied, tillage, and specific cultivar used by farmers.

A CTWN analysis is basically the evaluation of single-factor response to (i) CO_2 (360, 450, 540, 630, 720 ppm CO_2), (ii) temperature (−2, 0, +2, +4, +6, and +8C), (iii) rainfall (25%, 50%, 75%, 100%, 125%, 150%, 175%, 200% of ambient rainfall), and (iv) N fertilization (0, 30, 60, 90, 120, 150, 180, 210 kg N ha^{-1}), with the exception that response to CO_2 was done at two N levels . The results of CTWN analysis were important to understand the reasons for small climatic responses to CO_2 (caused by lack of N), temperature, rainfall (where elevated rainfall often reduced yield because it leached N), and N fertilization for these degraded soil sites. Figure 3 represents the main results of this analysis for a specific region: Koutiala, Mali. DSSAT's CERES-Maize and APSIM-Maize correctly showed very small response to elevated CO_2, which was small to nonexistent under zero N fertilizer applied (Fig. 3a and b). This was anticipated because maize is a C-4 species crop, and this confirmed the results from Durand et al. (2018). However, the simulated C-3 crops such as wheat and rice also showed much less response to rising CO_2 for N-limited crops compared to expected responses to CO_2 observed for well-fertilized crops, a response that is supported by literature (Nakagawa et al., 1994; Ziska et al., 1996). The APSIM and DSSAT maize models both showed yield reduction with rising temperature (Fig. 3c), but DSSAT's CERES-Maize model showed greater yield reductions (associated with model parameterization, i.e. grain growth rate reduces too strongly as a function of rising temperature). This response of temperature should be modified, based both on this intercomparison study

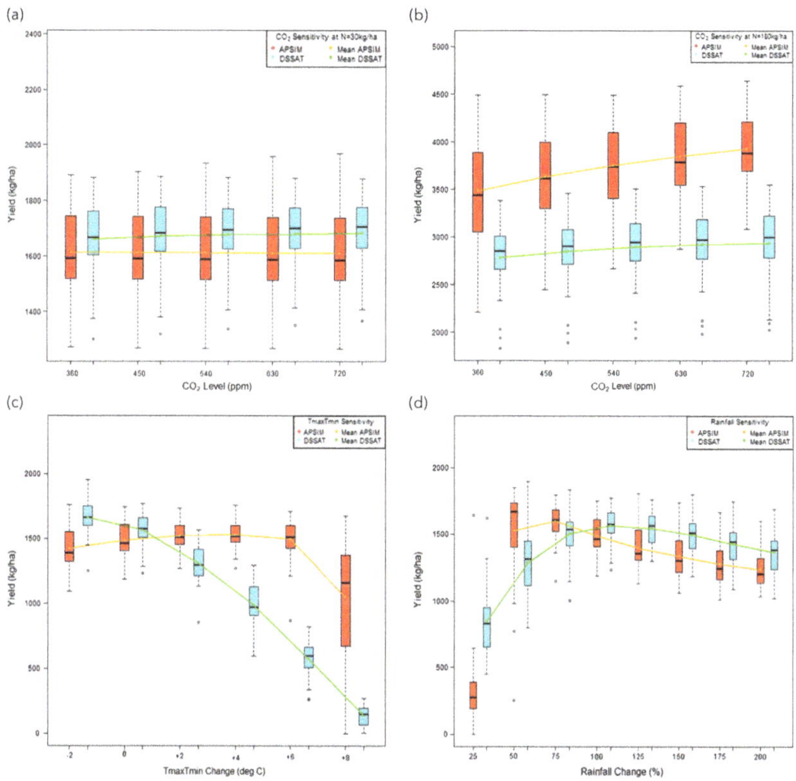

Figure 3 CO_2, temperature, water/rainfall, and nitrogen (CTWN) analysis of the crop models in Koutiala, Mali: (a) Response to elevated CO_2 under zero N fertilizer applied and (b) with 180 kg N.ha^{-1} applied, (c) response to temperature changes, (d) response to rainfall changes. Courtesy: CIWARA AgMIP team.

and on recent data on maize pollen viability, grain-set, and grain growth under heat stress observed by Lizaso et al. (2017, 2018). Yield response to elevated temperature for the wheat and rice models (DSSAT or APSIM) showed yield reductions with rising temperature, which were relatively similar across the two software systems, and deemed to be correct.

Rainfall responses of the different crops were quite revealing for the N-limited condition on degraded soils. When rainfall was very limiting, rising rainfall increased yields. However, more often, especially in West Africa and Southeast Africa with degraded soils, yield was reduced as rainfall increased above a given optimum rainfall (often less than ambient, Fig. 3d), because the increasing rainfall caused leaching of the precious little amount of N (Masikati et al., 2014). Another feature discovered with this modeling exercise was that rainfall was not simulated to be as limiting as expected. The reason 'in the crop models' is that the N was so limiting to LAI (often less than one to two units), that transpiration demand was low and water stress was simulated less than expected. Whether this is true in real fields is another question, but both models simulated that behavior, despite having totally different methods of transpiration simulation. As mentioned previously, DSSAT simulates transpiration relative to Priestly-Taylor ET methods and as a function of LAI (which is low in N-limited fields). APSIM simulates transpiration as a transpiration-efficiency function multiplied by daily biomass production (which is low in N-limited fields).

Following up on these findings, a new AgMIP initiative on low-input systems aims at comparing 29 existing soil-crop models to (i) evaluate their ability to simulate observed maize yield in diverse cropping systems across sub-Saharan Africa and (ii) to test model responses to CO_2, temperature, water, and nitrogen (N). The models are currently being calibrated and evaluated (for yield and biomass, but also soil water and soil nitrogen dynamics; Falconnier et al., 2019) with data from 2-year experiments in contrasting production environments: cool semiarid Rwanda, cool subhumid Ethiopia, hot semiarid Mali and Benin, and hot subhumid Ghana.

3.4 Importance of multi-model ensembles for climate impact and risk assessment

In the previous sections, we highlighted many studies carried out with model ensemble to assess the impact of climate change on crop productivity. Most of these examples are outcomes of the initiative 'Agricultural Model Intercomparison and Improvement Project (AgMIP)' (Rosenzweig et al., 2012). This initiative aims at simulating the end-to-end effects of climate change on crop productivity onward to economic productivity at site, regional, and global scales through model intercomparison. Model intercomparison is a useful approach to compare consistency and uncertainty in model simulations across the various available current crop-soil models. As described previously, current models have already been compared for some crops that are important in smallholders low-input systems, that is maize (Bassu et al., 2014), wheat (Asseng et al., 2013; Martre et al., 2015), and rice (Li et al., 2015). These intercomparison studies, though offering a first useful quantification of model simulation uncertainty linked with model structure and parameterization, focused on high-input, near-potential conditions. For example, in the maize intercomparison exercise, N fertilizer input ranged 61–255 kg/ha, sites were irrigated or had good rainfall, and grain yield ranged 5–11 t/ha. Furthermore, these intercomparison studies analyzed differences in model structure for crop growth processes mainly (e.g. simulation of net primary production as influenced by temperature and CO_2) but not the difference in structure for simulating soil processes. Different soil-model structures have however been scrutinized on their ability to reproduce water, C and N dynamics, that is the quality of SOC decomposition, N mineralization, nitrate leaching and denitrification estimates (Brilli et al., 2017), but again mostly in high-input conditions. Wallach and Thorburn (2017) with a special issue on models' uncertainty concluded that multi-model ensembles is a way to quantify model structural uncertainty. Also, Confalonieri et al. (2016) showed with an ensemble of 13 rice crop models that (i) differences in structure often resulted in similar predictions but also that (ii) similar structures can lead to large differences in model outputs. Hence, it seems important to conduct further research to better address the effect of climate change on crop productivity and assess with this diversity of models the range of potential futures as a response to climate change scenarios.

Most of the crop models in AgMIP are site-specific models requiring data on soils, crop, cultivar, management, and weather to simulate production at actual sites. However, global crop models are also employed, in which the site-specific nature is relegated to grid-positions over the entire globe, and the site-specific inputs per grid point are generated in a much more general way with 'data-products' (Ruane et al., 2017) So far, the AgMIP team members have an amazing history of collaborating, cooperating, sharing, and learning from each other, and have been supported by other local initiatives such as the MASCUR. The AgMIP modelers followed definite project protocols and plans (Ruane et al., 2014b), and reunite many crop modeling groups, including global crop modelers group, applying this methodology at a global scale. The gridded global modeling projects are aimed at addressing global production response under future climate change scenarios, considering the different fossil fuel-driven CO_2 pathways, as well as different GCMs used, and the different time slices of the future.

4 Conclusion and future trends

Crop models were originally designed to simulate crop growth and development as a response to climate, soils, crop management practices, and crop genetics. Most of them are mechanistic models aiming at representing the response of crop growth and development to temperature, CO_2, water, and nutrients. As a result, crop models are extensively used today to project the effect of climate change on agriculture. There are studies at different scales, from plot level to global level. But, this intensive use of modeling approaches, and in particular the use of the crop models, raises challenges that need to be addressed to increase the accuracy of their projections.

4.1 Future trends for agricultural data generation for climate change impact assessment

Most datasets, even from very detailed and controlled field crop experiments, do not contain all of the information required to run a crop model (i.e. initial soil moisture and nitrogen). Routine data such as row spacing, fertilizer application method, or depth of tillage operations may be overlooked when recording field management. This is especially true if data are collected on farm, or derived from farm surveys. A crop modeler often must fill in information, preferably from the data collector, but often based on the modeler's best agronomic knowledge of the modeled systems. For filling such management information, tools based on expert rules (i.e. QuadUI, dataMILL) have been developed within the framework of the AgMIP initiative. Also, in the case of weather data, gaps in the record are common. Models typically have no automated way of dealing with missing data and so the modeler must fill in these data using standard techniques. If gaps in temperature or radiation are less than a few days, data may be linearly interpolated through the gaps. If larger gaps occur, the data must be filled from nearby weather stations (Ruane et al., 2014a) or by using weather generators which use monthly statistics. Solar radiation data from satellite observations after 1983 have proven to be a reliable source for crop models (White et al., 2011b). However, as Corbeels et al. (2018) mentioned, crop modelers also need to be cautious about the use of climate-crop model ensembles in climate change impact assessments studies, as uncertainties coming from inputs data are very high.

Building on this experience, a data search and access platform for the Consultative Group for International Agricultural Research (CGIAR) is being developed to allow data from these research organizations to be accessible through the GARDIAN (gardian.bigdata.cgiar.org) web service. Datasets are currently being annotated with terms from using the Agronomy Ontology (Devare et al., 2016), Crop Ontology (Shrestha et al., 2010), and other ontologies so that data will be interpretable through semantic technologies. Methods to link CGIAR data to AgMIP translators are being developed to facilitate use of CGIAR data for modeling and other quantitative analyses. Also, developments of tools for data collection that use these standard vocabularies are underway (Devare et al., 2016).

4.2 Future trends for modeling major biotic and abiotic stresses for climate change impact

Though crop models were not originally designed to consider extreme events (Ewert et al., 2015), various recent efforts have been undertaken to improve them for response to heat stress (Maiorano et al., 2017), interactions of heat and drought stress (Webber et al., 2018b), and response to elevated atmospheric CO_2 concentrations under water stress (Durand et al., 2018), among others. Other stressors, such as atmospheric ozone which also reduces crop yields, are infrequently considered in crop models (Ainsworth, 2016). Of the models that do consider ozone damage, accelerated senescence and reduced photosynthesis or RUE are the most commonly included process responses to elevated ozone concentrations (Emberson et al., 2018).

Beyond the limited and variable representation of individual stressors, other challenges limit the use of crop models for climate change risk assessments. The first is the limited data available for testing the response to stressors across locations or production conditions. Rötter et al. (2018) reviewed recent crop modeling studies that tested model response to extreme events and found a relatively small number of cases that had based model improvements on observed experimental data. This is particularly troublesome when models are applied across wide spatial extents where the stressors and their combinations are expected to vary between regions are used, while their different effects according to the context/regions have not been yet properly tested. Some directions on how to consider four of the main climatic stressors (drought, heat, flooding, and frost risk) now anticipated to increase with climate change (IPCC, 2014) have been discussed in this chapter, highlighting their main mode of impacting crop growth and their representation in crop models. But limited interaction between experimental and modeling groups in experiment design can result in experiments being conducted that do not collect required datasets needed for modeling studies (Kersebaum et al., 2015). Another serious limitation to their use in climate risk assessments is because few models consider stress combinations. This largely results from the lack of scientific knowledge about stress combinations, as stresses are usually studied in isolation under control conditions, which differ substantially from field production conditions (Barnabás et al., 2008; Mittler, 2006; Suzuki et al., 2014). The work looking at multiple stress combinations focuses on understanding genetic basis of response and signaling pathways (Iyer et al., 2013; Rasmussen et al., 2013; Rivero et al., 2014; Rizhsky et al., 2004) which is difficult to convert to the level of physiological response needed in crop models, particularly when derived under controlled conditions. Stress combinations are of course not limited to the climatic stressors reviewed here, but include soil salinity and low fertility (which is more than N availability), and possibly ozone concentrations, and biotic stresses.

While recent years have seen model improvements for each of these factors (Emberson et al., 2018; Vanuytrecht and Thorburn, 2017; Adam et al., 2018; Webber et al., 2010), there is little if any evaluation of their performance under climatic stressors. Accounting for acclimation is another challenge in modeling the impacts of extreme events on crops. Acclimation refers to the fact that crops that have already been exposed to stress are often able to more rapidly enact protective mechanisms against damages, even when the exposure was to a different stressor. Despite the now widely reported importance of acclimation across crops and stressors (Bergjord Olsen et al., 2017; Bita and Gerats, 2013; Végh et al., 2018; Wang et al., 2012), crop models currently do not explicitly consider it (Rashid et al., 2018), although indirect effects such as reducing LAI, reducing specific leaf area, and increasing root under prior drought are examples of the crop models giving emergent acclimations. Smith and Dukes (2013) reviewed options to consider acclimation in Earth Systems' Models.

Further, particularly critical but difficult challenge is to account for pest, disease, and weed pressure and damage in crop model-based impact assessments. Pest and pathogen damages are now estimated to result in global crop yield losses of 22%, 30%, 23%, 17%, and 21% for wheat, rice, maize, potato, and soybean, respectively (Savary et al., 2019). As pests and diseases may also be expected to change with climate change, many questions exist regarding potential combined pest and climate change impacts for crops (Savary et al., 2018). Despite the significance of potential impacts, very few crop models consider pest and disease damage. The exceptions are often for a specific crop by pest by region combinations (e.g. Savary et al., 1990; Savary et al., 1997). More generic approaches are considered necessary for application in wide area impact studies, in which pests and diseases are grouped into generic types which have similar injury symptoms under similar production conditions and which can be simulated with quite generic pest modules (Savary et al., 2018). These authors propose that for each injury type by production condition, time varying reduction functions could be applied to affected processes. This is a more generic extension of earlier work, in which Boote et al (1983) proposed a typology of damages to crop states or processes and options for their inclusion in process-based crop models. Donatelli et al. (2017) outline a roadmap to guide the improvement of simulation of the damages to crop yields due to pest and diseases, including collection of better datasets for model development and improved integration between crop and pest and disease models, among others.

Another important component to consider when assessing the impact of climate change on agricultural production is to assess its impact on food quality. So far, little has been done on this subject, but Chenu et al. (2017) highlighted the negative effect of elevated CO_2 on wheat grain protein concentration and mentioned the work from Asseng and Milroy (2006) and Triboi et al. (2006) who modeled the environmental effects on grain protein concentration. Similarly, Martre et al. (2015) showed that the effect of climate change on grain protein content was linked to the simulation of total aboveground nitrogen. Asseng et al. (2019), with a multi-ensemble of wheat models, showed that higher temperature and changes in rainfall will negate the positive effect of CO_2 on wheat grain protein, but with an overall negative impact (−8.6%) on proteins yield, for most locations. This decline could be partly attributed to grain yield increase, and lower response of grain protein concentration to fertilization.

Also, efforts are underway in the crop modeling community to meet the need for developing new or adapting existing models for perennial crops and minor crops including vegetables. The DSSAT software suite has added perennial forages for grasses (Pequeno et al., 2018) and alfalfa (Malik et al., 2018) and has added some minor crops such as tomato (Boote et al., 2012), pigeonpea (Alderman et al., 2015), safflower (Singh et al., 2015), as well as snap bean, strawberry, quinoa, amaranthus, and chia (Boote, 2019, unpublished). The APSIM software suite also includes a number of crops and perennial forages (Keating et al., 2003). In addition, approaches can be found in the literature to help modelers adapt their models for new crops (Lamanda et al., 2012; Adam et al., 2013). Finally, both DSSAT and APSIM have full capabilities to run multiple crops in sequence, with carryover of crop residues, soil C, soil water, and soil inorganic N. This allows evaluation of C sequestration under long-term climate change scenarios, but these simulations have not been adequately employed by the modeling communities to this point and need further research. Corbeels et al. (2016) showed DSSAT responds to mulching, particularly when rooting depth is restricted, that is, when water is a critical limiting crop growth factor, but that some other important site-specific factors such as plough pan formation or improved soil structure are not modeled by DSSAT. No work has been done to explicitly test this carryover effect of crop residues under climate change scenarios.

Finally, a further challenge relates to the fact that few models consider the genetic basis of adaptive traits (e.g. early vigor, increased transpiration under high temperatures, increased assimilate partitioning to roots) that confer tolerance to extreme events (Reynolds and Langridge, 2016; Richards, 2006). Indeed, crop models offer great potential to aid in understanding in which environment a given adaptive trait will confer a yield benefit (Tardieu, 2012) or, perhaps more appropriately, assess the probability of a trait, in a particular system and location, conferring a yield benefit given the highly uncertain nature of the weather (Tardieu et al., 2018). However, most genetic coefficients in crop models do not consider correlation between traits and parameters, limiting their current validity in this setting (Boote et al., 2013). Combining process-based models information on quantitative trait loci (QTL) is one option to overcome this challenge (Chenu et al., 2009; Yin et al., 2003; Yin et al., 2005).

As the field of high-resolution phenotyping is rapidly developing with infield application, van Eeuwijk et al. (2019) highlight the need for coordinated efforts to ensure crop models are able to aid in the interpretation, as well as benefit from these new data sources.

5 Where to look for further information

5.1 Further reading

- For more information on calibration process for crop models, see He et al. (2017).
- For a general review of crop productivity responses to climate change factors of CO_2 and temperature, see Hatfield et al. (2008).
- For a review on water balance in DSSAT crop models, see Boote et al. (2009) and Ritchie (1998).
- For a review on how the DSSAT crop models address N uptake and N balance processes, see Godwin and Singh (1998) and Boote et al. (2008).

5.2 Key books/special issues

- Special issue of European Journal of Agronomy: 'Recent advances in crop modelling to support sustainable agricultural production and food security under global change' (Thorburn et al., 2018).
- Working with Dynamic Crop Models - 3rd Edition [WWW Document], n.d. URL https://www.elsevier.com/books/working-with-dynamic-crop-models/wallach/978-0-12-811756-9 (accessed 12 December 2018).
- Handbook of Climate Change and Agroecosystems: The Agricultural Model Intercomparison and Improvement Project (AgMIP) Integrated Crop and Economic Assessments – Joint Publication with the American Society of Agronomy, Crop Science Society of America, and Soil Science Society of America (In two parts).
- Special issue on uncertainty in crop model predictions: Wallach and Thorburn, 2017.

5.3 Major international research projects and conferences

- MASCUR https://macsur.eu/.
- AgMIP http://www.agmip.org/kql.
- Big data and crop modeling groups (CGIAR initiative): https://bigdata.cgiar.org/communities-of-practice/crop-modelling/.
- ICROPM conference, Berlin, 2016/ http://communications.ext.zalf.de/sites/crop-modelling/SitePages/iCROPM2016.aspx.

6 References

Adam, M., van Bussel, L. G. J., Leffelaar, P. A., van Keulen, H. and Ewert, F. 2011. Effects of modelling detail on simulated potential crop yields under a wide range of climatic conditions. *Ecological Modelling* 222, 131–43.

Adam, M., Wery, J., Leffelaar, P. A., Ewert, F., Corbeels, M., and Van Keulen, H. 2013. A systematic approach for re-assembly of crop models: an example to simulate pea growth from wheat growth. *Ecological Modelling* 250, 258–68. https://doi.org/10.1016/j.ecolmodel.2012.10.012.

Adam, M., Dzotsi, K. A., Hoogenboom, G., Traoré, P. C. S., Porter, C. H., Rattunde, H. F. W., Nebie, B., Leiser, W. L., Weltzien, E. and Jones, J. W. 2018. Modelling varietal differences in response to phosphorus in West African sorghum. *European Journal of Agronomy* 100, 35–43. https://doi.org/10.1016/j.eja.2018.04.001.

Adiku, S. G. K., MacCarthy, D. S., Hathie, I., Diancoumba, M., Freduah, B. S., Amikuzuno, J., Traore, P. C. S., Traore, S., Koomson, E., Agali, A., Lizaso, J. I., Fatondji, D., Adams, M., Tigana, L., Diarra, D. Z., N'Diaye, O. and Valdivia, R. O. 2015. Climate change impacts on West African agriculture: an integrated regional assessment (CIWARA). Ch. 2 In *Handbook of Climate Change and Agroecosystems: The Agricultural Model Intercomparison and Improvement Project (AgMIP)*. Ed. C. Rosenzweig and D. Hillel. Imperial College Press & American Society of Agronomy. https://www.ipar.sn/IMG/pdf/adiku20et20a0008.pdf.

Adnan, A. A., Jibrin, J. M., Kamara, A. Y., Abdulrahman, B. L. and Shaibu, A. S. 2017. Using CERES-Maize model to determine the nitrogen fertilization requirements of early maturing maize in the Sudan Savanna of Nigeria. *Journal of Plant Nutrition* 40, 1066–82. https://doi.org/10.1080/01904167.2016.1263330.

Ahmed, F., Rafii, M. Y., Ismail, M. R., Juraimi, A. S., Rahim, H. A., Asfaliza, R. and Latif, M. A. 2013. Waterlogging tolerance of crops: breeding, mechanism of tolerance, molecular approaches, and future prospects. *BioMed Research International* 2013, 10p.

Ainsworth, E. A. 2016. Understanding and improving global crop response to ozone pollution. *The Plant Journal* 90, 886–97. doi:10.1111/tpj.13298.

Akinseye, F. M., Adam, M., Agele, S. O., Hoffmann, M. P., Traore, P. C. S. and Whitbread, A. M. 2017. Assessing crop model improvements through comparison of sorghum (sorghum bicolor L. moench) simulation models: a case study of West African varieties. *Field Crops Research* 201, 19–31. https://doi.org/10.1016/j.fcr.2016.10.015.

Alderman, P. D., Boote, K. J., Jones, J. W. and Bhatia, V. S. 2015. Adapting the CSM-CROPGRO model for pigeonpea using sequential parameter estimation. *Field Crops Research* 181, 1–15.

Amouzou, K. A., Naab, J. B., Lamers, J. P. A. and Becker, M. 2018. CERES-Maize and CERES-Sorghum for modeling growth, nitrogen and phosphorus uptake, and soil moisture dynamics in the dry savanna of West Africa. *Field Crops Research* 217, 134–49. https://doi.org/10.1016/j.fcr.2017.12.017.

Araki, H., Hamada, A., Hossain, M. A. and Takahashi, T. 2012. Waterlogging at jointing and/or after anthesis in wheat induces early leaf senescence and impairs grain filling. *Field Crops Research* 137, 27–36.

Armstrong, W., Healy, M. T. and Lythe, S. 1983. Oxygen diffusion in pea. II. Oxygen concentrations in the primary pea root apex as affected by growth, the production of laterals and radial oxygen loss. *New Phytologist* 94, 549–59.

Asseng, S. and Milroy, S. P. 2006. Simulation of environmental and genetic effects on grain protein concentration in wheat. *European Journal of Agronomy, Modelling Quality Traits and Their Genetic Variability for Wheat* 25, 119–28. https://doi.org/10.1016/j.eja.2006.04.005.

Asseng, S., Foster, I. and Turner, N. C. 2011. The impact of temperature variability on wheat yields. *Global Change Biology* 17, 997–1012.

Asseng, S., Ewert, F., Rosenzweig, C., Jones, J. W., Hatfield, J. L., Ruane, A. C., Boote, K. J., Thorburn, P. J., Rötter, R. P., Cammarano, D., Brisson, N., Basso, B., Martre, P., Aggarwal, P. K., Angulo, C., Bertuzzi, P., Biernath, C., Challinor, A. J., Doltra, J., Gayler, S., Goldberg, R., Grant, R., Heng, L., Hooker, J., Hunt, L. A., Ingwersen, J., Izaurralde, R. C., Kersebaum, K. C., Müller, C., Naresh Kumar, S., Nendel, C., O'Leary, G., Olesen, J. E., Osborne, T. M., Palosuo, T., Priesack, E., Ripoche, D., Semenov, M. A., Shcherbak, I., Steduto, P., Stöckle, C., Stratonovitch, P., Streck, T., Supit, I., Tao, F., Travasso, M., Waha, K., Wallach, D., White, J. W., Williams, J. R. and Wolf, J. 2013. Uncertainty in simulating wheat yields under climate change. *Nature Climate Change* 3, 827–32. https://doi.org/10.1038/nclimate1916.

Asseng, S., Martre, P., Maiorano, A., Rötter, R. P., O'Leary, G. J., Fitzgerald, G. J., Girousse, C., Motzo, R., Giunta, F., Babar, M. A., Reynolds, M. P., Kheir, A. M. S., Thorburn, P. J., Waha, K., Ruane, A. C., Aggarwal, P. K., Ahmed, M., Balkovič, J., Basso, B., Biernath, C., Bindi, M., Cammarano, D., Challinor, A. J., Sanctis, G. D., Dumont, B., Rezaei, E. E., Fereres, E., Ferrise, R., Garcia-Vila, M., Gayler, S., Gao, Y., Horan, H., Hoogenboom, G., Izaurralde, R. C., Jabloun, M., Jones, C. D., Kassie, B. T., Kersebaum, K.-C., Klein, C., Koehler, A.-K., Liu, B., Minoli, S., Martin, M. M. S., Müller, C., Kumar, S. N., Nendel, C., Olesen, J. E., Palosuo, T., Porter, J. R., Priesack, E., Ripoche, D., Semenov, M. A., Stöckle, C., Stratonovitch, P., Streck, T., Supit, I., Tao, F., Velde, M. V. der, Wallach, D., Wang, E., Webber, H., Wolf, J., Xiao, L., Zhang, Z., Zhao, Z., Zhu, Y. and Ewert, F. 2019. Climate change impact and adaptation for wheat protein. *Global Change Biology* 25, 155–73. https://doi.org/10.1111/gcb.14481.

Barlow, K. M., Christy, B. P., O'Leary, G. J., Riffkin, P. A. and Nuttall, J. G. 2015. Simulating the impact of extreme heat and frost events on wheat crop production: a review. *Field Crops Research* 171, 109–19. https://doi.org/10.1016/j.fcr.2014.11.010.

Barnabás, B., Jäger, K. and Fehér, A. 2008. The effect of drought and heat stress on reproductive processes in cereals. *Plant, Cell & Environment* 31, 11–38.

Basso, B., Dumont, B., Maestrini, B., Shcherbak, I., Robertson, G. P., Porter, J. R., Smith, P., Paustian, K., Grace, P. R., Asseng, S., Bassu, S., Biernath, C., Boote, K. J., Cammarano, D., Sanctis, G. de, Durand, J.-L., Ewert, F., Gayler, S., Grant, R., Hyndman, D. W., Kent, J., Martre, P., Nendel, C., Priesack, E., Ripoche, D., Ruane, A. C., Sharp, J., Thorburn, P. J., Hatfield, J. L., Jones, J. W. and Rosenzweig, C. 2018. Soil organic carbon and nitrogen feedbacks on crop yields under climate change. *Agricultural and Environmental Letters* 3, 180026. https://doi.org/10.2134/ael2018.05.0026.

Bassu, S., Brisson, N., Durand, J.-L., Boote, K., Lizaso, J., Jones, J. W., Rosenzweig, C., Ruane, A. C., Adam, M., Baron, C., Basso, B., Biernath, C., Boogaard, H., Conijn, S., Corbeels, M., Deryng, D., De Sanctis, G., Gayler, S., Grassini, P., Hatfield, J., Hoek, S., Izaurralde, C., Jongschaap, R., Kemanian, A. R., Kersebaum, K. C., Kim, S.-H., Kumar, N. S., Makowski, D., Müller, C., Nendel, C., Priesack, E., Pravia, M. V., Sau, F., Shcherbak, I., Tao, F., Teixeira, E., Timlin, D. and Waha, K. 2014. How do various maize crop models vary in their responses to climate change factors? *Global Change Biology* 20, 2301–20. https://doi.org/10.1111/gcb.12520.

Bellocchi, G., Rivington, M., Donatelli, M. and Matthews, K. 2010. Validation of biophysical models: issues and methodologies. A review. *Agronomy for Sustainable Development* 30, 109-30. https://doi.org/10.1051/agro/2009001.

Bergjord Olsen, A. K., Persson, T., Wit, A., Nkurunziza, L., Sindhøj, E. and Eckersten, H. 2017. Estimating winter survival of winter wheat by simulations of plant frost tolerance. *Journal of Agronomy and Crop Science* 204, 62-73.

Bita, C. E. and Gerats, T. 2013. Plant tolerance to high temperature in a changing environment: scientific fundamentals and production of heat stress-tolerant crops. *Frontiers in Plant Science* 4. https://doi.org/10.3389/fpls.2013.00273.

Bondeau, A., Smith, P. C., Zaehle, S., Schaphoff, S., Lucht, W., Cramer, W., Gerten, D., Lotze-Campen, H., Müller, C., Reichstein, M. and Smith, B. 2007. Modelling the role of agriculture for the 20th century global terrestrial carbon balance. *Global Change Biology* 13, 679-706. https://doi.org/10.1111/j.1365-2486.2006.01305.x.

Boote, K. J., Jones, J. W., Mishoe, J. W. and Berger, R. D. 1983. Coupling pests to crop growth simulators to predict yield reductions. *Phytopathology* 73, 1581-7.

Boote, K. J., Hoogenboom, G., Jones, J. W. and Ingram, K. T. 2008. Modeling N-Fixation and its relationship to N uptake in the CROPGRO model. In *Quantifying and Understanding Plant Nitrogen Uptake for Systems Modeling*. Ed. L. Ma, L. Ahuja, and T. Bruulsema. Taylor & Francis Group LLC: Boca Raton, FL.

Boote, K. J., Sau, F., Hoogenboom, G. and Jones, J. W. 2009. Experience with water balance, evapotranspiration, and prediction of water stress effects in the CROPGRO Model. In *Response of Crops to Limited Water: Modeling Water Stress Effects on Plant Growth Processes, Volume 1 of Advances in Agricultural Systems Modeling*. Ed. L. R. Ahuja, V. R. Reddy, S. A. Saseendran, and Q. Yu. ASA-CSSA-SSSA: Madison, WI.

Boote, K. J., Allen, Jr., L. H., Vara Prasad, P. V. and Jones, J. W. 2010. Testing effects of climate change in crop models. In *Handbook of Climate Change and Agroecosystems*. Ed. D. Hillel and C. Rosenzweig. Imperial College Press: London, UK.

Boote, K. J., Rybak, M. R., Scholberg, J. M. S. and Jones, J. W. 2012. Improving the CROPGRO-Tomato model for predicting growth and yield response to temperature. *HortScience* 47, 1038-49.

Boote, K. J., Jones, J. W., White, J. W., Asseng, S. and Lizaso, J. I. 2013. Putting mechanisms into crop production models. *Plant, Cell & Environment* 36, 1658-72.

Boote, K. J., Vara Prasad, L. H., Allen, Jr., Singh, P. and Jones, J. W. 2018. Modeling sensitivity of grain yield to elevated temperature in the DSSAT crop models for peanut, soybean, dry bean, chickpea, sorghum, and millet. *European Journal of Agronomy* 100, 99-109.

Brilli, L., Bechini, L., Bindi, M., Carozzi, M., Cavalli, D., Conant, R., Dorich, C. D., Doro, L., Ehrhardt, F., Farina, R., Ferrise, R., Fitton, N., Francaviglia, R., Grace, P., Iocola, I., Klumpp, K., Léonard, J., Martin, R., Massad, R. S., Recous, S., Seddaiu, G., Sharp, J., Smith, P., Smith, W. N., Soussana, J.-F. and Bellocchi, G. 2017. Review and analysis of strengths and weaknesses of agro-ecosystem models for simulating C and N fluxes. *Science of the Total Environment* 598, 445-70. https://doi.org/10.1016/j.scitotenv.2017.03.208.

Brisson, N., Gary, C., Justes, E., Roche, R., Mary, B., Ripoche, D., Zimmer, D., Sierra, J., Bertuzzi, P., Burger, P., Bussière, F., Cabidoche, Y. M., Cellier, P., Debaeke, P., Gaudillère, J. P., Hénault, C., Maraux, F., Seguin, B. and Sinoquet, H. 2003. An overview of the crop model stics. *European Journal of Agronomy* 18, 309-32. https://doi.org/10.1016/S1161-0301(02)00110-7.

Busch, F. A. and Sage, R. F. 2017. The sensitivity of photosynthesis to O_2 and CO_2 concentration identifies strong Rubisco control above the thermal optimum. *New Phytologist* 213, 1036-51. https://doi.org/10.1111/nph.14258.

Challinor, A., Wheeler, T., Craufurd, P. and Slingo, J. 2005. Simulation of the impact of high temperature stress on annual crop yields. *Agricultural and Forest Meteorology* 135, 180-9.

Challinor, A. J., Watson, J., Lobell, D. B., Howden, S. M., Smith, D. R. and Chhetri, N. 2014. A meta-analysis of crop yield under climate change and adaptation. *Natural Climate Change* 4, 287.

Chenu, K., Chapman, S. C., Tardieu, F., McLean, G., Welcker, C. and Hammer, G. L. 2009. Simulating the yield impacts of organ-level quantitative trait loci associated with drought response in Maize-A 'Gene-to-Phenotype' Modeling approach. *Genetics* 83, 1507-23. https://doi.org/10.1534/genetics.109.105429.

Chenu, K., Porter, J. R., Martre, P., Basso, B., Chapman, S. C., Ewert, F., Bindi, M. and Asseng, S. 2017. Contribution of crop models to adaptation in wheat. *Trends in Plant Science* 22, 472-90. https://doi.org/10.1016/j.tplants.2017.02.003.

Christensen, J. H. and Christensen, O. B. 2003. Climate modelling: severe summertime flooding in Europe. *Nature* 421, 805.

Claessens, L., Antle, J. M., Stoorvogel, J. J., Valdivia, R. O., Thornton, P. K. and Herrero, M. 2012. A method for evaluating climate change adaptation strategies for small-scale farmers using survey, experimental and modeled data. *Agricultural Systems* 111, 85-95. https://doi.org/10.1016/j.agsy.2012.05.003.

Confalonieri, R., Bregaglio, S., Adam, M., Ruget, F., Li, T., Hasegawa, T., Yin, X., Zhu, Y., Boote, K., Buis, S., Fumoto, T., Gaydon, D., Lafarge, T., Marcaida, M., Nakagawa, H., Ruane, A. C., Singh, B., Singh, U., Tang, L., Tao, F., Fugice, J., Yoshida, H., Zhang, Z., Wilson, L. T., Baker, J., Yang, Y., Masutomi, Y., Wallach, D., Acutis, M. and Bouman, B. 2016. A taxonomy-based approach to shed light on the babel of mathematical models for rice simulation. *Environmental Modelling & Software* 85, 332-41. https://doi.org/10.1016/j.envsoft.2016.09.007.

Corbeels, M., McMurtrie, R. E., Pepper, D. A. and O'Connell, A. M. 2005. A process-based model of nitrogen cycling in forest plantations: part I. Structure, calibration and analysis of the decomposition model. *Ecological Modelling* 187, 426-48. https://doi.org/10.1016/j.ecolmodel.2004.09.005.

Corbeels, M., Chirat, G., Messad, S. and Thierfelder, C. 2016. Performance and sensitivity of the DSSAT crop growth model in simulating maize yield under conservation agriculture. *European Journal of Agronomy* 76, 41-53. https://doi.org/10.1016/j.eja.2016.02.001.

Corbeels, M., Berre, D., Rusinamhodzi, L. and Lopez-Ridaura, S. 2018. Can we use crop modelling for identifying climate change adaptation options? *Agricultural and Forest Meteorology* 256-7, 46-52. https://doi.org/10.1016/j.agrformet.2018.02.026.

Cromey, M. G., Wright, D. S. C. and Boddington, H. J. 1998. Effects of frost during grain filling on wheat yield and grain structure. *New Zealand Journal of Crop and Horticultural Science* 26, 279-90.

Deryng, D., Sacks, W. J., Barford, C. C. and Ramankutty, N. 2011. Simulating the effects of climate and agricultural management practices on global crop yield. *Global Biogeochemical Cycles* 25. https://doi.org/10.1029/2009GB003765.

Deryng, D., Conway, D., Ramankutty, N., Price, J. and Warren, R. 2014. Global crop yield response to extreme heat stress under multiple climate change futures.

Environmental Research Letters 9, 034011. https://doi.org/10.1088/1748-9326/9/3/034011.
Descheemaeker, K., Zijlstra, M., Masikati, P., Crespo, O. and Homann-Kee Tui, S. 2018. Effects of climate change and adaptation on the livestock component of mixed farming systems: a modelling study from semi-arid Zimbabwe. *Agricultural Systems* 159, 282–95. https://doi.org/10.1016/j.agsy.2017.05.004.
Devare, M., Aubert, C., Laporte, M.-A., Valette, L., Arnaud, E. and Buttigieg, P. L. 2016. Data-driven agricultural research for development – a need for data harmonization via semantics. In *7th International Conference on Biomedical Ontologies, ICBO'16, Vol. 1747 of CEUR Workshop Proceedings*. Ed. P. Jaiswal and R. Hoehndorf. Corvallis, OR, pp. 2.
Dingkuhn, M., Kouressy, M., Vaksmann, M., Clerget, B. and Chantereau, J. 2008. A model of sorghum photoperiodism using the concept of threshold-lowering during prolonged appetence. *European Journal of Agronomy* 28, 74–89. https://doi.org/10.1016/j.eja.2007.05.005.
Donatelli, M., Magarey, R. D., Bregaglio, S., Willocquet, L., Whish, J. P. and Savary, S. 2017. Modelling the impacts of pests and diseases on agricultural systems. *Agricultural Systems* 155, 213–24.
Durand, J.-L., Delusca, K., Boote, K., Lizaso, J., Manderscheid, R., Weigel, H. J., Ruane, A. C., Rosenzweig, C., Jones, J., Ahuja, L., Anapalli, S., Basso, B., Baron, C., Bertuzzi, P., Biernath, C., Deryng, D., Ewert, F., Gaiser, T., Gayler, S., Heinlein, F., Kersebaum, K. C., Kim, S.-H., Müller, C., Nendel, C., Olioso, A., Priesack, E., Villegas, J. R., Ripoche, D., Rötter, R. P., Seidel, S. I., Srivastava, A., Tao, F., Timlin, D., Twine, T., Wang, E., Webber, H. and Zhao, Z. 2018. How accurately do maize crop models simulate the interactions of atmospheric CO2 concentration levels with limited water supply on water use and yield? *European Journal of Agronomy* 100, 67–75.
Dzotsi, K. A., Jones, J. W., Adiku, S. G. K., Naab, J. B., Singh, U., Porter, C. H. and Gijsman, A. J. 2010. Modeling soil and plant phosphorus within DSSAT. *Ecological Modelling* 221, 2839–49. https://doi.org/10.1016/j.ecolmodel.2010.08.023.
Elliott, J., Kelly, D., Chryssanthacopoulos, J., Glotter, M., Jhunjhnuwala, K., Best, N., Wilde, M. and Foster, I. 2014. The parallel system for integrating impact models and sectors (pSIMS). *Environmental Modelling & Software* 62, 509–16. doi:10.1016/j.envsoft.2014.04.008.
Emberson, L. D., Pleijel, H., Ainsworth, E. A., van den Berg, M., Ren, W., Osborne, S., Mills, G., Pandey, D., Dentener, F., Büker, P., Ewert, F., Koeble, R. and Van Dingenen, R. 2018. Ozone effects on crops and consideration in crop models. *European Journal of Agronomy* 100, 19–34. https://doi.org/10.1016/j.eja.2018.06.002.
Ensminger, I., Busch, F. and Huner, N. P. A. 2006. Photostasis and cold acclimation: sensing low temperature through photosynthesis. *Physiologia Plantarum* 126, 28–44. https://doi.org/10.1111/j.1399-3054.2006.00627.x.
Esse, P. C., Buerkert, A., Hiernaux, P. and Assa, A. 2001. Decomposition of and nutrient release from ruminant manure on acid sandy soils in the Sahelian zone of Niger, West Africa. *Agriculture, Ecosystems & Environment* 83, 55–63. https://doi.org/10.1016/S0167-8809(00)00264-4.
Ewert, F., Rötter, R., Bindi, M., Webber, H., Trnka, M., Kersebaum, K., Olesen, J. E., van Ittersum, M., Janssen, S., Rivington, M., Semenov, M., Wallach, D., Porter, J., Stewart, D., Verhagen, J., Gaiser, T., Palosuo, T., Tao, F., Nendel, C., Roggero, P., Bartošová, L. and Asseng, S. 2015. Crop modelling for integrated assessment of

climate change risk to food production. *Environmental Modelling & Software* 72, 287-303.
Eyshi Rezaei, E., Webber, H., Gaiser, T., Naab, J. and Ewert, F. 2015a. Heat stress in cereals: mechanisms and modelling. *European Journal of Agronomy* 64, 98-113.
Eyshi Rezaei, E., Siebert, S. and Ewert, F. 2015b. Intensity of heat stress in winter wheat–phenology compensates for the adverse effect of global warming. *Environmental Research Letters* 10, 024012.
Eyshi Rezaei, E., Siebert, S. and Ewert, F. 2017. Climate and management interaction cause diverse crop phenology trends. *Agricultural and Forest Meteorology* 233, 55-70.
Fabre, D., Yin, X., Dingkuhn, M., Clément-Vidal, A., Roques, S., Rouan, L., Soutiras, A. and Luquet, D. 2019. Is triose phosphate utilization involved in the feedback inhibition of photosynthesis in rice under conditions of sink limitation? *Journal of Experimental Botany* 70, 5773-85. https://doi.org/10.1093/jxb/erz318.
Falconnier, G. N., Corbeels, M., Boote, K. J., Adam, M., Basso, B., Ruane, A. et al. 2019. Model intercomparison of maize response to climate change in low-input smallholder cropping systems. *3rd Agriculture and Climate Change Conference*, Budapest, Hungary, 24-26 March 2019.
Farquhar, G. D. and von Caemmerer, S. 1982. Modeling of photosynthetic response to environment. In *Encyclopedia of Plant Physiology*. Ed. O. L. Lange et al. New Series. Vol. 12B. Physiological Plant Ecology II. Springer-Verlag: Berlin, pp. 549-87.
Fatondji, D., Bationo, A., Tabo, R., Jones, J. W., Adamou, A. and Hassane, O. 2012. Water use and yield of millet under the Zai system: understanding the processes using simulation. In *Improving Soil Fertility Recommendations in Africa Using the Decision Support System for Agrotechnology Transfer (DSSAT)*. Springer Netherlands, pp. 77-100.
Faye, B., Webber, H., Naab, J. B., MacCarthy, D. S., Adam, M., Ewert, F., Lamers, J. P. A., Schleussner, C.-F., Ruane, A., Gessner, U., Hoogenboom, G., Boote, K., Shelia, V., Saeed, F., Wisser, D., Hadir, S., Laux, P. and Gaiser, T. 2018. Impacts of 1.5 versus 2.0°C on cereal yields in the West African Sudan Savanna. *Environmental Research Letters* 13, 034014. https://doi.org/10.1088/1748-9326/aaab40.
Fosu-Mensah, B. Y., MacCarthy, D. S., Vlek, P. L. G. and Safo, E. Y. 2012. Simulating impact of seasonal climatic variation on the response of maize (Zea mays L.) to inorganic fertilizer in sub-humid Ghana. *Nutrient Cycling in Agroecosystems* 94, 255-71. https://doi.org/10.1007/s10705-012-9539-4.
Frederiks, T. M., Christopher, J. T., Harvey, G. L., Sutherland, M. W. and Borrell, A. K. 2012. Current and emerging screening methods to identify post-head-emergence frost adaptation in wheat and barley. *Journal of Experimental Botany* 63, 5405-16. https://doi.org/10.1093/jxb/ers215.
Freduah, B. S., MacCarthy, D. S., Adam, M., Ly, M., Ruane, A. C., Timpong-Jones, E. C., Traore, P. S., Boote, K. J., Porter, C. and Adiku, S. G. K. 2019. Sensitivity of maize yield in smallholder systems to climate scenarios in semi-arid regions of West Africa: accounting for variability in farm management practices. *Agronomy* 9, 639. https://doi.org/10.3390/agronomy9100639.
Fuller, M. P., Fuller, A. M., Kaniouras, S., Christophers, J. and Fredericks, T. 2007. The freezing characteristics of wheat at ear emergence. *European Journal of Agronomy* 26, 435-41. https://doi.org/10.1016/j.eja.2007.01.001.
Gabaldón-Leal, C., Webber, H., Otegui, M. E., Slafer, G. A., Ordóñez, R. A., Gaiser, T., Lorite, I. J., Ruiz-Ramos, M. and Ewert, F. 2016. Modelling the impact of heat stress on maize yield formation. *Field Crops Research* 198, 226-37.

Gbegbelegbe, S., Cammarano, D., Asseng, S., Robertson, R., Chung, U., Adam, M., Abdalla, O., Payne, T., Reynolds, M., Sonder, K., Shiferaw, B. and Nelson, G. 2017. Baseline simulation for global wheat production with CIMMYT mega-environment specific cultivars. *Field Crops Research* 202, 122–35. https://doi.org/10.1016/j.fcr.2016.06.010.

Gijsman, A. J., Hoogenboom, G., Parton, W. J. and Kerridge, P. C. 2002. Modifying DSSAT crop models for low-input agricultural systems using a soil organic matter-residue module from CENTURY. *Agronomy Journal* 94, 462–74.

Godwin, D. C. and Singh, U. 1998. Nitrogen balance and crop response to nitrogen in upland and lowland cropping systems. In *Understanding Options for Agricultural Production*. Ed. G. Y. Tsuji, G. Hoogenboom and P. K. Thornton. Kluwer Academic Publishers: Dordrecht, The Netherlands, pp. 55–77.

Hasegawa, T., Lai, T., Yin, X., Zhu, Y., Boote, K., Baker, J., Bregaglio, S., Buis, S., Confalonieri, R., Fugice, J., Fumoto, T., Gaydon, D., Naresh Kumar, S., Lafarge, T., Marcaida, M., Masutomi, Y., Nakagawa, H., Oriol, P., Ruget, F., Singh, U., Tang, L., Tao, F., Wakatsuki, H., Wallach, D., Wang, Y., Wilson, L. T., Yang, L., Yang, Y., Yoshida, H., Zhang, Z. and Zhu, J. 2017. Causes of variation among rice models in yield response to CO_2 examined with Free-Air CO_2 Enrichment and growth chamber experiments. *Scientific Reports* 7. doi:10.1038/s41598-017-13582-y.

Hatfield, J., Boote, K., Fay, P., Hahn, L., Izaurralde, C., Kimball, B. A., Mader, T., Morgan, J., Ort, D., Polley, W., Thomson, A. and Wolfe, D. 2008. Agriculture. In *The Effects of Climate Change on Agriculture, Land Resources, Water Resources, and Biodiversity*. A Report by the U.S. Climate Change Science Program and the Subcommittee on Global Change Research. Washington, DC, USA, 362 pp.

He, D., Wang, E., Wang, J. and Robertson, M. J. 2017. Data requirement for effective calibration of process-based crop models. *Agricultural and Forest Meteorology* 234–35, 136–48. https://doi.org/10.1016/j.agrformet.2016.12.015.

Herzog, M., Striker, G. G., Colmer, T. D. and Pedersen, O. 2016. Mechanisms of waterlogging tolerance in wheat - a review of root and shoot physiology. *Plant Cell and Environment* 39, 1068–86. https://doi.org/10.1111/pce.12676.

Holzworth, D., Huth, N., deVoil, P., Zurcher, E., Herrmann, N., McLean, G., Chenu, K., van Oosterom, E., Snow, V., Murphy, C., Moore, A., Brown, H., Whish, J., Verrall, S., Fainges, J., Bell, L., Peake, A., Poulton, P., Hochman, Z., Thorburn, P., Gaydon, D., Dalgliesh, N., Rodriguez, D., Cox, H., Chapman, S., Doherty, A., Teixeira, E., Sharp, J., Cichota, R., Vogeler, I., Li, F., Wang, E., Hammer, G., Robertson, M., Dimes, J., Carberry, P., Hargreaves, J., MacLeod, N. C. M., Harsdorf, J., Wedgewood, S. and Keating, B. 2014. APSIM - Evolution towards a new generation of agricultural systems simulation. *Environmental Modeling and Software: Agricultural Systems Modeling & Software* 62, 327-50.

Hunt, L. A. and Boote, K. J. 1998. Data for model operation, calibration and evaluation. In *Understanding Options for Agricultural Production*. Ed. G. Y. Tsuji, G. Hoogenboom and P. K. Thornton. Kluwer Academic Publishers/ICASA: Dordrecht, pp. 9–40.

IPCC. 2014. *Climate Change 2014: Synthesis Report. Contribution of Working Groups I, II and III to the Fifth Assessment Report of the Intergovernmental Panel on Climate Change*. Ed. Core Writing Team, R. K. Pachauri and L. A. Meyer. IPCC: Geneva, Switzerland, 151pp.

Irfan, M., Hayat, S., Hayat, Q., Afroz, S. and Ahmad, A. 2010. Physiological and biochemical changes in plants under waterlogging. *Protoplasma* 241, 3–17. https://doi.org/10.1007/s00709-009-0098-8.

Iyer, N. J., Tang, Y. and Mahalingam, R. 2013. Physiological, biochemical and molecular responses to a combination of drought and ozone in Medicago truncatula. *Plant, Cell & Environment* 36, 706–20.

Izaurralde, R. C., Williams, J. R., McGill, W. B., Rosenberg, N. J. and Jakas, M. C. 2006. Simulating soil C dynamics with EPIC: model description and testing against long-term data. *Ecological Modelling* 192(3), 362–84.

Jagtap, S. S., Mornu, M. and Kang, B. T. 1993. Simulation of growth, development and yield of maize in the transition zone of Nigeria. *Agricultural Systems* 41, 215–29. https://doi.org/10.1016/0308-521X(93)90040-9.

Jamieson, P. D., Porter, J. R., Goudriaan, J., Ritchie, J. T., van Keulen, H. and Stol, W. 1998. A comparison of the models AFRCWHEAT2, CERES-wheat, Sirius, SUCROS2 and SWHEAT with measurements from wheat grown under drought. *Field Crops Research* 55, 23–44.

Jones, P. G. and Thornton, P. K. 2013. Generating downscaled weather data from a suite of climate models for agricultural modelling applications. *Agricultural Systems* 114, 1–5. https://doi.org/10.1016/j.agsy.2012.08.002.

Jones, J. W., Hoogenboom, G., Porter, C. H., Boote, K. J., Batchelor, W. D., Hunt, L. A., Wilkens, P. W., Singh, U., Gijsman, A. J. and Ritchie, J. T. 2003. The DSSAT cropping system model. *European Journal of Agronomy* 18, 235–65.

Jones, J. W., Naab, J., Fatondji, D., Dzotsi, K., Adiku, S. and He, J. 2012. Uncertainties in simulating crop performance in degraded soils and low input production systems. In *Improving Soil Fertility Recommendations in Africa Using the Decision Support System for Agrotechnology Transfer (DSSAT)*. Springer: Dordrecht, pp. 43–59. https://doi.org/10.1007/978-94-007-2960-5_4.

Jones, J. W., Antle, J. M., Basso, B., Boote, K. J., Conant, R. T., Foster, I., Godfray, H. C. J., Herrero, M., Howitt, R. E., Janssen, S., Keating, B. A., Munoz-Carpena, R., Porter, C. H., Rosenzweig, C. and Wheeler, T. R. 2017. Toward a new generation of agricultural system data, models, and knowledge products: state of agricultural systems science. *Agricultural Systems* 155, 269–88. https://doi.org/10.1016/j.agsy.2016.09.021.

Keating, B. A., Carberry, P. S., Hammer, G. L., Probert, M. E., Robertson, M. J., Holzworth, D., Huth, N. I., Hargreaves, J. N. G., Meinke, H., Hochman, Z., McLean, G., Verburg, K., Snow, V., Dimes, J. P., Silburn, M., Wang, E., Brown, S., Bristow, K. L., Asseng, S., Chapman, S., McCown, R. L., Freebairn, D. M. and Smith, C. J. 2003. An overview of APSIM, a model designed for farming systems simulation. *European Journal of Agronomy* 18(3–4).

Kersebaum, K., Boote, K., Jorgenson, J., Nendel, C., Bindi, M., Frühauf, C., Gaiser, T., Hoogenboom, G., Kollas, C. and Olesen, J. E. 2015. Analysis and classification of data sets for calibration and validation of agro-ecosystem models. *Environmental Modelling & Software* 72, 402–17.

Kimball, B. A., Boote, K. J., Hatfield, J. L., Ahuja, L. R., Stockle, C., Archontoulis, S., Baron, C., Basso, B., Bertuzzi, P., Constantin, J., Deryng, D., Dumont, B., Durand, J.-L., Ewert, F., Gaiser, T., Gayler, S., Hoffmann, M. P., Jiang, Q., Kim, S.-H., Lizaso, J., Moulin, S., Nendel, C., Parker, P., Palosuo, T., Priesack, E., Qi, Z., Srivastava, A., Stella, T., Tao, F., Thorp, K. R., Timlin, D., Twine, T. E., Webber, H., Willaume, M. and Williams, K. 2019. Simulation of maize evapotranspiration: an inter-comparison among 29 maize models. *Agricultural and Forest Meteorology* 271, 264–84. https://doi.org/10.1016/j.agrformet.2019.02.037.

Kristensen, K., Schelde, K. and Olesen, J. E. 2011. Winter wheat yield response to climate variability in Denmark. *The Journal of Agricultural Science* 149, 33–47.

Kuai, J., Zhou, Z., Wang, Y., Meng, Y., Chen, B. and Zhao, W. 2015. The effects of short-term waterlogging on the lint yield and yield components of cotton with respect to boll position. *European Journal of Agronomy* 67, 61–74. https://doi.org/10.1016/j.eja.2015.03.005.

Lamanda, N., Roux, S., Delmotte, S., Merot, A., Rapidel, B., Adam, M. and Wery, J. 2012. A protocol for the conceptualisation of an agro-ecosystem to guide data acquisition and analysis and expert knowledge integration. *European Journal of Agronomy* 38, 104–16. https://doi.org/10.1016/j.eja.2011.07.004.

Leemans, R. A. and Solomon, M. 1993. Modeling the potential change in yield and distribution of the earth's crops under a warmed climate. *Climate Research* 3, 79–96.

Li, T., Hasegawa, T., Yin, X., Zhu, Y., Boote, K., Adam, M., Bregaglio, S., Buis, S., Confalonieri, R., Fumoto, T., Gaydon, D., Marcaida, M., Nakagawa, H., Oriol, P., Ruane, A. C., Ruget, F., Singh, B.–, Singh, U., Tang, L., Tao, F., Wilkens, P., Yoshida, H., Zhang, Z. and Bouman, B. 2015. Uncertainties in predicting rice yield by current crop models under a wide range of climatic conditions. *Global Change Biology* 21, 1328–41. https://doi.org/10.1111/gcb.12758.

Lizaso, J. I., Ruiz-Ramos, M., Rodriguez, L., Gabaldon-Leal, C., Oliveira, J. A., Lorite, I. J., Rodriguez, A., Maddonni, G. A. and Otegui, M. E. 2017. Modeling the response of maize phenology, kernel set, and yield components to heat stress and heat shock with CSM-IXIM. *Field Crops Research* 214, 239–52.

Lizaso, J. I., Ruiz-Ramos, M., Rodriguez, L., Gabaldon-Leal, C., Oliveira, J. A., Lorite, I. J., Sanchez, D., Garcia, E. and Rodriguez, A. 2018. Impact of high temperatures in maize: phenology and yield components. *Field Crops Research* 216, 129–40.

Lobell, D. B. and Asseng, S. 2017. Comparing estimates of climate change impacts from process-based and statistical crop models. *Environmental Research Letters* 12, 015001. https://doi.org/10.1088/1748-9326/aa518a.

Lobell, D. B., Hammer, G. L., Chenu, K., Zheng, B., McLean, G. and Chapman, S. C. 2015. The shifting influence of drought and heat stress for crops in northeast Australia. *Global Change Biology* 21, 4115–27.

Lombardozzi, D. L., Smith, N. G., Cheng, S. J., Dukes, J. S., Sharkey, T. D., Rogers, A., Fisher, R. and Bonan, G. B. 2018. Triose phosphate limitation in photosynthesis models reduces leaf photosynthesis and global terrestrial carbon storage. *Environmental Research Letters* 13, 074025. https://doi.org/10.1088/1748-9326/aacf68.

MacCarthy, D. S., Vlek, P. L. G., Bationo, A., Tabo, R. and Fosu, M. 2010. Modeling nutrient and water productivity of sorghum in smallholder farming systems in a semi-arid region of Ghana. *Field Crops Research* 118, 251–8. https://doi.org/10.1016/j.fcr.2010.06.005.

Maiorano, A., Martre, P., Asseng, S., Ewert, F., Müller, C., Rötter, R. P., Ruane, A. C., Semenov, M. A., Wallach, D. and Wang, E. 2017. Crop model improvement reduces the uncertainty of the response to temperature of multi-model ensembles. *Field Crops Research* 202, 5–20.

Malik, A. I., Colmer, T. D., Lambers, H., Setter, T. L. and Schortemeyer, M. 2002. Short-term waterlogging has long-term effects on the growth and physiology of wheat. *New Phytologist* 153, 225–36.

Malik, W., Boote, K. J., Hoogenboom, G., Cavero, J. and Dechmi, F. 2018. Adapting the CROPGRO model to simulate alfalfa growth and yield. *Agronomy Journal* 110, 1777–90.

Manderscheid, R., Erbs, M. and Weigel, H. J. 2014. Interactive effects of free-air CO_2 enrichment and drought stress on maize growth. *European Journal of Agronomy* 52, 11-21.

Martre, P., Wallach, D., Asseng, S., Ewert, F., Jones, J. W., Rötter, R. P., Boote, K. J., Ruane, A. C., Thorburn, P. J., Cammarano, D., Hatfield, J. L., Rosenzweig, C., Aggarwal, P. K., Angulo, C., Basso, B., Bertuzzi, P., Biernath, C., Brisson, N., Challinor, A. J., Doltra, J., Gayler, S., Goldberg, R., Grant, R. F., Heng, L., Hooker, J., Hunt, L. A., Ingwersen, J., Izaurralde, R. C., Kersebaum, K. C., Müller, C., Kumar, S. N., Nendel, C., O'leary, G., Olesen, J. E., Osborne, T. M., Palosuo, T., Priesack, E., Ripoche, D., Semenov, M. A., Shcherbak, I., Steduto, P., Stöckle, C. O., Stratonovitch, P., Streck, T., Supit, I., Tao, F., Travasso, M., Waha, K., White, J. W. and Wolf, J. 2015. Multimodel ensembles of wheat growth: many models are better than one. *Global Change Biology* 21, 911-25. https://doi.org/10.1111/gcb.12768.

Masikati, P., Manschadi, A., van Rooyen, A. and Hargreaves, J. 2014. Maize-mucuna rotation: an alternative technology to improve water productivity in smallholder farming systems. *Agricultural Systems* 123, 62-70. https://doi.org/10.1016/j.agsy.2013.09.003.

Mittler, R. 2006. Abiotic stress, the field environment and stress combination. *Trends in Plant Science* 11, 15-19.

Moriondo, M., Giannakopoulos, C. and Bindi, M. 2011. Climate change impact assessment: the role of climate extremes in crop yield simulation. *Climatic Change* 104, 679-701.

Müller, C., Elliott, J., Chryssanthacopoulos, J., Arneth, A., Balkovic, J., Ciais, P., Deryng, D., Folberth, C., Glotter, M., Hoek, S., Iizumi, T., Izaurralde, R. C., Jones, C., Khabarov, N., Lawrence, P., Liu, W., Olin, S., Pugh, T. A. M., Ray, D. K., Reddy, A., Rosenzweig, C., Ruane, A. C., Sakurai, G., Schmid, E., Skalsky, R., Song, C. X., Wang, X., de Wit, A. and Yang, H. 2017. Global gridded crop model evaluation: benchmarking, skills, deficiencies and implications. *Geoscientific Model Development* 10, 1403-22. https://doi.org/10.5194/gmd-10-1403-2017.

Murray, M. B., Cape, J. N. and Fowler D. 1989. Quantification of frost damage in plant tissues by rates of electrolyte leakage. *New Phytologist* 113, 307-11.

Nagao, M., Minami, A., Arakawa, K., Fujikawa, S. and Takezawa, D. 2005. Rapid degradation of starch in chloroplasts and concomitant accumulation of soluble sugars associated with ABA-induced freezing tolerance in the moss Physcomitrella patens. *Journal of Plant Physiology* 162, 169-80. https://doi.org/10.1016/j.jplph.2004.06.012.

Nakagawa, H., Horie, T. and Kim, H. Y. 1994. Environmental factors affecting rice responses to elevated carbon dioxide concentrations. *International Rice Research Notes* 19, 45-6.

Olesen, J. E., Trnka, M., Kersebaum, K., Skjelvåg, A., Seguin, B., Peltonen-Sainio, P., Rossi, F., Kozyra, J. and Micale, F. 2011. Impacts and adaptation of European crop production systems to climate change. *European Journal of Agronomy* 34, 96-112.

Penning de Vries, F. W. T., Brunsting, A. H. M. and van Laar, H. H. 1974. Products, requirements and efficiency of biosynthesis: a quantitative approach. *Journal of Theoretical Biology* 45, 339-77.

Pequeno, D. N. L., Pedreira, C. G. S., Boote, K. J., Alderman, P. D. and Faria, A. F. G. 2018. Species-genotypic parameters of the CROPGRO Perennial Forage Model: implications for comparison of three tropical pasture grasses. *Grass and Forage Science* 73, 440-55.

Pickering, N. B., Jones, J. W. and Boote, K. J. 1995. Adapting SOYGRO V5.42 for prediction under climate change conditions. In *Climate Change and Agriculture: Analysis of Potential International Impacts*. Ed. C. Rosenzweig, J. W. Jones, and L. H. Allen, Jr. ASA Spec. Pub. No. 59, ASA-CSSA-SSSA: Madison, WI, pp. 77-98.

Porter, C. H., Villalobos, C., Holzworth, D., Nelson, R., White, J. W., Athanasiadis, I. N., Janssen, S., Ripoche, D., Cufi, J., Raes, D., Zhang, M., Knapen, R., Sahajpal, R., Boote, K. J. and Jones, J. W. 2014. Harmonization and translation of crop modeling data to ensure interoperability. *Environmental Modelling and Software* 62, 495-508. doi:10.1016/j.envsoft.2014.09.004.

Raes, D., Steduto, P., Hsiao, T. C. and Fereres, E. 2009. AquaCrop–the FAO crop model to simulate yield response to water: II. Main algorithms and software description. *Agronomy Journal* 101, 438-47.

Rashid, M. A., Andersen, M. N., Wollenweber, B., Zhang, X. and Olesen, J. E. 2018. Acclimation to higher VPD and temperature minimized negative effects on assimilation and grain yield of wheat. *Agricultural and Forest Meteorology* 248, 119-29. https://doi.org/10.1016/j.agrformet.2017.09.018.

Rasmussen, S., Barah, P., Suarez-Rodriguez, M. C., Bressendorff, S., Friis, P., Costantino, P., Bones, A. M., Nielsen, H. B. and Mundy, J. 2013. Transcriptome responses to combinations of stresses in Arabidopsis thaliana. *Plant Physiology* 112, 210773.

Ren, B., Zhang, J., Li, X., Fan, X., Dong, S., Liu, P. and Zhao, B. 2014. Effects of waterlogging on the yield and growth of summer maize under field conditions. *Canadian Journal of Plant Science* 94, 23-31. https://doi.org/10.4141/cjps2013-175.

Reynolds, M. and Langridge, P. 2016. Physiological breeding. *Current Opinion in Plant Biology* 31, 162-71.

Richards, R. A. 2006. Physiological traits used in the breeding of new cultivars for water-scarce environments. *Agricultural Water Management* 80, 197-211.

Richards, R., Rebetzke, G., Condon, A. and Van Herwaarden, A. 2002. Breeding opportunities for increasing the efficiency of water use and crop yield in temperate cereals. *Crop Science* 42, 111-21.

Rigby, J. R. and Porporato, A. 2008. Spring frost risk in a changing climate. *Geophysical Research Letters* 35, n/a-n/a. https://doi.org/10.1029/2008GL033955.

Rihan, H. Z. and Fuller, M. P. 2013. Frost hardiness expression and characterisation in Wheat at ear emergence. *Journal of Agronomy and Crop Science* 199, 66-74.

Ritchie, J. T. 1998. Soil water balance and plant water stress. In *Understanding Options for Agricultural Production*. Ed. G. Y. Tsuji, G. Hoogenboom and P. K. Thornton. Kluwer Academic Publishers: Dordrecht, The Netherlands, pp. 55-77.

Rivero, R. M., Mestre, T. C., Mittler, R., Rubio, F., Garcia-Sanchez, F. and Martinez, V. 2014. The combined effect of salinity and heat reveals a specific physiological, biochemical and molecular response in tomato plants. *Plant, Cell & Environment* 37, 1059-73.

Rizhsky, L., Liang, H., Shuman, J., Shulaev, V., Davletova, S. and Mittler, R. 2004. When defense pathways collide. The response of Arabidopsis to a combination of drought and heat stress. *Plant Physiology* 134, 1683-96.

Roberts, M. J., Braun, N. O., Sinclair, T. R., Lobell, D. B. and Schlenker, W. 2017. Comparing and combining process-based crop models and statistical models with some implications for climate change. *Environmental Research Letters* 12, 095010. https://doi.org/10.1088/1748-9326/aa7f33.

Robertson, R. D. 2017. *Mink: Details of a Global Gridded Crop Modeling System*. Washington, DC: International Food Policy Research Institute (IFPRI). http://ebrary.ifpri.org/cdm/ref/collection/p15738coll2/id/131406.

Rosenzweig, C. and Hillel, D. 2015. *Handbook of Climate Change and Agroecosystems: The Agricultural Model Intercomparison and Improvement Project (AgMIP) Integrated Crop and Economic Assessments – Joint Publication with American Society of Agronomy, Crop Science Society of America, and Soil Science Society of America (In 2 Parts). ICP Series on Climate Change Impacts, Adaptation, and Mitigation*. Imperial College Press. https://doi.org/10.1142/p970.

Rosenzweig, C., Jones, J. W., Hatfield, J. L., Ruane, A. C., Boote, K. J., Thorburn, P., Antle, J. M., Nelson, G. C., Porter, C., Janssen, S., Asseng, S., Basso, B., Ewert, F., Wallach, D., Baigorria, G. and Winter, J. M. 2012. The Agricultural Model Intercomparison and Improvement Project (AgMIP): protocols and pilot studies. *Agricultural and Forest Meteorology* 170, 166–82. http://dx.doi.org/10.1016/j.agrformet.2012.09.011.

Rosenzweig, C., Elliott, J., Deryng, D., Ruane, A. C., Müller, C., Arneth, A., Boote, K. J., Folberth, C., Glotter, M., Khabarov, N., Neumann, K., Piontek, F., Pugh, T. A. M., Schmid, E., Stehfest, E., Yang, H. and Jones, J. W. 2014. Global multi-model crop-climate impact assessment. *Proceedings of the National Academy of Sciences* 111(9), 3268–73. doi:10.1073/pnas.1222463110.

Rötter, R. P., Appiah, M., Fichtler, E., Kersebaum, K. C., Trnka, M. and Hoffmann, M. P. 2018. Linking modelling and experimentation to better capture crop impacts of agroclimatic extremes–A review. *Field Crops Research* 221, 142–56.

Rowhani, P., Lobell, D. B., Linderman, M. and Ramankutty, N. 2011. Climate variability and crop production in Tanzania. *Agricultural and Forest Meteorology* 151, 449–60. https://doi.org/10.1016/j.agrformet.2010.12.002.

Ruane, A. C., Winter, J. M., McDermid, S. P. and Hudson, N. I. 2014a. AgMIP climate data and scenarios for integrated assessment. In *Handbook of Climate Change and Agroecosystems, Series on Climate Change Impacts, Adaptation, and Mitigation*. Imperial College Press, pp. 45–78. https://doi.org/10.1142/9781783265640_0003.

Ruane, A. C., McDermid, S., Rosenzweig, C., Baigorria, G. A., Jones, J. W., Romero, C. C. and Cecil, L. D. 2014b. Carbon-temperature-water change analysis for peanut production under climate change: a prototype for the AgMIP Coordinated Climate-Crop Modeling Project (C3MP). *Global Change Biology* 20(2), 394–407. doi:10.1111/gcb.12412.

Ruane, A. C., Winter, J. M., McDermid, S. P. and Hudson, N. I. 2015. AgMIP climate data and scenarios for integrated assessment. In *Handbook of Climate Change and Agroecosystems: Agricultural Model Intercomparison and Improvement Project Integrated Crop and Economic Assessments* (Vol. 3 & 4). Ed. D. Hillel and C. Rosenzweig. ICP Series on Climate Change Impacts, Adaptation, and Mitigation. Imperial College Press.

Ruane, A. C., Rosenzweig, C., Asseng, S., Boote, K. J., Elliott, J., Ewert, F., Jones, J. W., Martre, P., McDermid, S. P., Müller, C., Snyder, A. and Thorburn, P. J. 2017. An AgMIP framework for improved agricultural representation in integrated assessment models. *Environmental Research Letters* 12, 125003. https://doi.org/10.1088/1748-9326/aa8da6.

Savary, S., De Jong, P. D., Rabbinge, R. and Zadoks, J. C. 1990. Dynamic simulation of groundnut rust, a preliminary model. *Agricultural Systems* 32, 113–41.

Savary, S., Willocquet, L. and Teng, P. S. 1997. Modelling sheath blight epidemics on rice tillers. *Agricultural Systems* 55, 359–84.

Savary, S., Nelson, A. D., Djurle, A., Esker, P. D., Sparks, A., Amorim, L., Bergamin Filho, A., Caffi, T., Castilla, N., Garrett, K., McRoberts, N., Rossi, V., Yuen, J. and Willocquet, L. 2018. Concepts, approaches, and avenues for modelling crop health and crop losses. *European Journal of Agronomy* 100, 4–18. https://doi.org/10.1016/j.eja.2018.04.003.

Savary, S., Willocquet, L., Pethybridge, S. J., Esker, P., McRoberts, N. and Nelson, A. 2019. The global burden of pathogens and pests on major food crops. *Nature Ecology & Evolution* 3, 430-9. https://doi.org/10.1038/s41559-018-0793-y.

Schleussner, C.-F., Deryng, D., Müller, C., Elliott, J., Saeed, F., Folberth, C., Liu, W., Wang, X., Pugh, T. A. and Thiery, W. 2018. Crop productivity changes in 1.5 C and 2 C worlds under climate sensitivity uncertainty. *Environmental Research Letters* 13, 064007.

Shaw, R. E. and Meyer, W. S. 2015. Improved empirical representation of plant responses to waterlogging for simulating crop yield. *Agronomy Journal* 107, 1711. https://doi.org/10.2134/agronj14.0625.

Shaw, R. E., Meyer, W. S., McNeill, A. and Tyerman, S. D. 2013. Waterlogging in Australian agricultural landscapes: a review of plant responses and crop models. *Crop and Pasture Science* 64, 549–62. https://doi.org/10.1071/CP13080.

Shrestha, R., Arnaud, E., Mauleon, R., Senger, M., Davenport, G. F., Hancock, D., Morrison, N., Bruskiewich, R. and McLaren, G. 2010. Multifunctional crop trait ontology for breeders' data: field book, annotation, data discovery and semantic enrichment of the literature. *AoB Plants* 2010.

Singh, S., Boote, K. J., Angadi, S. V., Grover, K., Begna, S. and Auld, D. 2015. Adapting the CROPGRO model to simulate growth and yield of spring safflower in semiarid conditions. *Agronomy Journal* 108, 64–72.

Smith, N. G. and Dukes, J. S. 2013. Plant respiration and photosynthesis in global-scale models: incorporating acclimation to temperature and CO2. *Global Change Biology* 19(1), 45–63.

Stöckle, C. O., Donatelli, M. and Nelson, R. 2003. CropSyst, a cropping systems simulation model. *European Journal of Agronomy* 18, 289–307. https://doi.org/10.1016/S1161-0301(02)00109-0.

Suzuki, N., Rivero, R. M., Shulaev, V., Blumwald, E. and Mittler, R. 2014. Abiotic and biotic stress combinations. *New Phytologist* 203, 32–43.

Tardieu, F. 2012. Any trait or trait-related allele can confer drought tolerance: just design the right drought scenario. *Journal of Experimental Botany* 63, 25–31.

Tardieu, F., Simonneau, T. and Muller, B. 2018. The physiological basis of drought tolerance in crop plants: a scenario-dependent probabilistic approach. *Annual Review of Plant Biology* 69, 733–59.

Taylor, C. M., Belušić, D., Guichard, F., Parker, D. J., Vischel, T., Bock, O., Harris, P. P., Janicot, S., Klein, C. and Panthou, G. 2017. Frequency of extreme Sahelian storms tripled since 1982 in satellite observations. *Nature* 544, 475–8. https://doi.org/10.1038/nature22069.

Tedeschi, L. O., Cavalcanti, L. F. L., Fonseca, M. A., Herrero, M. and Thornton, P. K. 2014. The evolution and evaluation of dairy cattle models for predicting milk production: an agricultural model intercomparison and improvement project (AgMIP) for livestock. *Animal Production Science* 54, 2052–67. https://doi.org/10.1071/AN14620.

Thorburn, P. J., Boote, K. J., Nendel, C., Rötter, R. P. and Ewert, F. 2018. Recent advances in crop modelling to support sustainable agricultural production and food security under global change. *European Journal of Agronomy* 100, 1–3. https://doi.org/10.1016/j.eja.2018.10.009.

Traore, B., Descheemaeker, K., van Wijk, M. T., Corbeels, M., Supit, I. and Giller, K. E. 2017. Modelling cereal crops to assess future climate risk for family food self-sufficiency in southern Mali. *Field Crops Research* 201, 133–45. https://doi.org/10.1016/j.fcr.2016.11.002.

Triboi, E., Martre, P., Girousse, C., Ravel, C. and Triboi-Blondel, A.-M. 2006. Unravelling environmental and genetic relationships between grain yield and nitrogen concentration for wheat. *European Journal of Agronomy* 25, 108–18. https://doi.org/10.1016/j.eja.2006.04.004.

Uehara, G. and Tsuji, G. Y. 1993. The IBSNAT project. In *Systems Approaches for Agricultural Development*. Ed. F. P. de Vries, P. Teng and K. Metselaar. Springer: Dordrecht.

van Eeuwijk, F. A., Bustos-Korts, D., Millet, E. J., Boer, M. P., Kruijer, W., Thompson, A., Malosetti, M., Iwata, H., Quiroz, R., Kuppe, C., Muller, O., Blazakis, K. N., Yu, K., Tardieu, F. and Chapman, S. C. 2019. Modelling strategies for assessing and increasing the effectiveness of new phenotyping techniques in plant breeding. *Plant Science The 4th International Plant Phenotyping Symposium* 282, 23–39. https://doi.org/10.1016/j.plantsci.2018.06.018.

van Oosterom, E. J., Carberry, P. S., Hargreaves, J. N. G. and O'Leary, G. J. 2001. Simulating growth, development, and yield of tillering pearl millet: II. Simulation of canopy development. *Field Crops Research* 72, 67–91. https://doi.org/10.1016/S0378-4290(01)00165-4.

Vanuytrecht, E. and Thorburn, P. J. 2017. Responses to atmospheric CO_2 concentrations in crop simulation models: a review of current simple and semicomplex representations and options for model development. *Global Change Biology* 23(5), 1806–20. doi:10.1111/gcb.13600.

Végh, B., Marček, T., Karsai, I., Janda, T. and Darkó, É. 2018. Heat acclimation of photosynthesis in wheat genotypes of different origin. *South African Journal of Botany* 117, 184–92.

Wallach, D. and Thorburn, P. J. 2017. Estimating uncertainty in crop model predictions: current situation and future prospects. *European Journal of Agronomy* 88, A1–A7. https://doi.org/10.1016/j.eja.2017.06.001.

Wang, E., Smith, C. J., Bond, W. J., and Verburg, K. 2005. Estimations of vapour pressure deficit and crop water demand in APSIM and their implications for prediction of crop yield, water use, and deep drainage. *Australian Journal of Agricultural Research* 55, 1227–40. https://doi.org/10.1071/AR03216.

Wang, X., Cai, J., Liu, F., Jin, M., Yu, H., Jiang, D., Wollenweber, B., Dai, T. and Cao, W. 2012. Pre-anthesis high temperature acclimation alleviates the negative effects of post-anthesis heat stress on stem stored carbohydrates remobilization and grain starch accumulation in wheat. *Journal of Cereal Science* 55, 331–6. https://doi.org/10.1016/j.jcs.2012.01.004.

Wang, E., Martre, P., Zhao, Z., Ewert, F., Maiorano, A., Rötter, R. P., Kimball, B. A., Ottman, M. J., Wall, G. W. and White, J. W. 2017. The uncertainty of crop yield projections is reduced by improved temperature response functions. *Nature Plants* 3, 17102.

Webber, H. A., Madramootoo, C. A., Bourgault, M., Horst, M. G., Stulina, G. and Smith, D. L. 2010. Adapting the CROPGRO model for saline soils: the case for a common bean crop. *Irrigation Science* 28(4), 317–29. doi:10.1007/s00271-009-0189-5.

Webber, H., Gaiser, T. and Ewert, F. 2014. What role can crop models play in supporting climate change adaptation decisions to enhance food security in Sub-Saharan Africa? *Agricultural Systems* 127, 161–77.

Webber, H., Ewert, F., Olesen, J. E., Müller, C., Fronzek, S., Ruane, A. C., Bourgault, M., Martre, P., Ababaei, B., Bindi, M., Ferrise, R., Finger, R., Fodor, N., Gabaldón-Leal, C., Gaiser, T., Jabloun, M., Kersebaum, K.-C., Lizaso, J. I., Lorite, I. J., Manceau, L., Moriondo, M., Nendel, C., Rodríguez, A., Ruiz-Ramos, M., Semenov, M. A., Siebert, S., Stella, T., Stratonovitch, P., Trombi, G. and Wallach, D. 2018a. Diverging importance of drought stress for maize and winter wheat in Europe. *Nature Communications* 9, 4249.

Webber, H., White, J., Kimball, B., Ewert, F., Asseng, S., Rezaei, E. E., Pinter, P., Hatfield, J. L., Reynolds, M. P., Ababaei, B., Bindi, M., Doltra, J., Ferrise, R., Kage, H., Kassie, B., Kersebaum, K. C., Luig, A., Olesen, J. E., Semenov, M. A., Stratonovitch, P., Ratjen, A., LaMorte, R., Leavitt, S. W., Hunsaker, D., Wall, G. W. and Martre, P. 2018b. Physical robustness of canopy temperature models for crop heat stress simulation across environments and production conditions. *Field Crops Research* 216, 75–88.

Wheeler, T. R., Craufurd, P. Q., Ellis, R. H., Porter, J. R. and Prasad, P. V. 2000. Temperature variability and the yield of annual crops. *Agriculture, Ecosystems & Environment* 82, 159–67.

White, J. W., Hoogenboom, G., Kimball, B. A. and Wall, G. W. 2011a. Methodologies for simulating impacts of climate change on crop production. *Field Crops Research* 124, 357–68.

White, Jeffery W., Hoogenboom, G., Wilkens, P. W., Stackhouse, P. W. and Hoell, J. M. 2011b. Evaluation of satellite-based, modeled-derived daily solar radiation data for the continental United States. *Agronomy Journal* 103(4), 1242–51.

White, J. W., Hunt, L. A., Boote, K. J., Jones, J. W., Koo, J., Kim, S., Porter, C. H., Wilkens, P. W. and Hoogenboom, G. 2013. Integrated description of agricultural field experiments and production: the ICASA version 2.0 data standards. *Computers and Electronics in Agriculture* 96, 1–12.

Wiengweera, A. and Greenway, H. 2004. Performance of seminal and nodal roots of wheat in stagnant solution: K+ and P uptake and effects of increasing O_2 partial pressures around the shoot on nodal root elongation. *Journal of Experimental Botany* 55, 2121–9.

Wilkinson, M. D., et al. 2016. The FAIR guiding principles for scientific data management and stewardship. *Science Data* 3, 160018. doi:10.1038/sdata.2016.18.

Yin, X., Stam, P., Kropff, M. J. and Schapendonk, A. H. C. M. 2003. Crop modeling, QTL mapping, and their complementary role in plant breeding. *Agronomy Journal* 95, 90–8.

Yin, X., Struik, P. C., van Eeuwijk, F. A., Stam, P. and Tang, J. 2005. QTL analysis and QTL-based prediction of flowering phenology in recombinant inbred lines of barley. *Journal of Experimental Botany* 56, 967–76.

Zheng, B., Chapman, S. C., Christopher, J. T., Frederiks, T. M. and Chenu, K. 2015. Frost trends and their estimated impact on yield in the Australian wheatbelt. *Journal of Experimental Botany* 66, 3611–23.

Ziska, L. H., Weerakoon, W., Namuco, O. S. and Pamplona, R. 1996. The influence of nitrogen on the elevated CO_2 response in field-grown rice. *Australian Journal of Plant Physiology* 23, 45–52.

Part 2

The contribution of agriculture to climate change

Chapter 4

Quantifying the role of livestock in climate change

Julie Wolf, USDA-ARS, USA

1 Introduction

The livestock sector has been increasingly newsworthy lately for its many negative impacts on the environment. Recent work has attributed five of eight recent bird species extinctions to deforestation in South America (Barkham, 2018), where forest clearing for grazing and soy crops has resumed (INPE, 2016; Mooney, 2016). Another hurricane has caused large quantities of swine manure to overflow from storage lagoons in North Carolina (Dove, 2018). Human disease caused by antibiotic-resistant bacteria has been traced, via the consumption of meat, to farm antibiotic use (Liu et al., 2018; McKenna, 2018). Despite this frequent news coverage, the impacts of the livestock sector on climate change, ecosystem services, and the broader environment are difficult to quantify. Because livestock also contribute positively to human economic livelihood and food security (Herrero et al., 2013; Herrero and Thornton, 2013), the pros and cons associated with the sector must be thoroughly explored. Therefore, this chapter explores some of the many facets of livestock's contributions to climate

http://dx.doi.org/10.19103/AS.2020.0064.08
Published by Burleigh Dodds Science Publishing Limited, 2020.

change and the difficulties involved in quantifying them, with a closer look into one area where scientific consensus has not been reached–the contribution of livestock methane emissions to changing atmospheric methane concentrations over the last few decades.

The keeping of livestock has contributed to environmental change since the earliest beginnings of animal housing and domestication. Wild ruminant mammals may have been housed, fed, and slaughtered within human settlements as early as the ninth century BCE (Stiner et al., 2014). Early pastoralism spread rapidly between 5000 and 2000 BCE, and methane emissions from increasing populations of ruminant livestock are likely to have contributed to rising global atmospheric methane concentrations over those years (Fuller et al., 2011). Domestication, breeding improvements, and use of animals for transportation, draft power, and production of eggs and dairy soon followed. Global data regarding the impacts of these early, rapid developments on climate and environment are rare, but the changes associated with the ongoing replacement of wild mammals in livestock have been studied in sub-Saharan Africa (Hempson et al., 2017) and additional assumptions can be made based on current knowledge.

In addition to ruminant methane emissions, the gathering and storing of fodder, restriction of animal movement and location, and changes in temporal and spatial grazing patterns are likely to have altered local carbon and nutrient cycles and fire regimes, because the spatial distribution of animals and manure, redistribution of nutrients across landscapes, and amounts of standing and consumed plant biomass are all affected (Hempson et al., 2017). The use of livestock draft power for soil cultivation must also have contributed to the loss of soil organic carbon in newly plowed lands (Sanderman et al., 2017), although this role has not been explicitly modeled. Grazing on semiarid lands is also associated with soil carbon loss (Sanderman et al., 2017); as crop cultivation expanded, the most productive lands were diverted from grazing to cropland, pushing grazing to semiarid lands in some regions (Asner et al., 2004) and likely decreasing soil carbon storage there.

After the first spread of livestock across Old World human settlements (Fuller et al., 2011), domesticated animals and husbandry practices continued to develop and spread. During the Industrial Revolution, however, rapid and drastic changes arose. The use of animals for traction and transportation began to be replaced by fossil fuels, and inorganic fertilizers increasingly took the place of manure on cropland soils; these changes short-circuit the relationship between livestock and crop cultivation from reciprocal to a more one-way flow of feed nutrients to livestock operations.

The last few decades have seen new and rapid changes from traditional or pastoral to large-scale 'industrial' livestock husbandry. Technology has greatly improved growth rates and productivity through breeding, feed optimization, and use of preventative and growth-promoting antibiotics. Long-term treatment of manure in pits or lagoons has also become more prevalent, taking the place of deposition on pasture or cropland (MacDonald et al., 2009; MacDonald and McBride, 2009) and animals perform less draft work (Xue et al., 2014) than in earlier decades. With globalization of trade, the local and regional distributions of livestock operations have shifted, and demand for livestock products has increased in many regions. Because these recent changes have large impacts on the livestock sector's greenhouse gas emissions, land-use extent and land-use change, cycling of nutrients, and other environmental factors, the contributions of livestock to climate change must be frequently reassessed in detail.

These changes in livestock are often associated with much greater greenhouse gas emissions per animal, particularly enteric fermentation methane emissions from meat and dairy cattle, which tend to be the largest source among the greenhouse gas emissions from livestock (Gerber et al., 2013; Herrero et al., 2013). Manure management methane is a much smaller source overall, but is likely to be significantly larger than previously expected in developed regions (Wolf et al., 2017), so detailed examination of this source is also merited.

While per-animal emissions have increased with these recent changes, productivity is also much higher (Capper, 2011). This often leads to lower emission intensities (greenhouse gas emissions per unit production). While lower emission intensities will be a crucial goal in mitigating greenhouse gas emissions from the sector, they are not a guarantee of sustainability, because additional 'economic, social or environmental externalities' may accompany higher productivity (Smith, 2018). Therefore, careful and broad consideration of benefits and impacts is needed to guide consumer choices and policy. Such guidance has been lacking; for example, increased dairy productivity in North America has been subsidized instead of allowed to track market demand, leading to excess production and waste (ABC News, 2018; Investor's Business Daily, 2018) and canceling out the achieved reductions in emissions intensity. The livestock sector's rapid growth and change (along with that of the entire food supply chain) is largely based on intensification and centralization. In industrialized countries, fewer and fewer people are involved with modern livestock production, resulting in general unawareness of how large the sector has grown, of how it functions, and of its broader impacts on climate and the environment.

2 Changes in the extent, management, and efficiency of the livestock sector

The environmental and climate impacts of the livestock sector are due in part to its recent rapid growth to an enormous scale. Consider the global biomasses of livestock, humans, and wild terrestrial mammals (Fig. 1, Table 1) presented in recent work from Bar-On et al. (2018): currently, the global biomass of mammalian livestock is 67% larger than that of humans and 757% larger than that of all wild mammals, and the global biomass of domesticated poultry

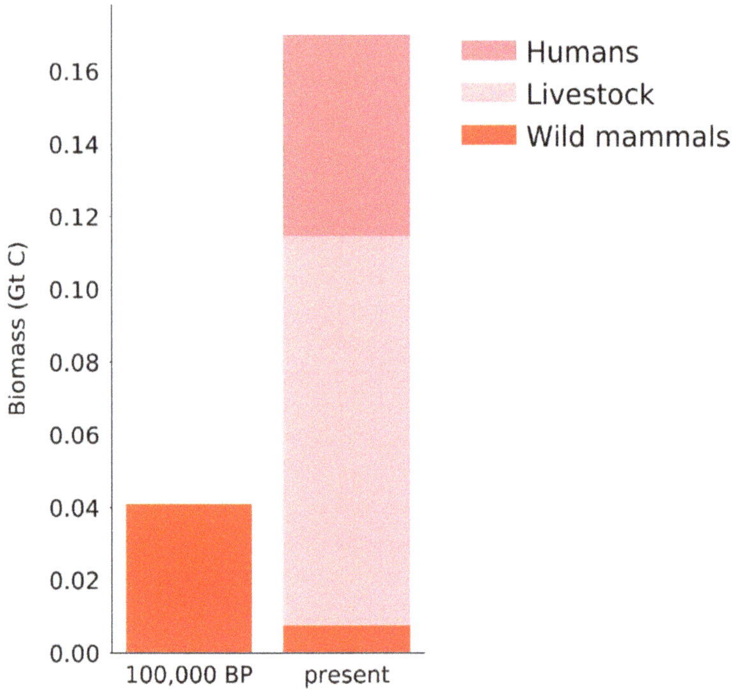

Figure 1 The impact of human civilization on the biomass of mammals. The biomass of wild mammals, livestock (dominated by cattle), and humans before human civilization and at present. Values are based on estimates presented in detail in the relevant sections for humans and livestock, wild mammals and pre-human biomass. Reproduced from Bar-On et al. (2018, Figure S5), distributed under Creative Commons Attribution-Non-Commercial No Derivatives License 4.0 (https://creativecommons.org/licenses/by/4.0/).

Table 1 Estimated current total biomass of humans, mammalian and avian livestock, and wild mammals and birds

Group	Total present-day biomass (Gt C)[a]
Humans	0.06
Mammalian livestock	0.1
Wild mammals	0.007
Domestic poultry	0.005
Wild birds	0.002

[a] data from Bar-On et al. (2018).

is 150% larger than that of all wild birds. Before modern human civilizations arose, the total biomass of wild mammals contained 0.04 gigatons of carbon; today the Earth supports 318% more mammalian biomass, most of which is domesticated cattle (Bar-On et al., 2018, supplemental data, shown here in Fig. 1). In sub-Saharan Africa, the loss of wildlife and replacement with domesticated animals has had large impacts on broader ecosystem processes, by changing plant community composition, decreasing herbivore trait diversity, altering fire regimes, decreasing nutrient diffusion capacity across landscapes, and other impacts (Hempson et al., 2017).

Global estimates indicate that the area of land used specifically for pasture and/or grazing increased from 1.4 to 3.7 million square kilometers between 800 and 1700 CE, and jumped to 29.6 million square kilometers by 1992 (Pongratz et al., 2008). Grazed area has continued to rise in subsequent years (Alkemade et al., 2013), now occupying 22–30% of the 130 million square kilometers of global ice-free land area, with an approximate third of cropland also used to grow feed (Herrero et al., 2016; Hooke and Martín-Duque, 2012; Phelps and Kaplan, 2017). In the United States, a total of 41% of land area is used to feed livestock–ca.10% for growing feed crops and the remainder for grazing cattle (Merrill and Leatherby, 2018).

Globally, the livestock sector has become bifurcated: it is increasingly intensive, technical, and productive in wealthier parts of the globe but remains largely extensive and low-input in poorer regions. This is reflected by the lower cattle body mass and higher percent of total livestock intake from grazing in some regions (Table 2), with some exceptions. Extensive cattle grazing is still widely practiced in some wealthier nations, for example, Australia and the United States. In the United States, grazing is concentrated in the arid western states, but very few animals spend their entire lives there (USDA, 2013); most are moved to be 'finished' on feedlots where they rapidly put on weight (Hayek and Garrett, 2018). Nevertheless, these extensive grazing systems bear some similarities to those in poorer nations. Where arable land is used to grow crops, grazing is pushed to the most marginal soils of drylands (savannah, shrublands, and deserts) or to grasslands with the most variable climates, so that crops can be cultivated on more productive lands (Asner et al., 2004). These lower-quality grazing lands produce feed of lower digestibility, leading to higher methane emissions intensities, slower growth rates, lower feed-use efficiency, and higher land-use extent (explored in Section 5.3). In suboptimal lands, removal of large quantities of plant biomass by grazers tends to degrade the land: via desertification in arid areas, desertification and/or woody encroachment in semiarid areas, and large releases of biomass carbon and further deforestation in humid tropical areas (Asner et al., 2004).

In contrast to extensive livestock husbandry, the foundations of intensive operations are improved, fast-growing breeds, minimal expenditures of animal

Table 2 Average meat cattle body mass and percent of total livestock intake from grazing

Region:	Africa	E.-S. E. Asia	E. Europe, W. Asia, and Central Asia	Latin America	Oceania	S. Asia	US and Canada	W. Europe
Average body mass of meat cattle (kg)[a]	173	259	351	323	376	110	420	401
Proportion of intake from grazing (%)[a]	81.2	58.7	33.2	71.9	88.8	48.8	46.7	29.7

[a] from Wolf et al. (2017; Tables 2 and 4).

Table 3 Comparison of impacts associated with extensive and intensive livestock management

	Extensive (grazing) ruminants	Intensive (housed and fed) ruminants and monogastrics
Products and benefits:	High-quality food (protein, fat iron, calcium) from human-inedible plant material	Highly productive, high feed-use efficiency)
	Manure may increase rangeland soil organic carbon	By-product feeds (e.g. DGGS, oilseed cakes) reduce waste from the biofuels and oilcrop industries
		Manure may be used as fertilizer and may increase cropland soil organic carbon
		Manure can be a source of usable biogas (via anaerobic digestion)
		Intensification greatly reduces land-use extent
Greenhouse gas emissions:	High enteric fermentation methane emissions on lower-quality forage	Lower enteric fermentation methane emissions on highly digestible feed (ruminants only, no enteric fermentation in poultry or swine)
	High carbon dioxide emissions associated with land-use change for pasture creation	Emissions from crop feed production, for example, nitrous oxide from fertilizers
		Higher manure CH4 emissions from liquid manure management (unless anaerobic digestion or capping used)
	Manure may increase rangeland soil nitrous oxide emissions	Manure and sludge may increase cropland soil nitrous oxide emissions
Land use and land-use change:	High land-use extent	Land used for crop feed growth
	High land-use change (e.g. deforestation) in some areas	Land used for manure management/ treatment
	Potential overgrazing, erosion, desertification, or woody encroachment	
	Grazing may decrease soil organic carbon on arid/ marginal grazing lands	
	Potential altered fuel load and fire regime	

(*Continued*)

Table 3 (*Continued*)

	Extensive (grazing) ruminants	Intensive (housed and fed) ruminants and monogastrics
Pollution, acidification, and/or eutrophication:		Large quantities of concentrated manure have high potential for eutrophication, contamination, etc.
		Fertilizers, herbicides, and pesticides used to grow feed may contribute to pollution, acidification, and/or eutrophication
Water use:		High water usage in some manure management systems; irrigation water used for some crop feeds
Biodiversity impacts:	Compete with wildlife and wildlands for habitat and feed	Potential disease vectors to and from wildlife
	Potential positive (e.g. bird habitat, grassland health) and/or negative impacts (e.g. deforestation, loss of ecosystem services)	Potential for growth hormones and other compounds to negatively affect wildlife
Energy use:	Energy use for processing and transport	Energy use for fertilizer and feed production
		Energy use for temperature/environmental control
		Energy use for processing and transport
Other:		Human-edible crops fed to livestock
		Potential contamination of soils, surface waters, and/or groundwater with harmful bacteria, residual medicines, growth hormones, etc.
		Concentrated animals receiving low-dose antibiotics–conducive to development of antibiotic-resistant 'super' bacteria
		Manure management pits and lagoons--bad odors, visually repellant, decrease quality of life for neighbors

energy (e.g. to stay warm or to move around), feed that is high in energy and digestibility (grains, soy, and, increasingly, by-products of biofuel production), high-tech feed supplements and/or growth-enhancing compounds, and ever-larger operations for economic efficiency (via economies of scale). Housed cattle and monogastrics (poultry and swine) fed on highly digestible crop-derived feeds have high growth rates and feed-use efficiencies, and these intensive systems use very little land compared to extensive grazing systems. Such large operations accumulate vast quantities of manure, often too much to be applied to adjoining cropland. Aside from beef cattle feedlots, which tend to store manure in dry piles, these intensive operations increasingly use long-term liquid manure management strategies based on pits or anaerobic lagoons, which emit much more methane than dry systems.

It is clear that the drastic increase in total animal biomass must impact the Earth. Recent growth of both grazed and housed animal populations, as well as recent changes associated with modern intensification and centralization (Table 3), are likely to have multiple impacts on climate change and the environment. At issue here is how to put numbers on these impacts, so that we can mitigate them, add their true costs to the price of livestock products, and/or otherwise change policy and behavior, to hopefully avoid catastrophic climate change (IPCC - SR15, 2018).

3 Livestock methane, global atmospheric concentrations, and the global methane cycle

Methane is the second most important greenhouse gas, responsible for more global warming (ca. 20% of current total warming) than all other gases except for carbon dioxide (Kirschke et al., 2013). Atmospheric concentrations of methane remained below ~700 ppb for ca. 800 000 years, but began a rapid increase around the time of the Industrial Revolution (Fig. 2), and the growth is thought to be due to anthropogenic (due to human activity) sources such as agriculture, fossil fuels, biomass burning, and wastes (Dlugokencky et al.,

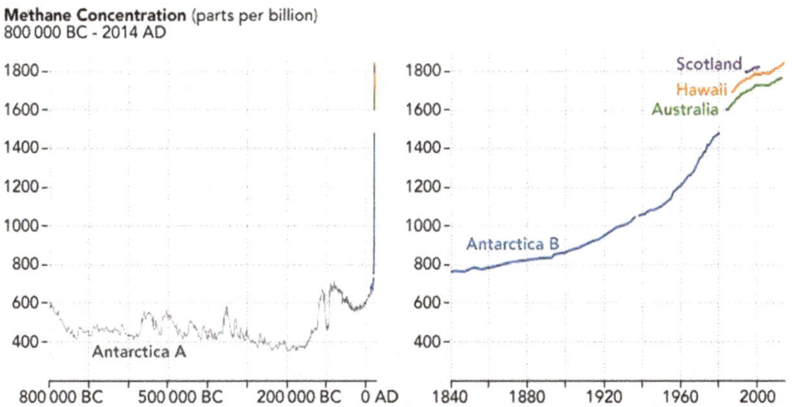

Figure 2 A combination of historical ice core data and air monitoring instruments reveals a consistent trend: global atmospheric methane concentrations have risen sharply in the past 2000 years. (NASA Earth Observatory image by Joshua Stevens, using data from the EPA., available at https://earthobservatory.nasa.gov/Features/MethaneMatters).

2011). Atmospheric concentrations are now ca. 1800 ppb, 2.5 times greater than preindustrial concentrations (Dlugokencky et al., 2011; Saunois et al., 2016). Methane has more warming potential per ton than carbon dioxide, but a much shorter turnover time in the atmosphere, and so represents an important means to limit climate change in the near future (Dlugokencky et al., 2011). In order to do so, however, the sources of methane emissions must be well understood so that best practices can be characterized and employed.

Domesticated (i.e. cattle, buffalo, yaks, goats, sheep, camels, alpacas, llamas, and reindeer) and wild (e.g. bison, deer, antelope, giraffes) ruminants have specialized guts where methanogenic bacteria help break down feed through fermentation. This specialization allows ruminants to extract nutrition more thoroughly from fibrous feeds that are high in cellulose and hemicellulose, creating high-quality meat and/or milk from human-inedible plant material. This process, called enteric fermentation, also results in emissions of methane, and so represents a trade-off (Table 3). The most recent assessments indicate that 'the production of animal products generates the majority of food-related [greenhouse gas] emissions (72-78% of total agricultural emissions) ...' (Springmann et al., 2018b). Livestock are thought to be the largest single anthropogenic source of methane emissions to the atmosphere (Fig. 3); the only larger single source is thought to be natural wetlands (Ghosh et al., 2015).

Methane sources are separated as anthropogenic (fossil fuel extraction, livestock enteric fermentation, submerged rice paddies, wastes, livestock manure, and biomass burning) or natural (natural wetlands, ocean exchange, geologic activity). The isotopic signatures of carbon and hydrogen in methane emissions may vary by source, depending on the process that generated them, which are not strictly anthropogenic or natural. These processes include (i) biogenic, produced by methanogenic bacteria in anaerobic environments such as natural wetland soils, rice paddy soils, ruminant guts, termite guts, and stockpiles of waste and manure; (ii) geologic; (iii) pyrogenic, produced by the burning of biomass and biofuels (Sapart et al., 2012), or (iv) thermogenic, resulting from extraction or release of fossil fuel methane (Schaefer et al., 2016). Variations in isotope signatures can also arise within sources from a process, for example, methane emissions from livestock vary with the proportions of C3 and C4 plant-based feeds in their diet (Klevenhusen et al., 2010). Therefore, while the mean isotopic signatures of methane produced by different processes vary, there is often considerable overlap in their ranges (Sherwood et al., 2017).

The methane cycle also includes sinks (Fig. 3b)–the dominant sink being reaction with hydroxyl (OH) radicals in the troposphere (Rigby et al., 2017). The overall atmospheric concentration of methane can be measured annually, and changes must relate to the difference between total emissions and total sinks (technically, mass mixing ratios, not volumetric concentrations, are measured; but 'concentration' will be used here). But attributing atmospheric methane to

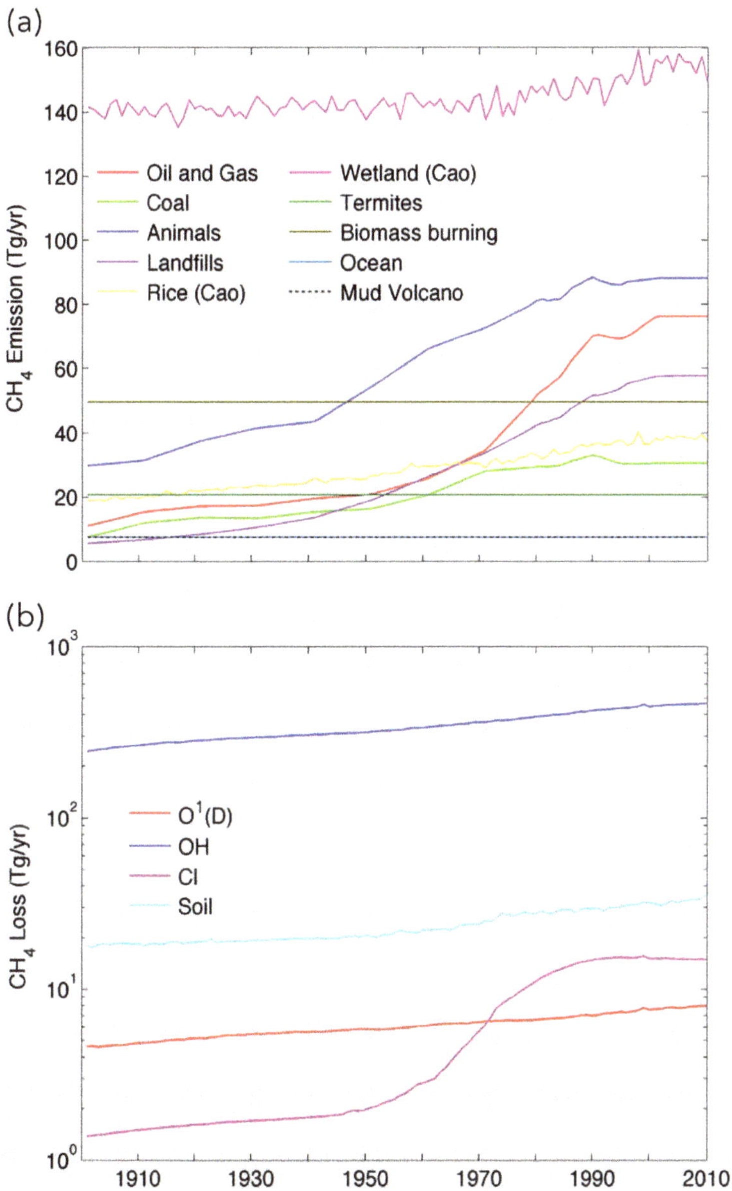

Figure 3 (a) Time series of methane emission inventory estimates from different categories during simulation [wetland and rice emissions data from a method developed by Cao, M., Gregson, K., and Marshall, S]. (b) Chemical loss of methane calculated using ACTM [(Atmospheric General Circulation Model-based Chemistry Transport Model)] simulation with initial emissions (E_{ini}) [note logarithmic scale in this panel] (Ghosh et al., 2015). This work is distributed under the Creative Commons Attribution 3.0 License (https://creativecommons.org/licenses/by/3.0/).

a single sector is not so simple. The multiple and variable sources of methane rapidly mix into the atmosphere, and their overlapping isotopic signatures do not permit clear separation (Sherwood et al., 2017). Additional methods are needed to quantify sectoral methane emissions, particularly the large livestock source which, unlike methane from natural wetlands, is under our control and therefore has potential for mitigation.

4 Quantifying enteric fermentation and manure management methane emissions

It is difficult to quantify gaseous methane emissions from the diverse global herds of livestock and their manure. Emissions from individual animals or small quantities of manure can be directly measured in the lab or with specialized equipment. Emissions from barns, whole herds, larger quantities of manure managed on-farm, or areas where livestock are the dominant source of methane, may be measured using surface, near-surface (tower, drone, or aircraft), or satellite measurements over a very small area. Surface, near-surface, and/or satellite measurements of atmospheric concentrations over regions and/or the entire globe can be made and sources can be attributed to various sectors with some effort and uncertainty. Lastly, inventory-based estimates can be assembled based on observations and modeling of per-animal emissions estimates. Each method has significant limitations and uncertainty involved, reviewed by Lassey (2007) and Hristov et al. (2018).

4.1 Bottom-up measurements

Bottom-up measurements involve making actual measurements of methane quantities emitted over a set period of time from individual animals and individual manure management locations, sufficiently representing variation to upscale to the population or condition of interest (e.g. emissions on a certain diet formulation). Methods for direct bottom-up measurements of enteric fermentation methane include placing individual animals in respiration chambers or hoods, which can influence animal feeding and other behavior, and the SF6 tracer technique, which is also associated with concerns and uncertainties (Hristov et al., 2018; Lassey, 2007; Olander et al., 2014). These measurements involve significant effort and cost, making it difficult to make measurements over long time periods and on large numbers of animals. Bottom-up measurements of manure management methane involve laboratory-based incubations and/or field observations (Owen and Silver, 2015; Pratt et al., 2015) and have similar limitations in terms of representing different animals and diets, multiple manure management strategies, locations, and conditions, and in tracking manure management emissions from production through ultimate

Published by Burleigh Dodds Science Publishing Limited, 2020.

fate. For example, sludge removal from anaerobic lagoons may occur as rarely as once every 5–10 or more years (Chastain and Henry, 2002a,b). Most of these variable details regarding animals, diets, and manure management have large impacts on total emission quantities, so extrapolating measurements brings considerable uncertainty.

For enteric methane quantification, detailed feed intake quantity and quality data should also be collected, so that the methane conversion factor or methane yield (Y_m, the energy contained in methane emissions per unit feed intake energy, which can be converted to mass of methane emitted per mass of dry matter intake) can be characterized (Lassey, 2007). The quantification of Y_m better allows measurements from individual animals to be extrapolated to herds and larger populations consuming similar diets. Analogously, for manure management methane, the biochemical methane potential (B_o) provides an estimate of methane production per mass of manure volatile solids (IPCC, 2006; Owen and Silver, 2015; Pratt et al., 2015). Along with estimates of the amount of volatile solids excreted in manure, B_o can be used to upscale experimental observations to larger herds and populations.

4.2 Bottom-up inventory-based estimations

Bottom-up inventory-based estimations rely on the direct measurements described above, along with models developed from them, to derive emissions factors that describe the average amount of a gas emitted per animal in a year. Annex 1 parties to the United Nations Framework Convention on Climate Change (UNFCCC) quantify their annual greenhouse gas emissions, including those from livestock, following International Panel on Climate Change Guidelines for National Greenhouse Gas Inventories (IPCC, 1997, 2000, 2006). The IPCC guidelines are the principal method for conducting bottom-up inventory-based estimates of livestock emissions. The resulting National Inventory Submissions can be downloaded in a uniform spreadsheet Common Reporting Format (CRF) that includes emission estimates for enteric fermentation and manure management methane and nitrous oxide emissions, as well as much (but not all) of the information used to calculate these quantities (UNFCCC, 2017).

Bottom-up inventories can be simple or more complex. The simpler Tier 1 estimates are calculated by multiplying annual populations of each animal type by default emission factors for each species and/or type of animals (e.g. dairy cows, broiler chickens) that are constructed separately for different global regions. The emission factors are derived using sets of mathematical formulae, developed through empirical and energy-requirements modeling, that relate regional livestock qualities and management to emissions quantity using the methane yield (Y_m) or biochemical methane potential (B_o) (IPCC, 2006; Lassey,

2007). Tier 1 estimates only require the user to know the various populations of livestock in the country or region of interest, and these numbers are available from the Food and Agriculture Organization of the United Nations (FAO, 2014), where precalculated Tier 1 estimates of emissions are also available for download.

At the Tier 2 level of bottom-up emission inventories, instead of default emission factors, the user constructs tailored emission factors for subsets of animal populations and manure management scenarios in the country or area of interest. This level of evaluation is appropriate for larger/more important emissions categories. The first step for Tier 2 assessments is to assemble detailed national or subnational activity data–this includes (i) populations of livestock subcategories such as age cohorts or housed and grazing cattle subgroups, and (ii) detailed information regarding factors that influence emissions, such as feed intake quantity and quality; milk production quantity; amounts of energy needed for growth, maintenance, draft work, foraging, and pregnancy; and utilization of various manure management systems. The energy-requirements modeling equations used to derive the default Tier 1 emission factors are instead calculated with this more detailed information. The emission factors created are then multiplied by corresponding subgroup populations to calculate emissions for each subgroup. These equations therefore allow for revision of emission factors as animal traits and management change.

A nation may also develop more complex and detailed modeling–the Tier 3 level–for one or more important categories of its livestock, such as the Cattle Enteric Fermentation Model (CEFM) developed by the US Environmental Protection Agency (EPA, 2016)–but the IPCC guidelines do not provide methods for this level of assessment.

In the IPCC 2006 guidelines, the default values for emission factors, activity data, and regional values for methane yield and biochemical methane potential, are based on literature reflecting earlier decades. For example, data sources listed for tables in Annexes 10.A1 and 10A.2 in the guidelines (IPCC, 2006) were published between 1976 and 2004, with most from the 1980s and 1990s. In many regions, this information no longer reflects the current state of livestock. For example, in many industrialized or industrializing nations, management of manure in pits or lagoons has become much more prevalent. Because methane emissions from anaerobic lagoons are estimated to be nearly twice the magnitude of those from dry systems per unit of manure input, these changes will have a large impact in nations using these systems at high rates, such as the United States (Fig. 4) and so must be taken into account in new bottom-up inventories. Recent work has also shown that these methods underestimate actual emissions for most recent types of manure management practices (Owen and Silver, 2015). Frequent reassessment of activity data and

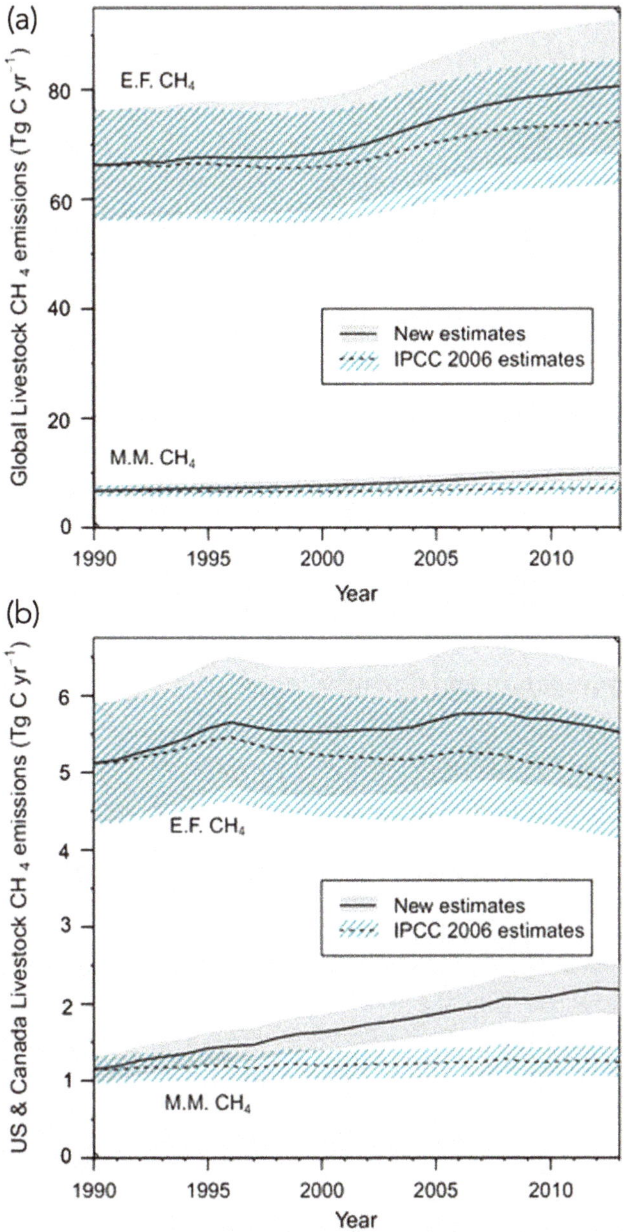

Figure 4 Comparison of enteric fermentation (EF) and manure management (MM) methane emissions estimated using IPCC 2006 and revised emission factors resulting from this study, for the globe (a), and the US and Canada region (b) (Wolf et al., 2017). Distributed under Creative Commons Attribution 4.0 International License (http://creativecommons.org/licenses/by/4.0/)

emissions factors is called for (Owen and Silver, 2015; VanderZaag et al., 2013; Wolf et al., 2017).

4.3 Top-down studies

Top-down studies of greenhouse gas concentrations and/or emissions are informed by atmospheric measurements made at or near the Earth's surface (surface station, flux tower, aircraft, drone, or similar), and/or from space (satellite). Using atmospheric methane concentration data with a 'forward' atmospheric transport model can solve for emissions, while 'inverse' modeling of sources and sinks runs atmospheric transport models with expected or 'prior' estimates of sources and sinks (often based on bottom-up inventories), compares the results to observations, and refines source/sink configurations to best reproduce observations (Hein et al., 1997; Houweling et al., 1999). Inverse modeling may also use the isotopic signature of detected methane to help assign emissions to specific sources (Sherwood et al., 2017). Focus on areas where there is good spatial separation of sectors and sources (Miller et al., 2013) and tracking co-emitted gases or tracers, such as ethane for fossil fuels emissions (Schaefer et al., 2016), can also help refine attribution to different source sectors.

4.4 Disaggregation and downscaling

Disaggregation and downscaling of bottom-up emission estimates to finer resolution can be accomplished in various ways, such as (i) using state or county livestock populations and spatial boundaries to place emissions on a map, and/or (ii) using gridded landcover/land-use maps, along with national, state, or county boundaries, to place emissions where the source populations are thought to be. Spatial discrepancies among and between multiple top-down and bottom-up assessments help evaluate findings and direct future efforts. While bottom-up estimates are often larger than top-down, especially for natural sources (Saunois et al., 2016), for livestock in the United States, bottom-up inventory estimates based on IPCC guidelines are much lower than top-down estimates (Hristov et al., 2014; Miller et al., 2013, 2014; Turner et al., 2015; Wecht et al., 2014a,b). Disaggregation and comparison of spatially downscaled emissions helped evaluate these discrepancies, which are explored below in the case study.

5 Broader assessments and considerations

Methane from enteric fermentation and manure management is not the only greenhouse gas emission directly associated with the production of livestock products. Several methods to quantify impacts in broader terms, as well as additional considerations, are described below in order of increasing scope.

5.1 Global warming potential and associated issues

In addition to methane, nitrous oxide, ammonia, and volatile organic compounds are also emitted from livestock urine and manure, and carbon dioxide is emitted through animal respiration and decomposition of manure organic carbon. Usually only methane and nitrous oxide are quantified as the others are considered negligible or neutral, and methane often takes the forefront. When multiple gases are assessed, their combined impact on warming can be summarized using the gases' global warming potentials (GWP). The GWP represents the amount of heat energy that 1 ton of emitted gas will absorb over a set period of time, often 100 years, relative to the amount of heat energy that 1 ton of carbon dioxide would absorb (US EPA, 2016). Therefore, multiplying the quantity of an emitted gas by its GWP converts the emissions to carbon dioxide equivalents (CO_2eq). The use of GWPs is complicated by the fact that greenhouse gases differ greatly in both their warming potential and lifetime in the atmosphere, so that the time period considered matters. Recommended GWPs may also change, because of reassessment and/or because atmospheric concentrations have changed enough to alter the heat trapping behavior of further emissions (US EPA, 2016). Recent work suggests that the use of GWPs underestimate the warming caused by non-CO_2 greenhouse gases because they do not account for (i) the cooling effect of aerosols in the last century and (ii) the different atmospheric dynamics of gases like methane, compared to those of CO_2. Therefore, the use of GWPs may not adequately reflect livestock contributions to historical and current warming (Allen et al., 2016; Reisinger and Clark, 2018). Using carbon cycle-climate modeling, Reisinger and Clark (2018) attributed 19% of climate warming in 2010 to direct livestock non-CO_2 emissions, much more than the 7% estimate that is obtained using 100-year GWPs. Because this type of modeling is not possible for everyone in the research community, reporting of actual quantities of emissions along with their CO_2eq values may be the best strategy for now.

5.2 Respired carbon dioxide and global losses of carbon fixation and storage

In most accounting frameworks, the carbon dioxide emitted through animal respiration and manure decomposition are considered a balanced return of recent photosynthetically fixed carbon to the atmosphere, and therefore are usually omitted from analysis and reporting. This view has been challenged by Goodland (2013, 2014), who rightly points out that:

> In recent decades, respiration has increased exponentially with livestock production–while intensified livestock and feed production has been accompanied by large-scale deforestation and forest-burning, huge increases in volatilization of

> soil carbon, and dramatic declines in Earth's photosynthetic capacity ... As a result, either carbon dioxide in livestock respiration–or its reflection in carbon absorption forgone on land used for livestock and feed production–should be counted as emissions ... (Goodland, 2014).

The carbon in direct carbon dioxide emissions from respiration and manure can be calculated as the difference between total intake carbon and production of methane and milk or eggs (if any), assuming a roughly constant standing herd biomass (Fig. 5). For global livestock of all types and species 2011, this quantity of carbon is 2681 of the total 2826 teragrams of carbon in intake (Fig. 5). Because this quantity only reflects the carbon in the emitted carbon dioxide, it must be multiplied by the molecular weight of carbon dioxide (44) divided by the atomic weight of carbon (12), leading to 9830 teragrams of carbon dioxide emissions. Doing the same for methane (multiplying methane carbon quantities by (12 + 4)/12) results in 119 teragrams of methane emitted. The carbon dioxide emissions add ca. 2.5-3 times the amount of warming potential due to the methane emissions estimated for 2011, even using the highest value for methane's GWP (US EPA, 2016).

This issue may seem negligible in the light of large carbon dioxide emissions from fossil fuel combustion and the greater warming potential of methane. However, a recent study 'suggests that land use halves the amount of carbon that is potentially stored in terrestrial biomass' (Erb et al., 2018)–around 450–466 petagrams of the globe's potential carbon storage are missing due to anthropogenic land use, land management, and deforestation since preindustrial times (not including changes in soil carbon storage). Where the climate is appropriate, all land not covered with infrastructure, crops, or forests (54.3 million square kilometers of area) was attributed to grazing in this study, which is higher than several current estimates of ca. 30 million square kilometers (see Section 1). This may be because the estimate is cumulative over the last several centuries. This 'grassland and grazing land' area is associated with 54–60% of total missing carbon storage (Erb et al., 2018, Extended Data Table 1). Cropland is responsible for an additional 28–31%, of which one-third is estimated to be used for crop feeds. These numbers suggest that 64–70% of the total missing carbon storage in biomass, ca. 288–322 petagrams of carbon, is solely due to the livestock sector. Even if the actual amount of grazed area is less than 54.3 million square kilometers, the estimate of 30 million square kilometers suggests that a minimum of 42%, 189 petagrams of carbon, can be attributed to livestock. These enormous quantities of lost carbon storage and photosynthetic potential lend support to Goodland's (2014) assertion above, that livestock's carbon dioxide emissions, or the associated 'foregone carbon absorption' on grazed and feed crop lands, should be included in estimates of the sector's climate impacts. In fact, carbon dioxide emission quantities might not adequately account for the impacts on carbon storage over longer time

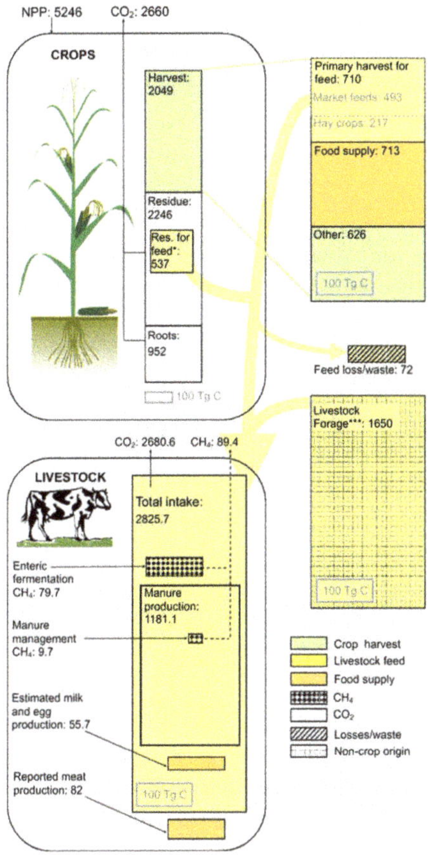

Figure 5 Revised global crop and livestock carbon budget for 2011, showing the consumption of biomass carbon from crop harvests and grazing and the fate of that carbon. The units of all represented quantities and fluxes are teragrams of carbon (Tg C) per year. The area of each box is proportional to the quantity of carbon (with total crop net primary productivity (NPP) shown at one fourth the magnitude of the harvest, forage, and livestock fluxes). Carbon quantities were calculated from inventory-based estimates of livestock populations and modeled quantities of carbon in feed intake, manure production, methane emissions, and milk and egg production. Yellow boxes and arrows represent carbon in livestock feed from crop and grazing sources. Orange boxes represent edible food (for humans) produced by the crop and livestock sectors. Solid and dashed black arrows show fluxes of carbon to and from the atmosphere: fixation (via photosynthesis) or emission (via respiration) of carbon dioxide, and emissions of methane (via enteric fermentation and manure management), respectively. Carbon emitted as carbon dioxide from livestock respiration is calculated by subtraction of all other carbon fluxes from total intake carbon, assuming constant total livestock biomass from year to year. All non-harvested crop biomass carbon, and all manure carbon not emitted as methane, is assumed to be decomposed and respired as carbon dioxide within the same year as production. From Wolf et al. (2017). Distributed under Creative Commons Attribution 4.0 International License (http://creativecommons.org/licenses/by/4.0/).

periods, because they result not only from 'foregone absorption' but also from decreased turnover time of remaining biomass due to (i) replacing trees with shorter-lived plants and (ii) harvesting of biomass (Erb et al., 2018). Additional work to formalize accounting of these quantities is merited.

5.3 Carbon footprint and life cycle analysis

When multiple non-CO_2 greenhouse gases are emitted, carbon footprint (CF) analysis can be conducted. This can be as simple as using the GWP to summarize emissions of several gases from a single process (e.g. livestock husbandry), or can also account for associated emissions up- and down-stream during the 'lifetime' of production. In the case of livestock, this might include nitrous oxide emissions from fertilizer used to grow feed, fossil fuels used to grow and transport the feed and to transport, process, and refrigerate the livestock products. For CF studies, system boundaries must be considered carefully and stated clearly so that comparisons can be made within and among studies (e.g. of different production systems, regions, livestock types, and even to non-livestock food sources). A life cycle analysis (LCA) could also be conducted, which follows International Organization for Standardization (ISO) guidelines (ISO, 2006a,b) to quantify multiple environmental impacts. LCAs are often just a broad CF (e.g. Gerber et al., 2010), but for livestock, it often includes one or more categories of impacts in addition to direct greenhouse gas emissions, such as carbon dioxide emissions due to land-use change, land costs/land-use extent, water usage relative to local water scarcity (Poore and Nemecek, 2018), fossil fuel/energy usage, eutrophication, pollution, and/or acidification potential, and/or impacts on biodiversity/wildlife (often related to land-use extent). Sometimes economic indicators are also included (Boehm et al., 2018). A recent meta-analysis showed large regional variability in the impacts of livestock grazing on loss of biodiversity (Alkemade et al., 2013). For EU27 nations, livestock were shown to dominate multiple impacts of the whole agricultural sector–responsible for more than three-fourths of total agricultural greenhouse gas emissions, reductions in terrestrial biodiversity, soil acidification, air pollution, and water pollution (Leip et al., 2015).

Depending on the system boundaries, CF analyses involving ruminant production usually find that enteric fermentation methane is the largest source, by far, of greenhouse gas emissions (or the largest contributor to emissions intensity) among cradle-to-farmgate, upstream, and downstream-associated emissions (Hyland et al., 2016). If the analysis is broadened to include emissions from land-use change and emission intensities, this may change. Using results from a publicly available global CF/LCA model (MacLeod et al., 2018), beef production from different regions and production systems can be compared (Fig. 6, not all regions shown). Enteric fermentation methane was

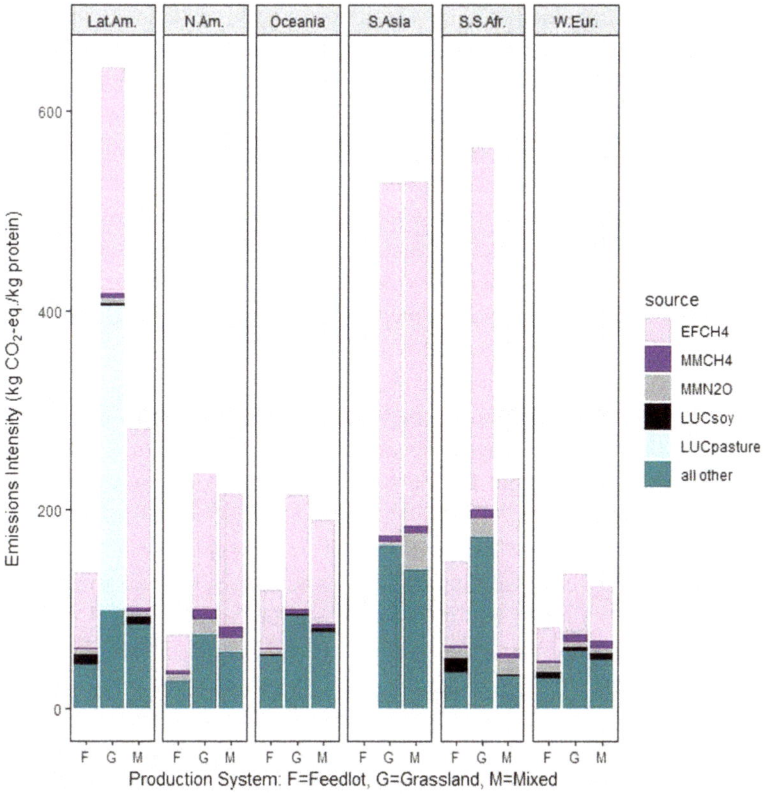

Figure 6 Comparison of beef cattle emission intensities by cumulative source, for feedlot, grassland-based, and mixed operations in Latin America, North America, Oceania, South Asia, sub-Saharan Africa, and West Europe. Emission source categories are: EFCH4, enteric fermentation methane; MMCH4, manure management methane; MMN2O, manure management nitrous oxide; LUCsoy, land-use change for soy cropping; LUCpasture, land-use change for pasture creation; and all other sources combined. Data taken from Global Livestock Environmental Assessment Model (GLEAM) (FAO, 2018).

the largest single contributor to emissions intensity in most beef systems and regions, except for grass-fed systems in Latin America; here, forest destruction for pasture contributes a large share of emissions and significantly raises the emissions intensity. This comprehensive model, the Global Livestock Assessment Model (GLEAM), is a CF model (but is called a LCA, however; see below). It models emission intensities from all major types of livestock products, regions, and production systems, and provides downscaled input information (e.g. livestock populations and management data). The model and results for data year 2010 are available for download (FAO, 2018). Output from the model shows that meat produced in feedlot systems has lower emissions intensity in every region where it is available (Fig. 6), as expected.

The results from GLEAM highlight the strength of CF/LCA, which is to capture and represent differences between production systems, regions, different types of livestock, or even between products of livestock and non-livestock origin. For example, animals that are housed intensively do not contribute further to deforestation for pasture expansion, as is currently occurring in the Amazon (Eaton, 2018). However, housed animals do consume more crop-derived feeds which may also contribute to deforestation (e.g. the 'land use change for soy cropping' category in Fig. 6). These crop-derived feeds could also, theoretically, alleviate human hunger, if such food could reach the people that need it, and was sufficiently familiar and palatable to those people; this is not yet clear (e.g. White and Hall, 2017, 2018; Emery, 2018; Meerbeek and Svenning, 2018; Springmann et al., 2018a).

The production efficiency associated with modernized, industrial husbandry relates to efficiency in animal growth per unit input (e.g. feed), which may translate to more profit for the farmer/rancher. However, this efficiency may also be associated with inefficiencies that are hidden in 'the commons' because the true cost is not assessed. Some possible examples include organic manure carbon that is lost to the atmosphere as carbon dioxide and methane instead of contributing to soil organic carbon, and facilitation of antibiotic-resistant bacteria and other disease agents (e.g. Liu et al., 2018). These costs are difficult to quantify, may have many conflicting aspects, and/or may not directly contribute to climate change, but still should be considered. For example, manure storage lagoons can be visually and olfactorily repellant and can decrease quality of life for nearby residents, so most people innately think of them as negative (until its grocery shopping time, when the lower prices of intensive meat are taken for granted). However, a recent meta-analysis suggests that, from a global warming standpoint, the soil organic carbon sequestration benefits of applying manure to cropland are largely negated by stimulation of soil nitrous oxide emissions (although improvements in soil structure, water- and nutrient-holding capacity were not analyzed) (Zhou et al., 2017). This finding therefore lends support for utilization of long-term manure storage and treatment despite their offensiveness, especially if use of methane capture or anaerobic digestion technology were to become more common. Overall, these complexities recommend a broad view of the environmental and climate impacts, as well as of the benefits associated with livestock.

The public, and particularly conservation-oriented people, often perceive low-intensity farming to be better for biodiversity (Balmford et al., 2012). However, using a novel framework to look at the covariance of multiple impacts, the opposite was strongly supported (Balmford et al., 2018). These authors examined externalities per unit production in a way analogous to emission intensities. Their meta-analysis of field experiments and LCAs reveals that per unit of production, high-yield, intensive, non-organic systems have lower greenhouse gas emissions, land use (and concomitant biodiversity impacts), nitrogen and phosphorus use, and soil impacts. While intensive livestock production appears by all measures so far to have lower impacts than intensive, there is still no avoiding the fact that these same measures demonstrate that even the most efficiently produced livestock products have vastly larger impacts than non-livestock food products. A meta-analysis of unprecedented comprehensiveness, covering multiple impacts of grain and protein staple foods in various production systems (Fig. 7; note different emissions intensity units than in Fig. 6), concludes that:

> ... for all environmental indicators and nutritional units examined, plant-based foods have the lowest environmental impacts; eggs, dairy, pork, poultry, non-trawling fisheries, and non-recirculating aquaculture have intermediate impacts; and ruminant meat has impacts ~100 times those of plant-based foods ... dietary shifts towards low-impact foods and increases in agricultural input use efficiency would offer larger environmental benefits than would switches from conventional agricultural systems to alternatives such as organic agriculture or grass-fed beef (Clark and Tilman, 2017).

We might also consider the economic, ethical, and social issues related to the sector, such as animal welfare and human quality of life issues (e.g., workers'

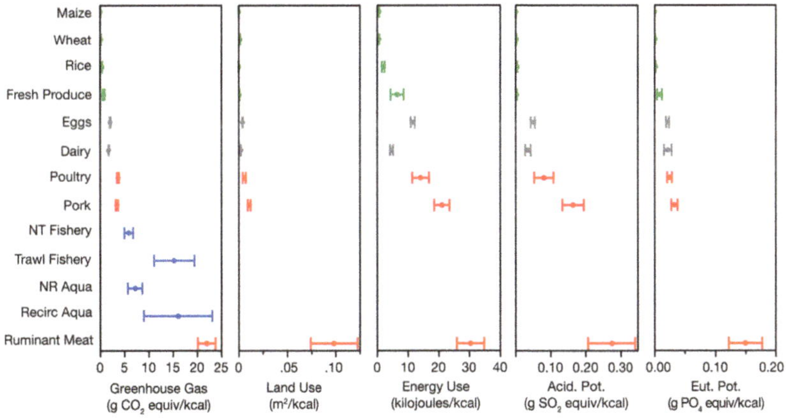

Figure 7 Environmental impacts of broad groups of foods per kilocalorie. The environmental indicators examined are greenhouse gas emissions, land use, energy use, acidification potential (Acid. Pot.) and eutrophication potential (Eut. Pot.). Bars show means and standard errors. Plant-based foods are in green; dairy and eggs are in grey; meats are in red; and seafood is in blue. Data from foods grown in greenhouses are not included when plotting this figure. Trawl Fishery = bottom-trawling fisheries; NT Fishery = all other fisheries (e.g. line, purse net, seine net, etc.); Recirc Aqua = recirculating aquaculture; NR Aqua = non-recirculating aquaculture (e.g. pond, net pen, flow-through, etc.). (Clark and Tilman, 2017) Distributed under Creative Commons Attribution 3.0 International License (http://creativecommons.org/licenses/by/3.0/).

rights and adequate salary, use of antibiotics/development of resistant bacteria, human health, etc.), but these issues fall partly outside the realm of climate change and agriculture, and analysis must stop somewhere. It would be good to, however, add one more consideration to future assessments–that of vulnerability.

> ... if grain prices rise in line with some predictions, the ability to convert poor-quality forage efficiently into milk or meat might be another attribute for which it will be worth [the] trouble to develop diagnostic [genetic] markers (Petherick, 2012).

The issue of grain feed prices is one of several potential vulnerabilities of high-intensity high-yield livestock agriculture (as well as of modern agriculture in general). The goals of breeding and crop research, for the most part, have been to obtain steady and high yields with high levels of inputs. Therefore, we are losing both the genetic diversity and the knowledge and infrastructure needed for alternative husbandry practices, and replacing them with a low number of high-yielding genotypes (both crop and livestock) and an extreme reliance on high-quality crop feeds, centralization, cheap fossil fuels and fertilizers, medications and growth promoters, and other technology. The supply of all these things is vulnerable to climate change, political or economic upheaval, and/or other unpredicted circumstances. More and larger anaerobic manure lagoons are at increasing risk of failure due to extreme events (e.g. hurricanes), as are the animals themselves (heatwaves, droughts, untimely snowstorms, bioterrorism/disease outbreaks). Centralized management of manure leads to too much phosphorus in places where it is not needed, while future shortages of phosphate fertilizer seem quite possible (Withers et al., 2018). The qualities of a resilient system may be different from those of highly productive systems, so this kind of planning will take lots of creativity and foresight.

6 Case study: evaluating livestock's contributions to global methane concentrations

Starting in 1999, annual increases in atmospheric methane concentrations 'plateaued' (Dlugokencky et al., 2003; Kai et al., 2011), but after 2006, concentrations resumed their annual increases (Fig. 2), with concurrent changes in the ^{13}C isotopic signature of atmospheric methane (Nisbet et al., 2016; Schaefer et al., 2016). These dynamics have been studied in detail to better understand the methane cycle and to attribute the changes to specific sectors or sinks that could be targeted for mitigation. However, it seems likely that there are multiple causes. Contributions of the livestock sector to the resumption of growth of atmospheric methane concentrations may be substantial (Wolf et al., 2017), but consensus has not been reached.

The sum of bottom-up estimates of methane emissions from all sectors is significantly larger than top-down estimates, although anthropogenic emissions, including livestock, exhibit better agreement at the global level (Saunois et al., 2016). Ongoing measurements, collaboration, and comparison of results from direct measurement, modeling, and estimation methods allow the community to zero in on a consensus. Almost every emissions sector has been suggested as the cause of the atmospheric methane dynamics over the last few decades (growth-plateau-growth), including decreased emissions from rice agriculture (Kai et al., 2011), agricultural sources (rice and/or livestock) in the Northern hemisphere tropics and subtropics (Schaefer et al., 2016), wetland responses to climate in the Southern hemisphere and Northern and Southern tropics (Nisbet et al., 2016), increased fossil fuel emissions with concurrent decreased biomass burning (Hausmann et al., 2016), increased livestock methane emissions (Wolf et al., 2017), and decreased biomass burning emissions alone (Worden et al., 2017). To further complicate the matter, two recent works implicate changes in the strength of the tropospheric OH sink for methane, finding evidence for increasing sink strength until 2004 followed by declining sink strength (Rigby et al., 2017; Turner et al., 2017; reviewed by Prather and Holmes, 2017). The concurrent dynamics of methane emission sources in these years are still unclear, with one group suggesting that emissions grew over the whole period, but the rate of growth slowed in the mid-2000s (Rigby et al., 2017), and the other suggesting another possibility–that methane emissions decreased over the period, but OH sink decreased more–which can also explain the observations (Turner et al., 2017). A third group does not find evidence to support changes in the OH sink, and instead supports increasing wetland, agricultural, and fossil fuel methane emissions causing the rise since 2007 (Thompson et al., 2018). These developments illustrate the complexity of greenhouse gas fluxes between multiple sources and sinks, and that there is still a lot of uncertainty in the magnitude and sometimes even the direction of the impact.

Hristov et al. (2018) point out that global cattle populations increased, while global methane concentrations remained stable from 1999–2006, and that the observed increases in global methane since 2006 cannot therefore be linked to cattle. However, the changes in multiple global sources and sinks described above do not allow clear attribution (Turner et al., 2017) and do not establish any expectations for livestock emissions dynamics. In addition, multiple changes *within* the global livestock system have also occurred over this time period. While global cattle populations did increase, (i) in the United States, cattle numbers remained stable (USDA, 2013), but body mass increased (Wolf et al., 2017); (ii) in Latin America, where per-animal methane emissions are higher, cattle populations increased greatly (FAO, 2014); iii) (intake of by-product feeds such as distillers grains with solubles (DGGS) increased in the United States and in other countries (Fig. 8) (Wolf et al., 2017), which is likely to decrease methane

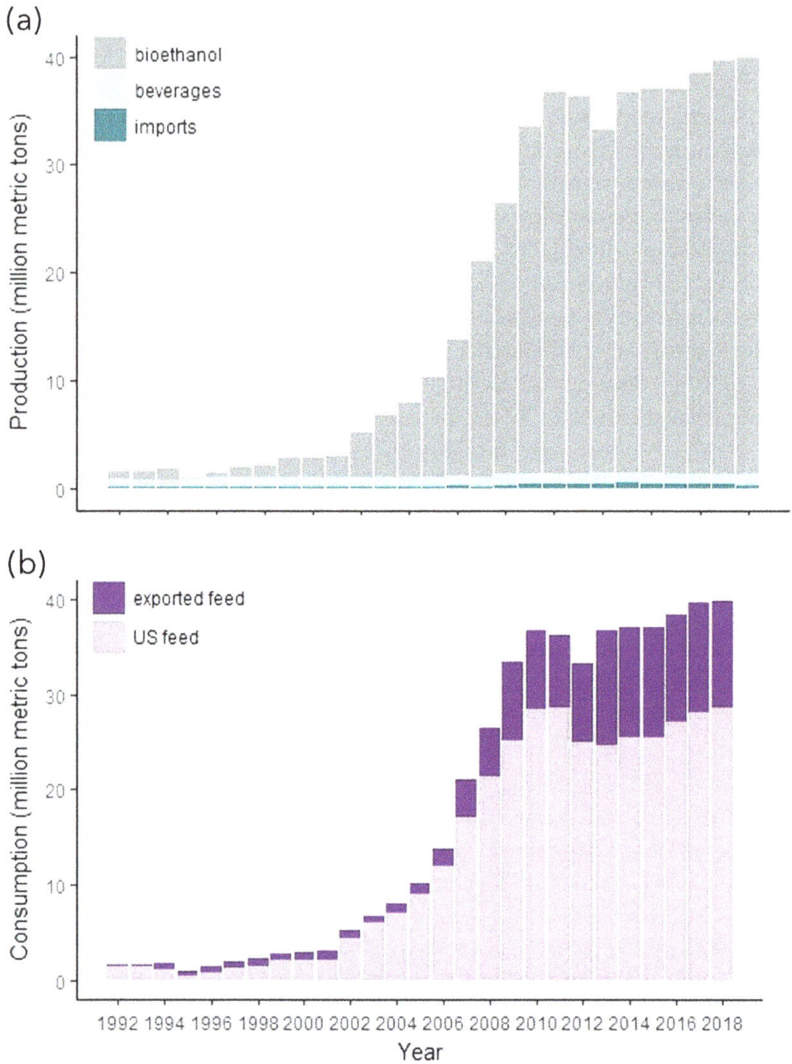

Figure 8 Supply and disappearance of dried distillers grain with solubles (DGGS). Sources of distillers grains (a), and use of those distillers grains for feed in the United States and exported to other nations (b); both show steep growth in the 2000s as the biofuels industry expanded with legislation in the United States (data from USDA ERS–U. S. Bioenergy Statistics, 2018).

emissions (but may not, depending on diet formulation, and may increase nitrous oxide emissions) (Benchaar et al., 2013; Hales et al., 2013; Hünerberg et al., 2013, 2014; McGinn et al., 2009; Penner et al., 2009); (iv) manure management methane emissions increased greatly over this time period, with some regional variability, due in large part to increased use of anaerobic lagoon management of manure, which is poorly characterized both for the rates of usage (Key et al., 2011; Livestock and Poultry Environmental Learning Center, 2018) and default emission factors (Owen and Silver, 2017; Wolf et al., 2017). Therefore, changes in global cattle population alone cannot be used to explain dynamics in the livestock sector's methane emissions over this time period.

These findings do not help discern the trajectory of livestock methane emissions, but they do not negate that methane concentrations have more than doubled in recent decades, or that bottom-up inventories suggest increasing livestock emissions over this time period. Wolf et al. (2017) summed the changes in annual livestock methane emissions over time by latitude (Table 4, Fig. 9) and found the largest increases between 30N and the equator (Northern tropics), which agrees with the conclusions of Schaefer et al. (2016) as well as with multiple global models summarized by Bergamaschi et al. (2013). However, inverse modeling depends on accurate, spatially disaggregated bottom-up inventories, and these are still being refined. For example, Hristov et al. (2017) re-evaluated and spatially disaggregated livestock methane emissions to a fine-scale in the contiguous United States. Their work revealed significant differences from the global gridded Emission Database for Global Atmospheric Research (European Commission, 2012), which has been the go-to emissions database used as priors in inverse modeling assessments of emissions from different sectors. More work is needed to better understand these dynamics, and the need continues for frequent updated activity data and emissions factors, revised inventory-based assessments, and more detailed spatial disaggregation of livestock emissions.

7 Conclusions

Quantification of the livestock sector's impact on climate change is not straightforward. Even for a single emitted gas, methane, best estimates must be based on comparisons of multiple bottom-up and top-down methods, and frequent reassessment is warranted. While there is still disagreement, I believe that livestock are a major contributor to recent increases in atmospheric methane, along with changes in other methane source sectors and sinks. Although overall methane emissions have increased in recent years, emissions intensity has decreased. This decrease in greenhouse gas emissions per unit of food production is due to modernizing livestock production systems, which have made large gains in productivity per animal and per unit of feed consumed. Nevertheless, even highly productive livestock systems have major impacts on climate change and the broader environment, and it is likely that we underestimate these impacts by using the GWP of emitted gases, and by neglecting to consider the loss of carbon storage and photosynthetic capacity

Table 4 Temporal changes in annual livestock CH_4 emissions by latitudinal zone (Note: Tg CH_4 C and Tg CH_4 both shown to facilitate comparisons to other studies) (Wolf et al., 2017). Distributed under Creative Commons Attribution 4.0 International License (http://creativecommons.org/licenses/by/4.0/)

Time period:	90N-60N	60N-30N	30N-equator	Equator-30S	30S-60S	60S-90S
			Tg CH_4 C-yr^{-1}			
2003-2010	−0.08 ± 0.15	1.7 ± 8.96	5.18 ± 9.87	2.01 ± 5.26	−0.16 ± 1.37	0
2000-2013	−0.12 ± 0.16	2.1 ± 8.97	5.59 ± 9.39	4.23 ± 5.19	0.28 ± 1.34	0
			Tg CH_4-yr^{-1}			
2003-2010	−0.11 ± 0.21	2.27 ± 11.95	6.91 ± 13.17	2.69 ± 7.02	−0.21 ± 1.83	0
2000-2013	−0.16 ± 0.21	2.8 ± 11.97	7.45 ± 12.51	5.64 ± 6.92	0.38 ± 1.78	0

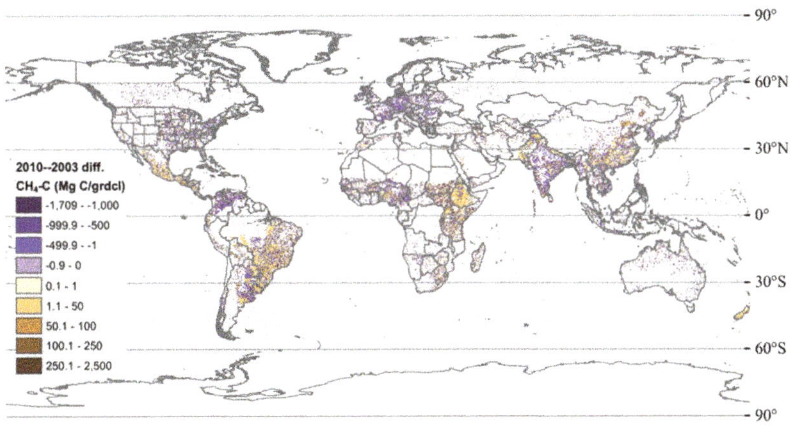

Figure 9 Change in total livestock methane emissions between 2003 and 2010 (2010 [minus] 2003 emissions [in Mg C]), per 0.05 × 0.05° gridcell (Wolf et al., 2017). Distributed under Creative Commons Attribution 4.0 International License (http://creativecommons.org/licenses/by/4.0/).

in vast areas of land used to feed livestock. Life cycle assessments and other methods are evolving to better account for the multiple impacts of livestock per unit of food production, economic return, and/or other benefit. These methods include impacts in multiple areas, such as greenhouse gas emissions, water usage and quality, energy usage, biodiversity losses, eutrophication, pollution, contamination, and land-use change. Studies employing these methods uniformly indicate that over many of these impact areas, the climate impacts of plant-based foods are miniscule relative to those of livestock foods. Among livestock products, the impacts are uniformly rated by LCAs in this order: eggs and dairy products < pork and poultry meat < beef and lamb meat.

While measurements, models, and other analyses do not convey certainty, multiple assessments indicate that increasing growth from the livestock sector is contributing to potentially catastrophic climate change and loss of biodiversity. However, as livestock operations become more intensive, emissions intensity and land-use extent decrease. Continued assessment of the sector, both detailed (e.g. changes in manure management methane emissions) and broad (e.g. displacement of global carbon storage in plant biomass, production system vulnerabilities, trade-offs) should guide public policy and consumer choices.

8 Where to look for further information

The NOAA's Global Greenhouse Gas Reference Network offers the Carbon Tracker, which provides maps, data, and accounting of global carbon dioxide, methane, and other carbon fluxes: https://www.esrl.noaa.gov/gmd/ccgg/carbontracker/.

The FAO's Global Livestock Assessment Model (GLEAM) is a GIS framework that simulates the biophysical processes and activities along livestock supply chains under a life cycle assessment approach. Input data, disaggregated and downscaled output data and maps for the year 2010 are available for download via: http://www.fao.org/gleam/en/.

The FAO's Livestock Environmental Assessment and Performance Partnership (LEAP) is developing publicly available guidelines for assessment of livestock sector impacts on biodiversity, water use, and soil carbon: http://www.fao.org/partnerships/leap/overview/the-partnership/en/.

The Cool Farm Alliance offers the Cool Farm Tool, a free online calculator of greenhouse gas, water, and biodiversity effects related to individual farm-scale livestock management: https://coolfarmtool.org/.

The Carnegie Mellon University offers a free economic input-output life cycle assessment: http://www.eiolca.net/index.html.

9 References

ABC News. 2018. What's behind the government's unprecedented move to buy $50M in milk [WWW Document]. *ABC News*. Available at: https://abcnews.go.com/Politics/us-government-buying-50-million-milk-dairy-farmers/story?id=57180332 (accessed on 20 October 2018).

Alkemade, R., Reid, R. S., Berg, M. van den, Leeuw, J. de and Jeuken, M. 2013. Assessing the impacts of livestock production on biodiversity in rangeland ecosystems. *Proc. Natl. Acad. Sci. U.S.A.* 110(52), 20900–5. doi:10.1073/pnas.1011013108.

Allen, M. R., Fuglestvedt, J. S., Shine, K. P., Reisinger, A., Pierrehumbert, R. T. and Forster, P. M. 2016. New use of global warming potentials to compare cumulative and short-lived climate pollutants. *Nat. Clim. Change* 6(8), 773–6. doi:10.1038/nclimate2998.

Asner, G. P., Elmore, A. J., Olander, L. P., Martin, R. E. and Harris, A. T. 2004. Grazing systems, ecosystem responses, and global change. *Annu. Rev. Environ. Resour.* 29(1), 261–99. doi:10.1146/annurev.energy.29.062403.102142.

Balmford, A., Green, R. and Phalan, B. 2012. What conservationists need to know about farming. *Proc. R. Soc. B-Biol. Sci.* 279(1739), 2714–24. doi:10.1098/rspb.2012.0515.

Balmford, A., Amano, T., Bartlett, H., Chadwick, D., Collins, A., Edwards, D., Field, R., Garnsworthy, P., Green, R., Smith, P., Waters, H., Whitmore, A., Broom, D. M., Chara, J., Finch, T., Garnett, E., Gathorne-Hardy, A., Hernandez-Medrano, J., Herrero, M., Hua, F., Latawiec, A., Misselbrook, T., Phalan, B., Simmons, B. I., Takahashi, T., Vause, J., Ermgassen, E. zu and Eisner, R. 2018. The environmental costs and benefits of high-yield farming. *Natl. Sustain.* 1(9), 477–85. doi:10.1038/s41893-018-0138-5.

Barkham, P. 2018. Eight bird species are first confirmed avian extinctions this decade. *The Guardian*.

Bar-On, Y. M., Phillips, R. and Milo, R. 2018. The biomass distribution on Earth. *Proc. Natl. Acad. Sci. U.S.A.* 115(25), 6506–11. doi:10.1073/pnas.1711842115 29784790.

Benchaar, C., Hassanat, F., Gervais, R., Chouinard, P. Y., Julien, C., Petit, H. V. and Massé, D. I. 2013. Effects of increasing amounts of corn dried distiller grains with solubles in dairy cow diets on methane production, ruminal fermentation, digestion,

N balance, and milk production. *J. Dairy Sci.* 96(4), 2413-27. doi:10.3168/jds.2012-6037.
Bergamaschi, P., Houweling, S., Segers, A., Krol, M., Frankenberg, C., Scheepmaker, R. A., Dlugokencky, E., Wofsy, S. C., Kort, E. A., Sweeney, C., Schuck, T., Brenninkmeijer, C., Chen, H., Beck, V. and Gerbig, C. 2013. Atmospheric CH_4 in the first decade of the 21st century: Inverse modeling analysis using SCIAMACHY satellite retrievals and NOAA surface measurements. *J. Geophys. Res. Atmos.* 118(13), 7350-69. doi:10.1002/jgrd.50480.
Boehm, R., Wilde, P. E., Ver Ploeg, M., Costello, C. and Cash, S. B. 2018. A comprehensive life cycle assessment of greenhouse gas emissions from U.S. household food choices. *Food Policy* 79, 67-76. doi:10.1016/j.foodpol.2018.05.004.
Capper, J. L. 2011. The environmental impact of beef production in the United States: 1977 compared with 2007. *J. Anim. Sci.* 89(12), 4249-61. doi:10.2527/jas.2010-3784.
Chastain, J. P. and Henry, S. 2002a. Management of lagoon and storage structures for swine manure. In: *Confined Animal Manure Managers Swine Training Manual*. Clemson University Cooperative Extension.
Chastain, J. P. and Henry, S. 2002b. Management of lagoon and storage structures for dairy manure. In: *Confined Animal Manure Managers Dairy Training Manual*. Clemson University Cooperative Extension.
Clark, M. and Tilman, D. 2017. Comparative analysis of environmental impacts of agricultural production systems, agricultural input efficiency, and food choice. *Environ. Res. Lett.* 12(6). doi:10.1088/1748-9326/aa6cd5, 064016.
Dlugokencky, E. J., Houweling, S., Bruhwiler, L., Masarie, K. A., Lang, P. M., Miller, J. B. and Tans, P. P. 2003. Atmospheric methane levels off: temporary pause or a new steady-state? *Geophys. Res. Lett.* 30(19), 1992. doi:10.1029/2003GL018126.
Dlugokencky, E. J., Nisbet, E. G., Fisher, R. and Lowry, D. 2011. Global atmospheric methane: budget, changes and dangers. *Philos. Trans. A Math. Phys. Eng. Sci.* 369(1943), 2058-72. doi:10.1098/rsta.2010.0341.
Dove, R. 2018. Perspective | I saw Florence sending millions of gallons of animal feces flooding across North Carolina. *Washington Post*.
Eaton, S. 2018. Tropical forests are flipping from storing carbon to releasing it. *The Nation*.
Emery, I. 2018. Without animals, US farmers would reduce feed crop production. *Proc. Natl. Acad. Sci. U.S.A.* 115(8), E1703-. doi:10.1073/pnas.1720760115.
EPA. 2016. Inventory of U.S. greenhouse gas emissions and sinks: 1990-2014 [WWW Document]. *US EPA*. Available at: https://www.epa.gov/ghgemissions/inventory-us-greenhouse-gas-emissions-and-sinks-1990-2014 (accessed on 29 May 2017).
Erb, K. H., Kastner, T., Plutzar, C., Bais, A. L. S., Carvalhais, N., Fetzel, T., Gingrich, S., Haberl, H., Lauk, C., Niedertscheider, M., Pongratz, J., Thurner, M. and Luyssaert, S. 2018. Unexpectedly large impact of forest management and grazing on global vegetation biomass. *Nature* 553(7686), 73-6. doi:10.1038/nature25138.
European Commission. 2012. EDGAR: the emissions database for global atmospheric research, part III [WWW Document]. *Greenhouse Gas Emissions*. Available at: http://edgar.jrc.ec.europa.eu/docs/IEA_PARTIII.pdf (accessed on 03 November 2016).
FAO. 2014. Food and Agriculture Organization of the United Nations Statistics Division (FAOSTAT) [WWW Document]. Available at: http://faostat.fao.org/ (accessed on 01 December 2015).

FAO. 2018. Results | Global Livestock Environmental Assessment Model (GLEAM) [WWW Document]. *Food and Agriculture Organization of the United Nations*. Available at: http://www.fao.org/gleam/results/en/ (accessed on 30 October 2018).

Fuller, D. Q., van Etten, J., Manning, K., Castillo, C., Kingwell-Banham, E., Weisskopf, A., Qin, L., Sato, Y.-I. and Hijmans, R. J. 2011. The contribution of rice agriculture and livestock pastoralism to prehistoric methane levels: an archaeological assessment. *Holocene* 21(5), 743–59. doi:10.1177/0959683611398052.

Gerber, P., Vellinga, T., Opio, C., Henderson, B. and Steinfeld, H. 2010. *Greenhouse Gas Emissions from the Dairy Sector: a Life Cycle Assessment*.

Gerber, P., Steinfeld, H., Henderson, B., Mottet, A., Opio, C., Dijkman, J., Falcucci, A. and Tempio, G. 2013. *Tackling Climate Change through Livestock – A Global Assessment of Emissions and Mitigation Opportunities*. FAO, 115p.

Ghosh, A., Patra, P. K., Ishijima, K., Umezawa, T., Ito, A., Etheridge, D. M., Sugawara, S., Kawamura, K., Miller, J. B., Dlugokencky, E. J., Krummel, P. B., Fraser, P. J., Steele, L. P., Langenfelds, R. L., Trudinger, C. M., White, J. W. C., Vaughn, B., Saeki, T., Aoki, S. and Nakazawa, T. 2015. Variations in global methane sources and sinks during 1910–2010. *Atmos. Chem. Phys.* 15(5), 2595–612. doi:10.5194/acp-15-2595-2015.

Goodland, R. 2013. Lifting livestock's long shadow. *Natl. Clim. Change* 3(1), 2. doi:10.1038/nclimate1755.

Goodland, R. 2014. A fresh look at livestock greenhouse gas emissions and mitigation potential in Europe. *Glob. Change Biol.* 20(7), 2042–4. doi:10.1111/gcb.12454.

Hales, K. E., Cole, N. A. and MacDonald, J. C. 2013. Effects of increasing concentrations of wet distillers grains with solubles in steam-flaked, corn-based diets on energy metabolism, carbon-nitrogen balance, and methane emissions of cattle. *J. Anim. Sci.* 91(2), 819–28. doi:10.2527/jas.2012-5418.

Hausmann, P., Sussmann, R. and Smale, D. 2016. Contribution of oil and natural gas production to renewed increase in atmospheric methane (2007–2014): top–down estimate from ethane and methane column observations. *Atmos. Chem. Phys.* 16(5), 3227–44. doi:10.5194/acp-16-3227-2016.

Hayek, M. N. and Garrett, R. D. 2018. Nationwide shift to grass-fed beef requires larger cattle population. *Environ. Res. Lett.* 13(8). doi:10.1088/1748-9326/aad401, 084005.

Hein, R., Crutzen, P. J. and Heimann, M. 1997. An inverse modeling approach to investigate the global atmospheric methane cycle. *Glob. Biogeochem. Cycles* 11(1), 43–76. doi:10.1029/96GB03043.

Hempson, G. P., Archibald, S. and Bond, W. J. 2017. The consequences of replacing wildlife with livestock in Africa. *Sci. Rep.* 7(1), 17196. doi:10.1038/s41598-017-17348-4.

Herrero, M. and Thornton, P. K. 2013. Livestock and global change: emerging issues for sustainable food systems. *Proc. Natl. Acad. Sci. U.S.A.* 110(52), 20878–81. doi:10.1073/pnas.1321844111.

Herrero, M., Havlík, P., Valin, H., Notenbaert, A., Rufino, M. C., Thornton, P. K., Bluemmel, M., Weiss, F., Grace, D. and Obersteiner, M. 2013. Biomass use, production, feed efficiencies, and greenhouse gas emissions from global livestock systems. *Proc. Natl. Acad. Sci. U.S.A.* 110(52), 20888–93. doi:10.1073/pnas.1308149110.

Herrero, M., Henderson, B., Havlík, P., Thornton, P. K., Conant, R. T., Smith, P., Wirsenius, S., Hristov, A. N., Gerber, P., Gill, M., Butterbach-Bahl, K., Valin, H., Garnett, T. and Stehfest, E. 2016. Greenhouse gas mitigation potentials in the livestock sector. *Nat. Clim. Change* 6(5), 452–61. doi:10.1038/nclimate2925.

Hooke, R. L. and Martín-Duque, J. F. 2012 Land transformation by humans: a review. *GSA Today* 12(12), 4–10. doi:10.1130/GSAT151A.1.
Houweling, S., Kaminski, T., Dentener, F., Lelieveld, J. and Heimann, M. 1999. Inverse modeling of methane sources and sinks using the adjoint of a global transport model. *J. Geophys. Res.* 104(D21), 26137–60. doi:10.1029/1999JD900428.
Hristov, A. N., Johnson, K. A. and Kebreab, E. 2014. Livestock methane emissions in the United States. *Proc. Natl. Acad. Sci. U.S.A.* 111(14), E1320–. doi:10.1073/pnas.1401046111.
Hristov, A. N., Harper, M., Meinen, R., Day, R., Lopes, J., Ott, T., Venkatesh, A. and Randles, C. A. 2017. Discrepancies and uncertainties in bottom-up gridded inventories of livestock methane emissions for the contiguous United States. *Environ. Sci. Technol.* 51(23), 13668–77. doi:10.1021/acs.est.7b03332.
Hristov, A. N., Kebreab, E., Niu, M., Oh, J., Bannink, A., Bayat, A. R., Boland, T. M., Brito, A. F., Casper, D. P., Crompton, L. A., Dijkstra, J., Eugène, M., Garnsworthy, P. C., Haque, N., Hellwing, A. L. F., Huhtanen, P., Kreuzer, M., Kuhla, B., Lund, P., Madsen, J., Martin, C., Moate, P. J., Muetzel, S., Muñoz, C., Peiren, N., Powell, J. M., Reynolds, C. K., Schwarm, A., Shingfield, K. J., Storlien, T. M., Weisbjerg, M. R., Yáñez-Ruiz, D. R. and Yu, Z. 2018. Symposium review: uncertainties in enteric methane inventories, measurement techniques, and prediction models. *J. Dairy Sci.* 101(7), 6655–74. doi:10.3168/jds.2017-13536.
Hünerberg, M., McGinn, S. M., Beauchemin, K. A., Okine, E. K., Harstad, O. M. and McAllister, T. A. 2013. Effect of dried distillers grains plus solubles on enteric methane emissions and nitrogen excretion from growing beef cattle. *J. Anim. Sci.* 91(6), 2846–57. doi:10.2527/jas.2012-5564.
Hünerberg, M., Little, S. M., Beauchemin, K. A., McGinn, S. M., O'Connor, D., Okine, E. K., Harstad, O. M., Kröbel, R. and McAllister, T. A. 2014. Feeding high concentrations of corn dried distillers' grains decreases methane, but increases nitrous oxide emissions from beef cattle production. *Agric. Syst.* 127, 19–27. doi:10.1016/j.agsy.2014.01.005.
Hyland, J. J., Styles, D., Jones, D. L. and Williams, A. P. 2016. Improving livestock production efficiencies presents a major opportunity to reduce sectoral greenhouse gas emissions. *Agric. Syst.* 147, 123–31. doi:10.1016/j.agsy.2016.06.006.
INPE. 2016. PRODES estima 7.989 km^2 de desmatamento por corte raso na Amazônia em 2016 [WWW Document]. *Braz. Natl. Inst. Space Res. Minist. Sci. Technol. Innov. Commun.* Available at: http://www.inpe.br/noticias/noticia.php?Cod_Noticia=4344 (accessed on 24 October 2018).
Investor's Business Daily. 2018. Dairy subsidies have created a mountain of surplus cheese [WWW Document]. *Investor's Business Daily*. Available at: https://www.investors.com/politics/editorials/dairy-subsidies-government-farm-programs-surplus-cheese/ (accessed on 20 October 2018).
IPCC. 1997. Volume 3 (reference manual), Chapter 4: agriculture. In: *IPCC Good Practice Guidance and Uncertainty Management in National Greenhouse Gas Inventories*. OECD/ODCE, Paris.
IPCC. 2000. *Good Practice Guidance for Uncertainty Management in Natinal Greenhouse Gas Inventories*. Kanagawa, Japan.
IPCC. 2006. *2006 IPCC Guidelines for National Greenhouse Gas Inventories, Volume 4: Agriculture, Forestry and Other Land Use*. Institute for Global Environmental Strategies, Kanagawa, Japan.

IPCC - SR15. 2018. [WWW Document]. Available at: https://www.ipcc.ch/report/sr15/ (accessed on 25 October 2018).

ISO. 2006a. ISO 14040: environmental management – life cycle assessment – principles and framework [WWW Document]. Available at: https://www.iso.org/obp/ui/#iso:std:iso:14040:ed-2:v1:en (accessed on 28 October 2018).

ISO. 2006b. ISO 14044: environmental management – life cycle assessment – requirements and guidelines [WWW Document]. Available at: https://www.iso.org/obp/ui/#iso:std:iso:14044:ed-1:v1:en (accessed on 28 October 2018).

Kai, F. M., Tyler, S. C., Randerson, J. T. and Blake, D. R. 2011. Reduced methane growth rate explained by decreased Northern Hemisphere microbial sources. *Nature* 476(7359), 194–7. doi:10.1038/nature10259.

Key, N., McBride, W. D., Ribaudo, M. and Sneeringer, S. 2011. Trends and developments in hog manure management: 1998–2009 [WWW Document]. Available at: https://www.ers.usda.gov/publications/pub-details/?pubid=44580 (accessed on 27 October 2018).

Kirschke, S., Bousquet, P., Ciais, P., Saunois, M., Canadell, J. G., Dlugokencky, E. J., Bergamaschi, P., Bergmann, D., Blake, D. R., Bruhwiler, L., Cameron-Smith, P., Castaldi, S., Chevallier, F., Feng, L., Fraser, A., Heimann, M., Hodson, E. L., Houweling, S., Josse, B., Fraser, P. J., Krummel, P. B., Lamarque, J.-F., Langenfelds, R. L., Le Quéré, C., Naik, V., O'Doherty, S., Palmer, P. I., Pison, I., Plummer, D., Poulter, B., Prinn, R. G., Rigby, M., Ringeval, B., Santini, M., Schmidt, M., Shindell, D. T., Simpson, I. J., Spahni, R., Steele, L. P., Strode, S. A., Sudo, K., Szopa, S., van der Werf, G. R., Voulgarakis, A., van Weele, M., Weiss, R. F., Williams, J. E. and Zeng, G. 2013. Three decades of global methane sources and sinks. *Nat. Geosci.* 6(10), 813–23. doi:10.1038/ngeo1955.

Klevenhusen, F., Bernasconi, S. M., Kreuzer, M. and Soliva, C. R. 2010. Experimental validation of the Intergovernmental Panel on Climate Change default values for ruminant-derived methane and its carbon-isotope signature. *Anim. Prod. Sci.* 50(3), 159. doi:10.1071/AN09112.

Lassey, K. R. 2007. Livestock methane emission: from the individual grazing animal through national inventories to the global methane cycle. *Agric. For. Meteorol.* 142(2–4), 120–32. doi:10.1016/j.agrformet.2006.03.028.

Leip, A., Billen, G., Garnier, J., Grizzetti, B., Lassaletta, L., Reis, S., Simpson, D., Sutton, M. A., de Vries, W., Weiss, F. and Westhoek, H. 2015. Impacts of European livestock production: nitrogen, sulphur, phosphorus and greenhouse gas emissions, land-use, water eutrophication and biodiversity. *Environ. Res. Lett.* 10(11). doi:10.1088/1748-9326/10/11/115004, 115004.

Liu, C. M., Stegger, M., Aziz, M., Johnson, T. J., Waits, K., Nordstrom, L., Gauld, L., Weaver, B., Rolland, D., Statham, S., Horwinski, J., Sariya, S., Davis, G. S., Sokurenko, E., Keim, P., Johnson, J. R. and Price, L. B. 2018. Escherichia coli ST131-H22 as a foodborne uropathogen. *mBio* 9(4), e00470-18. doi:10.1128/mBio.00470-18.

Livestock and Poultry Environmental Learning Center. 2018. Animal agriculture in the U.S. - trends in production and manure management - eXtension [WWW Document]. Available at: https://articles.extension.org/pages/74481/animal-agriculture-in-the-us-trends-in-production-and-manure-management (accessed on 27 October 2018).

MacDonald, J. and McBride, W. 2009. The transformation of U.S. livestock agriculture: scale, efficiency, and risks [WWW Document]. USDA Economic Research Service

- EIB43. Available at: http://www.ers.usda.gov/publications/eib-economic-information-bulletin/eib43.aspx (accessed on 29 January 2014).

MacDonald, J., Ribaudo, M., Livingston, M., Beckman, J. and Huang, W.-Y. 2009. *Manure Use for Fertilizer and for Energy: Report to Congress USDA ERS - AP-037*.

MacLeod, M. J., Vellinga, T., Opio, C., Falcucci, A., Tempio, G., Henderson, B., Makkar, H., Mottet, A., Robinson, T., Steinfeld, H. and Gerber, P. J. 2018. Invited review: a position on the Global Livestock Environmental Assessment Model (GLEAM). *Animal* 12(2), 383-97. doi:10.1017/S1751731117001847.

McGinn, S. M., Chung, Y. -H., Beauchemin, K. A., Iwaasa, A. D. and Grainger, C. 2009. Use of corn distillers' dried grains to reduce enteric methane loss from beef cattle. *Can. J. Anim. Sci.* 89(3), 409-13. doi:10.4141/CJAS08133.

McKenna, M. 2018. The hidden link between farm antibiotics and human illness. *Wired*.

Meerbeek, K. V. and Svenning, J. C. 2018. Causing confusion in the debate about the transition toward a more plant-based diet. *Proc. Natl. Acad. Sci. U.S.A.* 115(8), E1701-2. doi:10.1073/pnas.1720738115.

Merrill, D. and Leatherby, L. 2018. *Here's How America Uses Its Land*. Bloomberg Press.

Miller, S. M., Wofsy, S. C., Michalak, A. M., Kort, E. A., Andrews, A. E., Biraud, S. C., Dlugokencky, E. J., Eluszkiewicz, J., Fischer, M. L., Janssens-Maenhout, G., Miller, B. R., Miller, J. B., Montzka, S. A., Nehrkorn, T. and Sweeney, C. 2013. Anthropogenic emissions of methane in the United States. *Proc. Natl. Acad. Sci. U.S.A.* 110(50), 20018-22. doi:10.1073/pnas.1314392110.

Miller, S. M., Michalak, A. M. and Wofsy, S. C. 2014. Linking methane emissions inventories with atmospheric observations. *Proc. Natl. Acad. Sci. U.S.A.* 111(14), E1321. doi:10.1073/pnas.1401703111.

Mooney, C. 2016. Destruction of the Amazon is speeding up – just when the planet can least afford it. *Washington Post*.

Nisbet, E. G., Dlugokencky, E. J., Manning, M. R., Lowry, D., Fisher, R. E., France, J. L., Michel, S. E., Miller, J. B., White, J. W. C., Vaughn, B., Bousquet, P., Pyle, J. A., Warwick, N. J., Cain, M., Brownlow, R., Zazzeri, G., Lanoisellé, M., Manning, A. C., Gloor, E., Worthy, D. E. J., Brunke, E. -G., Labuschagne, C., Wolff, E. W. and Ganesan, A. L. 2016. Rising atmospheric methane: 2007-2014 growth and isotopic shift. *Glob. Biogeochem. Cycles* 30(9), 1356-70. doi:10.1002/2016GB005406.

Olander, L. P., Wollenberg, E., Tubiello, F. N. and Herold, M. 2014. Synthesis and review: advancing agricultural greenhouse gas quantification. *Environ. Res. Lett.* 9(7), 075003. doi:10.1088/1748-9326/9/7/075003.

Owen, J. J. and Silver, W. L. 2015. Greenhouse gas emissions from dairy manure management: a review of field-based studies. *Glob. Change Biol.* 21(2), 550-65. doi:10.1111/gcb.12687.

Owen, J. J. and Silver, W. L. 2017. Greenhouse gas emissions from dairy manure management in a Mediterranean environment. *Ecol. Appl.* 27(2), 545-59. doi:10.1002/eap.1465.

Penner, G. B., Yu, P. and Christensen, D. A. 2009. Effect of replacing forage or concentrate with wet or dry distillers' grains on the productivity and chewing activity of dairy cattle. *Anim. Feed Sci. Technol.* 153(1-2), 1-10. doi:10.1016/j.anifeedsci.2009.05.006.

Petherick, A. 2012. Light is cast on a long shadow. *Natl. Clim. Change* 2(10), 705-6. doi:10.1038/nclimate1703.

Phelps, L. N. and Kaplan, J. O. 2017. Land use for animal production in global change studies: defining and characterizing a framework. *Glob. Change Biol.* 23(11), 4457-71. doi:10.1111/gcb.13732.

Pongratz, J., Reick, C., Raddatz, T. and Claussen, M. 2008. A reconstruction of global agricultural areas and land cover for the last millennium. *Glob. Biogeochem. Cycles* 22(3). doi:10.1029/2007GB003153.

Poore, J. and Nemecek, T. 2018. Reducing food's environmental impacts through producers and consumers. *Science* 360(6392), 987-92. doi:10.1126/science.aaq0216.

Prather, M. J. and Holmes, C. D. 2017. Overexplaining or underexplaining methane's role in climate change. *Proc. Natl. Acad. Sci. U.S.A.* 114(21), 5324-6. doi:10.1073/pnas.1704884114.

Pratt, C., Redding, M., Hill, J. and Jensen, P. D. 2015. Does manure management affect the latent greenhouse gas emitting potential of livestock manures? *Waste Manag.* 46, 568-76. doi:10.1016/j.wasman.2015.08.019.

Reisinger, A. and Clark, H. 2018. How much do direct livestock emissions actually contribute to global warming? *Glob. Change Biol.* 24(4), 1749-61. doi:10.1111/gcb.13975.

Rigby, M., Montzka, S. A., Prinn, R. G., White, J. W. C., Young, D., O'Doherty, S., Lunt, M. F., Ganesan, A. L., Manning, A. J., Simmonds, P. G., Salameh, P. K., Harth, C. M., Mühle, J., Weiss, R. F., Fraser, P. J., Steele, L. P., Krummel, P. B., McCulloch, A. and Park, S. 2017. Role of atmospheric oxidation in recent methane growth. *Proc. Natl. Acad. Sci. U.S.A.* 114(21), 5373-7. doi:10.1073/pnas.1616426114.

Sanderman, J., Hengl, T. and Fiske, G. J. 2017. Soil carbon debt of 12,000 years of human land use. *Proc. Natl. Acad. Sci. U.S.A.* 114(36), 9575-80. doi:10.1073/pnas.1706103114.

Sapart, C. J., Monteil, G., Prokopiou, M., Wal, R. S. W. van de, Kaplan, J. O., Sperlich, P., Krumhardt, K. M., Veen, C. van der, Houweling, S., Krol, M. C., Blunier, T., Sowers, T., Martinerie, P., Witrant, E., Dahl-Jensen, D. and Röckmann, T. 2012. Natural and anthropogenic variations in methane sources during the past two millennia. *Nature* 490(7418), 85-8. doi:10.1038/nature11461.

Saunois, M., Bousquet, P., Poulter, B., Peregon, A., Ciais, P., Canadell, J. G., Dlugokencky, E. J., Etiope, G., Bastviken, D., Houweling, S., Janssens-Maenhout, G., Tubiello, F. N., Castaldi, S., Jackson, R. B., Alexe, M., Arora, V. K., Beerling, D. J., Bergamaschi, P., Blake, D. R., Brailsford, G., Brovkin, V., Bruhwiler, L., Crevoisier, C., Crill, P., Curry, C., Frankenberg, C., Gedney, N., Höglund-Isaksson, L., Ishizawa, M., Ito, A., Joos, F., Kim, H.-S., Kleinen, T., Krummel, P., Lamarque, J.-F., Langenfelds, R., Locatelli, R., Machida, T., Maksyutov, S., McDonald, K. C., Marshall, J., Melton, J. R., Morino, I., O'Doherty, S., Parmentier, F.-J. W., Patra, P. K., Peng, C., Peng, S., Peters, G. P., Pison, I., Prigent, C., Prinn, R., Ramonet, M., Riley, W. J., Saito, M., Schroeder, R., Simpson, I. J., Spahni, R., Steele, P., Takizawa, A., Thornton, B. F., Tian, H., Tohjima, Y., Viovy, N., Voulgarakis, A., Weele, M. van, Werf, G. van der, Weiss, R., Wiedinmyer, C., Wilton, D. J., Wiltshire, A., Worthy, D., Wunch, D. B., Xu, X., Yoshida, Y., Zhang, B., Zhang, Z. and Zhu, Q. 2016. The global methane budget: 2000-2012. *Earth Syst. Sci. Data* 8, 697-751. doi:10.5194/essd-2016-25.

Schaefer, H., Fletcher, S. E. M., Veidt, C., Lassey, K. R., Brailsford, G. W., Bromley, T. M., Dlugokencky, E. J., Michel, S. E., Miller, J. B., Levin, I., Lowe, D. C., Martin, R. J., Vaughn, B. H. and White, J. W. C. 2016. A 21st-century shift from fossil-fuel to

biogenic methane emissions indicated by $^{13}CH_4$. *Science* 352, 80–4. doi:10.1126/science.aad2705.
Sherwood, O. A., Schwietzke, S., Arling, V. A. and Etiope, G. 2017. Global inventory of gas geochemistry data from fossil fuel, microbial and burning sources, version 2017. *Earth Syst. Sci. Data* 9, 639–56. doi:10.5194/essd-9-639-2017.
Smith, P. 2018. Managing the global land resource. *Proc. R. Soc. B* 285, 20172798. doi:10.1098/rspb.2017.2798.
Springmann, M., Clark, M., Mason-D'Croz, D., Wiebe, K., Bodirsky, B. L., Lassaletta, L., Vries, W. de, Vermeulen, S. J., Herrero, M., Carlson, K. M., Jonell, M., Troell, M., DeClerck, F., Gordon, L. J., Zurayk, R., Scarborough, P., Rayner, M., Loken, B., Fanzo, J., Godfray, H. C. J., Tilman, D., Rockström, J. and Willett, W. 2018a. Options for keeping the food system within environmental limits. *Nature* 562(7728), 519–25. doi:10.1038/s41586-018-0594-0.
Springmann, M., Clark, M. and Willett, W. 2018b. Feedlot diet for Americans that results from a misspecified optimization algorithm. *Proc. Natl. Acad. Sci. U.S.A.* 115(8), E1704–5. doi:10.1073/pnas.1721335115.
Stiner, M. C., Buitenhuis, H., Duru, G., Kuhn, S. L., Mentzer, S. M., Munro, N. D., Pöllath, N., Quade, J., Tsartsidou, G. and Özbaşaran, M. 2014. A forager-herder trade-off, from broad-spectrum hunting to sheep management at Aşıklı Höyük, Turkey. *Proc. Natl. Acad. Sci. U.S.A.* 111(23), 8404–9. doi:10.1073/pnas.1322723111.
Thompson, R. L., Nisbet, E. G., Pisso, I., Stohl, A., Blake, D., Dlugokencky, E. J., Helmig, D. and White, J. W. C. 2018. Variability in atmospheric methane from fossil fuel and microbial sources over the last three decades. *Geophys. Res. Lett.* 45(20) doi:10.1029/2018GL078127.
Turner, A. J., Jacob, D. J., Wecht, K. J., Maasakkers, J. D., Lundgren, E., Andrews, A. E., Biraud, S. C., Boesch, H., Bowman, K. W., Deutscher, N. M., Dubey, M. K., Griffith, D. W. T., Hase, F., Kuze, A., Notholt, J., Ohyama, H., Parker, R., Payne, V. H., Sussmann, R., Sweeney, C., Velazco, V. A., Warneke, T., Wennberg, P. O. and Wunch, D. 2015. Estimating global and North American methane emissions with high spatial resolution using GOSAT satellite data. *Atmos. Chem. Phys.* 15(12), 7049–69. doi:10.5194/acp-15-7049-2015.
Turner, A. J., Frankenberg, C., Wennberg, P. O. and Jacob, D. J. 2017. Ambiguity in the causes for decadal trends in atmospheric methane and hydroxyl. *Proc. Natl. Acad. Sci. U.S.A.* 114(21), 5367–72. doi:10.1073/pnas.1616020114.
United Nations Framework Convention on Climate Change. 2017. National inventory submissions 2017 [WWW Document]. Available at: http://unfccc.int/national_reports/annex_i_ghg_inventories/national_inventories_submissions/items/10116.php (accessed on 27 May 2017).
USDA. 2013. Quickstats 2.0 [WWW Document]. *National Agricultural Statistics Service*. Available at: http://quickstats.nass.usda.gov/ (accessed on 11.26.13).
USDA ERS. 2018. U.S. bioenergy statistics [WWW Document]. Available at: https://www.ers.usda.gov/data-products/us-bioenergy-statistics/us-bioenergy-statistics/#Coproducts (accessed on 10.11.18).
US EPA. 2016. Understanding global warming potentials [WWW Document]. *US EPA*. Available at: https://www.epa.gov/ghgemissions/understanding-global-warming-potentials (accessed on 10.28.18). US EPA.
VanderZaag, A. C., MacDonald, J. D., Evans, L., Verge, X. P. C. and Desjardins, R. L. 2013. Towards an inventory of methane emissions from manure management that is

responsive to changes on Canadian farms. *Environ. Res. Lett.* 8(3). doi:10.1088/1748-9326/8/3/035008, 035008.

Wecht, K. J., Jacob, D. J., Frankenberg, C., Jiang, Z. and Blake, D. R. 2014a. Mapping of North American methane emissions with high spatial resolution by inversion of SCIAMACHY satellite data. *J. Geophys. Res. Atmos.* 119(12), 7741–56. doi:10.1002/2014JD021551.

Wecht, K. J., Jacob, D. J., Sulprizio, M. P., Santoni, G. W., Wofsy, S. C., Parker, R., Boesch, H. and Worden, J. 2014b. Spatially resolving methane emissions in California: constraints from the CalNex aircraft campaign and from present (GOSAT, TES) and future (TROPOMI, geostationary) satellite observations. *Atmos. Chem. Phys.* 14(15), 8173–84. doi:10.5194/acp-14-8173-2014.

White, R. R. and Hall, M. B. 2017. Nutritional and greenhouse gas impacts of removing animals from US agriculture. *Proc. Natl. Acad. Sci. U.S.A.* 114(48), E10301–8. doi:10.1073/pnas.1707322114.

White, R. R. and Hall, M. B. 2018. Reply to Van Meerbeek and Svenning, Emery, and Springmann et al.: clarifying assumptions and objectives in evaluating effects of food system shifts on human diets. *Proc. Natl. Acad. Sci. U.S.A.* 115(8), E1706–8. doi:10.1073/pnas.1720895115.

Withers, P., Doody, D. and Sylvester-Bradley, R. 2018. Achieving sustainable phosphorus use in food systems through circularisation. *Sustainability* 10(6), 1804. doi:10.3390/su10061804.

Wolf, J., Asrar, G. R. and West, T. O. 2017. Revised methane emissions factors and spatially distributed annual carbon fluxes for global livestock. *Carbon Balance Manag.* 12(1), 16. doi:10.1186/s13021-017-0084-y.

Worden, J. R., Bloom, A. A., Pandey, S., Jiang, Z., Worden, H. M., Walker, T. W., Houweling, S. and Röckmann, T. 2017. Reduced biomass burning emissions reconcile conflicting estimates of the post-2006 atmospheric methane budget. *Nat. Commun.* 8(1), 2227. doi:10.1038/s41467-017-02246-0.

Xue, B., Wang, L. Z. and Yan, T. 2014. Methane emission inventories for enteric fermentation and manure management of yak, buffalo and dairy and beef cattle in China from 1988 to 2009. *Agric. Ecosyst. Environ.* 195, 202–10. doi:10.1016/j.agee.2014.06.002.

Zhou, M., Zhu, B., Wang, S., Zhu, X., Vereecken, H. and Brüggemann, N. 2017. Stimulation of N_2O emission by manure application to agricultural soils may largely offset carbon benefits: a global meta-analysis. *Glob. Change Biol.* 23(10), 4068–83. doi:10.1111/gcb.13648.

Chapter 5

The role of crop cultivation in contributing to climate change

Sonali Shukla McDermid and David Kanter, New York University, USA

1 Introduction

While there has been much research emphasis on quantifying climate change impacts on food, fuel, and fiber production (e.g. Rosenzweig et al., 2014), agriculture itself is a major contributor to global environmental change (Smith et al., 2013a,b). Agriculture has been identified as a major driver of Earth, exceeding key 'planetary boundaries', including biogeochemical cycling, biodiversity, water availability, quality, and among others (Campbell et al., 2017; Conijn et al., 2018). It is thus imperative that as we develop and adapt agricultural systems to efficiently provide food and nutrients for nine to ten billion people by 2050, we also consider and quantify agriculture's environmental and climate ramifications in an effort to reduce or eliminate deleterious impacts and foster more responsible and restorative practices.

Our agricultural footprint now occupies nearly 40% of the ice-free land surface (Fig. 1) and has experienced large gains in productivity over the latter half of the twentieth century, equating to a near doubling of production (Burney et al., 2010). These gains have resulted from a host of technological developments, including improved crop varieties, access to groundwater irrigation, and

http://dx.doi.org/10.19103/AS.2020.0064.09

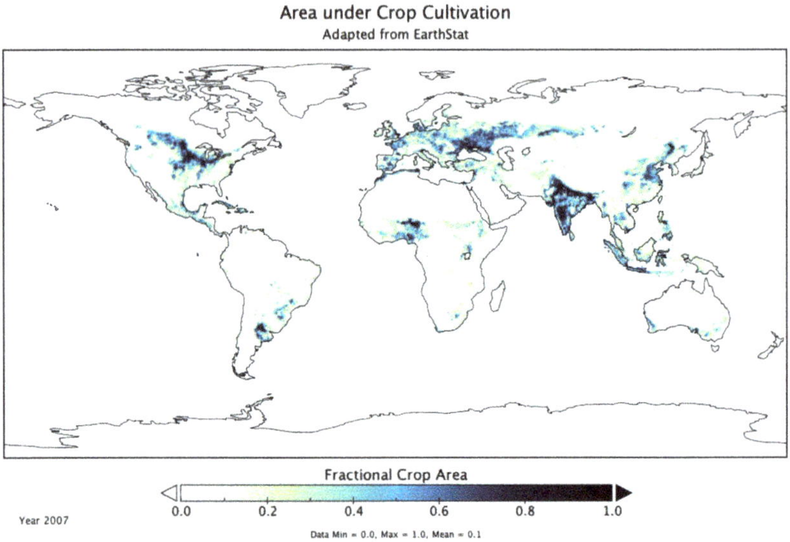

Figure 1 Fractional area under crop cultivation. Data used for figure was obtained from EarthStat, based on the work of Ramankutty et al. (2008).

mechanization. Perhaps among the most important developments, however, has been the advent of relatively cheap mineral fertilizer production, which has dramatically spurred crop growth and improved harvest indices (Erisman et al., 2008). These developments have transformed our previously extensive agricultural systems into ones that are now also intensive, producing vast quantities of biomass that require substantial inputs (Burney et al., 2010).

Similar to other sectoral and industrial developments, this increasing intensity of agriculture has been accompanied by a large increase in emissions of greenhouse gases (GHG). To date, the global food system, including landuse change, contribute nearly one-quarter of global GHG emissions–on par with power generation (Vermeulen et al., 2012b), making these the two biggest sectoral contributions. A majority of these emissions, ~80-86%, are concentrated on agricultural production activities, including livestock and crop cultivation (Vermeulen et al., 2012b).

Cropland emissions were approximately 1780 Tg of carbon dioxide equivalent (CO_2-eq) in 2016 (FAO, 2018). Studies centered on the year 2000 indicate that the highest emission contributions arise from drained peatlands at higher latitudes and Indonesia, and rice paddy production and fertilizer applications, particularly across South, Southeast, and East Asia, exemplified in Fig. 2 (Carlson et al., 2016).

There is however utility in identifying all the contributions from cropland management in order to aid in mitigation and 'climate-smart' agricultural

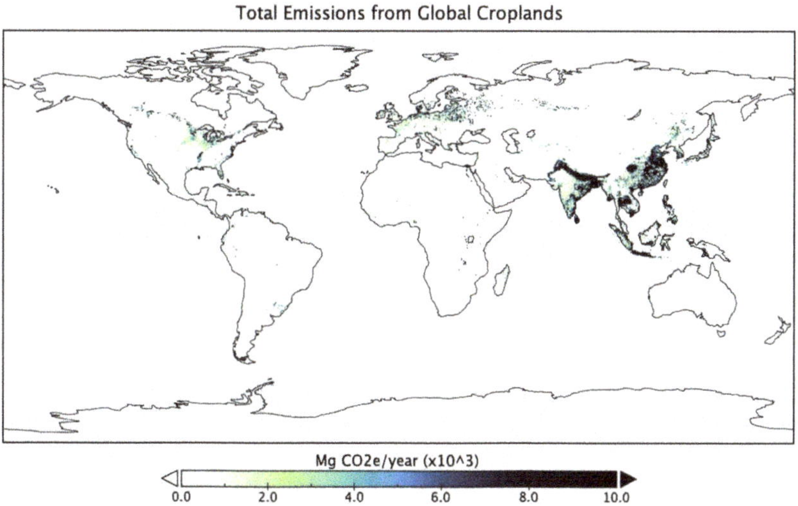

Figure 2 Adapted from Figure 1 of Carlson et al. (2016). Total cropland emissions from each grid cell in Mg CO_2-eq/year. Plot on left shows emissions by latitude for peatlands (green), fertilizer (purple), and rice (orange).

development (discussed further in Section 3). We herein focus specifically on the various contributions of crop cultivation to GHG emissions inclusive of crop residue management; cultivated organic soils and crop-induced soil organic matter (SOM) changes; nutrient applications; rice cultivation; and bioenergy. The relative decadal contribution of these emissions is shown in Fig. 3, summarized by the Intergovernmental Panel on Climate Change 5th Assessment Report (Smith et al., 2014). For each of these contributions, we discuss current trends, prospective developments, and outstanding uncertainties. We then provide a general discussion of possible future trajectories and drivers of GHG emissions from cropping systems, and their outstanding uncertainties, which must be considered as we move towards designing 'climate-smart' and sustainable agricultural systems. Finally, we situate sustainable crop cultivation in the broader context of sustainable development and how the evolution of how we grow our food will have implications not just for the climate, but also for the state of aquatic and terrestrial ecosystems, as well as for social and economic well-being.

2 Greenhouse gas (GHG) emissions from crop residues

Agricultural biomass burning (as a form of residue management) is a significant source of GHGs, primarily methane and nitrous oxide, as well as other aerosols and pollutants (i.e. precursors to ozone formation and black carbon) that can compromise air quality (Sivakumar, 2007). Using fire to manage agricultural

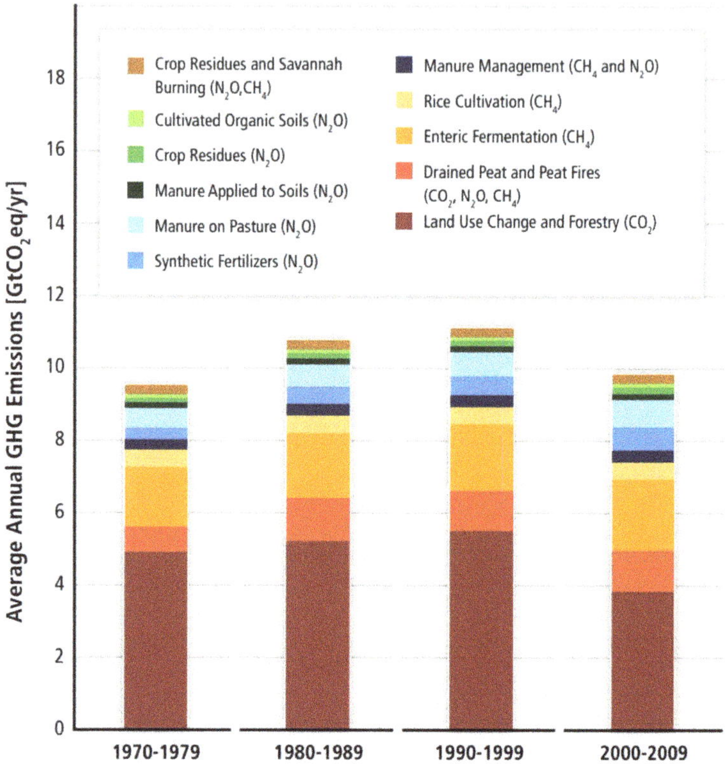

Figure 3 Adapted from Figure 11.2 of Smith et al. (2014), IPCC AR5, Working Group 3, Agriculture, Forestry and Other Landuse. AFOLU emissions for the last four decades. For the agricultural subsectors, emissions are shown for separate categories, based on FAOSTAT, (2013). Emissions from crop residues, manure applied to soils, manure left on pasture, cultivated organic soils, and synthetic fertilizers are typically aggregated to the category 'agricultural soils' for IPCC reporting. For the Forestry and Other Land Use (FOLU), subsector data are from the Houghton bookkeeping model results (Houghton et al., 2012). Emissions from drained peat and peat fires are, for the 1970s and the 1980s, from JRC/PBL (2013), derived from Hooijer et al. (2010) and van der Werf et al. (2006) and for the 1990s and the 2000s, from FAOSTAT (2013).

systems, including those based on maize, paddy rice, wheat, and sugarcane, can serve multiple purposes. These include returning nutrients to the soil (particularly in nutrient-limited areas) and quickly, efficiently clearing fields to prepare them for harvest (e.g. in the case of sugarcane) or the next crop in rotation (Lin et al., 2012). In total, agricultural waste burning comprises ~3% of global fire carbon emissions, most notably across Asia and the Americas (Smith et al., 2014; Tubiello et al., 2014). However, agricultural residue burning contributed 29 Mt CO_2-eq (in 2011), which was 0.5% of total agricultural GHG emissions and of which non-Annex I countries comprised two-thirds (Tubiello et al., 2014; Smith, 2014). These annual emissions have stayed relatively constant over the last half of the twentieth century, but have slightly increased in the beginning of the twenty-first century and continued growth is expected (Fig. 3; Tubiello et al., 2014; Smith et al., 2014).

The integration of crop residues into soils as a nutrient source for future growing seasons and their subsequent decomposition can also lead to GHG emissions, specifically N_2O (U. S. EPA, 2012; FAO, 2018). Global N_2O emissions from this source increased over 20% during 2006–2016 (FAO, 2018). Though it may also lead to increased soil biological activity, decreased risk of soil erosion, and increased soil organic carbon (SOC) sequestration, particularly in certain types of tropical soils (Chivenge et al., 2007; Powlson et al., 2014). More detail on this N_2O dynamic is provided in Section 4.

3 Cultivated organic soils and soil organic matter in cropping systems

3.1 Changes in soil organic matter content in agricultural soils

Soil organic matter (SOM), usually taken as SOC, is critical to biogeochemical cycling, water availability, and productivity of agro-ecosystems (Lal, 2011). In addition to sequestering carbon, SOC comprises the largest terrestrial carbon stock (~1500 PgC; (FAO and ITPS, 2015; Lal, 2010), lends to the soil its nutrient holding capacity, and can act as a highly effective buffer against soil acidification. However, conversions from natural ecosystems and ongoing agricultural land management, including deforestation, the ploughing of native grasslands, and soil overturning and disturbance between cropping cycles, can substantially decrease SOM. Nearly 25% of historic global soil carbon losses have resulted from deforestation, largely for agricultural and resource production (Smith et al., 2016). For example, the clearing of forest for agricultural and/or grazing land can reduce the presence of aggregates and destroy the soil structure (Wei et al., 2015), which can thereby increase organic matter mineralization and enhance CO_2 emissions. Tillage can also lead to soil compaction, compromising soil structure, which makes precious agricultural top soils prone to wind and water erosion and carbon oxidation (Conant et al., 2007).

Nearly 50 million km^2 of soils are currently being managed to produce food, fuel, and fiber (Sanderman et al., 2017). Total land cover changes, cultivation, and SOC losses have collectively contributed ~214 Gt carbon over the historical period (Zomer et al., 2017), an amount comparable to fossil fuel burning (Lal, 2004). In total, it is estimated that upward of one-third of the world's soils are currently degraded (FAO and ITPS, 2015). SOC losses, estimated in Fig. 4, have occurred both in highly productive agroecosystems–such

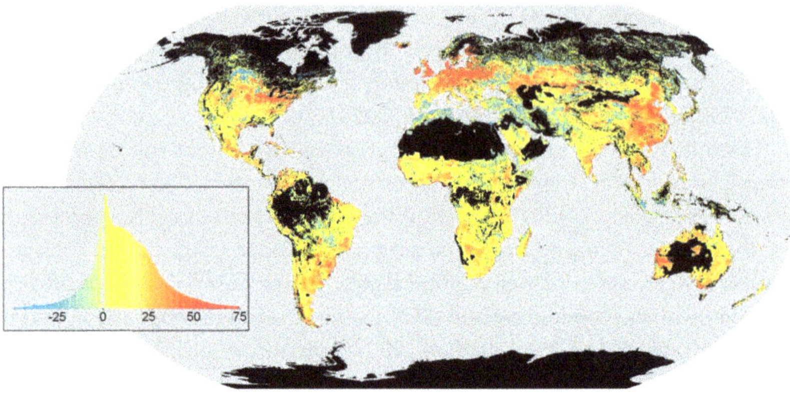

Figure 4 Adapted from Figure 1, Sanderman et al. (2017). Global distribution of modeled SOC change in the top 2 m. Legend is presented as histogram of SOC loss (Mg C·ha^{-1}), with positive values indicating loss and negative values depicting net gains in SOC.

as the North American Great Plains and Midwest, the European Union, and eastern China–and in low-input systems across the semi-arid tropics. In high production, industrial agricultural zones, much of these SOC losses are driven by cultivation, particularly mechanized ploughing (e.g. North American Great Plains) to prepare for the next crop rotation, or the loss of above-ground biomass through deforestation, which limits the recycling of organic matter (Lal, 2011; Smith, 2008). In other, low-input regions, the reasons for SOC losses and overall soil degradation are varied, and can include deforestation, suboptimal land management, or overgrazing, particularly in already-marginal environments (Sanderman et al., 2017).

3.2 Conversion of peatlands to croplands

Of particular concern to cropland GHG emissions are those originating from drained peatlands, saturated, low-oxygen environments comprised of heterogeneous mixtures of decomposed accumulated plant material. Peatlands comprise no more than 3% of Earth's land area, mostly in boreal and subarctic climate zones. However, they are extremely important to the global carbon balance as they house between 20% and 30% of global soil carbon (nearly 528 Pg; Leifeld and Menichetti, 2018; Murdiyarso et al., 2010). Upwards of 10% of global peatlands have been drained and/or converted to accommodate agricultural production, both crop and pasture. In doing so, previously protected carbon- and nitrogen-rich soils are exposed to the atmosphere and allowed to oxidize to CO_2 and N_2O. Fires set to peatlands, often for land management and conversion in the tropics, may also contribute to the release of CO_2, CO, and CH_4 (Leifeld and Menichetti, 2018). In total, degraded peatlands, inclusive of all agricultural landuses such as cropland, grazing lands, and managed forests, contribute ~1.91 Gt equivalent CO_2 emissions annually, and the majority of contributions originate across tropical peatland areas, the EU, and Russia, in that order (Fig. 5; Joosten, 2010; Leifeld and Menichetti, 2018). Tropical forest peatland areas, particularly Indonesia, have undergone a rapid expansion of agriculture over the last few decades. By 2006, nearly 47% of the then identified peatlands in Southeast Asia were either drained or burned to facilitate agricultural expansion, primarily for oil palm or wood pulp products (Austin et al., 2015; Frolking et al., 2011).

3.3 The role of cropland expansion into tropical peatlands: the case of oil palm

Oil palm has become an increasingly in-demand agricultural commodity, known for its wide variety of uses, relatively cheap production, and substantial economic returns that have made its cultivation widely popular across the forested tropics (Fig. 6) (Carlson et al., 2013; Leifeld and Menichetti, 2018; Murdiyarso et al., 2010; Pirker et al., 2016). Palm oil is used in the production and manufacturing of a wide variety of products, including processed foods, cosmetics, household cleaners, and biofuels. While oil palm is expanding across the tropics (Vijay et al., 2016), Malaysia and Indonesia together comprise nearly 85% of the total global crude palm oil production (Austin et al., 2015; Murdiyarso et al., 2010).

Though profitable, land conversions for oil palm displace natural vegetation, and soils rich in carbon and thus oil palm contribute significantly to global carbon emissions (Carlson et al., 2013; Carlson and Curran, 2014). Rapid oil palm development has led to nearly 75% of Indonesia's CO_2-eq

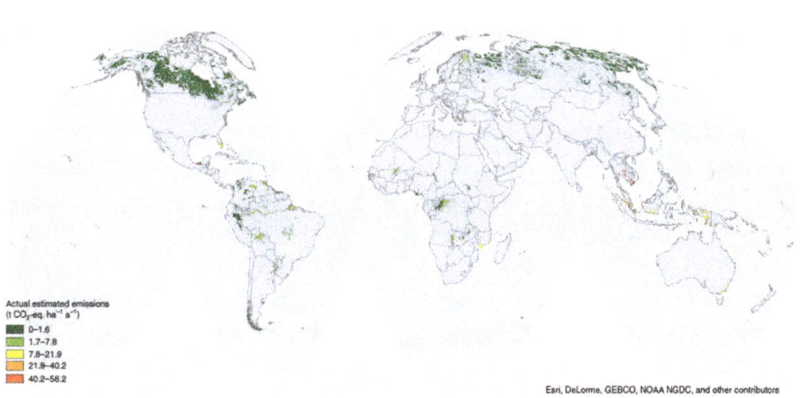

Figure 5 Adapted from Figure 2 of Leifeld et al. (2018). Global peatland distribution and estimated annual actual emissions from peatland degradation.

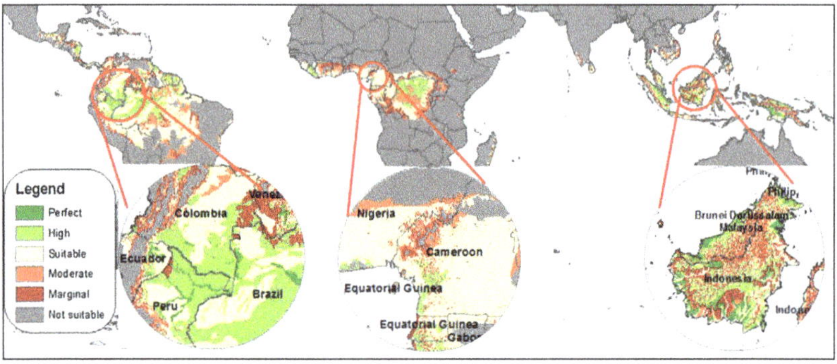

Figure 6 Adapted from Figure 1 of Pirker et al. (2016). Global suitability map and zoom into three oil palm focus areas (from left to right): The Amazon region; the coast of Central Africa; and the island of Borneo.

emissions resulting from plantation-related land conversions (Austin et al., 2015; Carlson et al., 2013), and these emissions further comprise 2–9% of all tropical landuse-related carbon emissions up to 2010 (Carlson and Curran, 2014). Nearly 87% of the total area under oil palm occurs on mineral soils, and contribute nearly 61–73% of oil palm emissions (Carlson et al., 2013). Oil palm emissions from Kalimantan, Indonesia, alone are estimated to be between 0.038 and 0.045 GtC/year from 2000 to 2010 (Carlson and Curran, 2014). However, there is an increasing threat of oil palm plantation expansion into carbon-rich peatlands in Indonesia and elsewhere (Austin et al., 2015). A large fraction of recent, at-risk peatland can be found in Indonesia, which comprises the largest proportion of carbon in tropical peatlands (Page et al., 2011a,b).

A continuation of rapid oil palm development is anticipated in Indonesia (Austin et al., 2015; Carlson et al., 2013; Dohong et al., 2018; Page et al., 2011a,b; Austin et al., 2015; Carlson et al., 2013; Dohong et al., 2018), and nearly one-fifth of this expansion could require the conversion of peatland areas in Kalimantan, and in total, could contribute 18–22% of equivalent CO_2 emissions (Carlson et al., 2013). Additionally, oil palm production is expected to expand across African and Central and South American tropical forested regions due to globally cheap production costs and reliably high economic returns (Wijedasa et al., 2018). Austin et al. (2015) estimates emissions trends of 128.4–211.4 $MtCO_2$/year under business-as-usual expansion scenarios based on current trends. However, a substantial amount of tropical areas that are suitable for oil palm production do or could fall under protected land status, or are currently designated for other uses, which could limit the overall global expansion of oil palm notwithstanding major policy changes (Pirker et al., 2016).

3.4 Potential future soil carbon trajectories and outstanding uncertainties

Despite these cropland carbon losses, or perhaps because of them, the advent of the 2015 Paris climate agreement has initiated interest in leveraging agricultural soils to sequester meaningful amounts of anthropogenic carbon (Chabbi et al., 2017). Although varied, estimates for potential SOC storage in global croplands are between 0.90 and 1.85 Pg C/yr across intensively cultivated regions (Zomer et al., 2017), and more if areas of severe land degradation are rehabilitated (Lal, 2015). In particular, the French Ministry of Agriculture has launched 4p1000 initiative on Soil for Food Security and Climate, which has committed to sequestering 3.5 Gt carbon annually in agricultural soils (Chabbi et al., 2017; Minasny et al., 2017). Not only would this aid global climate mitigation efforts and emission targets defined by Nationally Determined Contributions (NDCs), but such SOC gains would also help to revitalize soils, restoring fertility, nutrient cycling, structure, water holding capacity, and may also considerably aid in boosting crop yields, particularly in areas with outstanding yield gaps (Paustian et al., 2016; Smith, 2008; Soussana et al., 2017).

However, there exist outstanding uncertainties surrounding both measuring existing SOC stocks and estimating their potential gains via deliberate management (Van Groenigen et al., 2017; Zomer et al., 2017). First, obtaining representative data on the geographic distribution of SOC remains challenging, and there exists much variation between datasets, particularly over long time frames and across the developing world, where management data can also be sparse (FAO and ITPS, 2015; Harden et al., 2017; Sanderman and Baldock, 2010). This has implications for predicting the amount of carbon such soils could store and for how long, given regional climate and management pressures (Sanderman and Baldock, 2010). Additionally, soils hold finite amount of carbon directly related to their organic matter inputs and subsequent decomposition, which can vary seasonally and interannually (Sommer and Bossio, 2014). Building soil carbon stocks requires the continuous inputs of organic matter, a serious management consideration, which stand to also be potentially impacted by climate change and more specifically changes in surface temperature and precipitation variability (Zomer et al., 2017).

Furthermore, managing agroecosystems for soil carbon storage may incite new trade-offs for certain domains. For example, putting a price on carbon, and incentivizing management practices that sequester carbon such as reduced- or no-till cropping systems, could contribute to farmers' household income as well as soil health improvements (Soussana et al., 2017), but could also present challenges for regional food security or cropping system adaptation to climate changes (FAO and ITPS, 2015; Zomer et al., 2017). While higher carbon prices could facilitate enhanced carbon storage, they may not necessarily lead to full utilization of the soil's total carbon storage potential and/or may take a fairly long time to do so (Grace et al., 2012; Sanderman and Baldock, 2010; Soussana et al., 2017). Furthermore, effective and long-term carbon storage would require the provisioning of specialized equipment for reduced-tillage systems to farmers that could incur high transaction costs and/or levels of institutional capacity (Grace et al., 2012). Reaching the maximum potential carbon storage is also nitrogen limited, and could demand an additional mineral nitrogen application of 100 Tg/year, or double the amount that is currently sequestered by agricultural systems (Van Groenigen et al., 2017). While there is scope to recover current nitrogen losses (and thereby limit nitrogen pollution) to facilitate these demands, these losses are not distributed evenly across agricultural regions and ensuring their full uptake by crops remains a significant management challenge (Van Groenigen et al., 2017).

4 Nutrient applications for global croplands

Nitrogen (N) is an essential input across all cropping systems. However, its inefficient use - driven by a lack of synchronization between plant N demand and soil N availability - can lead to N losses, which exacerbate a range of environmental impacts. These losses can intensify if the ratio of soil N is disproportionate to other key plant nutrients, particularly phosphorus and potassium (Kanter and Brownlie, 2019). A key impact of N losses to the environment is climate change, which results from the emission of nitrous oxide (N_2O), the third most abundantly emitted GHG globally in terms of CO_2 equivalents and the largest remaining threat to the stratospheric ozone layer (Ciais et al., 2013; WMO, 2014). In 2016, N_2O emissions from crop cultivation amounted to three million tons N_2O-N, equivalent to 1.25 billion tons CO_2 (FAO, 2018). These emissions have several sources: crop residue burning (<1% of N_2O crop emissions) and integration (17%), the cultivation of organic soils (10%), and the application of manure (15%) and synthetic fertilizer (57%). The latter is by far the dominant source, constituting approximately 5% of global emissions from the agricultural sector (FAO, 2018). Indeed, isotopic analysis recently identified synthetic fertilizer use as the main driver of rising atmospheric N_2O concentrations (Park et al., 2012). These concentrations have increased from pre-industrial levels of 270 ppbv to 328 ppbv in 2016 (Butler and Montzka, 2017). The rate of increase has also gone up with time, from approximately 0.15 ppbv year^{-1} 100 years ago to 0.85 ppbv year^{-1} over 2001-2015 (Wells et al., 2018).

The underlying biogeochemical processes that lead to soil N_2O emissions from manure, fertilizer, and crop residue integration are incomplete nitrification (the conversion of ammonium (NH_4^+) to nitrate (NO_3^-)), and incomplete denitrification (the conversion of NO_3^- to atmospheric dinitrogen (N_2)).

Nitrification dominates in soils with high NH_4^+ concentrations and O_2 levels (the latter determined by soil water content), while denitrification dominates in soils with high NO_3^- concentrations, low O_2 levels, and a ready supply of organic carbon (Robertson anad Groffman, 2007). As a result, crop N_2O emissions are characterized by hot spots and hot moments (Groffman et al., 2009), meaning that they are often concentrated in brief periods and small areas where conditions are optimal (e.g., high soil moisture after springtime N application). Warmer and wetter conditions associated with a changing climate will enhance the conditions for soil N_2O emissions, acting as a positive feedback to climate change.

In terms of the regional distribution of crop N_2O emissions, the Asian agricultural sector accounts for over 50% (FAO, 2018). Recent findings confirm significant increases in emissions from the Asian agricultural sector, likely driven by an increase in synthetic N fertilizer use due to increased food demand and changing diets toward higher meat consumption (Saikawa et al., 2014; Wells et al., 2018). Indeed, FAOSTAT data shows that China and India were responsible for 43% of global synthetic N fertilizer consumption in 2016.

Despite the increasing contributions of crop N_2O emissions to climate change, there are studies that suggest that increased synthetic fertilizer use over the twentieth century – perhaps the hallmark of agricultural intensification – has provided significant climate benefits due to higher crop yields. Higher yields mean that less land has had to be put into cultivation as global population has ballooned, avoiding 161 billion tons of carbon emissions over 1850–2005 according to one study, equivalent to 34% of the total carbon emissions over that period (Burney et al., 2010).

Looking ahead is somewhat a challenge, as most projections and scenarios present future N_2O emissions for the entire agricultural sector without disaggregating the various contributors, including crop cultivation (Bouwman et al., 2013; Davidson, 2012; van Vuuren et al., 2011). One of the few exceptions is the UNEP N_2O report (UNEP, 2013), which breaks down emissions from synthetic fertilizer, manure, and other sources. In a business-as-usual scenario derived from FAO assumptions on future livestock and crop demand (Alexandratos and Bruinsma, 2012), N_2O emission levels from crops are expected to increase from almost 50% of 2016 values to 4.4 million tons of N_2O-N by 2050. Over 80% of this comes from fertilizer, with the remainder coming from manure. Herculean efforts in improving nitrogen-use efficiency (15% increase by 2050), increasing manure recycling (100% increase by 2050), and changing diets toward WHO guidelines could reduce emissions to 2.4 million tons in N_2O-N by 2050. Figure 7 shows model projections of agricultural N inputs based on the main environmental scenarios framework in use today (the Shared Socioeconomic Pathways (SSPs); Mogollon et al., 2018; Riahi et al., 2017). More about the underlying assumptions in the SSPs and others can be found in Section 7.

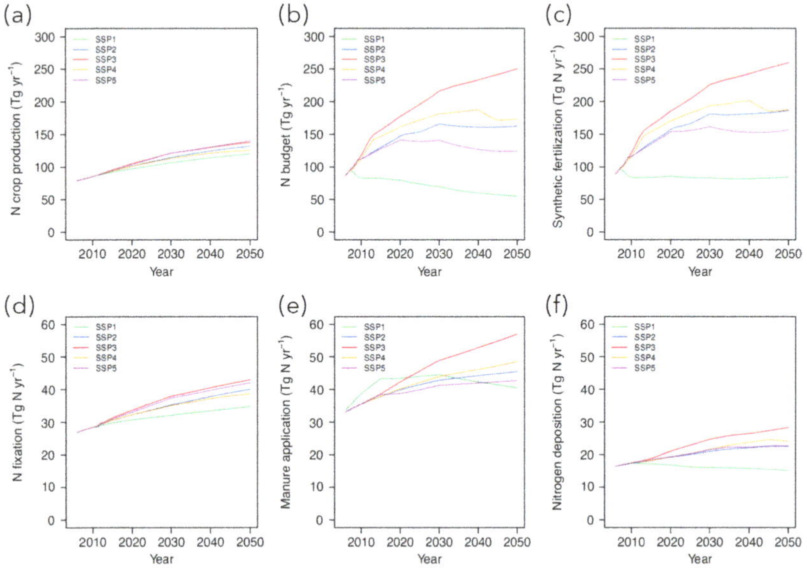

Figure 7 Adapted from Figure 2 of Mogollon et al. (2018). (a) Global N production, (b) global N budget; and global N inputs in croplands from (c) synthetic fertilizers, (d) fixation, (e) manure, and (f) deposition from 2006 until 2050. The green lines represent SSP1, the blue lines represent SSP2, the red lines represent SSP3, the orange lines represent SSP4, and the purple lines represent SSP5. Note the different Y axis for panels (a), (b), (c) vs panels (d), (e), (f).

There are two central uncertainties when estimating crop N_2O emissions. Both can be evaluated through the lens of the most basic (Tier 1) emission factors (EFs) developed by the Intergovernmental Panel on Climate Change (Eggleston et al., 2006). Tier 1 EFs are used to indirectly estimate GHG emissions from a range of production and consumption data, usually at national or international scales. In the case of crop N_2O emissions, annual synthetic fertilizer or manure consumption data is multiplied by an EF to estimate annual emission fluxes. What underpins the uncertainty here is that N_2O EFs assume a linear relationship between N application rates and N_2O emissions, with a 1% EF applied to synthetic N fertilizer rates to estimate direct emissions (e.g. for every 100 kg N applied, 1 kg of N_2O is emitted). However, recent studies are finding nonlinear relationships, suggesting that N_2O emissions per ha are lower than EFs at low N application rates, and higher at high N application rates - likely due to the greater excess N unused by crops, which is then available to be emitted as N_2O (Gerber et al., 2016; Shcherbak et al., 2014). The second major uncertainty is the estimation of indirect crop N_2O emissions, that is, emissions after N has already been lost the environment in other forms such as NO_3^- or NH_3. Recent studies suggest that the IPCC EFs are biased low, especially the 0.75% EF for indirect N_2O from leached NO_3^- (Turner et al., 2015).

5 Greenhouse gas (GHG) emissions from rice production

Rice occupies >10% of arable lands, and serves as a dietary staple for nearly two billion people globally. The rice economy also supports the household income of ~150 million households worldwide (Kritee et al., 2018; van Groenigen et al., 2013), making it perhaps a singularly important crop that directly supports and/or feeds the most people. The majority of rice is typically transplanted and grown in flooded paddies, producing anaerobic environments. While rice does not need to be flooded, it is a water-tolerant crop and flooding typically aids in weed suppression (Berkhout et al., 2015). However, these conditions give rise to methanogenic populations that can contribute significantly to global carbon cycling (Kirschke et al., 2013).

In particular, rice-based farming systems constitute ~2.5% of global GHG contributions (Kritee et al., 2018), 10% of anthropogenic CH_4 emissions (van Groenigen et al., 2013), and comprise approximately 32% of GHG emissions from global croplands (Carlson et al., 2016). A majority of these rice-system emissions originate from South, Southeast, and East Asia (Kirschke et al., 2013), and India alone is thought to contribute ~27% of total rice-based CH_4 emissions. It is thus thought that curbing CH_4 emissions from paddy production could comprise significant contributions toward countries' NDCs. New water management regimes with intermittent flooding, such as alternate wetting and drying (AWD), have potential to minimize rice-system CH_4 emissions and promote soil aeration (Lampayan et al., 2015; Linquist et al., 2015). In South Indian rice-based systems, the inclusion of AWD could reduce CH_4 emissions by ~20–40% (Oo et al., 2018).

However, there exist potential trade-offs between CH_4 and N_2O emissions, the latter resulting from oxidized soil conditions which can prevail when rice flooding is reduced (Kritee et al., 2018; Oo et al., 2018). It is possible that, under some conditions, these aggregate nitrous oxide emissions from rice-based systems with alternative water management have a lower global warming potential relative to the CH_4 emissions from their conventionally flooded counterparts (van Groenigen et al., 2013). However, it has been recently shown that there is still much uncertainty surrounding the adequate and reliable measurement of N_2O emissions from rice paddies under a variety of water management conditions, which in some cases have been found to be 30–45 times higher when improved estimates are obtained (Kritee et al., 2018). In some cases, even intermittent flooding can lead to elevated risks of N_2O production relative to continuous flooding conditions due to strong bouts of microbial activity, which enhances mineralization and nitrification-denitrification processes (Fig. 8).

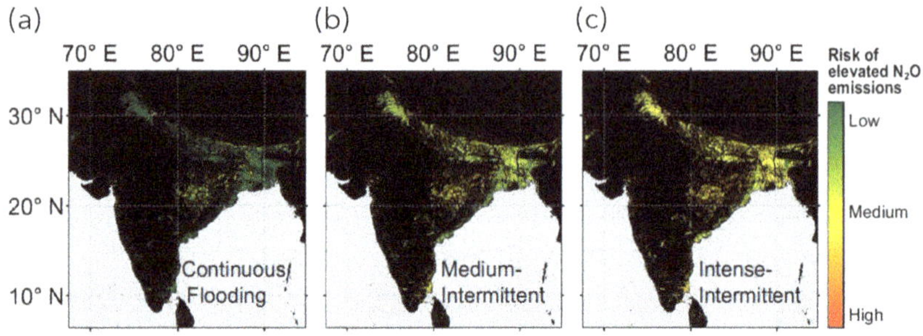

Figure 8 Adapted from Figure 2 of Kritee et al. (2018). Qualitative risk of elevated N_2O emissions from the Indian subcontinent under three flooding scenarios: continuous (a), medium-intermittent (b), and intense-intermittent (c) flooding.

As such, there is a need for not just improved estimation methods of CH_4 emissions alone, but also improved estimation of the multiple co-variables that drive GHG emissions from rice systems and longer time series of the relevant and necessary data (Kritee et al., 2018). Augmenting these concerns, future climate conditions, may further challenge mitigating methane emissions from rice production. Although rising atmospheric CO_2 concentrations could promote increased yields, it also enhances CH_4 emissions at a faster rate, thereby enhancing the GHG intensity of rice production (van Groenigen et al., 2013). Likewise, rising surface temperatures also enhance GHG intensity via negative impacts on yield, and the combination of these two forcing factors leads to a near doubling of rice production GHG intensity by the end of the century in the absence of adaptation (van Groenigen et al., 2013).

6 Greenhouse gas (GHG) emissions from bioenergy systems

Bioenergy is an energy source derived from biomass grown on agricultural lands which can be used as fuel in a number of sectors, from transport to electricity production. Bioenergy is seen as a key mitigation strategy in pathways aiming to limit global average temperature increases to 2°C or 1.5°C; the central assumption being that the amount of CO_2 emitted from the combustion of biomass will be re-sequestered in future years, making it climate neutral. However, this assumption does not hold when GHG emissions associated with harvest, transport and conversion, landuse change, and land management are taken into account (Crutzen et al., 2008; Searchinger et al., 2009). In this chapter, we focus on emissions from crop cultivation, specifically N_2O from synthetic N fertilizer use and CO_2 from soil C disturbance.

N_2O emissions from synthetic N fertilizer use are one of the largest GHG sources across many biofuel production cycles. Depending on the crop, fertilization rate, measurement period, and other management factors, emission rates can vary by as much as 700% for the same site (Smith et al., 2013b). N_2O emissions alone can transform bioenergy systems from net GHG sinks to net sources - offsetting the climate benefits of transitioning away from fossil fuels - particularly for first-generation crop-based biofuels such as corn that have high N requirements (Crutzen et al., 2008). For second-generation biofuels with lower N requirements, excessive N application rates can significantly reduce the climate benefits of the system in addition to exacerbating other N-related impacts, such as nitrate leaching. For example, in a study on a switchgrass system in the US Midwest, fertilizing at a rate of 196 kg N ha^{-1} led to a mitigation capacity of approximately three tons CO_2 ha^{-1} yr^{-1} compared to almost six tons CO_2 ha^{-1} yr^{-1} when fertilized at a rate of 56 kg N ha^{-1} (Ruan et al., 2016).

Furthermore, shifting land previously used for agricultural or other purposes to biomass production can lead to a considerable soil C 'debt'. For example, an uncropped ecosystem or pasture will likely have significant amounts of C stored above and below ground, a large portion of which could be lost to the atmosphere during the conversion to a bioenergy system (Gelfand et al., 2011). Consequently, if such a system is to produce a net climate benefit, it will need to have C accumulation rates greater than those in the preexisting system, through C-storing management practices such as no-tillage and carbon capture and storage (discussed further below) or the use of crops that tend to sequester more soil C, such as perennials (West and Post, 2002). Indeed, perennial crops like switchgrass and miscanthus have been shown to provide both soil C and N benefits, due to lower N requirements and the year-round growth which extends the period for soil N uptake, thereby reducing potential N_2O emissions (Robertson et al., 2017). Moreover, the absence of tillage allows for N accumulation in SOM, which can also increase soil C storage capacity given fixed C:N ratios.

A key add-on to traditional bioenergy systems that has emerged in recent years is the combination with carbon capture and storage, commonly known as BECCS. If effective, it qualifies as a negative emissions technique, as it extracts CO_2 from the atmosphere during plant growth and subsequently stores it underground. It also provides a renewable energy source, therefore reducing fossil fuel demand. However, this technology is still being tested and concerns exist if it became widely deployed as part of a highly ambitious mitigation pathway consistent with a 1.5°C or 2°C target. This is due to the risk of higher N_2O emissions from more land under fertilization, and the landuse needs which could threaten food production and forest ecosystems with potentially higher C storage capacity (Harper et al., 2018; Rogelj et al., 2018). Very high-yielding bioenergy crops would be required to repay the C debt from landuse conversion in a reasonable timeframe.

Consequently, looking ahead, bioenergy systems could be an important contributor to climate mitigation over the course of this century or a net GHG source whose emissions need to be offset elsewhere. The key determinant will be the stringency of national and international climate policies, which will subsequently determine how much energy and land is needed and put aside for bioenergy production, the crops used, and the N management practices and technologies deployed (Smith et al., 2013a). One recent study estimates that production of second-generation biofuels could range from close to zero to over 20 billion tons in 2100 depending on the development and climate trajectory (Popp et al., 2017; Fig. 9). For perspective, total global grain production in 2016/2017 was 2.7 billion tons (USDA, 2018). Other social, political, and economic factors will also play a role, including the evolution of land tenure and regulation, trade and technology development (Dornburg et al., 2010). From an N_2O perspective, bioenergy systems could emit less <1 to >18 million tons N_2O-N by 2100 depending on the scenario, meaning that bioenergy could remain a relatively trivial source or become the most significant source of anthropogenic N_2O emissions at some point in this century (Davidson and Kanter, 2014). N management practices such as the use of N inhibitors could marginally limit these emissions, but minimizing N fertilizer use overall

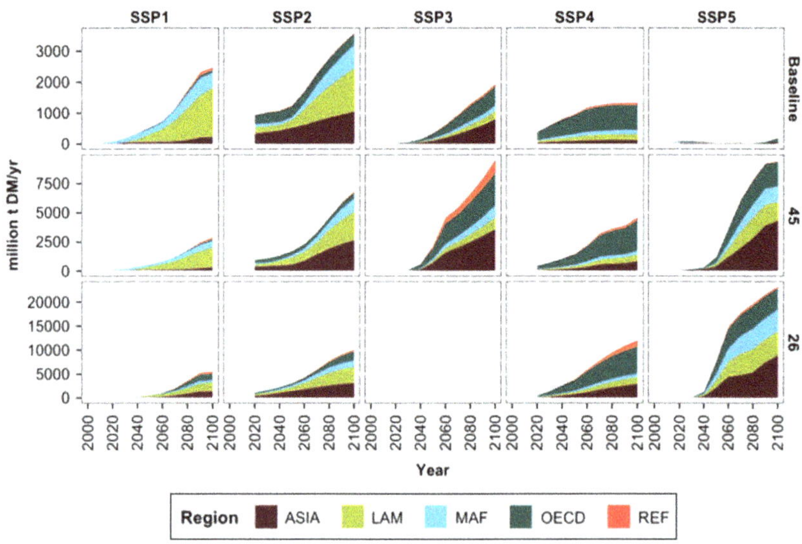

Figure 9 Adapted from Supplementary Information Figure 9 of Popp et al. (2017). Regional production of dedicated second-generation bioenergy crops of the five SSP marker scenarios for the baseline (upper row), RCP4.5 (middle row) and RCP2.6 (lower row) cases. Data is shown for five aggregate regions: (1) OECD90 countries (OECD), (2) reforming economies of Eastern Europe and the Former Soviet Union (REF), (3) countries of the Middle East and Africa (MAF), (4) countries of Latin America and the Caribbean (LAM), and (5) Asian countries with the exception of the Middle East, Japan, and Former Soviet Union states (ASIA).

will have to be an important component of maximizing the climate benefits of bioenergy systems (Ogle et al., 2016).

7 Future trends

Future trajectories of agriculture and cropping system development trends must balance the need to feed ~ten billion by 2050 with minimizing environmental and climate impacts (Godfray et al., 2010; Vermeulen et al., 2012b). It has been highlighted that crop intensification over the twentieth century, particularly in industrialized or rapidly developing nations, may have actually averted higher historical GHG emission trends (Burney et al., 2010). This may have been partly due to technology trends that produced higher yields, and prevented crop expansion onto marginal lands. Indeed, it has been suggested that those areas displaying high GHG intensities (CO_2eq/kcal produced), in addition to high gross GHG emissions, should be the focus of emerging policy measures to reduce GHGs from agriculture (Carlson et al., 2016). Such areas include widespread rice-growing regions across South, Southeast, and East Asia, and peatlands, particularly in Indonesia and the Maritime continent, which could benefit from either improvements in crop technology or potential crop switching. However, in some of these locations, particularly rice-growing regions, much of the crop calories produced are delivered to the food system for direct human consumption and constitute important forms of regional food security (Carlson et al., 2016; Cassidy et al., 2013), despite their emissions intensity. This contrasts sharply with industrialized, intensive commodity production areas, where major crops are processed for a variety of end uses, such as animal feed or biofuels, which may or may not eventuate in consumed food calories. As such, future agricultural policy measures targeting GHG emissions and intensities should carefully evaluate how these quantities are normalized, and the role regional crop production systems play in food security and the provision of livelihoods.

In the next two sections, we briefly consider how agricultural and cropping system development may be incorporated alongside larger, global multi-sectoral development trends, global GHG concentration trajectories, and the sustainable development goals (SDGs) in order to optimize across food security and environmental preservation.

7.1 Considering future trends of cropland greenhouse gas (GHG) emissions using Representative Agricultural Pathways (RAPs)

Future projections of agricultural GHG emissions can be calculated either by extrapolating current trends or by envisioning different trajectories or scenarios that more fully capture the range of uncertainty in combined policy, socioeconomic trends, and biophysical feedbacks space. For five

IPCC assessment reports, the climate research community has explored the range of climate responses via scenarios, first with the Special Report on Emissions Scenarios (SRES; Nakicenovic et al., 2000) and subsequently with the Representative Concentration Pathways (RCPs), which describe both aggressive and more moderate GHG mitigation as well as business-as-usual scenarios (van Vuuren et al., 2011). In tandem, the Integrated Assessment Modeling community has recently blended these RCPs with scenarios of global development known as SSPs, which provide socioeconomic narratives for reaching different climate forcing and adaptation levels by 2100 (O'Neill et al., 2017). Each of the five SSPs is defined by different trajectories in major socioeconomic, demographic, technological, lifestyle, policy, institutional, and other trends. They encompass a range of futures that span the societal challenges associated with mitigating and adapting to climate change (Riahi et al., 2017). These futures were expanded for landuse and agriculture, with storylines differentiated by level of landuse regulation, agricultural productivity, dietary preferences, trade patterns, globalization, and climate mitigation approaches (Popp et al., 2017). For example, SSP 1 assumes a large increase in crop productivity with rapid diffusion of best practices and ambitious climate mitigation, while SSP 3 assumes low technology development with a very limited role for agriculture in climate mitigation. This has important implications for GHG emissions trajectories even independent of different RCPs, though emissions from crop cultivation are not reported separately.

Much of cropland GHG contributions arise from management norms and choices, which may be greatly influenced by policies and economic factors at multiple governance scales (from local market access to international commodity pricing; Antle et al., 2014; Valdivia et al., 2015). As such, there is a need to consider how cropland management, and resulting GHG emissions, respond to agricultural development trends as informed by both scenarios of biophysical and socioeconomic change.

To address these needs for agricultural development scenarios, protocols were recently developed under the auspices of the Agricultural Model Intercomparison and Improvement Project (AgMIP; Rosenzweig et al., 2013) to establish regional and global Representative Agricultural Pathways (RAPs; Antle et al., 2014; Valdivia et al., 2015). Figure 10 details the AgMIP regional integrated assessment modeling framework and demonstrates how RAPs are linked to both the RCPs and SSPs, while also incorporating highly calibrated (crop) production system models ('PSM' on Fig. 10) to better resolve changes specific to particular cropping systems. Thus, RAPs comprise of a set of scenarios for agricultural development that are consistent with global pathways of change, but can provide a more highly resolved understanding of how these larger-scale forcings can affect changes in regional agriculture, inclusive of changes in preferred or dominant crop varieties, farming systems, and their

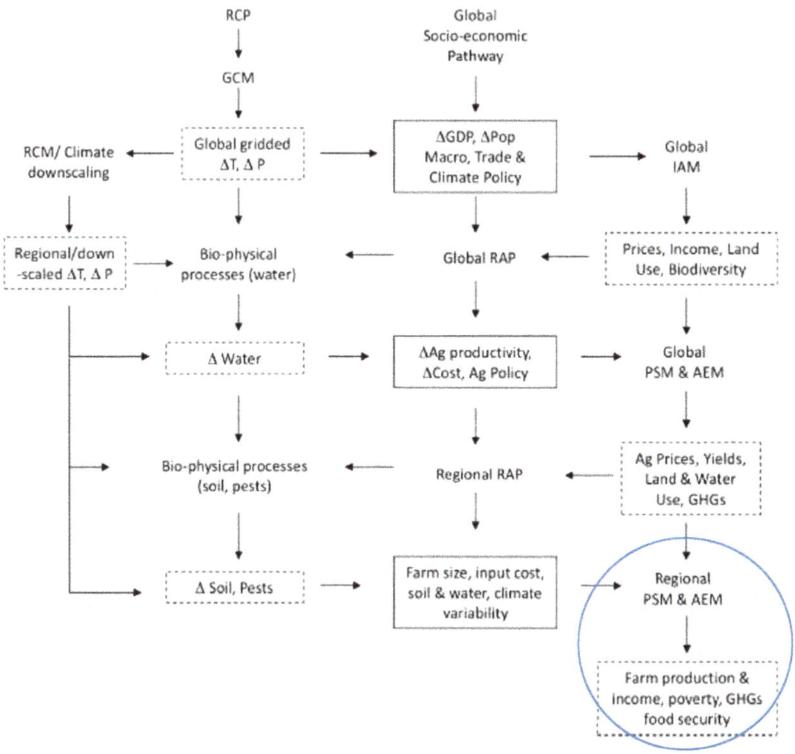

Figure 10 Adapted from Figure 3 of Valdivia et al. (2015), *Handbook of Climate Change and Agroecosystems*. An elaboration of AgMIP's global and regional integrated assessment modeling framework. Moving from top to bottom represents different geographic scales (global to regional to local), and the three columns of the figure represent biophysical models and data (left-hand column), biophysical and socioeconomic pathways (center), and impact models (right-hand column). Note: IAM = integrated assessment model; RAP = representative agricultural pathway; PSM = biophysical production system model; AEM = agricultural economic model; solid boxes indicate variables determined by global socioeconomic pathways and RAPs, dashed boxes indicate model outputs.

management (e.g. labor, seed, fertilizer, water resources, etc.; Valdivia et al., 2015). Such detailed information is critical for improved impacts assessments, and assessments of cropland GHG contributions resulting from varied scenarios of future global environmental and economic change (see blue circle on Fig. 10).

7.2 Crop cultivation and sustainable development

It should be noted that future trajectories of agricultural GHG emissions and the different technical, economic, and social dynamics that underpin each of them do not exist in a vacuum. The evolution of the agricultural sector and the broader food system will be a key determinant of the success of the SDGs agreed to by the United Nations in 2015, with the potential for trade-offs and synergies. The sum aim of the SDGs is to improve social, economic, and environmental well-being, with 17 goals and 169 targets ranging from gender equality to protecting aquatic and terrestrial ecosystems, all by 2030. Goal 2 is the most directly related to crop cultivation as it aims to end hunger and within this, target 2.4 calls for 'sustainable food production systems' by 2030 'that strengthen capacity for adaptation to climate change, extreme weather, drought, flooding and other disasters and that progressively improve land and soil quality' (UN, 2015). This target has important interactions with several other goals. For example, there is a clear link to Goal 13 with its focus on climate action–specifically targets 13.2 and 13.3, which call for the integration of climate measures into national policies and further development of knowledge and capacity on climate mitigation and adaptation, respectively. But the interactions are more extensive than just climate. Sustainable crop cultivation–producing more nutritious food while minimizing the environmental and health impacts–could make important contributions to achieving several of the other SDGs by improving food security (Goal 1: ending poverty; Goal 3: healthy lives; Goal 5: gender equality) and reducing air and water pollution (Goal 6: clean water and sanitation; Goal 7: affordable and clean energy; Goal 14: protect life below water; Goal 15: protect life on land).

However, another target within Goal 2, target 2.3, calls for a doubling of agricultural productivity, particularly for small-scale producers. This could have synergies with other SDGs, including ending poverty (Goal 1) and providing affordable and clean energy (Goal 7). Yet if implemented following traditional agronomic approaches (i.e. increased use of synthetic fertilizer and pesticides, water resources, heavy machinery etc.), target 2.3 could have precisely the opposite effect of what target 2.4 is aiming to achieve in many areas–namely a negative impact on health and environmental outcomes (ICSU, 2017). Consequently, how crop cultivation develops over the next decade and beyond will have a significant influence over the fulfillment of a range of SDG goals and targets. Indeed, targets 2.3 and 2.4 encapsulate one of the central tensions in sustainable development–improving social and economic outcomes while attempting to minimize the environmental impacts. Tools are emerging to help decision-makers across a range of scales and settings to better understand and address the trade-offs and synergies involved in the transition toward more sustainable crop cultivation (Kanter et al., 2018). These tools will be key to developing effective and equitable strategies to address this multi-objective optimization problem.

8 Conclusion

Global population is expected to pass nine billion by 2050 (Godfray et al., 2010), in tandem with increasing trends in meat consumption and per capita energy demand. As a consequence, crop cultivation–for food, feed, and energy–will also need to grow and intensify in order to meet these demands. Furthermore, climate change threatens harsher growing conditions, such as increased heat and water stress, and potentially stronger weather shocks, which will make it harder to meet these needs (Godfray et al., 2010; Tigchelaar et al., 2018; Vermeulen et al., 2012a). However, crop cultivation is a key contributor to agricultural GHG emissions, and thus climate change overall, via a range of land and production management practices. Consequently, transitioning to more sustainable and climate-smart agricultural practices is essential to both reducing the carbon footprint of crop cultivation and increasing its resilience to climate change. Looking ahead, future research must place a strong emphasis on improving GHG measurement from heterogeneous croplands across all scales–from field to farm, regional to global. More exact, frequent, and consistent measurements would allow for the implementation and monitoring of climate mitigation and adaptation measures better tailored to specific geographic contexts, and inform the development of more accurate climate and crop models that can help humanity better plan for future developments. If managed with an eye toward minimizing its multiple impacts and maximizing ecosystem services, crop cultivation could be a positive contributor not only to climate mitigation and adaptation, but also to achieving sustainable development writ large.

9 Where to look for further information

Further work exploring the topic of greenhouse gas emissions from cropping systems can consult the inventories and databases of the following:

- UN Food and Agriculture Organization (http://www.fao.org/faostat/en/#data);
- the EarthStat database from the University of Minnesota and University of British Columbia (http://www.earthstat.org/), which also contains detailed information on landuse, crop distributions, fertilizer applications, yield gaps and other key management data and the Global Carbon Project (https://www.globalcarbonproject.org/), which is also compiling data on global methane and nitrous oxide budgets.

10 References

Alexandratos, N. and Bruinsma, J. 2012. *World Agriculture Towards 2030/2050: The 2012 Revision*. FAO, Rome, Italy.

Antle, J. M., Stoorvogel, J. J. and Valdivia, R. O. 2014. New parsimonious simulation methods and tools to assess future food and environmental security of farm populations. *Philos. Trans. R Soc. Lond. B Biol. Sci.* 369, 1–15. https://doi.org/10.1098/rstb.2012.0280.

Austin, K. G., Kasibhatla, P. S., Urban, D. L., Stolle, F. and Vincent, J. 2015. Reconciling oil palm expansion and climate change mitigation in Kalimantan, Indonesia. *PLoS ONE* 10, e0127963. https://doi.org/10.1371/journal.pone.0127963.

Berkhout, E., Glover, D. and Kuyvenhoven, A. 2015. On-farm impact of the System of Rice Intensification (SRI): Evidence and knowledge gaps. *Agric. Syst.* 132, 157–66. https://doi.org/10.1016/J.AGSY.2014.10.001.

Bouwman, A. F., Beusen, A. H. W., Griffioen, J., Van Groenigen, J. W., Hefting, M. M., Oenema, O., Van Puijenbroek, P. J. T. M., Seitzinger, S., Slomp, C. P., Stehfest, E., Galloway, J., Tiedje, J., Yamulki, S., Harrison, R., Goulding, K., Webster, C., Saad, A., Conrad, R., McClain, M., Böhlke, J.-K., Wanty, R., Tuttle, M., Delin, G., Landon, M., Zhang, Y.-C., Slomp, C., Broers, H., Passier, H., Cappellen, P. Van, Mayer, P., Reynolds, S., McCutchen, M., Canfield, T., Heuvel, R. Van den, Bakker, S., Jetten, M., Hefting, M., Groffman, P., Gold, A., Jacinthe, P.-A., Bouwman, A., Goldewijk, K. K., Hoek, K. Van der, Beusen, A., Vuuren, D. Van, Willems, W., Rufinoe, M., Stehfest, E., Boyer, E., Alexander, R., Parton, W., Li, C., Butterbach-Bahl, K., Donner, S., Skaggs, R., Grosso, S. Del, Bouwman, A., Beusen, A., Billen, G., Bouwman, A., Kram, T., Goldewijk, K. K., Fekete, B., Wisser, D., Kroeze, C., Mayorga, E., Bouwman, A., Wollheim, W., Vörösmarty, C., Vitousek, P., Menge, D., Reed, S., Cleveland, C., Cleveland, C., Drecht, G. Van, Bouwman, A., Knoop, J., Beusen, A., Meinardi, C., Vidon, P., Hill, A., IPCC, Seitzinger, S., Harrison, J., Böhlke, J., Bouwman, A., Lowrance, R., Peterson, B., Tobias, C., Drecht, G., Emery, K., Orr, W., Rittenberg, S., Gruber, N., Sarmiento, J., Codispoti, L., Brandes, J., Christensen, J., Devol, A., Naqvi, S., Pearl, H., Yoshinari, T., Søderlund, R., Svensson, B., Canfield, D., Glazer, A., Falkowski, P., Hofstra, N., Bouwman, A., Ju, X., Liu, X., Zhang, F., Roelcke, M., Ducharne, A., Griffioen, J., Brunt, R., Vasak, S., Gun, J. Van der, Drecht, G. Van, Bouwman, A., Boyer, E., Green, P., Siebert, S., Beusen, A., Bouwman, A., Heuberger, P., Drecht, G. Van, Hoek, K. Van Der, Bodirsky, B., Popp, A., Weindl, I., Dietrich, J., Rolinski, S., Scheiffele, L., Schmitz, C., Lotze-Campen, H., Groffman, P., Heinen, M., Rivett, M., Buss, S., Morgan, P., Smith, J., Bemment, C., Postma, D., Boesen, C., Kristiansen, H., Larsen, F., Hill, A., Stehfest, E., Bouwman, A., Bouwman, A., Boumans, L., Batjes, N., Ryan, M., Müller, C., Di, H., Cameron, K., Farquharson, R., Baldock, J., Corazza, M., Vilain, G., Garnier, J., Tallec, G., Tournebize, J., DeSimone, J., Macrae, M., Bourbonniere, R., Wrage, N., Velthof, G., Beusichem, M. Van and Oenema, O. 2013. Global trends and uncertainties in terrestrial denitrification and N_2O emissions. *Philos. Trans. R. Soc. B Biol. Sci.* 368. https://doi.org/10.1098/rstb.2013.0112.

Burney, J. A., Davis, S. J. and Lobell, D. B. 2010. Greenhouse gas mitigation by agricultural intensification. *Proc. Natl. Acad. Sci.* 107, 12052–7. https://doi.org/10.1073/pnas.0914216107.

Butler, J. H. and Montzka, S. A. 2017. The NOAA Annual Greenhouse Gas Index (AGGI).

Campbell, B. M., Beare, D. J., Bennett, E. M., Hall-Spencer, J. M., Ingram, J. S. I., Jaramillo, F., Ortiz, R., Ramankutty, N., Sayer, J. A. and Shindell, D. 2017. Agriculture production as a major driver of the Earth system exceeding planetary boundaries. *Ecol. Soc.* 22, art8. https://doi.org/10.5751/ES-09595-220408.

Carlson, K. M. and Curran, L. M. 2014. Refined carbon accounting for oil palm agriculture: disentangling potential contributions of indirect emissions and smallholder farmers. *Carbon Management* 4, 347-9. https://doi.org/10.4155/cmt.13.39.

Carlson, K. M., Curran, L. M., Asner, G. P., Pittman, A. M., Trigg, S. N. and Marion Adeney, J. 2013. Carbon emissions from forest conversion by Kalimantan oil palm plantations. *Nat. Clim. Chang.* 3, 283-7. https://doi.org/10.1038/nclimate1702.

Carlson, K. M., Gerber, J. S., Mueller, N. D., Herrero, M., Macdonald, G. K., Brauman, K. A., Havlik, P., O'connell, C. S., Johnson, J. A., Saatchi, S. and West, P. C. 2016. Greenhouse gas emissions intensity of global croplands. *Nat. Clim. Chang.* 7, 63-71. https://doi.org/10.1038/NCLIMATE3158.

Cassidy, E. S., West, P. C., Gerber, J. S. and Foley, J. A. 2013. Redefining agricultural yields: from tonnes to people nourished per hectare. *Environ. Res. Lett.* https://doi.org/10.1088/1748-9326/8/3/034015.

Chabbi, A., Lehmann, J., Ciais, P., Loescher, H. W., Cotrufo, M. F., Don, A., SanClements, M., Schipper, L., Six, J., Smith, P. and Rumpel, C. 2017. Aligning agriculture and climate policy. *Nat. Clim. Chang.* 7, 307-9. https://doi.org/10.1038/nclimate3286.

Chivenge, P. P., Murwira, H. K., Giller, K. E., Mapfumo, P. and Six, J. 2007. Long-term impact of reduced tillage and residue management on soil carbon stabilization: Implications for conservation agriculture on contrasting soils. *Soil Tillage Res.* 94, 328-37. https://doi.org/10.1016/j.still.2006.08.006.

Ciais, P., Sabine, C., Bala, G., Bopp, L., Brovkin, V., Canadell, J., Chhabra, A., DeFries, R., Galloway, J., Heimann, M., Jones, C., Le Quéré, C., Myneni, R. B., Piao, S. and Thornton, P. 2013. Carbon and other biogeochemical cycles. In: Stocker, T. F., Qin, D., Plattner, G.-K., Tignor, M., Allen, S. K., Boschung, J., Nauels, A., Xia, Y., Bex, V. and Midgley, P. M. (Eds), *Climate Change 2013: The Physical Science Basis. Contribution of Working Group I to the Fifth Assessment Report of the Intergovernmental Panel on Climate Change.* Cambridge University Press, Cambridge, United Kingdom and New York, NY, USA.

Conant, R. T., Easter, M., Paustian, K., Swan, A. and Williams, S. 2007. Impacts of periodic tillage on soil C stocks: A synthesis. *Soil Tillage Res.* https://doi.org/10.1016/j.still.2006.12.006.

Conijn, J. G., Bindraban, P. S., Schroder, J. J. and Jongschaap, R. E. E. 2018. Can our global food system meet food demand within planetary boundaries? *Agric. Ecosyst. Environ.* 251, 244-56. https://doi.org/10.1016/j.agee.2017.06.001.

Crutzen, P. J., Mosier, A. R., Smith, K. A. and Winiwarter, W. 2008. N_2O release from agro-biofuel production negates global warming reduction by replacing fossil fuels. *Atmos. Chem. Phys.* 8, 389-95. https://doi.org/DOI 10.5194/acp-8-389-2008.

Davidson, E. A. 2012. Representative concentration pathways and mitigation scenarios for nitrous oxide. *Environ. Res. Lett.* 7. https://doi.org/Artn 02400510.1088/1748-9326/7/2/024005.

Davidson, E. A. and Kanter, D. 2014. Inventories and scenarios of nitrous oxide emissions. *Environ. Res. Lett.* 9. https://doi.org/Artn 10501210.1088/1748-9326/9/10/105012.

Dohong, A., Aziz, A. A. and Dargusch, P. 2018. Carbon emissions from oil palm development on deep peat soil in Central Kalimantan Indonesia. *Anthropocene* 22, 31-39. https://doi.org/10.1016/J.ANCENE.2018.04.004.

Dornburg, V., van Vuuren, D., van de Ven, G., Langeveld, H., Meeusen, M., Banse, M., van Oorschot, M., Ros, J., van den Born, G. J., Aiking, H., Londo, M., Mozaffarian, H., Verweij,

P., Lysen, E. and Faaij, A. 2010. Bioenergy revisited: Key factors in global potentials of bioenergy. *Energy Environ. Sci.* 3, 258–67. https://doi.org/10.1039/b922422j.

Eggleston, S., Buendia, L., Miwa, K., Ngara, T. and Tanabe, K. 2006. *IPCC Guidelines for National Greenhouse Gas Inventories*. Intergovernmental Panel on Climate Change, Kanagawa, Japan.

Erisman, J. W., Sutton, M. A., Galloway, J., Klimont, Z. and Winiwarter, W. 2008. How a century of ammonia synthesis changed the world. *Nat. Geosci.* 1, 636–9. https://doi.org/10.1038/ngeo325.

FAO. 2018. FAOSTAT Statistics Database.

FAO and ITPS. 2015. *Status of the World's Soil Resources (SWSR) – Main Report*. Rome.

Frolking, S., Talbot, J., Jones, M. C., Treat, C. C., Kauffman, J. B., Tuittila, E.-S. and Roulet, N. 2011. Peatlands in the Earth's 21st century climate system. *Environ. Rev.* 19, 371–96. https://doi.org/10.1139/a11-014.

Gelfand, I., Zenone, T., Jasrotia, P., Chen, J. Q., Hamilton, S. K. and Robertson, G. P. 2011. Carbon debt of Conservation Reserve Program (CRP) grasslands converted to bioenergy production. *Proc. Natl. Acad. Sci. U. S. A.* 108, 13864–9. https://doi.org/10.1073/pnas.1017277108.

Gerber, J. S., Carlson, K. M., Makowski, D., Mueller, N. D., de Cortazar-Atauri, I. G., Havlik, P., Herrero, M., Launay, M., O'Connell, C. S., Smith, P. and West, P. C. 2016. Spatially explicit estimates of N2O emissions from croplands suggest climate mitigation opportunities from improved fertilizer management. *Glob. Chang. Biol.* 22, 3383–94. https://doi.org/10.1111/gcb.13341.

Godfray, H. C. J., Beddington, J. R., Crute, I. R., Haddad, L., Lawrence, D., Muir, J. F., Pretty, J., Robinson, S., Thomas, S. M. and Toulmin, C. 2010. Food security: the challenge of feeding 9 billion people. *Science* 327, 812–18. https://doi.org/10.1126/science.1185383.

Grace, P. R., Antle, J., Aggarwal, P. K., Ogle, S., Paustian, K. and Basso, B. 2012. Soil carbon sequestration and associated economic costs for farming systems of the Indo-Gangetic Plain: A meta-analysis. *Agric. Ecosyst. Environ.* 146, 137–46. https://doi.org/10.1016/j.agee.2011.10.019.

Groffman, P. M., Butterbach-Bahl, K., Fulweiler, R. W., Gold, A. J., Morse, J. L., Stander, E. K., Tague, C., Tonitto, C. and Vidon, P. 2009. Challenges to incorporating spatially and temporally explicit phenomena (hotspots and hot moments) in denitrification models. *Biogeochemistry* 93, 49–77. https://doi.org/10.1007/s10533-008-9277-5.

Harden, J. W., Hugelius, G., Ahlström, A., Blankinship, J. C., Bond-Lamberty, B., Lawrence, C. R., Loisel, J., Malhotra, A., Jackson, R. B., Ogle, S., Phillips, C., Ryals, R., Todd-Brown, K., Vargas, R., Vergara, S. E., Cotrufo, M. F., Keiluweit, M., Heckman, K. A., Crow, S. E., Silver, W. L., DeLonge, M. and Nave, L. E. 2017. Networking our science to characterize the state, vulnerabilities, and management opportunities of soil organic matter. *Glob Chang Biol.* 24, e705–18. https://doi.org/10.1111/gcb.13896.

Harper, A. B., Powell, T., Cox, P. M., House, J., Huntingford, C., Lenton, T. M., Sitch, S., Burke, E., Chadburn, S. E., Collins, W. J., Comyn-Platt, E., Daioglou, V., Doelman, J. C., Hayman, G., Robertson, E., van Vuuren, D., Wiltshire, A., Webber, C. P., Bastos, A., Boysen, L., Ciais, P., Devaraju, N., Jain, A. K., Krause, A., Poulter, B. and Shu, S. J. 2018. Land-use emissions play a critical role in landbased mitigation for Paris climate targets. *Nat. Commun.* 9. https://doi.org/ARTN 293810.1038/s41467-018-05340-z.

Hooijer, A., Page, S., Canadell, J. G., Silvius, M., Kwadijk, J., Wösten, H., and Jauhiainen, J. 2010. Current and future CO_2 emissions from drained peatlands in Southeast Asia. *Biogeosciences* 7, 1505–1514. https://doi.org/10.5194/bg-7-1505-2010.

Houghton, R. A., House, J. I., Pongratz, J., Van Der Werf, G. R., DeFries, R. S., Hansen, M. C., et al. 2012. Carbon emissions from land use and land-cover change. *Biogeosciences* (9, 5125–5142. https://doi.org/10.5194/bg-9-5125-2012.

ICSU. 2017. *A Guide to SDG Interactions: from Science to Implementation*. International Council for Science, Paris, France.

Joosten, H. 2010. *The Global Peatland CO_2 picture Peatland Status and Drainage Related Emissions in All Countries of the World*. Wetlands International. https://doi.org/10.1137/S1064827501399006.

Kanter, D. R. and Brownlie, W. J. 2019. Joint nitrogen and phosphorus managemetn for sustainable development and climate goals. *Environ. Sci. Policy* 92, 1–8.

Kanter, D. R., Musumba, M., Wood, S. L. R., Palm, C., Antle, J., Balvanera, P., Dale, V. H., Havlik, P., Kline, K. L., Scholes, R. J., Thornton, P., Tittonell, P. and Andelman, S. 2018. Evaluating agricultural trade-offs in the age of sustainable development. *Agric. Syst.* 163, 73–88. https://doi.org/10.1016/j.agsy.2016.09.010.

Kirschke, S., Bousquet, P., Ciais, P., Saunois, M., Canadell, J. G., Dlugokencky, E. J., Bergamaschi, P., Bergmann, D., Blake, D. R., Bruhwiler, L., Cameron-Smith, P., Castaldi, S., Chevallier, F., Feng, L., Fraser, A., Heimann, M., Hodson, E. L., Houweling, S., Josse, B., Fraser, P. J., Krummel, P. B., Lamarque, J.-F., Langenfelds, R. L., Le Quéré, C., Naik, V., O'Doherty, S., Palmer, P. I., Pison, I., Plummer, D., Poulter, B., Prinn, R. G., Rigby, M., Ringeval, B., Santini, M., Schmidt, M., Shindell, D. T., Simpson, I. J., Spahni, R., Steele, L. P., Strode, S. A., Sudo, K., Szopa, S., van der Werf, G. R., Voulgarakis, A., van Weele, M., Weiss, R. F., Williams, J. E. and Zeng, G. 2013. Three decades of global methane sources and sinks. *Nat. Geosci.* 6, 813–23. https://doi.org/10.1038/ngeo1955.

Kritee, K., Nair, D., Zavala-Araiza, D., Proville, J., Rudek, J., Adhya, T. K., Loecke, T., Esteves, T., Balireddygari, S., Dava, O., Ram, K., Abhilash, S. R., Madasamy, M., Dokka, R. V, Anandaraj, D., Athiyaman, D., Reddy, M., Ahuja, R. and Hamburg, S. P. 2018. High nitrous oxide fluxes from rice indicate the need to manage water for both long- and short-term climate impacts. *Proc. Natl. Acad. Sci. U. S. A.* 201809276. https://doi.org/10.1073/pnas.1809276115.

Lal, R. 2004. Agricultural activities and the global carbon cycle. *Nutr. Cycl. Agroecosyst.* 70, 103–16. https://doi.org/10.1023/B:FRES.0000048480.24274.0f.

Lal, R. 2010. Enhancing eco-efficiency in agro-ecosystems through soil carbon sequestration. *Crop Sci.* 50, S-120-S-131. https://doi.org/10.2135/cropsci2010.01.0012.

Lal, R. 2011. Sequestering carbon in soils of agro-ecosystems. *Food Policy* 36, S33--S39. https://doi.org/10.1016/j.foodpol.2010.12.001.

Lal, R. 2015. Restoring Soil Quality to Mitigate Soil Degradation. *Sustainability* 7, 5875–95. https://doi.org/10.3390/su7055875.

Lampayan, R. M., Rejesus, R. M., Singleton, G. R. and Bouman, B. A. M. 2015. Adoption and economics of alternate wetting and drying water management for irrigated lowland rice. *Field Crop. Res.* 170, 95–108. https://doi.org/10.1016/J.FCR.2014.10.013.

Leifeld, J. and Menichetti, L. 2018. The underappreciated potential of peatlands in global climate change mitigation strategies. *Nat. Commun.* 9(1), 1071. https://doi.org/10.1038/s41467-018-03406-6.

Lin, H.-W., Jin, Y., Giglio, L., Foley, J. A. and Randerson, J. T. 2012. Evaluating greenhouse gas emissions inventories for agricultural burning using satellite observations of active fires. *Ecol. Appl.* 22, 1345–64. https://doi.org/10.1890/10-2362.1.

Linquist, B. A., Anders, M. M., Adviento-Borbe, M. A. A., Chaney, R. L., Nalley, L. L., da Rosa, E. F. F. and van Kessel, C. 2015. Reducing greenhouse gas emissions, water use, and grain arsenic levels in rice systems. *Glob. Chang. Biol.* 21, 407–17. https://doi.org/10.1111/gcb.12701.

Minasny, B., Malone, B. P., McBratney, A. B., Angers, D. A., Arrouays, D., Chambers, A., Chaplot, V., Chen, Z.-S., Cheng, K., Das, B. S., Field, D. J., Gimona, A., Hedley, C. B., Hong, S. Y., Mandal, B., Marchant, B. P., Martin, M., McConkey, B. G., Mulder, V. L., O'Rourke, S., Richer-de-Forges, A. C., Odeh, I., Padarian, J., Paustian, K., Pan, G., Poggio, L., Savin, I., Stolbovoy, V., Stockmann, U., Sulaeman, Y., Tsui, C.-C., Vågen, T.-G., van Wesemael, B. and Winowiecki, L. 2017. Soil carbon 4 per mille. *Geoderma* 292, 59–86. https://doi.org/10.1016/J.GEODERMA.2017.01.002.

Mogollon, J. M., Lassaletta, L., Beusen, A. H. W., van Grinsven, H. J. M., Westhoek, H. and Bouwman, A. F. 2018. Assessing future reactive nitrogen inputs into global croplands based on the shared socioeconomic pathways. *Environ. Res. Lett.* 13. https://doi.org/ARTN 04400810.1088/1748-9326/aab212.

Murdiyarso, D., Hergoualc'h, K. and Verchot, L. V, 2010. Opportunities for reducing greenhouse gas emissions in tropical peatlands. *Proc. Natl. Acad. Sci.* 107, 19655–60 SPECIAL. https://doi.org/10.1073/pnas.0911966107/-/DCSupplemental.

Nakicenovic, J., Davis, G., Vries, B. d., Fenhann, J. V., Gaffin, S., Gregory, K., Grübler, A., Jung, T. Y., Kram, T., Rovere, E. L., Michaelis, L., Mori, S., Morita, T., Pepper, W., Pitcher, H., Price, L., Riahi, K., Roehrl, A., Rogner, H. H., Sankovski, A., Alcamo, N. and Press, C. U. 2000. *Special Report on Emissions Scenarios: Special Report of Working Group III of the Intergovernmental Panel on Climate Change*. Intergovernmental Panel on Climate Change.

Ogle, S. M., McCarl, B. A., Baker, J., Del Grosso, S. J., Adler, P. R., Paustian, K. and Parton, W. J. 2016. Managing the nitrogen cycle to reduce greenhouse gas emissions from crop production and biofuel expansion. *Mitig. Adapt. Strateg. Glob. Chang.* 21, 1197–212. https://doi.org/10.1007/s11027-015-9645-0.

O'Neill, B. C., Oppenheimer, M., Warren, R., Hallegatte, S., Kopp, R. E., Pörtner, H. O., Scholes, R., Birkmann, J., Foden, W., Licker, R., Mach, K. J., Marbaix, P., Mastrandrea, M. D., Price, J., Takahashi, K., Van Ypersele, J. P. and Yohe, G. 2017. IPCC reasons for concern regarding climate change risks. *Nat. Clim. Chang.* 7. https://doi.org/10.1038/nclimate3179.

Oo, A. Z., Sudo, S., Inubushi, K., Mano, M., Yamamoto, A., Ono, K., Osawa, T., Hayashida, S., Patra, P. K., Terao, Y., Elayakumar, P., Vanitha, K., Umamageswari, C., Jothimani, P. and Ravi, V. 2018. Methane and nitrous oxide emissions from conventional and modified rice cultivation systems in South India. *Agric. Ecosyst. Environ.* 252, 148–58. https://doi.org/10.1016/J.AGEE.2017.10.014.

Page, S. E., Morrison, R., Malins, C., Hooijer, A., Rieley, J. O. and Jauhiainen, J. 2011a. Review of peat surface greenhouse gas emissions from oil palm plantations in Southeast Asia. Indirect effects of biofuel production series.

Page, S. E., Rieley, J. O. and Banks, C. J. 2011b. Global and regional importance of the tropical peatland carbon pool. *Glob. Chang. Biol.* 17, 798–818. https://doi.org/10.1111/j.1365-2486.2010.02279.x.

Park, S., Croteau, P., Boering, K. A., Etheridge, D. M., Ferretti, D., Fraser, P. J., Kim, K. R., Krummel, P. B., Langenfelds, R. L., van Ommen, T. D., Steele, L. P. and Trudinger, C. M. 2012. Trends and seasonal cycles in the isotopic composition of nitrous oxide since 1940. *Nat. Geosci.* 5, 261-65. https://doi.org/10.1038/Ngeo1421.

Paustian, K., Lehmann, J., Ogle, S., Reay, D., Robertson, G. P. and Smith, P. 2016. Climate-smart soils. *Nature* 532, 49-57. https://doi.org/10.1038/nature17174.

Pirker, J., Mosnier, A., Kraxner, F., Havlík, P. and Obersteiner, M. 2016. What are the limits to oil palm expansion? *Glob. Environ. Chang.* 40, 73-81. https://doi.org/10.1016/J.GLOENVCHA.2016.06.007.

Popp, A., Calvin, K., Fujimori, S., Havlik, P., Humpenoder, F., Stehfest, E., Bodirsky, B. L., Dietrich, J. P., Doelmann, J. C., Gusti, M., Hasegawa, T., Kyle, P., Obersteiner, M., Tabeau, A., Takahashi, K., Valin, H., Waldhoff, S., Weindl, I., Wise, M., Kriegler, E., Lotze-Campen, H., Fricko, O., Riahi, K. and van Vuuren, D. P. 2017. Land-use futures in the shared socio-economic pathways. *Glob. Environ. Chang. Policy Dimens.* 42, 331-45. https://doi.org/10.1016/j.gloenvcha.2016.10.002.

Powlson, D. S., Stirling, C. M., Jat, M. L., Gerard, B. G., Palm, C. A., Sanchez, P. A. and Cassman, K. G. 2014. Limited potential of no-till agriculture for climate change mitigation. *Nat. Clim. Chang.* 4, 678-83. https://doi.org/10.1038/nclimate2292.

Ramankutty, N., Evan, A. T., Monfreda, C. and Foley, J. A. (2008). Farming the planet: 1. Geographic distribution of global agricultural lands in the year 2000. *Global Biogeochemical Cycles, 22*(1). https://doi.org/10.1029/2007GB002952.

Riahi, K., van Vuuren, D. P., Kriegler, E., Edmonds, J., O'Neill, B. C., Fujimori, S., Bauer, N., Calvin, K., Dellink, R., Fricko, O., Lutz, W., Popp, A., Cuaresma, J. C., Samir, K. C., Leimbach, M., Jiang, L. W., Kram, T., Rao, S., Emmerling, J., Ebi, K., Hasegawa, T., Havlik, P., Humpenoder, F., da Silva, L. A., Smith, S., Stehfest, E., Bosetti, V., Eom, J., Gernaat, D., Masui, T., Rogelj, J., Strefler, J., Drouet, L., Krey, V., Luderer, G., Harmsen, M., Takahashi, K., Baumstark, L., Doelman, J. C., Kainuma, M., Klimont, Z., Marangoni, G., Lotze-Campen, H., Obersteiner, M., Tabeau, A. and Tavoni, M. 2017. The Shared Socioeconomic Pathways and their energy, land use, and greenhouse gas emissions implications: An overview. *Glob. Environ. Chang. Policy Dimens.* 42, 153-68. https://doi.org/10.1016/j.gloenvcha.2016.05.009.

Robertson, G. P. and Groffman, P. M. 2007. Nitrogen transformations. In: Paul, E. A. (Ed.), *Soil Microbiology, Biochemistry, and Ecology*. Springer, pp. 341-64.

Robertson, G. P., Hamilton, S. K., Barham, B. L., Dale, B. E., Izaurralde, R. C., Jackson, R. D., Landis, D. A., Swinton, S. M., Thelen, K. D. and Tiedje, J. M. 2017. Cellulosic biofuel contributions to a sustainable energy future: Choices and outcomes. *Science* 356. https://doi.org/ARTN eaal232410.1126/science.aal2324.

Rogelj, J., Shindell, D., Jiang, K., Fifita, S., Forsters, P., Ginzberg, V., Handa, C., Kheshgi, H., Kobayashi, S., Kriegler, E., Mundaca, L., Séférian, R. and Vilariño, M. V, 2018. Mitigation pathways compatible with 1.5C in the context of sustainable development. In: Masson-Delmotte, V., Zhai, P., Pörtner, H. O., Roberts, D., Skea, J., Shukla, P. R., Pirani, A., Moufouma-Okia, W., Péan, C., Pidcock, R., Connors, S., Matthews, J. B. R., Chen, Y., Zhou, X., Gomis, M. I., Lonnoy, E., Maycock, T., Tignor, M. and Waterfield, T. (Eds), *IPCC, 2018: Global Warming of 1.5C. An IPCC Special Report on the Impacts of Global Warming of 1.5C above Pre-industrial Levels and Related Global Greenhouse Gas Emissions Pathways, in the Context of Strengthening the Global Response to the Threat of Climate Change.*

Rosenzweig, C., Jones, J. W., Hatfield, J. L., Ruane, A. C., Boote, K. J., Thorburn, P., Antle, J. M., Nelson, G. C., Porter, C., Janssen, S., Asseng, S., Basso, B., Ewert, F., Wallach, D., Baigorria, G. and Winter, J. M. 2013. The Agricultural Model Intercomparison and Improvement Project (AgMIP): Protocols and pilot studies. *Agric. For. Meteorol.* 170, 166–82. https://doi.org/10.1016/j.agrformet.2012.09.011.

Rosenzweig, C., Elliott, J., Deryng, D., Ruane, A. C., Müller, C., Arneth, A., Boote, K. J., Folberth, C., Glotter, M., Khabarov, N., Neumann, K., Piontek, F., Pugh, T. A., Schmid, M. E., Stehfest, E., Yang, H. and Jones, J. W. 2014. Assessing agricultural risks of climate change in the 21st century in a global gridded crop model intercomparison. *Proc. Natl. Acad. Sci. U. S. A.* 111, 3268–73. https://doi.org/10.1073/pnas.1222463110.

Ruan, L. L., Bhardwaj, A. K., Hamilton, S. K. and Robertson, G. P. 2016. Nitrogen fertilization challenges the climate benefit of cellulosic biofuels. *Environ. Res. Lett.* 11. https://doi.org/Artn 06400710.1088/1748-9326/11/6/064007.

Saikawa, E., Prinn, R. G., Dlugokencky, E., Ishijima, K., Dutton, G. S., Hall, B. D., Langenfelds, R., Tohjima, Y., Machida, T., Manizza, M., Rigby, M., O'Doherty, S., Patra, P. K., Harth, C. M., Weiss, R. F., Krummel, P. B., van der Schoot, M., Fraser, P. J., Steele, L. P., Aoki, S., Nakazawa, T. and Elkins, J. W. 2014. Global and regional emissions estimates for N_2O. *Atmos. Chem. Phys.* 14, 4617–41. https://doi.org/10.5194/acp-14-4617-2014.

Sanderman, J. and Baldock, J. A. 2010. Accounting for soil carbon sequestration in national inventories: a soil scientist's perspective. *Environ. Res. Lett.* 5, 034003. https://doi.org/10.1088/1748-9326/5/3/034003.

Sanderman, J., Hengl, T. and Fiske, G. J. 2017. Soil carbon debt of 12,000 years of human land use. *Proc. Natl. Acad. Sci.* https://doi.org/10.1073/pnas.1706103114.

Searchinger, T. D., Hamburg, S. P., Melillo, J., Chameides, W., Havlik, P., Kammen, D. M., Likens, G. E., Lubowski, R. N., Obersteiner, M., Oppenheimer, M., Robertson, G. P., Schlesinger, W. H. and Tilman, G. D. 2009. Fixing a critical climate accounting error. *Science* 326, 527–8. https://doi.org/10.1126/science.1178797.

Shcherbak, I., Millar, N. and Robertson, G. P. 2014. Global metaanalysis of the nonlinear response of soil nitrous oxide (N_2O) emissions to fertilizer nitrogen. *Proc. Natl. Acad. Sci. U. S. A.* 111, 9199–204. https://doi.org/10.1073/pnas.1322434111.

Sivakumar, M. V. K. 2007. Interactions between climate and desertification. *Agric. For. Meteorol.* 142, 143–55. https://doi.org/10.1016/j.agrformet.2006.03.025.

Smith, P. 2008. Land use change and soil organic carbon dynamics. *Nutr. Cycl. Agroecosyst.* 81, 169–78. https://doi.org/10.1007/s10705-007-9138-y.

Smith, P., Ashmore, M. R., Black, H. I. J., Burgess, P. J., Evans, C. D., Quine, T. A., Thomson, A. M., Hicks, K. and Orr, H. G. 2013a. REVIEW: The role of ecosystems and their management in regulating climate, and soil, water and air quality. *J. Appl. Ecol.* 50, 812–29. https://doi.org/10.1111/1365-2664.12016.

Smith, P., Bustamante, M., Ahammad, H., Clark, H., Dong, H., Elsiddig, E. A., Haberl, H., Harper, R., House, J., Jafari, M., Masera, O., Mbow, C., Ravindranath, N. H., Rice, C. W., Robledo Abad, C., Romanovskaya, A., Sperling, F. and Tubiello, F. 2013b. Agriculture, forestry and other land use. In: Edenhofer, O., Pichs-Madruga, R. and Sokona, Y. (Eds), *Mitigation of Climate Change – Working Group III Contribution to the Fifth Assessment Report of the Intergovernmental Panel on Climate Change*. Cambridge University Press.

Smith, P., House, J. I., Bustamante, M., Sobocká, J., Harper, R., Pan, G., West, P. C., Clark, J. M., Adhya, T., Rumpel, C., Paustian, K., Kuikman, P., Cotrufo, M. F., Elliott, J. A.,

McDowell, R., Griffiths, R. I., Asakawa, S., Bondeau, A., Jain, A. K., Meersmans, J. and Pugh, T. A. M. 2016. Global change pressures on soils from land use and management. *Glob. Chang. Biol.* 22, 1008–28. https://doi.org/10.1111/gcb.13068.
Smith P., Bustamante, M., Ahammad, H., Clark, H., Dong, H., Elsiddig, E. A., Haberl, H., Harper, R., House, J., Jafari, M., Masera, O., Mbow, C., Ravindranath, N. H., Rice, C. W., Robledo Abad, C., Romanovskaya, A., Sperling, F. and Tubiello, F. 2014: Agriculture, Forestry and Other Land Use (AFOLU). In: Edenhofer, O., Pichs-Madruga, R., Sokona, Y., Farahani, E., Kadner, S., Seyboth, K., Adler, A., Baum, I., Brunner, S., Eickemeier, P., Kriemann, B., Savolainen, J., Schlömer, S., von Stechow, C., Zwickel, T. and Minx, J. C. (Eds), *Climate Change 2014: Mitigation of Climate Change. Contribution of Working Group III to the Fifth Assessment Report of the Intergovernmental Panel on Climate Change.* Cambridge University Press, Cambridge, United Kingdom and New York, NY, USA.
Sommer, R. and Bossio, D. 2014. Dynamics and climate change mitigation potential of soil organic carbon sequestration. *J. Environ. Manag.* 144, 83–7. https://doi.org/10.1016/j.jenvman.2014.05.017.
Soussana, J.-F., Lutfalla, S., Ehrhardt, F., Rosenstock, T., Lamanna, C., Havlík, P., Richards, M., Wollenberg, E. (Lini), Chotte, J.-L., Torquebiau, E., Ciais, P., Smith, P. and Lal, R. 2017. Matching policy and science: Rationale for the '4 per 1000 – soils for food security and climate' initiative. *Soil Tillage Res.* 188, 3–15. https://doi.org/10.1016/J.STILL.2017.12.002.
Tigchelaar, M., Battisti, D. S., Naylor, R. L. and Ray, D. K. 2018. Future warming increases probability of globally synchronized maize production shocks. *Proc. Natl. Acad. Sci.* 201718031. https://doi.org/10.1073/PNAS.1718031115.
Tubiello, F. N., Salvatore, M., Cóndor Golec, R. D., Ferrara, A., Rossi, S., Biancalani, R., Federici, S., Jacobs, H. and Flammini, A. 2014. Agriculture, forestry and other land use emissions by sources and removals by sinks. *FAO – e Food Agric. Organ. United Nations* 2, 4–89.
Turner, K. G., Anderson, S., Gonzales-Chang, M., Costanza, R., Courville, S., Dalgaard, T., Dominati, E., Kubiszewski, I., Ogilvy, S., Porfirio, L., Ratna, N., Sandhu, H., Sutton, P. C., Svenning, J.-C., Turner, G. M., Varennes, Y.-D., Voinov, A. and Wratten, S. 2015. A review of methods, data, and models to assess changes in the value of ecosystem services from land degradation and restoration. *Ecol. Modell.* 319, 190–207. https://doi.org/10.1016/j.ecolmodel.2015.07.017.
UN. 2015. *Transforming Our World: The 2030 Agenda for Sustainable Development.* United Nations, New York, NY.
UNEP. 2013. *Drawing Down N_2O To Protect Climate and the Ozone Layer.* https://doi.org/978-92-807-3358-7.
USDA. 2018. *World Agricultural Production, Circular Series.* Washington DC.
US EPA. 2012. *Global Anthropogenic Non-CO_2 Greenhouse Gas Emissions: 1990-2030.* Washington DC. https://doi.org/EPA 430-R-12-006.
Valdivia, R. O., Antle, J. M., Rosenzweig, C., Ruane, A. C., Vervoort, J., Ashfaq, M., Hathie, I., Tui, S. H.-K., Mulwa, R., Nhemachena, C., Ponnusamy, P., Rasnayaka, H. and Singh, H. 2015. Representative agricultural pathways and scenarios for regional integrated assessment of climate change impacts, vulnerability, and adaptation. In: Rosenzweig, C. and Hillel, D. (Eds), *Handbook of Climate Change and Agroecosystems: The Agricultural Model Intercomparison and Improvement Project (AgMIP) Integrated Crop and Economic Assessments.* Imperial College Press, London, UK, pp. 101–45. https://doi.org/10.1142/9781783265640_0005.

Van Der Werf, G. R., Randerson, J. T., Giglio, L., Collatz, G. J., Kasibhatla, P. S. and Arellano, A. F. 2006. Interannual variability in global biomass burning emissions from 1997 to 2004. *Atmos. Chem. Phys.* 6, 3423–41. https://doi.org/10.5194/acp-6-3423-2006

van Groenigen, K. J., van Kessel, C. and Hungate, B. A. 2013. Increased greenhouse-gas intensity of rice production under future atmospheric conditions. *Nat. Clim. Chang.* 3, 288–91. https://doi.org/10.1038/nclimate1712.

Van Groenigen, J. W., Kessel, C. Van, Hungate, B. A., Oenema, O., Powlson, D. S. and Van Groenigen, K. J. 2017. Sequestering soil organic carbon: a nitrogen dilemma. *Environ. Sci. Technol.* 51, 4738–9. https://doi.org/10.1021/acs.est.7b01427.

van Vuuren, D. P., Edmonds, J., Kainuma, M., Riahi, K., Thomson, A., Hibbard, K., Hurtt, G. C., Kram, T., Krey, V., Lamarque, J. F., Masui, T., Meinshausen, M., Nakicenovic, N., Smith, S. J. and Rose, S. K. 2011. The representative concentration pathways: an overview. *Clim. Change* 109, 5–31. https://doi.org/10.1007/s10584-011-0148-z.

Vermeulen, S. J., Aggarwal, P. K., Ainslie, A., Angelone, C., Campbell, B. M., Challinor, A. J., Hansen, J. W., Ingram, J. S. I., Jarvis, A., Kristjanson, P., Lau, C., Nelson, G. C., Thornton, P. K. and Wollenberg, E. 2012a. Options for support to agriculture and food security under climate change. *Environ. Sci. Policy* 15, 136–44. https://doi.org/10.1016/J.ENVSCI.2011.09.003.

Vermeulen, S. J., Campbell, B. M. and Ingram, J. S. I. 2012b. Climate Change and Food Systems. *Annu. Rev. Environ. Resour.* 37, 195–222. https://doi.org/10.1146/annurev-environ-020411-130608.

Vijay, V., Pimm, S. L., Jenkins, C. N., Smith, S. J., Walker, W., Soto, C., Trigg, S., Gaveau, D., Lawrence, D. and Rodrigues, H. 2016. The Impacts of Oil Palm on Recent Deforestation and Biodiversity Loss. *PLoS ONE* 11, e0159668. https://doi.org/10.1371/journal.pone.0159668.

Wei, X., Shao, M., Gale, W. and Li, L. 2015. Global pattern of soil carbon losses due to the conversion of forests to agricultural land. *Sci. Rep.* 4, 4062. https://doi.org/10.1038/srep04062.

Wells, K. C., Millet, D. B., Bousserez, N., Henze, D. K., Griffis, T. J., Chaliyakunnel, S., Dlugokencky, E. J., Saikawa, E., Xiang, G., Prinn, R. G., O'Doherty, S., Young, D., Weiss, R. F., Dutton, G. S., Elkins, J. W., Krummel, P. B., Langenfelds, R. and Steele, L. P. 2018. Top-down constraints on global N2O emissions at optimal resolution: application of a new dimension reduction technique. *Atmos. Chem. Phys.* 18, 735–56. https://doi.org/10.5194/acp-18-735-2018.

West, T. O. and Post, W. M. 2002. Soil organic carbon sequestration rates by tillage and crop rotation: A global data analysis. *Soil Sci. Soc. Am. J.* 66, 1930–46. https://doi.org/DOI 10.2136/sssaj2002.1930.

Wijedasa, L. S., Sloan, S., Page, S. E., Clements, G. R., Lupascu, M. and Evans, T. A. 2018. Carbon emissions from South-East Asian peatlands will increase despite emission-reduction schemes. *Glob. Chang. Biol.* 24, 4598–613. https://doi.org/10.1111/gcb.14340.

WMO. 2014. *Scientific Assessment of Ozone Depletion: 2014, Global Ozone Research and Monitoring Project*. Report No. 55. Geneva, Switzerland.

Zomer, R. J., Bossio, D. A., Sommer, R. and Verchot, L. V. 2017. Global sequestration potential of increased organic carbon in cropland soils. *Sci. Rep.* 7, 15554. https://doi.org/10.1038/s41598-017-15794-8.

Chapter 6

The role of agricultural expansion, land cover and land-use change in contributing to climate change

Catherine E. Scott, University of Leeds, UK

1 Introduction

Land-use change has accompanied the arrival and movement of human populations into and between regions of the Earth for thousands of years. The dominant effect of human arrival is the removal of forests to provide land that can be used for agriculture. Reliable land-use surveys exist only from the mid-twentieth century onwards, so the Earth's vegetation distribution must be reconstructed prior to that. On geological time scales, the knowledge of climatic conditions and indicators in the fossil record enable a reconstruction of natural vegetation across the globe.

Reconstructing land cover during the period of more substantial human influence (i.e. the past several thousand years) presents considerable challenges and is often based on estimates of the human population as well as the assumption that the amount of land 'used' per person for agricultural purposes has remained broadly similar over time (e.g. Ramankutty and Foley 1999; Klein Goldewijk 2001; Pongratz et al. 2008; Klein Goldewijk et al. 2011).

An alternative approach (Boserup 1965; Ruddiman 2003; Kaplan et al. 2009) considers the possibility that the amount of land required per capita may have declined substantially over time due to the intensification of agricultural

http://dx.doi.org/10.19103/AS.2020.0064.10

practises. Combining this approach with the estimates of population change results in much larger areas of land being under agricultural use over the past two millennia (Fig. 1; Kaplan et al. 2011).

Up until the Industrial Revolution in the eighteenth century, there was widespread deforestation across the temperate regions of Asia, Europe and North America on the land that was considered most suitable for farming (Williams 2003). At the start of the twentieth century, deforestation rates in tropical regions began to accelerate (FAO 2012), particularly in South and Central America, Southeast Asia and Central Africa.

Forest clearance in the tropics still occurs predominantly to acquire the land that is suitable for agriculture with more than 80% of new agricultural land acquired across the tropics between 1980 and 2000 coming from the clearance of intact or disturbed forests (Gibbs et al. 2010); however, the specific commodities driving agricultural clearance vary from region to region within the tropics.

In South and Central America, beef cattle ranching has been the dominant driver of 'modern' forest clearance (e.g. Grainger 1993; Fearnside 2005; Gibbs et al. 2010). During the 1960s, 1970s and 1980s, road building and financial incentives by the Brazilian government encouraged deforestation of the Amazon to create pasture and cattle ranches (Carvalho et al. 2002; Fearnside 2005). This process often involves clearance of the forest followed by burning

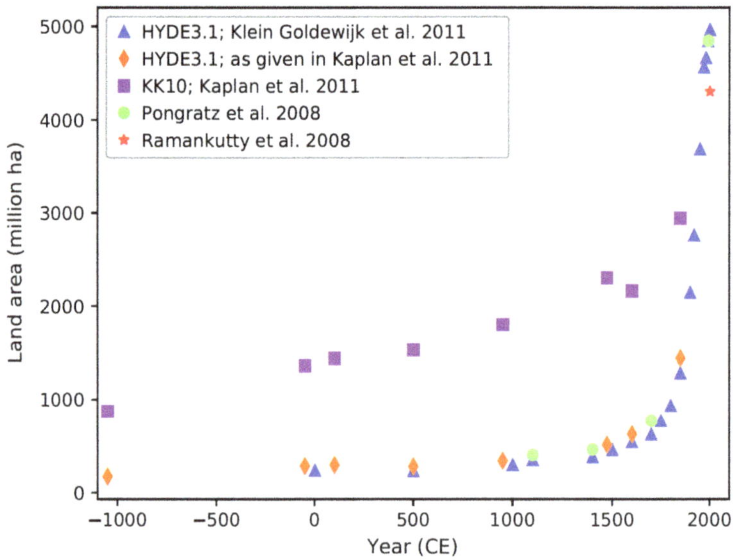

Figure 1 Land area occupied by agricultural activities (i.e. cropland and pastureland combined) in historical reconstructions up to the present day (Pongratz et al. 2008; Ramankutty et al. 2008; Kaplan et al. 2011; Klein Goldewijk et al. 2011).

to remove any residual trees. Until the 1990s, Brazilian beef was usually sold domestically, but in the early 2000s, international demand for Brazilian beef partly drove a spike in deforestation rates between 2002 and 2004 (Nepstad et al. 2006).

During the 1970s, an oil embargo prompted a rapid expansion of biofuel crop growth in South America, specifically the growth of sugar cane to produce ethanol. In 1977, the Brazilian government mandated that all gasoline must be blended with ethanol. This mandate is still in place and the current minimum blend level is set at 27% ethanol. As well as directly driving forest clearance, the growth of bioenergy crops can indirectly lead to deforestation if it displaces food production, which then moves onto forested land.

During the 1990s and 2000s the growth of soybeans also began to contribute substantially to the clearance of Amazon forests. Rather than being directly consumed by humans, soybean crop is mainly used to feed cattle, pigs and chicken. In the late 1990s, new cultivars enabled farmers to grow soybeans in regions that had not previously been climatically suitable, leading to a rapid expansion of soy farms into the Amazon forest (Fearnside 2001; Nepstad et al. 2006). In response to growing environmental concerns, a moratorium was announced by the exporters and processors of soybeans stating that they would not buy crops grown on farmland within the Brazilian Amazon that had been deforested since June 2006. Since its implementation, the soy moratorium appears to have been successful, with most new soy expansion occurring on previously cleared land (Rudorff et al. 2011; Gibbs et al. 2015) and has contributed to the overall decline in Brazilian deforestation rates (Hansen et al. 2013).

In Southeast Asia, extensive forest loss has been driven by food and fuel crop growth, as well as rubber and timber production (Gibbs et al. 2008; Miettinen et al. 2011). Malaysia and Indonesia now produce more than 80% of the world's palm oil (USDA-FAS 2019) through both industrial scale and smallholder plantations (Schoneveld et al. 2019). Oil palms grow only in humid, tropical conditions but are extremely efficient producers of oil compared to other crops (e.g. soybean, sunflower, rapeseed). The major environmental issue associated with oil palm growth is the conversion of old-growth or peat forests which contain dense carbon stocks, 6% of tropical peatlands in Southeast Asia had been converted to oil palm plantations by the early 2000s (Koh et al. 2011). In 2011, the Indonesian government imposed a moratorium on new oil palm and timber plantations on peatlands or primary forests, but its effectiveness remains unclear (Busch et al. 2015).

In Africa, forests are cleared to provide wood fuel and to make way for smallholder agriculture, but information about the scale and extent of deforestation is less robust than for South America or Southeast Asia, both due to a lower reporting capacity and the challenges associated with detecting small-scale forest clearance by satellites (Malhi et al. 2013). The lack of industrial-scale land clearance for agriculture means that deforestation rates in Africa have remained lower than those in South America or Southeast Asia throughout the 1990s and 2000s (Achard et al. 2002; Mayaux et al. 2013). Other pressures on tropical forests include mineral mining (e.g. to obtain gold, copper, tin), coal mining and oil drilling which have been particularly prevalent in parts of South America and Africa (Grainger 1993).

Monitoring changes to land cover and rates of forest loss relies on either country-level statistics of forest area (e.g. FAO 2018) or, in recent years, remote sensing from both airborne (e.g. Asner 2009; Saatchi et al. 2011a) and satellite instruments (e.g. Defries et al. 2000; Hansen et al. 2003, 2013; Saatchi et al. 2011b). At the global scale, natural forests were being lost at a rate of 10.6 million ha per year during the 1990s. Between 2010 and 2015 this rate slowed to approximately 6.5 million ha lost per year (FAO 2015). For the 2010–2015 period, the overall rate of forest area change was estimated as a net loss of 3.3 million ha per year (FAO 2015); this is lower than the rate of direct forest loss due to extensive afforestation, particularly in China (Fang et al. 1998, 2001; Wang et al. 2007), and the natural expansion of forests onto previously managed lands.

2 Impacts of land-use change on climate

When land cover or land use is changed, the fluxes of carbon, energy and water between the land-surface and the atmosphere can be altered substantially (Fig. 2; Bonan 2008). Changes to these fluxes can be broken down into radiative (i.e. energy) and non-radiative (i.e. carbon and water) effects. The following sections outline our understanding of the way that these fluxes change, and estimates of the extent to which this has occurred due to agricultural expansion.

2.1 Carbon emission

Plants take in carbon, in the form of carbon dioxide (CO_2), from the atmosphere. Approximately half of this carbon is returned to the atmosphere during respiration, while the other half is fixed as plant biomass during photosynthesis. The metabolic activity of ecosystems influences the amount of carbon in the atmosphere, with a marked seasonal cycle evident in measurements of atmospheric CO_2 concentrations.

The total amount of carbon stored in terrestrial ecosystems is uncertain: the tropical forests of Latin America, sub-Saharan Africa and Asia are estimated to contain between 247 and 553 petagrams of carbon (PgC), temperate forests between 159 and 292 PgC and boreal forests between 395 and 559 PgC (Dixon et al. 1994; Prentice et al. 2001; Saatchi et al. 2011b). As a result of the carbon stored in the trees, a considerable emission of carbon can be associated

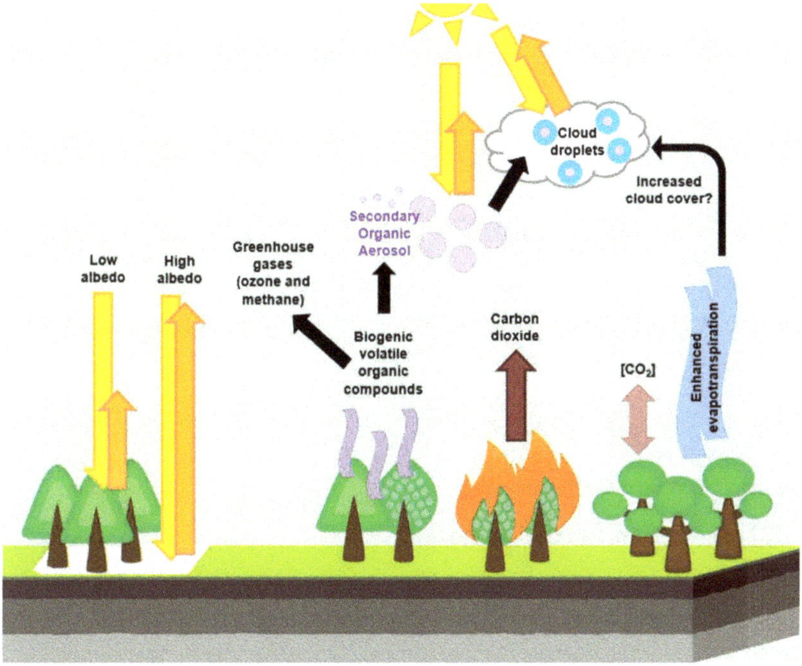

Figure 2 Schematic of the processes described in Section 2 of this chapter.

with the process of land-use and land cover change (LULCC), predominantly through the decay and burning of vegetation when forests are converted to agricultural land. Conversely, the process of afforestation will remove carbon dioxide from the atmosphere, generating a carbon sink.

Estimating the net flux of carbon to the atmosphere as a result of LULCC is challenging. Three methods are often used: (1) a 'bookkeeping' method that combines country-level forest area data with regionally averaged biomass values (e.g. Houghton 2003; Baccini et al. 2012); (2) dynamic global vegetation models (DGVMs) that simulate the fluxes of carbon between the land surface and the atmosphere; and (3) regional analyses based on satellite data (e.g. van der Werf et al. 2010).

While fluxes of carbon within the carbon cycle are conventionally referred to as an amount of carbon 'C', anthropogenic emissions are more commonly referred to in terms of a quantity of carbon dioxide 'CO_2', the value of which will be a factor of 3.67 (44 ÷ 12) higher for the same amount of carbon. For the most recent decade available (2008–2017), annual emissions from LULCC are estimated to be 1.5 ± 0.7 PgC (equivalent to 5.5 PgCO_2; Le Quéré et al. 2018), approximately 14% of the total annual carbon emission from anthropogenic activities. Despite this annual source of carbon emission being generated due to LULCC, globally the land remains a carbon sink because vegetation removes more carbon dioxide from the atmosphere than that is put back (Jia et al. 2019).

The challenges posed by quantifying LULCC emissions in the present day are further exacerbated prior to the mid-twentieth century where the changes in land-cover must be inferred (see Section 1). It is estimated that since the year 1750 CE, the process of land-use change has resulted in a total emission of almost 200 PgC into the atmosphere, compared to an estimated 27 PgC emitted due to LULCC between 800 CE and 1750 CE (Pongratz et al. 2009; Ciais et al. 2013). However, Kaplan et al. (2011) suggest that over 300 PgC could have been emitted as a result of agricultural land clearance by 1850 CE due to the larger area occupied in their estimates (Fig. 1).

2.2 Surface energy fluxes

2.2.1 Reflection of solar radiation

As well as influencing the atmospheric concentration of CO_2, the presence of large-scale vegetation also affects the energy balance at the Earth's surface. Forests are generally darker in colour than other land surface types, particularly cultivated vegetation such as cropland. This dark surface means that forested land absorbs most of the shortwave radiation that it receives from the Sun. The ratio of reflected to incident shortwave solar radiation is known as the albedo. A very bright surface, for example, fresh snow, would have an albedo of around 0.9, whereas dark surfaces like the ocean have a much lower albedo of around 0.1. Forests typically have a very low albedo of 0.08–0.19 (Betts and Ball 1997; Monteith and Unsworth 2008), whereas grass or cropland have a slightly higher albedo of 0.15–0.26 (Monteith and Unsworth 2008).

In the boreal region (above 60°N), snow covers the land surface for several months of the year. If this snow cover is lying on short vegetation, it will completely cover it and the surface albedo will be very high, for example, 0.75 (Betts and Ball 1997). However, if coniferous boreal trees are present, they will protrude from the snow, lowering the albedo to 0.1–0.15 (Leonard and Eschner 1968; Robinson and Kukla 1985; Thomas and Rowntree 1992; Betts and Ball 1997).

For an evergreen forest in a snow-free region, the albedo will be relatively constant year round. However, for the deciduous forests occupying temperate and boreal regions, the albedo of the forest varies according to the time of year and whether or not the trees are in leaf. In the winter months, the albedo of a deciduous forest will be mainly controlled by the underlying surface, and the process of leaf-out tends to increase the albedo of a deciduous forest by 20–50% (Hollinger et al. 2010; Richardson et al. 2013).

The overall climate impact of land-cover change is explored in Section 3, but the conversion of forest to cropland or other agricultural land will generally result in an increase in the surface albedo. The LULCC that has occurred since 1750 is therefore considered to have increased the overall albedo of the Earth's land, resulting in an overall cooling effect on the climate. This is quantified as a radiative forcing of approximately −0.18 W m^{-2} (Myhre et al. 2013).

2.2.2 Evapotranspiration and hydrological impacts

The presence of vegetation also mediates the transfer of water from the land surface to the atmosphere via evapotranspiration. Evapotranspiration is the sum of physical evaporation from soils and surfaces in the canopy, and biological transpiration. During transpiration, water that has been taken up from the soil via plant roots is lost through the stomata on leaves.

Higher rates of evapotranspiration have been measured above forests than other land-cover types (Spracklen et al. 2018). During the dry season, trees are able to sustain high evapotranspiration rates because their long roots, when compared to other vegetation, facilitate access to deep soil water (Nepstad et al. 1994; Canadell et al. 1996). Evapotranspiration plays such a strong role in hydrological cycling that parcels of air travelling over forests have been shown to produce at least twice as much rainfall as the air that has passed over little vegetation (Spracklen et al. 2012).

By modulating water fluxes, forests may also alter the distribution of low-level clouds. Observational studies have reached conflicting conclusions on the impact of deforestation on cloud cover. Wang et al. (2009) found that shallow clouds formed preferentially over patches of land in the Amazon that had been deforested, while Teuling et al. (2017) saw a strong increase in cloud cover over forested regions of western Europe.

As well as being an important part of the hydrological cycle, the process of evapotranspiration plays a vital role in the Earth's surface energy balance. The energy associated with a change from liquid water into water vapour during evapotranspiration is referred to as latent heat. The other important energy flux from the surface to the atmosphere is sensible heat, which refers directly to the change in atmospheric temperature. The ratio of sensible to latent heat is known as the Bowen ratio (Bowen 1926); by controlling evapotranspiration, vegetation can affect the Bowen ratio and influence local temperatures. Conversion of forests to cropland or grassland is therefore likely to reduce evapotranspiration rates, altering the Bowen ratio and potentially increasing local temperatures.

2.3 Emission of reactive gases from vegetation

In addition to storing carbon, controlling the reflectivity of the land surface and mediating the transfer of moisture to the atmosphere, vegetation present on the land surface can influence the atmospheric concentrations of a number of climatically important non-CO_2 greenhouse gases and particles.

The vegetation emits biogenic volatile organic compounds (BVOCs) into the air and the BVOCs include isoprene (with chemical formula C_5H_8), monoterpenes ($C_{10}H_{16}$) and sesquiterpenes ($C_{15}H_{24}$), with their emission rates dependent upon plant species, temperature, sunlight levels and atmospheric CO_2 concentration. Approximately 500 Tg of isoprene is emitted annually by vegetation, with around 100 Tg of monoterpenes and 30 Tg of sesquiterpenes (Went 1960; Rasmussen and Went 1965; Sanadze and Kursanov 1966; Guenther et al. 1991, 2012).

Producing BVOCs requires a large investment of energy from plants. This investment suggests that there is some form of advantage to be gained by their emission. Potential benefits to the plant include: enhancing resilience abiotic stress (e.g. temperature, light and oxidative damage; Loreto and Velikova 2001; Vickers et al. 2009), preventing the establishment of competing plants (Muller 1966), altering the climate (e.g. temperature, precipitation and cloud cover; for example, Spracklen et al. 2008; Paasonen et al. 2013; Scott et al. 2018a), allowing belowground signalling (e.g. Rasmann et al. 2005), or reducing insect and herbivore attack (e.g. Oh et al. 1967; Kessler and Baldwin 2001; Amin et al. 2013).

Once emitted into the atmosphere, BVOCs undergo a series of chemical reactions to give a wide range of products. One consequence of these atmospheric reactions is the formation of biogenic particles. By scattering incoming solar radiation, and acting as seeds for cloud droplet formation (thereby increasing the brightness of clouds), biogenic particles are likely to have a cooling effect on the global climate (Scott et al. 2014; Zhu et al. 2017, 2019). Recent research suggests that the products of monoterpene oxidation may be involved in the very first stages of new particle formation in the atmosphere, a process that was previously thought to rely on the presence of human-made pollution (Gordon et al. 2016; Kirkby et al. 2016).

Due to the complex atmospheric chemistry in which BVOCs participate, their presence can alter the concentration of some non-CO_2 greenhouse gases (ozone (O_3) and methane (CH_4)) which have a warming effect on the climate (Unger 2014). When evaluated over a 100-year time period, CH_4 is estimated to be between 28 and 34 times more effective at warming the climate than CO_2 (Myhre et al. 2013). The ability of O_3 to warm the climate is dependent on its location in the atmosphere, so it is not possible to quantify in quite the same way as CH_4. However, the rise in O_3 concentrations in the lower atmosphere due to anthropogenic activity is thought to be the third largest contributor to climate change of all greenhouse gases (behind CO_2 and CH_4; Myhre et al. 2013).

The results of the chemical reactions that BVOCs participate in depend upon the concentration of other gases in the atmosphere. While BVOCs can react directly with O_3, decreasing its concentration, under certain conditions (in the presence of sufficient nitrogen oxides), the emission of BVOCs contributes to the production of O_3, leading to an overall increase in O_3 concentration (Monks et al. 2015).

Unger (2014) found that LULCC since the year 1850 had resulted in a net cooling effect on climate through decreases in BVOC emission and therefore O_3 and CH_4 concentrations. In contrast, Scott et al. (2018b), found that the warming effect associated with a reduction in biogenic particles outweighed the cooling effect due to a reduction in O_3 and CH_4, leading to a 10% enhancement of the overall warming due to deforestation.

3 Estimating the impacts of land-use change on climate

The climate impact of land-use change is difficult to isolate from observations, so computer models can be used to explore the effects of converting one land cover type to another in idealised experiments (e.g. Lean and Warrilow 1989; Bonan et al. 1992; Betts 2000; Claussen et al. 2001; Bounoua et al. 2002; Snyder et al. 2004; Feddema et al. 2005; Gibbard et al. 2005; Bala et al. 2007; Davin and de Noblet-Ducoudré 2010; Pongratz et al. 2010; Arora and Montenegro 2011; Swann et al. 2012; Hallgren et al. 2013).

Modelling studies find that the overall impact of deforestation on climate is latitude dependent. Bala et al. (2007) found that the net climate impact of simulated global forest removal was a temperature reduction of −0.3°C. Removing tropical forest led to a global mean warming (+0.7°C) due to a reduction in evapotranspiration and high carbon storage in the tropics, whereas the removal of boreal forests resulted in a global mean cooling (−0.8°C) due to the dominance of the surface albedo effect. Davin and de Noblet-Ducoudré (2010) also found that, when looking only at the biogeophysical impacts, simulated global forest removal generated a cooling (−1°C) with the albedo effect dominant at high latitudes, and the effects of reduced surface roughness and evapotranspiration dominant in the tropics.

In temperate regions, the balance between competing biogeophysical effects is less clear than for either boreal or tropical forests (Claussen et al. 2001). Bala et al. (2007) simulated a global annual mean cooling of −0.04°C for total temperate deforestation. For the northern hemisphere (NH) alone, Snyder et al. (2004) obtained a larger annual mean cooling of −1.1°C and found that the effect was seasonally dependent, with temperate deforestation causing a cooling during the local winter and a warming during the summer. Swann et al. (2012) found a global mean temperature change of between −0.4°C and +0.1°C due to northern mid-latitude afforestation, but also simulated a northward shift of tropical precipitation belts and drying of the southern Amazon.

All the previously mentioned studies used models to examine idealised deforestation or afforestation scenarios. Estimating the climatic impact of historical land-use change combines the challenges associated with understanding the climate impact of specific land-cover transitions, with the challenge of reconstructing historical land-use change. Pongratz et al. (2010) estimated that LULCC since 1850 has resulted in a biogeochemical warming of 0.16–0.18°C and biogeophysical cooling of −0.03°C, giving a combined overall warming.

4 Role of the land sector in climate change mitigation

The 2015 Paris Agreement on Climate commits signatories to 'hold the increase in the global average temperature to well below 2°C above pre-industrial levels' and to 'pursue efforts to limit the temperature increase to 1.5°C'. Globally, anthropogenic CO_2 emissions are still rising, and emissions from fossil fuel combustion have increased from 3.1 ± 0.2 PgC (11.4 $PgCO_2$) per year during the 1960s to an average of 9.4 ± 0.5 PgC (35 $PgCO_2$) per year between 2008 and 2017 (Le Quéré et al. 2018). Future emission projections that are able to limit the global temperature increase to 1.5°C tend to require that global CO_2 emissions peak around 2020 and reach net zero by mid-century (Rogelj et al. 2018).

Achieving net zero emissions means that any remaining CO_2 emissions are balanced by processes that remove CO_2 from the atmosphere. This can be achieved in a number of ways, but the most frequently cited strategies are an increase in afforestation or reforestation and the large-scale deployment of bioenergy with carbon capture and storage (BECCS). BECCS involves the deliberate growth of crops that can be burned to generate energy, and the subsequent burial of the CO_2 emitted during their combustion; estimates suggest that BECCS could remove around 3 PgC (11 $PgCO_2$) per year by 2100, requiring up to 0.7 billion ha of land (Smith et al. 2016).

By 2050, the world's population is projected to reach 9 billion and the Food and Agriculture Organisation predicts that a 70% increase in food production will be required (Food and Agriculture Organisation of the United Nations (FAO) 2009). This is unlikely to be achieved by improving agricultural yields or intensifying livestock production alone, suggesting that the expansion of the current agricultural land area would occur (Bajzelj et al. 2014). Taking the requirements for food and bioenergy production together gives a complex set of possible future land-use change scenarios; a combination of increasing population, potentially increasing meat consumption and a requirement for BECCS will place enormous demands on global land that may not be sustainable (Benton et al. 2018). Over 75% of current agricultural land is used to raise livestock; future dietary changes, such as a reduction in meat consumption, therefore have the potential to reduce the amount of land required (Stehfest et al. 2009), as well as the greenhouse gas emissions associated with agriculture (Springmann et al. 2016).

In conjunction with the Paris Agreement on Climate, countries around the world prepared Nationally Determined Contributions (NDCs) to state what they would do to mitigate, and adapt to, future climate change. Initial assessments of the NDCs suggest that countries are currently expecting approximately one-quarter of their mitigation targets by 2025–2030 to be met by the land-use sector, through reduced deforestation and increased afforestation (Forsell et al. 2016; Grassi et al. 2017).

4.1 Reducing deforestation

The UN-REDD (Reducing Emissions from Deforestation and forest Degradation) programme, and the extension REDD+, aims to reduce forest loss in developing countries by introducing financial mechanisms to benefit countries that preserve the carbon stocks in their forests (e.g. Angelsen and Wetrtz-Kanounnikoff 2008). Reducing deforestation rates by 50% by 2050 (relative to rates observed in the 1990s), and maintaining them at that level until 2100 would avoid the direct release of approximately 50 PgC (Gullison et al. 2007), equivalent to five years of fossil fuel carbon emissions. Using a dynamic global vegetation model, Gumpenberger et al. (2010) found that tropical carbon stocks in 2100 decreased by 35 and 134 PgC, relative to 2012, under a continued deforestation scenario, whereas under a forest protection scenario tropical carbon stocks could be increased by between 7 and 121 PgC.

4.2 Increasing reforestation, restoration and afforestation

While preserved or increased forest cover would enhance CO_2 sequestration and storage, forests also exert the biogeochemical and biogeophysical impacts discussed in sections 2 and 3; as such the overall climatic impact of modifications to forest area will be complex and location specific.

Pongratz et al. (2011) found that the majority of historical anthropogenic LULCC in temperate and boreal regions has occurred on the most productive land, thereby generating higher than average (i.e. for a particular latitude) CO_2 emissions. Subsequently, reforestation of these areas could potentially induce a cooling effect from CO_2 sequestration that would outweigh any warming effect due to an albedo increase. Arora and Montenegro (2011) found that gradually replacing cropland in an Earth system model with forests reduced the simulated global mean temperature at the end of the twenty-first century by 0.45°C because the impact of increased carbon sequestration outweighed the warming from biogeophysical effects.

Using photo-interpretation of satellite imagery, Bastin et al. (2019) identified 0.9 billion ha of land globally that could support forests. Their analysis suggested that these additional forests could potentially store 205 PgC, but did not consider the biogeophysical impacts of the additional forests and may have overestimated the capacity for above-ground and soil carbon increases (Friedlingstein et al. 2019; Lewis et al. 2019; Veldman et al. 2019).

4.3 Growth of crops for bioenergy

Since vegetation takes in CO_2 during photosynthesis (see Section 2.1), it is theoretically possible to generate carbon-neutral energy by harvesting and burning crops and other plants. This is often done using fast-growing perennial grasses such as Miscanthus (also known as silvergrass) or coppicing short-rotation trees such as poplar, willow and eucalyptus. While the use of bioenergy crops can potentially displace fossil fuels and therefore reduce carbon emissions, bioenergy crops grown on former high carbon forest or peatland may struggle to repay the carbon debt associated with their initial establishment (Harper et al. 2018). If combined with carbon capture and storage (CCS) the global mitigation potential of BECCS is estimated to be up to around 3 PgC per year (Smith et al. 2016; Jia et al. 2019).

While motivated by the need to reduce carbon emissions, the large-scale growth of bioenergy crops could also have substantial impacts on surface energy fluxes and therefore local climate. Modelling studies indicate that the expansion of perennial bioenergy crops onto land previously used to grow annual crops would lead to an increase in both surface albedo and evapotranspiration, and therefore a localised cooling effect (Georgescu et al. 2011).

Many bioenergy crops emit higher levels of isoprene than the food crops they may have replaced. Ashworth et al. (2012) found that, while the impacts on global climate were negligible, replacing food crops with oil palm and short-rotation coppice resulted in localised increases in both O_3 and secondary organic aerosol concentrations.

5 Future land-use trajectories

As discussed in Section 3, computer models are often used to explore the impact of changes in land cover on the climate. The same approach is taken to assess future potential climate change as a result of different levels of greenhouse gas, and other anthropogenic, emissions.

A set of Shared Socioeconomic Pathways (SSPs; Riahi et al. 2017) has been developed that describe five different future narratives for society. The SSPs reflect a range of levels of possible global challenges around mitigation and adaptation to climate change (Table 1), but do not include

Table 1 Summary of the Shared Socioeconomic Pathways (Riahi et al. 2017) and their implications for future land-use change (Popp et al. 2017)

Shared Socioeconomic Pathway	Overview of pathway	Description of agriculture and land-use sector
SSP1	Sustainability: Taking the Green Road (*low challenges to mitigation and adaptation*)	Strong land-use regulation. Improvements in agricultural productivity, low meat diets and low growth in food consumption. Full participation of land-use sector in mitigation.
SSP2	Middle of the Road (*medium challenges to mitigation and adaptation*)	Medium land-use regulation. Material-intensive consumption and medium meat consumption. Slow decline in deforestation rate. Partial participation of land-use sector in mitigation.
SSP3	Regional Rivalry: A Rocky Road (*high challenges to mitigation and adaptation*)	Limited regulation. Resource intensive consumption. Continued deforestation. Limited participation of land-use sector in mitigation.
SSP4	Inequality: A Road Divided (*low challenges mitigation but high challenges to adaptation*)	Uneven regulation. High deforestation rates in low-income countries. Unequal consumption. Partial participation of land-use sector in mitigation.
SSP5	Fossil-fuelled Development: Taking the Highway (*high challenges to mitigation but low challenges to adaptation*)	Medium regulation. Material-intensive consumption and meat-rich diets. Slow decline in deforestation rate. Full participation of land-use sector in mitigation.

specific climate policies. Integrated assessment models are then used to realise the SSPs in the context of different levels of climate change along multiple Representative Concentration Pathways (RCPs; Moss et al. 2010; van Vuuren et al. 2011). Each RCP reaches a specific level of anthropogenic radiative forcing by 2100, for example, RCP6.0 reaches a level of +6.0 W m^{-2} and RCP1.9 reaches +1.9 W m^{-2}.

Land-use is of great importance in these pathways because of its potential to contribute to continued CO_2 emissions, and capacity to remove CO_2 from the atmosphere. The pace of land-use change described in these pathways over the twenty-first century is much more rapid than that has been seen historically. The median annual global carbon emission from land-use change across the SSPs is 0.8 PgC (3 $PgCO_2$) in 2030, 0.5 PgC (1.9 $PgCO_2$) in 2050 and −0.2 PgC (−0.7 $PgCO_2$) in 2100 (Jia et al. 2019).

In the resulting matrix of SSP and RCP combinations (O'Neill et al. 2016), the change in the global forested area in 2100 (relative to 2010) varies by over 2 billion ha (Fig. 3), equivalent to half of the present-day forested land area. The greatest increases in the forested area are seen in pathways that follow SSP1, which includes strong regulation of the land-sector, increased agricultural productivity, low food waste, and a shift towards lower meat consumption. Accordingly, SSP1 scenarios see reductions in the amount of land used for pasture, and cropland in some realisations (Fig. 3), which allows for forests to regenerate naturally on abandoned land and deliberate afforestation. In the pathways with the most stringent climate target (i.e. RCP1.9) afforestation is a sink of −0.6 PgC (−2.4 $PgCO_2$) per year by 2100 (median value across all SSPs in five IAMs). However, the warming biophysical impacts (i.e. decreased albedo) of forest expansion under some pathways (i.e. RCP4.5) may outweigh the cooling induced by carbon uptake (Davies-Barnard et al. 2014).

The area used for cropland includes both food and energy crops, here the decline in demand for food crops in most SSP1 scenarios is offset by an

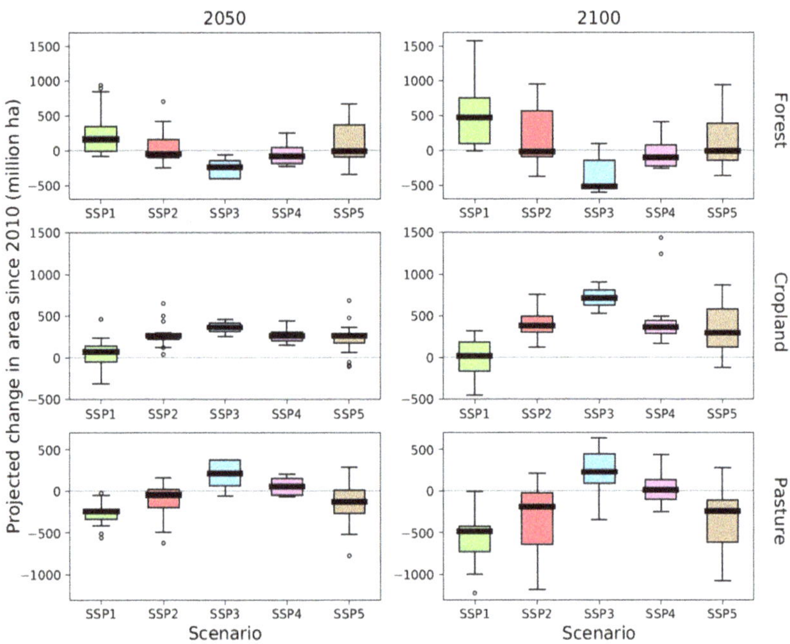

Figure 3 Projected change in land area covered by forest (top panels), cropland (central panels) and pasture (lower panels) in 2050 (left) and 2100 (right), relative to the year 2010, under five Shared Socioeconomic Pathways (Riahi et al. 2017). Each SSP is realised with five integrated assessment models (AIM, GCAM, IMAGE, MESSAGE-GLOBIOM and REMIND/MAgPIE; as described in Popp et al. 2017) under five representative concentration pathways (van Vuuren et al. 2011), data obtained from the IAMC Scenario Explorer (Huppmann et al. 2018).

increased demand for biofuel crop growth. In the pathways with the most stringent climate target, BECCS is a sink of −4 PgC (−14.9 $PgCO_2$) per year by 2100. The greatest increases in cropland and pasture, coupled with a decline in the forested area, are seen in SSP3 scenarios (Fig. 3), where deforestation occurs due to limited regulation and continued competition for resources.

Rather than being predictions, the RCPs and SSPs explore possible future scenarios and are being used by the global climate science community during the Coupled Model Intercomparison Project Phase 6 (CMIP6) to determine the impacts on climate, beyond carbon emissions, associated with different trajectories for the coming century.

6 Future trends and conclusion

Agriculture is currently the main driver of global forest loss, which occurs mainly in the tropics, with land cleared predominantly to grow crops or raise livestock (Gibbs et al. 2010). The specific drivers of land-use change vary regionally and temporally, however, with livestock and soybean growth currently dominating in South America and oil palm growth dominating in Southeast Asia.

Although reconstructing historical land-use change is challenging (Pongratz et al. 2009; Kaplan et al. 2011), it is estimated to have resulted in the emission of around 200 PgC (730 $PgCO_2$) since the year 1750 with current annual emissions estimated at 1.5 ± 0.7 PgC (or 5.5 $PgCO_2$, approximately 14% of total anthropogenic emissions; Le Quéré et al. 2018). Aside from the direct emission of CO_2, the process of land-use change has substantial impacts on the fluxes of energy, moisture and volatile gases between the land surface and the atmosphere.

Computer simulations can be used to assess the overall impact of a particular land-use change; modelling studies find that the overall impact of forests and land-use change on climate is latitude dependent. Forests at high northern latitudes exert an overall warming impact on climate due to their low albedo (Betts 2000; Davin and de Noblet-Ducoudré 2010), whereas forests in tropical latitudes sequester huge quantities of carbon, giving them an overall cooling impact on the climate (Bala et al. 2007). The climate impacts of forests at temperate latitudes are less clear (Claussen et al. 2001; Swann et al. 2012), but recent observational studies indicate that temperate forests may have a stronger cooling impact on climate than model simulations have previously suggested.

When combined, the agriculture and land-use sector contribute around one-quarter of current anthropogenic GHG emissions, highlighting the crucial role that land management practices must play in climate change mitigation over the coming decades. Meeting the commitments outlined in the Paris Agreement on Climate, to 'hold the increase in the global average temperature to well below 2°C above pre-industrial levels' and to 'pursue efforts to limit the temperature increase to 1.5°C', is likely to require global CO_2 emissions to reach net zero by around 2050. Achieving net zero will necessitate a complete, or almost complete, elimination of all GHG emissions from agriculture and the land sector, with any remaining emissions being balanced by processes that remove carbon dioxide from the atmosphere.

In addition to reducing direct GHG emissions from agriculture, reducing the amount of land used for food production could enable the natural regeneration, or deliberate replanting, of forests. Scenarios designed to allow the climate science community to explore the impacts of different global futures indicate that a very wide range of land-use trajectories are possible, depending on levels of regulation and cooperation between regions. By 2100, global forested land varies by around 2 billion ha between different realisations of five SSPs combined with RCPs. Current and future research is exploring the impacts of such substantial land-use changes using fully coupled Earth System Models. These models will enable scientists to go beyond quantifying carbon emissions and diagnose the impacts of land-cover change on surface reflectivity, evapotranspiration and the composition of the atmosphere.

7 Where to look for further information

- 'Ecological Climatology' by Gordon Bonan (2016) provides an excellent overview of interactions between the land-surface, atmosphere and climate.
- Global Forest Watch provides up-to-date information on rates of land-cover change around the world: https://www.globalforestwatch.org/.
- Intergovernmental Panel on Climate Change (IPCC) published a Special Report on Climate Change and Land (2019) to collate the latest scientific evidence on the role of LULCC in climate change: https://www.ipcc.ch/report/srccl/. Chapter 2 in particular discusses the links between land-use and climate (Jia et al. 2019).

8 References

Achard, F., Eva, H. D., Stibig, H. J., Mayaux, P., Gallego, J., Richards, T. and Malingreau, J. P. 2002. Determination of deforestation rates of the world's humid tropical forests. *Science* 297(5583), 999-1002. doi:10.1126/science.1070656.

Amin, H. S., Russo, R. S., Sive, B., Richard Hoebeke, E., Dodson, C., McCubbin, I. B., Gannet Hallar, A. and Huff Hartz, K. E. 2013. Monoterpene emissions from bark beetle infested *Engelmann spruce* trees. *Atmospheric Environment* 72(0), 130-3. doi:10.1016/j.atmosenv.2013.02.025.

Angelsen, A. and Wetrtz-Kanounnikoff, S. 2008. What are the key design issues for REDD and the criteria for assessing options? In: Angelsen, A. (Ed.), *Moving Ahead with REDD*. CIFOR, pp. 11-22.

Arora, V. K. and Montenegro, A. 2011. Small temperature benefits provided by realistic afforestation efforts. *Nature Geoscience* 4(8), 514–8. doi:10.1038/ngeo1182.

Ashworth, K., Folberth, G., Hewitt, C. N. and Wild, O. 2012. Impacts of near-future cultivation of biofuel feedstocks on atmospheric composition and local air quality. *Atmospheric Chemistry and Physics* 12(2), 919–39. doi:10.5194/acp-12-919-2012.

Asner, G. P. 2009. Tropical forest carbon assessment: integrating satellite and airborne mapping approaches. *Environmental Research Letters* 4(3). doi:10.1088/1748-9326/4/3/034009.

Baccini, A., Goetz, S. J., Walker, W. S., Laporte, N. T., Sun, M., Sulla-Menashe, D., Hackler, J., Beck, P. S. A., Dubayah, R., Friedl, M. A., Samanta, S. and Houghton, R. A. 2012. Estimated carbon dioxide emissions from tropical deforestation improved by carbon-density maps. *Nature Climate Change* 2(3), 182–5. doi:10.1038/nclimate1354.

Bajzelj, B., Richards, K. S., Allwood, J. M., Smith, P., Dennis, J. S., Curmi, E. and Gilligan, C. A. 2014. Importance of food-demand management for climate mitigation. *Nature Climate Change* 4(10), 924–9. doi:10.1038/nclimate2353.

Bala, G., Caldeira, K., Wickett, M., Phillips, T. J., Lobell, D. B., Delire, C. and Mirin, A. 2007. Combined climate and carbon-cycle effects of large-scale deforestation. *Proceedings of the National Academy of Sciences of the United States of America* 104(16), 6550–5. doi:10.1073/pnas.0608998104.

Bastin, J. F., Finegold, Y., Garcia, C., Mollicone, D., Rezende, M., Routh, D., Zohner, C. M. and Crowther, T. W. 2019. The global tree restoration potential. *Science* 365(6448), 76–9. doi:10.1126/science.aax0848.

Benton, T. G., Bailey, R., Froggatt, A., King, R., Lee, B. and Wellesley, L. 2018. Designing sustainable landuse in a 1.5°C world: the complexities of projecting multiple ecosystem services from land. *Current Opinion in Environmental Sustainability* 31, 88–95. doi:10.1016/j.cosust.2018.01.011.

Betts, R. A. 2000. Offset of the potential carbon sink from boreal forestation by decreases in surface albedo. *Nature* 408(6809), 187–90. doi:10.1038/35041545.

Betts, A. K. and Ball, J. H. 1997. Albedo over the boreal forest. *Journal of Geophysical Research: Atmospheres* 102(D24), 28901–9. doi:10.1029/96JD03876.

Bonan, G. B. 2008. Forests and climate change: forcings, feedbacks, and the climate benefits of forests. *Science* 320(5882), 1444–9. doi:10.1126/science.1155121.

Bonan, G. B., Pollard, D. and Thompson, S. L. 1992. Effects of boreal forest vegetation on global climate. *Nature* 359(6397), 716–8. doi:10.1038/359716a0.

Boserup, E. 1965. *The Conditions of Agricultural Growth: The Economics of Agrarian Change Under Population Pressure*. G. Allen & Unwin, London, UK.

Bounoua, L., DeFries, R., Collatz, G. J., Sellers, P. and Khan, H. 2002. Effects of land cover conversion on surface climate. *Climatic Change* 52(1–2), 29–64.

Bowen, I. S. 1926. The ratio of heat losses by conduction and by evaporation from any water surface. *Physical Review* 27(6), 779–87. doi:10.1103/PhysRev.27.779.

Busch, J., Ferretti-Gallon, K., Engelmann, J., Wright, M., Austin, K. G., Stolle, F., Turubanova, S., Potapov, P. V., Margono, B., Hansen, M. C. and Baccini, A. 2015. Reductions in emissions from deforestation from Indonesia's moratorium on new oil palm, timber, and logging concessions. *Proceedings of the National Academy of Sciences of the United States of America* 112(5), 1328–33. doi:10.1073/pnas.1412514112.

Canadell, J., Jackson, R. B., Ehleringer, J. B., Mooney, H. A., Sala, O. E. and Schulze, E. D. 1996. Maximum rooting depth of vegetation types at the global scale. *Oecologia* 108(4), 583–95. doi:10.1007/BF00329030.

Carvalho, G. O., Nepstad, D., McGrath, D., del Carmen Vera Diaz, M., Santilli, M. and Barros, A. C. 2002. Frontier expansion in the Amazon: balancing development and sustainability. *Environment: Science and Policy for Sustainable Development* 44(3), 34–44. doi:10.1080/00139150209605606.

Ciais, P., Sabine, C., Bala, G., Bopp, L., Brovkin, V., Canadell, J., Chhabra, A., DeFries, R., Galloway, J., Heimann, M., Jones, C., Le Quere, C., Myneni, R. B., Piao, S. and Thornton, P. 2013. Carbon and other biogeochemical cycles. In: Stocker, T. F., Qin, D., Plattner, G.-K., Tignor, M., Allen, S. K., Boschung, J., Nauels, A., Xia, Y., Bex, V. and Midgley, P. M. (Eds), *Climate Change 2013: The Physical Science Basis. Contribution of Working Group I to the Fifth Assessment Report of the Intergovernmental Panel on Climate Change*. Cambridge University Press, Cambridge, UK and New York, NY.

Claussen, M., Brovkin, V. and Ganopolski, A. 2001. Biogeophysical versus biogeochemical feedbacks of large-scale land cover change. *Geophysical Research Letters* 28(6), 1011–4. doi:10.1029/2000GL012471.

Davies-Barnard, T., Valdes, P. J., Singarayer, J. S., Pacifico, F. M. and Jones, C. D. 2014. Full effects of land use change in the representative concentration pathways. *Environmental Research Letters* 9(11). doi:10.1088/1748-9326/9/11/114014.

Davin, E. L. and de Noblet-Ducoudré, N. 2010. Climatic impact of global-scale deforestation: radiative versus nonradiative processes. *Journal of Climate* 23(1), 97–112. doi:10.1175/2009JCLI3102.1.

Defries, R. S., Hansen, M. C., Townshend, J. R. G., Janetos, A. C. and Loveland, T. R. 2000. A new global 1-km dataset of percentage tree cover derived from remote sensing. *Global Change Biology* 6(2), 247–54. doi:10.1046/j.1365-2486.2000.00296.x.

Dixon, R. K., Solomon, A. M., Brown, S., Houghton, R. A., Trexier, M. C. and Wisniewski, J. 1994. Carbon pools and flux of global forest ecosystems. *Science* 263(5144), 185–90. doi:10.1126/science.263.5144.185.

Fang, J.-Y., Wang, G. G., Liu, G.-H. and Xu, S.-L. 1998. Forest biomass of China: an estimate based on the biomass-volume relationship. *Ecological Applications* 8(4), 1084–91.

Fang, J., Chen, A., Peng, C., Zhao, S. and Ci, L. 2001. Changes in forest biomass carbon storage in China between 1949 and 1998. *Science* 292(5525), 2320–2. doi:10.1126/science.1058629.

FAO. 2012. *State of the World's Forests*. United Nations, Rome, Italy.

FAO. 2015. *Global Forest Resources Assessment 2015: How Are the World's Forests Changing?* United Nations, Rome, Italy.

FAO. 2018. *The State of the World's Forests 2018 – Forest Pathways to Sustainable Development*. United Nations, Rome, Italy.

Fearnside, P. M. 2001. Soybean cultivation as a threat to the environment in Brazil. *Environmental Conservation* 28(1), 23–38. doi:10.1017/S0376892901000030.

Fearnside, P. M. 2005. Deforestation in Brazilian Amazonia: history, rates, and consequences. *Conservation Biology* 19(3), 680–8. doi:10.1111/j.1523-1739.2005.00697.x.

Feddema, J., Oleson, K., Bonan, G., Mearns, L., Washington, W., Meehl, G. and Nychka, D. 2005. A comparison of a GCM response to historical anthropogenic land cover change and model sensitivity to uncertainty in present-day land cover representations. *Climate Dynamics* 25(6), 581–609. doi:10.1007/s00382-005-0038-z.

Food and Agriculture Organisation of the United Nations (FAO). 2009. *How to Feed the World in 2050*. United Nations, Rome, Italy.

Forsell, N., Turkovska, O., Gusti, M., Obersteiner, M., Elzen, M. D. and Havlik, P. 2016. Assessing the INDCs' land use, land use change, and forest emission projections. *Carbon Balance and Management* 11(1), 26. doi:10.1186/s13021-016-0068-3.

Friedlingstein, P., Allen, M., Canadell, J. G., Peters, G. P. and Seneviratne, S. I. 2019. Comment on "the global tree restoration potential. *Science* 366(6463), eaay8060. doi:10.1126/science.aay8060.

Georgescu, M., Lobell, D. B. and Field, C. B. 2011. Direct climate effects of perennial bioenergy crops in the United States. *Proceedings of the National Academy of Sciences of the United States of America* 108(11), 4307–12. doi:10.1073/pnas.1008779108.

Gibbard, S., Caldeira, K., Bala, G., Phillips, T. J. and Wickett, M. 2005. Climate effects of global land cover change. *Geophysical Research Letters* 32(23), L23705. doi:10.1029/2005GL024550.

Gibbs, H. K., Johnston, M., Foley, J. A., Holloway, T., Monfreda, C., Ramankutty, N. and Zaks, D. 2008. Carbon payback times for crop-based biofuel expansion in the tropics: the effects of changing yield and technology. *Environmental Research Letters* 3(3). doi:10.1088/1748-9326/3/3/034001.

Gibbs, H. K., Ruesch, A. S., Achard, F., Clayton, M. K., Holmgren, P., Ramankutty, N. and Foley, J. A. 2010. Tropical forests were the primary sources of new agricultural land in the 1980s and 1990s. *Proceedings of the National Academy of Sciences of the United States of America* 107(38), 16732–7.

Gibbs, H. K., Rausch, L., Munger, J., Schelly, I., Morton, D. C., Noojipady, P., Soares-Filho, B., Barreto, P., Micol, L. and Walker, N. F. 2015. Environment and development. Brazil's soy moratorium. *Science* 347(6220), 377–8. doi:10.1126/science.aaa0181.

Gordon, H., Sengupta, K., Rap, A., Duplissy, J., Frege, C., Williamson, C., Heinritzi, M., Simon, M., Yan, C., Almeida, J., Tröstl, J., Nieminen, T., Ortega, I. K., Wagner, R., Dunne, E. M., Adamov, A., Amorim, A., Bernhammer, A. K., Bianchi, F., Breitenlechner, M., Brilke, S., Chen, X., Craven, J. S., Dias, A., Ehrhart, S., Fischer, L., Flagan, R. C., Franchin, A., Fuchs, C., Guida, R., Hakala, J., Hoyle, C. R., Jokinen, T., Junninen, H., Kangasluoma, J., Kim, J., Kirkby, J., Krapf, M., Kürten, A., Laaksonen, A., Lehtipalo, K., Makhmutov, V., Mathot, S., Molteni, U., Monks, S. A., Onnela, A., Peräkylä, O., Piel, F., Petäjä, T., Praplan, A. P., Pringle, K. J., Richards, N. A., Rissanen, M. P., Rondo, L., Sarnela, N., Schobesberger, S., Scott, C. E., Seinfeld, J. H., Sharma, S., Sipilä, M., Steiner, G., Stozhkov, Y., Stratmann, F., Tomé, A., Virtanen, A., Vogel, A. L., Wagner, A. C., Wagner, P. E., Weingartner, E., Wimmer, D., Winkler, P. M., Ye, P., Zhang, X., Hansel, A., Dommen, J., Donahue, N. M., Worsnop, D. R., Baltensperger, U., Kulmala, M., Curtius, J. and Carslaw, K. S. 2016. Reduced anthropogenic aerosol radiative forcing caused by biogenic new particle formation. *Proceedings of the National Academy of Sciences of the United States of America* 113(43), 12053–8. doi:10.1073/pnas.1602360113.

Grainger, A. 1993. The causes of deforestation. In: Grainger, A. (Ed.), *Controlling Tropical Deforestation*. Earthscan, London, UK, pp. 49–68.

Grassi, G., House, J., Dentener, F., Federici, S., den Elzen, M. and Penman, J. 2017. The key role of forests in meeting climate targets requires science for credible mitigation. *Nature Climate Change* 7(3), 220–6. doi:10.1038/nclimate3227.

Guenther, A. B., Monson, R. K. and Fall, R. 1991. Isoprene and monoterpene emission rate variability: observations with eucalyptus and emission rate algorithm development. *Journal of Geophysical Research* 96(D6), 10799–808. doi:10.1029/91JD00960.

Guenther, A. B., Jiang, X., Heald, C. L., Sakulyanontvittaya, T., Duhl, T., Emmons, L. K. and Wang, X. 2012. The Model of Emissions of Gases and Aerosols from Nature version 2.1 (MEGAN2.1): an extended and updated framework for modeling biogenic emissions. *Geoscientific Model Development* 5(6), 1471-92. doi:10.5194/gmd-5-1471-2012.

Gullison, R. E., Frumhoff, P. C., Canadell, J. G., Field, C. B., Nepstad, D. C., Hayhoe, K., Avissar, R., Curran, L. M., Friedlingstein, P., Jones, C. D. and Nobre, C. 2007. Tropical forests and climate policy. *Science* 316(5827), 985-6. doi:10.1126/science.1136163.

Gumpenberger, M., Vohland, K., Heyder, U., Poulter, B., Macey, K., Rammig, A., Popp, A. and Cramer, W. 2010. Predicting pan-tropical climate change induced forest stock gains and losses - implications for REDD. *Environmental Research Letters* 5(1), 014013. doi:10.1088/1748-9326/5/1/014013.

Hallgren, W., Schlosser, C. A., Monier, E., Kicklighter, D., Sokolov, A. and Melillo, J. 2013. Climate impacts of a large-scale biofuels expansion. *Geophysical Research Letters* 40(8), 1624-30. doi:10.1002/grl.50352.

Hansen, M. C., DeFries, R. S., Townshend, J. R. G., Carroll, M., Dimiceli, C. and Sohlberg, R. A. 2003. Global percent tree cover at a spatial resolution of 500 meters: first results of the MODIS vegetation continuous fields algorithm. *Earth Interactions* 7(10), 1-15. doi:10.1175/1087-3562(2003)007<0001:GPTCAA>2.0.CO;2.

Hansen, M. C., Potapov, P. V., Moore, R., Hancher, M., Turubanova, S. A., Tyukavina, A., Thau, D., Stehman, S. V., Goetz, S. J., Loveland, T. R., Kommareddy, A., Egorov, A., Chini, L., Justice, C. O. and Townshend, J. R. 2013. High-resolution global maps of 21st-century forest cover change. *Science* 342(6160), 850-3. doi:10.1126/science.1244693.

Harper, A. B., Powell, T., Cox, P. M., House, J., Huntingford, C., Lenton, T. M., Sitch, S., Burke, E., Chadburn, S. E., Collins, W. J., Comyn-Platt, E., Daioglou, V., Doelman, J. C., Hayman, G., Robertson, E., van Vuuren, D., Wiltshire, A., Webber, C. P., Bastos, A., Boysen, L., Ciais, P., Devaraju, N., Jain, A. K., Krause, A., Poulter, B. and Shu, S. 2018. Land-use emissions play a critical role in land-based mitigation for Paris climate targets. *Nature Communications* 9(1), 2938. doi:10.1038/s41467-018-05340-z.

Hollinger, D. Y., Ollinger, S. V., Richardson, A. D., Meyers, T. P., Dail, D. B., Martin, M. E., Scott, N. A., Arkebauer, T. J., Baldocchi, D. D., Clark, K. L., Curtis, P. S., Davis, K. J., Desai, A. R., Dragoni, D., Goulden, M. L., Gu, L., Katul, G. G., Pallardy, S. G., Paw U, K. T., Schmid, H. P., Stoy, P. C., Suyker, A. E. and Verma, S. B. 2010. Albedo estimates for land surface models and support for a new paradigm based on foliage nitrogen concentration. *Global Change Biology* 16(2), 696-710. doi:10.1111/j.1365-2486.2009.02028.x.

Houghton, R. A. 2003. Revised estimates of the annual net flux of carbon to the atmosphere from changes in land use and land management 1850-2000. *Tellus B* 55(2), 378-90. doi:10.1034/j.1600-0889.2003.01450.x.

Huppmann, D., Kriegler, E., Krey, V., Riahi, K., Rogelj, J., Calvin, K., Humpenoeder, F., Popp, A., Rose, S. K., Weyant, J., Bauer, N., Bertram, C., Bosetti, V., Doelman, J., Drouet, L., Emmerling, J., Frank, S., Fujimori, S., Gernaat, D., Grubler, A., Guivarch, C., Haigh, M., Holz, C., Iyer, G., Kato, E., Keramidas, K., Kitous, A., Leblanc, F., Liu, J.-Y., Löffler, K., Luderer, G., Marcucci, A., McCollum, D., Mima, S., Sands, R. D., Sano, F., Strefler, J., Tsutsui, J., Van Vuuren, D., Vrontisi, Z., Wise, M. and Zhang, R. 2018. IAMC 1.5°C scenario explorer and data hosted by IIASA. Integrated Assessment Modeling Consortium & International Institute for Applied Systems Analysis.

Jia, G., Shevliakova, E., Artaxo, P., De Noblet-Ducoudré, N., Houghton, R., House, J., Kitajima, K., Lennard, C., Popp, A., Sirin, A., Sukumar, R. and Verchot, L. 2019. Land-climate interactions. In: Shukla, P. R., Skea, J., Calvo Buendia, E., Masson-Delmotte, V., Pörtner, H.-O., Roberts, D. C., Zhai, P., Slade, R., Connors, S., van Diemen, R., Ferrat, M., Haughey, E., Luz, S., Neogi, S., Pathak, M., Petzold, J., Portugal Pereira, J., Vyas, P., Huntley, E., Kissick, K., Belkacemi, M. and Malley, J. (Eds), *Climate Change and Land: an IPCC Special Report on Climate Change, Desertification, Land Degradation, Sustainable Land Management, Food Security, and Greenhouse Gas Fluxes in Terrestrial Ecosystems* (in press).

Kaplan, J. O., Krumhardt, K. M. and Zimmermann, N. 2009. The prehistoric and preindustrial deforestation of Europe. *Quaternary Science Reviews* 28(27–28), 3016–34. doi:10.1016/j.quascirev.2009.09.028.

Kaplan, J. O., Krumhardt, K. M., Ellis, E. C., Ruddiman, W. F., Lemmen, C. and Goldewijk, K. K. 2011. Holocene carbon emissions as a result of anthropogenic land cover change. *The Holocene* 21(5), 775–91. doi:10.1177/0959683610386983.

Kessler, A. and Baldwin, I. T. 2001. Defensive function of herbivore-induced plant volatile emissions in nature. *Science* 291(5511), 2141–4. doi:10.1126/science.291.5511.2141.

Kirkby, J., Duplissy, J., Sengupta, K., Frege, C., Gordon, H., Williamson, C., Heinritzi, M., Simon, M., Yan, C., Almeida, J., Tröstl, J., Nieminen, T., Ortega, I. K., Wagner, R., Adamov, A., Amorim, A., Bernhammer, A. K., Bianchi, F., Breitenlechner, M., Brilke, S., Chen, X., Craven, J., Dias, A., Ehrhart, S., Flagan, R. C., Franchin, A., Fuchs, C., Guida, R., Hakala, J., Hoyle, C. R., Jokinen, T., Junninen, H., Kangasluoma, J., Kim, J., Krapf, M., Kürten, A., Laaksonen, A., Lehtipalo, K., Makhmutov, V., Mathot, S., Molteni, U., Onnela, A., Peräkylä, O., Piel, F., Petäjä, T., Praplan, A. P., Pringle, K., Rap, A., Richards, N. A., Riipinen, I., Rissanen, M. P., Rondo, L., Sarnela, N., Schobesberger, S., Scott, C. E., Seinfeld, J. H., Sipilä, M., Steiner, G., Stozhkov, Y., Stratmann, F., Tomé, A., Virtanen, A., Vogel, A. L., Wagner, A. C., Wagner, P. E., Weingartner, E., Wimmer, D., Winkler, P. M., Ye, P., Zhang, X., Hansel, A., Dommen, J., Donahue, N. M., Worsnop, D. R., Baltensperger, U., Kulmala, M., Carslaw, K. S. and Curtius, J. 2016. Ion-induced nucleation of pure biogenic particles. *Nature* 533(7604), 521–6. doi:10.1038/nature17953.

Klein Goldewijk, K. K. 2001. Estimating global land use change over the past 300 years: the HYDE Database. *Global Biogeochemical Cycles* 15(2), 417–33. doi:10.1029/1999GB001232.

Klein Goldewijk, K., Beusen, A., Van Drecht, G. and De Vos, M. 2011. The HYDE 3.1 spatially explicit database of human-induced global land-use change over the past 12,000 years. *Global Ecology and Biogeography* 20(1), 73–86. doi:10.1111/j.1466-8238.2010.00587.x.

Koh, L. P., Miettinen, J., Liew, S. C. and Ghazoul, J. 2011. Remotely sensed evidence of tropical peatland conversion to oil palm. *Proceedings of the National Academy of Sciences of the United States of America* 108(12), 5127–32. doi:10.1073/pnas.1018776108.

Lean, J. and Warrilow, D. A. 1989. Simulation of the regional climatic impact of Amazon deforestation. *Nature* 342(6248), 411–3. doi:10.1038/342411a0.

Leonard, R. E. and Eschner, A. R. 1968. Albedo of intercepted snow. *Water Resources Research* 4(5), 931–5. doi:10.1029/WR004i005p00931.

Le Quéré, C., Andrew, R. M., Friedlingstein, P., Sitch, S., Hauck, J., Pongratz, J., Pickers, P. A., Korsbakken, J. I., Peters, G. P., Canadell, J. G., Arneth, A., Arora, V. K., Barbero, L.,

Bastos, A., Bopp, L., Chevallier, F., Chini, L. P., Ciais, P., Doney, S. C., Gkritzalis, T., Goll, D. S., Harris, I., Haverd, V., Hoffman, F. M., Hoppema, M., Houghton, R. A., Hurtt, G., Ilyina, T., Jain, A. K., Johannessen, T., Jones, C. D., Kato, E., Keeling, R. F., Goldewijk, K. K., Landschützer, P., Lefèvre, N., Lienert, S., Liu, Z., Lombardozzi, D., Metzl, N., Munro, D. R., Nabel, J. E. M. S., Nakaoka, S., Neill, C., Olsen, A., Ono, T., Patra, P., Peregon, A., Peters, W., Peylin, P., Pfeil, B., Pierrot, D., Poulter, B., Rehder, G., Resplandy, L., Robertson, E., Rocher, M., Rödenbeck, C., Schuster, U., Schwinger, J., Séférian, R., Skjelvan, I., Steinhoff, T., Sutton, A., Tans, P. P., Tian, H., Tilbrook, B., Tubiello, F. N., van der Laan-Luijkx, I. T., van der Werf, G. R., Viovy, N., Walker, A. P., Wiltshire, A. J., Wright, R., Zaehle, S. and Zheng, B. 2018. Global carbon budget 2018. *Earth System Science Data* 10(4), 2141–94. doi:10.5194/essd-10-2141-2018.

Lewis, S. L., Mitchard, E. T. A., Prentice, C., Maslin, M. and Poulter, B. 2019. Comment on "the global tree restoration potential. *Science* 366(6463), eaaz0388. doi:10.1126/science.aaz0388.

Loreto, F. and Velikova, V. 2001. Isoprene produced by leaves protects the photosynthetic apparatus against ozone damage, quenches ozone products, and reduces lipid peroxidation of cellular membranes. *Plant Physiology* 127(4), 1781–7. doi:10.1104/pp.010497.

Malhi, Y., Adu-Bredu, S., Asare, R. A., Lewis, S. L. and Mayaux, P. 2013. African rainforests: past, present and future. *Philosophical Transactions of the Royal Society of London. Series B, Biological Sciences* 368(1625), 20120312. doi:10.1098/rstb.2012.0312.

Mayaux, P., Pekel, J. F., Desclée, B., Donnay, F., Lupi, A., Achard, F., Clerici, M., Bodart, C., Brink, A., Nasi, R. and Belward, A. 2013. State and evolution of the African rainforests between 1990 and 2010. *Philosophical Transactions of the Royal Society of London. Series B, Biological Sciences* 368(1625), 20120300. doi:10.1098/rstb.2012.0300.

Miettinen, J., Shi, C. and Liew, S. C. 2011. Deforestation rates in insular Southeast Asia between 2000 and 2010. *Global Change Biology* 17(7), 2261–70. doi:10.1111/j.1365-2486.2011.02398.x.

Monks, P. S., Archibald, A. T., Colette, A., Cooper, O., Coyle, M., Derwent, R., Fowler, D., Granier, C., Law, K. S., Mills, G. E., Stevenson, D. S., Tarasova, O., Thouret, V., von Schneidemesser, E., Sommariva, R., Wild, O. and Williams, M. L. 2015. Tropospheric ozone and its precursors from the urban to the global scale from air quality to short-lived climate forcer. *Atmospheric Chemistry and Physics* 15(15), 8889–973. doi:10.5194/acp-15-8889-2015.

Monteith, J. L. and Unsworth, M. H. 2008. Microclimatology of Radiation (i). In: *Principles of Environmental Physics*. Academic Press, pp.86–99.

Moss, R. H., Edmonds, J. A., Hibbard, K. A., Manning, M. R., Rose, S. K., van Vuuren, D. P., Carter, T. R., Emori, S., Kainuma, M., Kram, T., Meehl, G. A., Mitchell, J. F., Nakicenovic, N., Riahi, K., Smith, S. J., Stouffer, R. J., Thomson, A. M., Weyant, J. P. and Wilbanks, T. J. 2010. The next generation of scenarios for climate change research and assessment. *Nature* 463(7282), 747–56. doi:10.1038/nature08823.

Muller, C. H. 1966. The role of chemical inhibition (allelopathy) in vegetational composition. *Bulletin of the Torrey Botanical Club* 93(5), 332–51. doi:10.2307/2483447.

Myhre, G., Shindell, D., Breon, F.-M., Collins, W., Fuglestvedt, J., Huang, J., Koch, D., Lamarque, J.-F., Lee, D., Mendoza, B., Nakajima, T., Robock, A., Stephens, G., Takemura, T. and Zhang, H. 2013. Anthropogenic and natural radiative forcing. In: Stocker, T. F., Qin, D., Plattner, G.-K., Tignor, M., Allen, S. K., Boschung, J., Nauels, A., Xia, Y., Bex, V. and Midgley, P. M. (Eds), *Climate Change 2013: The Physical*

Science Basis. Contribution of Working Group I to the Fifth Assessment Report of the Intergovernmental Panel on Climate Change. Cambridge University Press, Cambridge, UK and New York, NY.

Nepstad, D. C., de Carvalho, C. R., Davidson, E. A., Jipp, P. H., Lefebvre, P. A., Negreiros, G. H., da Silva, E. D., Stone, T. A., Trumbore, S. E. and Vieira, S. 1994. The role of deep roots in the hydrological and carbon cycles of Amazonian forests and pastures. *Nature* 372(6507), 666–9. doi:10.1038/372666a0.

Nepstad, D. C., Stickler, C. M. and Almeida, O. T. 2006. Globalization of the Amazon soy and beef industries: opportunities for conservation. *Conservation Biology: the Journal of the Society for Conservation Biology* 20(6), 1595–603. doi:10.1111/j.1523-1739.2006.00510.x.

Oh, H. K., Sakai, T., Jones, M. B. and Longhurst, W. M. 1967. Effect of various essential oils isolated from Douglas fir needles upon sheep and deer rumen microbial activity. *Applied Microbiology* 15(4), 777–84. doi:10.1128/AEM.15.4.777-784.1967.

O'Neill, B. C., Tebaldi, C., van Vuuren, D. P., Eyring, V., Friedlingstein, P., Hurtt, G., Knutti, R., Kriegler, E., Lamarque, J., Lowe, J., Meehl, G. A., Moss, R., Riahi, K. and Sanderson, B. M. 2016. The scenario model intercomparison project (ScenarioMIP) for CMIP6. *Geoscientific Model Development* 9(9), 3461–82. doi:10.5194/gmd-9-3461-2016.

Paasonen, P., Asmi, A., Petäjä, T., Kajos, M. K., Äijälä, M., Junninen, H., Holst, T., Abbatt, J. P. D., Arneth, A., Birmili, W., van der Gon, H. D., Hamed, A., Hoffer, A., Laakso, L., Laaksonen, A., Leaitch, W. R., Plass-Dülmer, C., Pryor, S. C., Räisänen, P., Swietlicki, E., Wiedensohler, A., Worsnop, D. R., Kerminen, V. and Kulmala, M. 2013. Warming-induced increase in aerosol number concentration likely to moderate climate change. *Nature Geoscience* 6(6), 438–42. doi:10.1038/ngeo1800.

Pongratz, J., Reick, C., Raddatz, T. and Claussen, M. 2008. A reconstruction of global agricultural areas and land cover for the last millennium. *Global Biogeochemical Cycles* 22(3), GB3018. doi:10.1029/2007GB003153.

Pongratz, J., Reick, C. H., Raddatz, T. and Claussen, M. 2009. Effects of anthropogenic land cover change on the carbon cycle of the last millennium. *Global Biogeochemical Cycles* 23(4), n/a-. doi:10.1029/2009GB003488.

Pongratz, J., Reick, C. H., Raddatz, T. and Claussen, M. 2010. Biogeophysical versus biogeochemical climate response to historical anthropogenic land cover change. *Geophysical Research Letters* 37(8), L08702. doi:10.1029/2010GL043010.

Pongratz, J., Reick, C. H., Raddatz, T., Caldeira, K. and Claussen, M. 2011. Past land use decisions have increased mitigation potential of reforestation. *Geophysical Research Letters* 38(15), L15701. doi:10.1029/2011GL047848.

Popp, A., Calvin, K., Fujimori, S., Havlik, P., Humpenöder, F., Stehfest, E., Bodirsky, B. L., Dietrich, J. P., Doelmann, J. C., Gusti, M., Hasegawa, T., Kyle, P., Obersteiner, M., Tabeau, A., Takahashi, K., Valin, H., Waldhoff, S., Weindl, I., Wise, M., Kriegler, E., Lotze-Campen, H., Fricko, O., Riahi, K. and Vuuren, D. Pv 2017. Land-use futures in the shared socio-economic pathways. *Global Environmental Change* 42, 331–45. doi:10.1016/j.gloenvcha.2016.10.002.

Prentice, I. C., Farquhar, G. D., Fasham, M. J. R., Goulden, M. L., Heimann, M., Jaramillo, V. J., Kheshgi, H. S., Le Quéré, C., Scholes, R. J. and Wallace, D. W. R. 2001. The carbon cycle and atmospheric carbon dioxide. In: Houghton, J. T., Ding, Y., Griggs, D. J., et al. *Climate Change 2001: the Physical Science Basis. Contribution of Working Group I to the Third Assessment Report of the Intergovernmental Panel on Climate Change*. Cambridge University Press, Cambridge, UK and New York, NY.

Ramankutty, N. and Foley, J. A. 1999. Estimating historical changes in global land cover: croplands from 1700 to 1992. *Global Biogeochemical Cycles* 13(4), 997–1027. doi:10.1029/1999GB900046.

Ramankutty, N., Evan, A. T., Monfreda, C. and Foley, J. A. 2008. Farming the planet: 1. Geographic distribution of global agricultural lands in the year 2000. *Global Biogeochemical Cycles* 22(1). doi:10.1029/2007GB002952.

Rasmann, S., Köllner, T. G., Degenhardt, J., Hiltpold, I., Toepfer, S., Kuhlmann, U., Gershenzon, J. and Turlings, T. C. 2005. Recruitment of entomopathogenic nematodes by insect-damaged maize roots. *Nature* 434(7034), 732–7. doi:10.1038/nature03451.

Rasmussen, R. A. and Went, F. W. 1965. Volatile organic material of plant origin in the atmosphere. *Proceedings of the National Academy of Sciences of the United States of America* 53(1), 215–20. doi:10.1073/pnas.53.1.215.

Riahi, K., van Vuuren, D. P., Kriegler, E., Edmonds, J., O'Neill, B. C., Fujimori, S., Bauer, N., Calvin, K., Dellink, R., Fricko, O., Lutz, W., Popp, A., Cuaresma, J. C., Kc, S., Leimbach, M., Jiang, L., Kram, T., Rao, S., Emmerling, J., Ebi, K., Hasegawa, T., Havlik, P., Humpenöder, F., Da Silva, L. A., Smith, S., Stehfest, E., Bosetti, V., Eom, J., Gernaat, D., Masui, T., Rogelj, J., Strefler, J., Drouet, L., Krey, V., Luderer, G., Harmsen, M., Takahashi, K., Baumstark, L., Doelman, J. C., Kainuma, M., Klimont, Z., Marangoni, G., Lotze-Campen, H., Obersteiner, M., Tabeau, A. and Tavoni, M. 2017. The Shared Socioeconomic Pathways and their energy, land use, and greenhouse gas emissions implications: an overview. *Global Environmental Change* 42, 153–68. doi:10.1016/j.gloenvcha.2016.05.009.

Richardson, A. D., Keenan, T. F., Migliavacca, M., Ryu, Y., Sonnentag, O. and Toomey, M. 2013. Climate change, phenology, and phenological control of vegetation feedbacks to the climate system. *Agricultural and Forest Meteorology* 169, 156–73. doi:10.1016/j.agrformet.2012.09.012.

Robinson, D. A. and Kukla, G. 1985. Maximum surface albedo of seasonally snow-covered lands in the northern hemisphere. *Journal of Climate and Applied Meteorology* 24(5), 402–11. doi:10.1175/1520-0450(1985)024<0402:MSAOSS>2.0.CO;2.

Rogelj, J., Shindell, D., Jiang, K., Fifita, S., Forster, P., Ginzburg, V., Handa, C., Kheshgi, H., Kobayashi, S., Kriegler, E., Mundaca, L., Séférian, R. and Vilariño, M. V. 2018. Mitigation pathways compatible with 1.5°C in the context of sustainable development. In: Masson-Delmotte, V., Zhai, P., Pörtner, H.-O., et al. (Eds), *Global Warming of 1.5°C. An IPCC Special Report on the Impacts of Global Warming of 1.5°C above Pre-industrial Levels and Related Global Greenhouse Gas Emission Pathways, in the Context of Strengthening the Global Response to the Threat of Climate Change, Sustainable Development, and Efforts to Eradicate Poverty* (in press).

Ruddiman, W. F. J. C. C. 2003. The anthropogenic greenhouse era began thousands of years ago. *Climatic Change* 61(3), 261–93. doi:10.1023/B:CLIM.0000004577.17928.fa.

Rudorff, B. F. T., Adami, M., Aguiar, D. A., Moreira, M. A., Mello, M. P., Fabiani, L., Amaral, D. F. and Pires, B. M. 2011. The soy moratorium in the Amazon biome monitored by remote sensing images. *Remote Sensing* 3(1), 185–202. doi:10.3390/rs3010185.

Saatchi, S., Marlier, M., Chazdon, R. L., Clark, D. B. and Russell, A. E. 2011a. Impact of spatial variability of tropical forest structure on radar estimation of aboveground biomass. *Remote Sensing of Environment* 115(11), 2836–49. doi:10.1016/j.rse.2010.07.015.

Saatchi, S. S., Harris, N. L., Brown, S., Lefsky, M., Mitchard, E. T., Salas, W., Zutta, B. R., Buermann, W., Lewis, S. L., Hagen, S., Petrova, S., White, L., Silman, M. and Morel, A. 2011b. Benchmark map of forest carbon stocks in tropical regions across three continents. *Proceedings of the National Academy of Sciences of the United States of America* 108(24), 9899–904. doi:10.1073/pnas.1019576108.

Sanadze, G. A. and Kursanov, A. L. 1966. On certain conditions of the evolution of the diene C_5H_8 from poplar leaves. *Soviet Plant Physiology* 13, 184–9.

Schoneveld, G. C., Ekowati, D., Andrianto, A. and van der Haar, S. 2019. Modeling peat- and forestland conversion by oil palm smallholders in Indonesian Borneo. *Environmental Research Letters* 14(1), (014006). doi:10.1088/1748-9326/aaf044.

Scott, C. E., Rap, A., Spracklen, D. V., Forster, P. M., Carslaw, K. S., Mann, G. W., Pringle, K. J., Kivekäs, N., Kulmala, M., Lihavainen, H. and Tunved, P. 2014. The direct and indirect radiative effects of biogenic secondary organic aerosol. *Atmospheric Chemistry and Physics* 14(1), 447–70. doi:10.5194/acp-14-447-2014.

Scott, C. E., Arnold, S. R., Monks, S. A., Asmi, A., Paasonen, P. and Spracklen, D. V. 2018a. Substantial large-scale feedbacks between natural aerosols and climate. *Nature Geoscience* 11(1), 44–8. doi:10.1038/s41561-017-0020-5.

Scott, C. E., Monks, S. A., Spracklen, D. V., Arnold, S. R., Forster, P. M., Rap, A., Äijälä, M., Artaxo, P., Carslaw, K. S., Chipperfield, M. P., Ehn, M., Gilardoni, S., Heikkinen, L., Kulmala, M., Petäjä, T., Reddington, C. L. S., Rizzo, L. V., Swietlicki, E., Vignati, E. and Wilson, C. 2018b. Impact on short-lived climate forcers increases projected warming due to deforestation. *Nature Communications* 9(1), 157. doi:10.1038/s41467-017-02412-4.

Smith, P., Davis, S. J., Creutzig, F., Fuss, S., Minx, J., Gabrielle, B., Kato, E., Jackson, R. B., Cowie, A., Kriegler, E., van Vuuren, D. P., Rogelj, J., Ciais, P., Milne, J., Canadell, J. G., McCollum, D., Peters, G., Andrew, R., Krey, V., Shrestha, G., Friedlingstein, P., Gasser, T., Grübler, A., Heidug, W. K., Jonas, M., Jones, C. D., Kraxner, F., Littleton, E., Lowe, J., Moreira, J. R., Nakicenovic, N., Obersteiner, M., Patwardhan, A., Rogner, M., Rubin, E., Sharifi, A., Torvanger, A., Yamagata, Y., Edmonds, J. and Yongsung, C. 2016. Biophysical and economic limits to negative CO_2 emissions. *Nature Climate Change* 6(1), 42–50. doi:10.1038/nclimate2870.

Snyder, P. K., Delire, C. and Foley, J. A. 2004. Evaluating the influence of different vegetation biomes on the global climate. *Climate Dynamics* 23(3–4), 279–302. doi:10.1007/s00382-004-0430-0.

Spracklen, D. V., Bonn, B. and Carslaw, K. S. 2008. Boreal forests, aerosols and the impacts on clouds and climate. *Philosophical Transactions. Series A, Mathematical, Physical, and Engineering Sciences* 366(1885), 4613–26. doi:10.1098/rsta.2008.0201.

Spracklen, D. V., Arnold, S. R. and Taylor, C. M. 2012. Observations of increased tropical rainfall preceded by air passage over forests. *Nature* 489(7415), 282–5. doi:10.1038/nature11390.

Spracklen, D. V., Baker, J. C. A., Garcia-Carreras, L. and Marsham, J. H. 2018. The effects of tropical vegetation on rainfall. *Annual Review of Environment and Resources* 43(1), 193–218. doi:10.1146/annurev-environ-102017-030136.

Springmann, M., Godfray, H. C., Rayner, M. and Scarborough, P. 2016. Analysis and valuation of the health and climate change cobenefits of dietary change. *Proceedings of the National Academy of Sciences of the United States of America* 113(15), 4146–51. doi:10.1073/pnas.1523119113.

Stehfest, E., Bouwman, L., van Vuuren, D. P., den Elzen, M. G. J., Eickhout, B. and Kabat, P. 2009. Climate benefits of changing diet. *Climatic Change* 95(1-2), 83-102. doi:10.1007/s10584-008-9534-6.

Swann, A. L. S., Fung, I. Y. and Chiang, J. C. 2012. Mid-latitude afforestation shifts general circulation and tropical precipitation. *Proceedings of the National Academy of Sciences of the United States of America* 109(3), 712-6. doi:10.1073/pnas.1116706108.

Teuling, A. J., Taylor, C. M., Meirink, J. F., Melsen, L. A., Miralles, D. G., van Heerwaarden, C. C., Vautard, R., Stegehuis, A. I., Nabuurs, G. J. and de Arellano, J. V. 2017. Observational evidence for cloud cover enhancement over Western European forests. *Nature Communications* 8, 14065. doi:10.1038/ncomms14065.

Thomas, G. and Rowntree, P. R. 1992. The boreal forests and climate. *Quarterly Journal of the Royal Meteorological Society* 118(505), 469-97. doi:10.1002/qj.49711850505.

Unger, N. 2014. Human land-use-driven reduction of forest volatiles cools global climate. *Nature Climate Change* 4(10), 907-10. doi:10.1038/nclimate2347.

USDA-FAS. 2019. *Oilseeds: World Markets and Trade*. Circular Series FOP July 2019. US Department of Agriculture - Foreign Agriculture Service, Washington DC.

van der Werf, G. R., Randerson, J. T., Giglio, L., Collatz, G. J., Mu, M., Kasibhatla, P. S., Morton, D. C., DeFries, R. S., Jin, Y. and van Leeuwen, T. T. 2010. Global fire emissions and the contribution of deforestation, savanna, forest, agricultural, and peat fires (1997-2009). *Atmospheric Chemistry and Physics* 10(23), 11707-35.

van Vuuren, D. P., Edmonds, J., Kainuma, M., Riahi, K., Thomson, A., Hibbard, K., Hurtt, G. C., Kram, T., Krey, V., Lamarque, J., Masui, T., Meinshausen, M., Nakicenovic, N., Smith, S. J. and Rose, S. K. 2011. The representative concentration pathways: an overview. *Climatic Change* 109(1-2), 5-31. doi:10.1007/s10584-011-0148-z.

Veldman, J. W., Aleman, J. C., Alvarado, S. T., Anderson, T. M., Archibald, S., Bond, W. J., Boutton, T. W., Buchmann, N., Buisson, E., Canadell, J. G., Dechoum, M. S., Diaz-Toribio, M. H., Durigan, G., Ewel, J. J., Fernandes, G. W., Fidelis, A., Fleischman, F., Good, S. P., Griffith, D. M., Hermann, J. M., Hoffmann, W. A., Le Stradic, S., Lehmann, C. E. R., Mahy, G., Nerlekar, A. N., Nippert, J. B., Noss, R. F., Osborne, C. P., Overbeck, G. E., Parr, C. L., Pausas, J. G., Pennington, R. T., Perring, M. P., Putz, F. E., Ratnam, J., Sankaran, M., Schmidt, I. B., Schmitt, C. B., Silveira, F. A. O., Staver, A. C., Stevens, N., Still, C. J., Strömberg, C. A. E., Temperton, V. M., Varner, J. M. and Zaloumis, N. P. 2019. Comment on "the global tree restoration potential". *Science* 366(6463), eaay7976. doi:10.1126/science.aay7976.

Vickers, C. E., Gershenzon, J., Lerdau, M. T. and Loreto, F. 2009. A unified mechanism of action for volatile isoprenoids in plant abiotic stress. *Nature Chemical Biology* 5(5), 283-91. doi:10.1038/nchembio.158.

Wang, S., Chen, J. M., Ju, W. M., Feng, X., Chen, M., Chen, P. and Yu, G. 2007. Carbon sinks and sources in China's forests during 1901-2001. *Journal of Environmental Management* 85(3), 524-37. doi:10.1016/j.jenvman.2006.09.019.

Wang, J., Chagnon, F. J., Williams, E. R., Betts, A. K., Renno, N. O., Machado, L. A., Bisht, G., Knox, R. and Bras, R. L. 2009. Impact of deforestation in the Amazon Basin on cloud climatology. *Proceedings of the National Academy of Sciences of the United States of America* 106(10), 3670-4. doi:10.1073/pnas.0810156106.

Went, F. W. 1960. Organic matter in the atmosphere, and its possible relation to petroleum formation. *Proceedings of the National Academy of Sciences of the United States of America* 46(2), 212-21. doi:10.1073/pnas.46.2.212.

Williams, M. 2003. *Deforesting the Earth*. The University of Chigaco Press.
Zhu, J., Penner, J. E., Lin, G., Zhou, C., Xu, L. and Zhuang, B. 2017. Mechanism of SOA formation determines magnitude of radiative effects. *Proceedings of the National Academy of Sciences of the United States of America* 114(48), 12685–90. doi:10.1073/pnas.1712273114.
Zhu, J., Penner, J. E., Yu, F., Sillman, S., Andreae, M. O. and Coe, H. 2019. Decrease in radiative forcing by organic aerosol nucleation, climate, and land use change. *Nature Communications* 10(1), 423. doi:10.1038/s41467-019-08407-7.

Chapter 7

Measuring and quantifying greenhouse gas emissions from agricultural activities

Mohammad Ibrahim Khalil, University College Dublin & Prudence College Dublin/GSustain, Ireland; Syed Faiz-ul Islam, University College Dublin, Ireland; Macdara O'Neill, University College Dublin & Teagasc, Ireland; and Bruce Osborne, University College Dublin, Ireland

1 Introduction

Formulation and adoption of policies and initiatives to reduce agricultural greenhouse gases (GHGs) are a global concern. Reporting of national estimates to regional and international agencies is mandatory, particularly for Annex-I countries, using mainly IPCC guidelines (IPCC, 2006), which have been updated recently (IPCC, 2019). However, precise measurements at local levels that require the replacement of IPCC default values are increasingly required

http://dx.doi.org/10.19103/AS.2020.0064.12

for more refined national inventories and the identification of regionally appropriate mitigation options. There is still a paucity of national measurement and accounting approaches that cover the detail and spatial resolution that would limit several of the often-associated errors in GHG estimations (Williams et al., 2012). Permanent pasture covers 77%, barren lands occupy 15.2% and cropland (with annual and perennial crops) 12.6% (FAO, 2015; IPCC, 2014), and this should determine the major approaches to be used for GHG measurements. The assessment of GHG fluxes in agricultural systems can broadly be categorized into two main measurement techniques: (a) chamber measurements and (b) micrometeorological measurements, or a combination of both (Denmead, 2008). Due to several reasons, chamber methodologies have been most commonly used for the last 30 years, and more than 95% of the thousands of published studies used chamber methodologies, particularly static chambers, for reporting GHG fluxes (Rochette, 2011).

Measurements of GHGs have now been reported from an extensive range of ecosystems and landscapes. Despite the fact that the manual (static) chamber technique is labour intensive and the measurements are subject to significant uncertainties, this provides a simple, low-cost approach that is easily portable or relocatable, as well as being universally acceptable for determining GHG emissions from agricultural soils. Whilst the automatic (dynamic) chamber method could minimize the temporal variability inherent in static chamber measurements, it has several limitations in terms of area/land use coverage and cost. In terms of the gas sampling strategy to be used with either static or dynamic chamber, this will likely vary depending on the GHG (N_2O, CH_4 and CO_2) under investigation, as well as with land use, plant row spacing and crop growth stage (e.g. Caprez et al., 2012). Though the analysis of gas samples should be done immediately, this is almost always never practical, so that the storage method and duration of storage also need to be considered.

The choice of instruments for the measurement of the three GHGs, either separately or together, will depend on the required precision and the budget. Gas chromatography (GC) has commonly been used for the analysis of GHGs, and several companies provide suitable instruments. Besides GC, GC-mass spectrometry (GC-MS), Fourier-transformed infrared spectroscopy (FTIR), nondispersive infrared (NDIR) spectroscopy, quantum cascade laser spectrometry (QCLS), cavity ring-down spectroscopy (CRDS) and photoacoustic methods can be used to measure GHGs.

Micrometeorological approaches have, however, largely become the method of choice for GHG measurements since the 1970s (Desjardins and Lemon, 1974), as they do not have the same limitations as chamber-based approaches and can capture fluxes over significantly larger areas. These rely on the physical principles governing the transfer of radiation, mass and momentum in the atmosphere, specifically, the interaction of the lower atmosphere with the surface of Earth (e.g. Monteith and Unsworth, 2013). However, this approach also has its own limitations and the high costs and complexity of operation limit its widespread use by farmers so that alternative and cheaper approaches, such as the surface renewal (SR) technique, are being explored (e.g. Hu et al., 2018). The appropriate deployment of either EC or chamber-based approaches in agricultural systems will depend on several factors, including cost, crop type, cultivation method and management intervention. As EC approaches are still largely focussed on CO_2 flux measurements, this will also depend on which of the three major GHGs is being investigated.

In this chapter, we provide guidelines on the various methodological approaches for measuring GHGs in agricultural systems, focussing largely on chamber and EC-based techniques for different agricultural activities. Among others, an overview of chamber, tracer, EC and other alternative techniques for measuring GHGs from livestock, housing and manure storage is given. Increasingly, it is becoming important to have assessments of all three GHGs (N_2O, CH_4 and CO_2), so this means that many studies will include a combination of the chamber and EC measurements until EC techniques for measuring N_2O and CH_4 become more widely available. We will, however, focus on the techniques/methods of choice for most current investigations and the basis for large monitoring projects such as the integrated carbon observation system (ICOS) (Franz et al., 2018), Fluxnet and Ameri-flux, and the further advancements that might be required for obtaining accurate measurements.

2 Greenhouse gas measurement techniques for arable farming

Arable farming includes either rainfed or with limited irrigation (except irrigated flooded/wetland rice) and drylands with low and erratic rainfall without irrigation. Croplands include the land under annual crops with or without cover/catch crops, meadows or pasture (arable ley), kitchen gardens and temporary fallow and comprises in total 31% of global agricultural land. Crop and livestock production systems are responsible for up to a third of total anthropogenic GHG emissions (Vermeulen et al., 2012) where land use change may play a key role in land-based emissions but lower than agricultural production emissions (Tubiello et al., 2015). In 2000, total global cropland GHG emissions were 1994–2172 Tg CO_2, representing 4.5–4.9% of anthropogenic emissions, but excluded emissions from the livestock sector, land-cover change and pre- and post-production emissions (IPCC, 2007). Among the total emissions, 48% are derived from flooded rice, 32% from peatland cultivation and 20% from N fertilizer application (Carlson et al., 2017).

Cereals and oil crops are the world's most prominent crops. Application of both organic and inorganic fertilizers/materials to croplands releases significant amounts of GHGs to the atmosphere. Nitrogen fertilizer additions to increase food production is the major driver of increased global N_2O emissions (Qin et al., 2012), and the estimated global demand by 2030 will be 135 Mt (Van Groenigen et al., 2010). This could contribute more to global warming (Snyder et al., 2014) unless mitigation measures are taken. The emissions grew in Asia by 5.3% per annum, in Africa by 2.0% and in the Americas by 1.5%, whereas emissions decreased in Europe (−1.8%).

In general, arable and drylands have been managed intensively (e.g. intensive tillage, application of fertilizers and other chemicals and irrigation), enhancing GHG emissions. However, relatively little is known about how the management practices that are used affect all GHGs, particularly CH_4 and CO_2, and their impact on GWP and the GHG budget. Hence, all GHGs should be measured and evaluated separately for a national to global inventory and for formulating mitigation measures.

Arable agriculture contributes directly and indirectly to GHG emissions and is the main focus of this subsection. These systems are also the main anthropogenic source of N_2O, largely because of N fertilizer use associated with the timing, amount and form used. However, CH_4 oxidation, as well as the SOC gain and loss processes, should also be addressed to obtain an overall GHG budget. There are various methods used to quantify the emissions though data on measured N_2O and CH_4 emission factors (EFs), as well as heterotrophic/soil respiration (and importantly precise accounting of SOC density, Khalil and Osborne, 2018), are limited. This means that most countries report EFs either using IPCC default (Tier 1) values, which fail to account for soil type, climate or location, and/or country-specific (Tier 2) factors. The latter are largely constrained by inadequate spatial and temporal sampling, as well as limited coverage of different land uses and soil types.

3 Chamber techniques for measuring emissions from arable farming: design and features

3.1 Chamber types

There are two types of chambers: (i) dynamic and (ii) static. With dynamic chambers, the headspace air is exchanged at a high rate and GHG fluxes are calculated from the difference in gas concentrations at the inlet and outlet of the chambers multiplied by the gas volume, taking into account the area which is covered by the chamber (Butterbach-Bahl et al., 1997). Whereas static chambers are fully gas tight, without the forced exchange of the headspace gas volume, dynamic chambers are usually vented to allow pressure equalization between the chamber headspace and the ambient air pressure (Xu et al., 2006; Butterbach-Bahl et al., 2016). Both types of chambers are invasive and can change the vertical diffusion of gases in the soil, the soil energy balance and the degree of turbulence above the soil (Rochette, 2011) and affect the accuracy of the results if not properly designed and used.

3.2 Chamber systems

3.2.1 Manual systems

The use of a standard chamber design for all plant types is not possible because different ecosystems require different chamber dimensions/sizes (Pavelka et al., 2018). However, chambers designed for small understory vegetation or forest soil may also be suitable for arable and grasslands having a similar vegetation height and/or where the chamber can be accommodated between two rows of a wider crop such as maize. Depending on the plant density and height, the same chamber or one with a wider size with options to raise the height as the vegetation grows can also be designed and used for gas sampling.

A chamber may be cylindrical, rectangular or square, and the size may differ among agricultural land use types (see later). However, it is important to follow some chamber design and measurement protocols, with the aim of minimizing any disadvantages of the approach. These include, for example, changes in temperature and pressure inside the chamber, radiation, precipitation, wind speed and site disturbance effects. This avoids any bias on chamber air concentration changes over time and produces comparable results across chamber types in different ecosystems. Some examples of chamber (static) design for dryland (arable and grasslands) and wetland (rice paddies) systems are shown in Fig. 1. The following guidelines and considerations should improve the quality of the measurements and should facilitate the comparison of results between studies across regions and landscapes. We also provide recommendations on the minimum requirements for using static chambers for obtaining reliable and accurate results.

3.2.2 Automated systems

Studies using manual chamber techniques may miss the main peak events after rainfall or fertilization, if not deployed appropriately and at the relevant times. Also, chambers can be deployed and sealed only for a limited duration during each measurement to minimize microclimatic modifications. Some of these issues can be resolved using automated chamber systems, capable of flux measurements every few hours (Smith and Dobbie, 2001; Grace et al., 2013). However, spatial coverage is likely to be even smaller than that for manual chambers, due to the cost of large numbers of chambers.

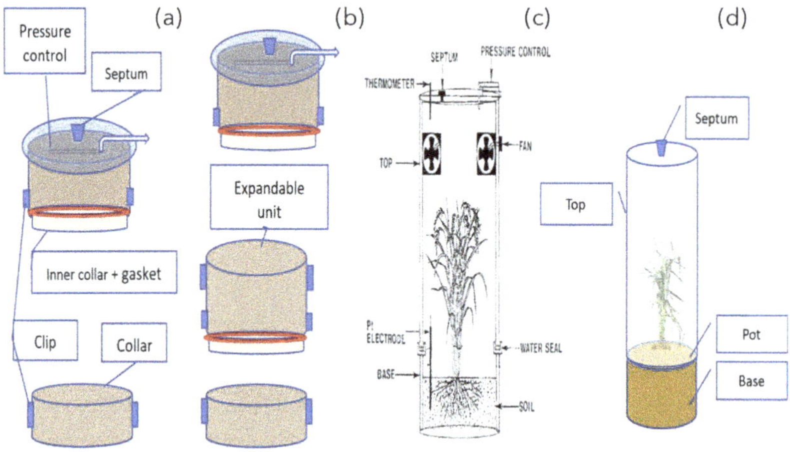

Figure 1 Schematic design ideas for a static chamber to collect gas samples from terrestrial ecosystems (a) base/collar (b) plus expandable option for both terrestrial and wetland ecosystems (c) inclusion of base/soil with fan and valve; and (d) laboratory and greenhouse conditions, with or without vegetation.

In both open and closed dynamic flow systems, tubing is required to sample air from the chamber to the gas analyser. The gas volume in the tube is replaced with a known gas concentration and directed to the chamber to maintain pressure equilibrium. The key differences are that one is open and another remains closed, and the frequencies of air samplings vary. Further information and operational details can be found elsewhere (Livingston and Hutchinson, 1995; Pumpanen et al., 2009). Typically, this system includes the chamber with or without a collar, a pump/sampling unit to transfer the air sample to a gas analyser and a controlling unit to operate the system, together with data storage and analysis capabilities.

A closed dynamic approach has commonly been used and recommended as a long-term (~20 years) optimum infrastructure (Fig. 2). The closed system

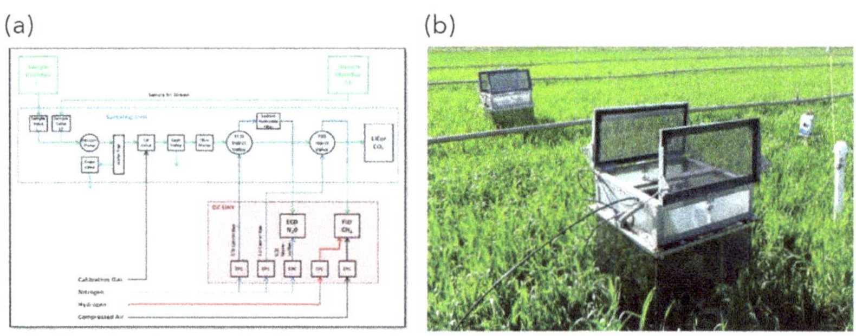

Figure 2 Schematic for the sample air path and carrier gases and calibration gas (a) and an automatic gas sampling chamber system connected directly to a GC (b) (From de Klein and Harvey, 2015).

integrates all required gas analysers in the field with an automatic *in situ* collection of air samples and analysis opportunities that capture diurnal and temporal variations but not spatial heterogeneity. In addition to a GC, an infrared gas analyser (IRGA) used for automatic chamber measurements can be connected to a single chamber. This requires sufficiently long sampling tubing for circulating chamber air via an outlet equipped with stopcocks to terminate IRGA measurements when required, together with inlet tubing. The inlet can be further used for syringe sampling (Hensen et al., 2006), which can also be linked to the sequential closing of the manual chambers.

In the next section, we will mainly focus on the use of static chamber techniques, which are commonly used for measuring GHG fluxes in agricultural fields and under laboratory/glasshouse conditions.

3.3 Chamber design and construction

3.3.1 Materials

Gas chamber materials should not react with any gases released from the soil and should not emit any contaminants into the atmosphere inside the chamber or pollute soils or vegetation. The commonly recommended chamber materials are stainless steel, aluminium, polyvinyl chloride (PVC), polyethene, polycarbonate and polymethyl methacrylate (Plexiglas®, acrylic sheet) (Parkin and Venterea, 2010), but non-metallic materials should generally be preferred. All other chamber materials such as seals, tubing, septa and venting should also be inert and rigid. The colour of the chamber influences the temperature inside the chamber. Generally, the outer colour of the chambers should be white (Pavelka et al., 2018) and the chamber should be insulated. If the chamber closure period is shorter (≤5 min), the use of an opaque chamber is recommended as it will ensure that the headspace air temperature does not increase as much as that in a transparent chamber, and thereby minimizes unintended warming effects inside the chamber. In case of the use of such transparent chambers, the whole chamber should be painted white. Alternatively, the chamber can also be covered with tin foil to block out sunlight. In all other cases such as gas collection from wetland rice paddies and measuring net ecosystem exchange (NEE), the use of transparent chambers is recommended (Minamikawa et al., 2015; Wang et al., 2013).

3.3.2 Dimensions

The chambers used should ideally be as large as practically possible in order to maximize the coverage of spatial variation. The minimum chamber size should be 0.2 m × 0.2 m or an equivalent ground area, which can be increased up to 2 m^2. For proper chamber design, the area-perimeter ratio (ratio of the chamber area, cm^2, and the length, cm of the chamber perimeter) should be ≥10 cm (Rochette and Eriksen-Hamel, 2008). This will reduce any relative error associated with a poor chamber seal as the diameter of a chamber increases. Though recommended a range of 0.1-0.4 m for arable crops (Butterbach-Bahl et al., 2016), the chamber height should be at least 0.3 m but could be higher in order to accommodate vegetation. The increase in the height of the chamber increases the minimum detectable flux but reduces environmental variability such as humidity, temperature or N_2O diffusion gradients within the soil profile. In contrast, a decrease in chamber height reduces the minimum detectable flux at the expense of greater variability.

Gas sampling from fields with tall plants (e.g. rice, wheat and maize) will require a higher chamber height, but this will lower the detectable fluxes, whilst increasing the closure period could reverse the situation. Rochette and Eriksen-Hamel (2008) indicated the ratio of chamber height (cm) to deployment time (h) should be ≥ 40 cm h^{-1}. But it is recommended to minimize the closure times as much as possible because longer deployment can increase the temperature inside the chamber. A better option is to use chamber extensions to increase the height of the chamber as the plant grows, but care needs to be taken with the seal between chamber extensions and the extent of mixing in the larger headspace volume, which can be compromised by the crop canopy.

To capture spatial variability, the recommended shape for the chamber is cylindrical, as it allows better mixing of the enclosed air. Often rectangular-shaped chambers, which may create dead space in the upper corners of the chamber where the enclosed air may not be properly mixed (Livingston and Hutchinson, 1995), have been used. In this case, the use of fans is recommended to ensure better mixing of the enclosed air. In row crops, chambers can be narrow and rectangular or square in size if the objective is to capture the flux gradient (FG) between plant rows, covering the whole inter-row spacing (Rochette and Hutchinson, 2005). Chambers are often custom made and should be designed in such a way that they cover the area of interest according to the required objectives.

3.3.3 Seals

The effective sealing of the chamber is an essential requirement for robust flux measurements. The aim of providing a gas-tight seal between the two components of the chamber is to prevent the leakage of gas out of the chamber, as well as stop external air from entering the chamber. To achieve an air-tight seal, a rubber gasket is usually placed between the chamber, and its base or a built-in trough is used which remains attached to the base, that holds water and acts as a seal between the two components (Christiansen et al., 2011; Hutchinson and Livingston, 2001; Parkin and Venterea, 2010). A study showed that gasket material must have a very low internal cross-sectional area for diffusion and must be pliable enough to form a good seal when compressed (Hutchinson and Livingston, 2001).

To achieve easy and accurate sealing, the use of precision-machined chamber components is recommended. To test the effectiveness of the seals, it is recommended to use a known concentration of reference gas inserted into the chamber, which is sealed to an impermeable surface and then to measure the rate of concentration change over time. Among the two types of sealing, the water seal is very effective and the recommended option. However, there are limitations in the use of water seals; for example, this approach needs a flat surface and the water may dry out quickly especially in a warm climate. This also requires the supply of clear water for each sampling, which may become dirty with algal growth as time progresses. However, most of these deficiencies can be overcome with proper preparation and careful monitoring, except for the possibility of N_2O dissolution in water (Roper et al., 2013) if the gas samplings are continued for a long time.

3.3.4 Venting

After a chamber (top chamber) is placed on its base, as well as during the gas sampling period, headspace pressure gradients can be formed inside the chamber (Christiansen et al., 2011). Thus, the inclusion of a vent that can be opened during sampling or when installing the gas chamber is often a recommended practice (Bain et al., 2005; Davidson et al., 2002; Hutchinson and Livingston, 2001; Hutchinson and Mosier, 1981; Xu et al., 2006). Such vents prevent pressure gradients between the interior and exterior of the chamber from influencing gas exchange. The vent tube is constructed from inert tubing and secured through the chamber with an appropriate gas-tight bulkhead fitting, which should satisfy several criteria: (i) the tube diameter (D) must be small enough to minimize diffusive losses but large enough to permit air - moving in response to pressure changes - to flow down the tube with a pressure loss no greater than 0.01 Pa, and (ii) the vent tube length (L) must be not less than that gives an internal volume five times greater than the volume of enclosed air displaced by the largest anticipated pressure wave (Hutchinson and Mosier, 1981).

3.3.5 Insulation and temperature control

Closure of non-insulated chambers exposed to direct sunlight can cause the headspace temperature to rise by 10–20°C within 20 min. Any significant temperature changes (> 5°C h^{-1}) can produce errors in the calculated fluxes (Rochette and Hutchinson, 2005; Parkin and Venterea, 2010). Studies have found that a 1°C change in chamber temperature could result in a 333 Pa change in chamber pressure (Xu et al., 2006), which also increased soil respiration by 10%

at 10°C in the absence of other limiting factors. Therefore, any increase in soil temperature may enhance CO_2 emissions, which in turn could alter the soil's oxygen status and indirectly affect N_2O and CH_4 production. The temperature change inside the chamber can be achieved by covering the chamber, outer walls and dorsal surface with either a reflective foil or an insulating material, or preferably a combination of both. To test if the proper insulation and temperate control are in place, air and soil temperatures inside and outside the chamber need to be measured and compared for the deployment duration before gas sampling.

3.3.6 Sampling port

To collect gas samples from the chamber, an inert and gas-tight sampling port is required. The most used materials for the sampling port are butyl rubber septa with syringe taps sealed to the chamber. To prevent any leaks, these materials should be changed at a regular interval.

3.3.7 Installation of fan

Proper mixing of headspace air during manual gas sampling is suggested to eliminate bias in the results (Christiansen et al., 2011; Rochette and Eriksen-Hamel, 2008; Rochette and Hutchinson, 2005). Mixing can be done with a syringe, but without using a fan it might still underestimate the soil gas fluxes. Therefore, the use of a fan is commonly recommended to eliminate possible bias from vertical gas concentration gradients (Pumpanen et al., 2004). The average air movement speed should be less than 0.5 m s^{-1} inside the chamber and ideally should be measured at four points at the half of the height of the chamber before the first sampling.

To prevent the unwanted suppression of gas diffusion from the soil into the chamber atmosphere, the fan has to be installed in such a way that the main airstream is not directed towards the soil but stays around the middle part of the chamber. It is recommended that two fans are used for tall or rectangular chambers on the opposite side walls to facilitate the homogenous mixture of headspace air. As an alternative to using a fan, a simple gas sampling manifold inside the headspace and built into the chamber cover can also be used to draw headspace air from four quadrants during sampling to minimize vertical gas concentration gradients (Parkin and Venterea, 2010).

3.3.8 Special consideration for a chamber with plants

Plants can have significant effects on CO_2, N_2O and CH_4 fluxes (Jørgensen et al., 2012; Pihlatie et al., 2005; Yu and Chen, 2009). An expandable chamber is a better option to accommodate all above-ground plant parts inside the chamber, as the plant grows in height, and this will reduce the possibility of the development of a dead volume. Depending on the size of the plants, it is commonly recommended that at least four plants are enclosed in the chamber, while studies in growth chambers in the laboratory may contain only one to two plants (Butterbach-Bahl et al., 2016; Islam et al., 2018). The use of an opaque chamber can block the incoming radiation, leading to stomatal closure and reduced photosynthesis (Hopkins and Hüner, 2009), which may ultimately reduce GHG fluxes (Yu and Chen, 2009; de Klein and Harvey, 2015). This problem can be overcome by using transparent chambers instead of an opaque chamber to enclose the plants. However, such measures will create a conflict between the need to control temperature changes in the chamber and any requirements to assess the contribution of photosynthesis. Therefore, researchers need to be mindful of these issues when designing experiments specifically to look at plant effects on GHG fluxes.

4 Chamber techniques for measuring emissions from arable farming: deployment and sampling techniques

4.1 Deployment

Chambers consisting of a paired 'base-chamber' (collar) design are recommended, where the chamber (top chamber) is placed and sealed onto a base previously inserted into the soil (Parkin and Venterea, 2010). Chamber bases should be installed long enough before measurement to allow a good chamber-to-soil seal, typically >24 h before the first sampling. In vegetated areas, installation of bases should be established for at least a week to a month before the gas sampling to allow root regrowth (Rochette and Hutchinson, 2005). Both chamber and collar must have the same dimensions for both manual and automated measurements, whether they include plants or not. Further points to consider include the minimization of the effects of rainfall and fertilizer addition and a reduction of the Venturi effect (e.g. Pavelka et al., 2018). Strategies in chamber deployment for air/gas sampling vary with land use type, plant canopy architecture and height over the growing period. In general, collar placement should cover heterogeneity in soil, vegetation, topography, seasonal changes (winter, spring, summer and autumn) and ecosystem-specific events such as re-wetting, freeze-thaw cycles and fertilization-harvest events. Some of these points are covered in Pavelka et al. (2018). The preferred approaches in chamber deployment are:

- From a statistical point of view, the random deployment of chambers would minimize both spatial and temporal variations.
- Collar insertion avoids root zone cutting and trenching effects.

- The depth of insertion will depend on the closure time (de Klein and Harvey, 2015), shallower for short-time (e.g. 3–5 cm) and deeper (e.g. 5–15 cm) for longer-duration sampling to carry the chamber load and avoid gas diffusion in or out of the chamber.
- A thin wire mesh along the inner boundary of the collar can be used to protect vegetation inside the chamber.
- Collars should be deployed in a transect with regular distances between adjacent points and distributed at random to cover field heterogeneity but conditional for an automated system.
- If required, the fall line direction for sloping terrain, cover patches and keep a utility boardwalk to prevent any disturbance to the surrounding area and vulnerable vegetation.
- Relocation of chambers is suggested when the soil water content within the chamber differs from the surroundings.
- A vent is needed to prevent pressure gradients during chamber deployment, particularly with manual systems (Bain et al., 2005). Where an inlet pipe is used, the inner part should be located close to the chamber lid and the top directed against the wind.
- A net or filter should be fixed to prevent incoming insects into the chamber. For an automated system having tubes longer than 10 m, the permeability of gases must be examined.
- For manual chamber measurements, at least five chambers should be installed and for automated at least three per variant (with or without plants) but more could be used to minimize spatial variability.
- An additional fan in all chamber types mixes the inside air to maintain a constant and homogeneous increase in the gas concentration (Christiansen et al., 2011).
- Auxiliary sensors are used or installed to record air temperature, pressure and relative humidity inside and outside the chamber to assess 'chamber effects'.

In addition, before the first measurements, the following points need to be considered (Bahn et al., 2009), depending on land use type and management practices:

Arable crops: For tilled crops (seasonal to annual mostly) sown or planted in rows (narrow and wide) with a distance of 10 and 20 cm (wheat, barley, paddy rice, etc.) and 40–75 cm (cotton, maize, sugarcane, etc.), the chamber diameter may vary, but it is recommended to deploy chambers that can cover at least two rows for narrow row spacing and smaller versions for inter-row placement (Fig. 3a and b). The chambers should be removed during harvest and/or before ploughing for the next crop. In case of limitations in space, single plants may

Figure 3 Details of chamber placement in a wheat field covering three rows (a), using a chamber of the same size placed in between rows in a maize field (b) and a transparent chamber placed in a rice field (c). The chambers placed in the wheat/maize field have extendable options to incorporate above ground vegetation in the either crop fields.

be removed after germination to ensure the measurements are subsequently associated with limited disturbance.

4.2 Gas sampling and collection

Among the agricultural GHGs, N_2O is a potent reactive gas, and the air sample container should be a standard glass tube/vial of a desirable size (>6 ml), depending on the amount of gas required for an analyser (e.g. GC, GC-MS, CRDS) and GC-IRMS for isotope studies (natural/enriched). The tube/vial should be leak proof, which may be achieved by screwing the cap on 'finger tight', followed by another quarter-turn, preferably with a rubber septum (butyl rubber and similar) or screw-on plastic caps with rubber septa. It is important to have an option for evacuation at <100 Pa several times (Fig. 4a), and where feasible to flush with inert gases (e.g. He, Ar and N_2). The septum used should frequently be replaced by a new one though pre-evacuated vials (vacutainer) are also available.

It is preferable that the vials are kept under water and to check for leakage through the examination of the water level inside and that one to two additional samples should be taken. A similar approach applies in order to keep the container gas tight to prevent sample loss during storage until analysis. For short-term storage, good-quality gas-tight syringes, although

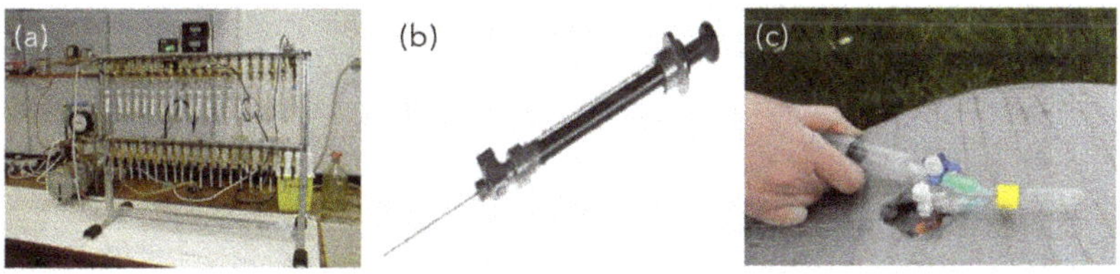

Figure 4 Illustration of a multiple evacuation system (a), sample-lock gas-tight syringe (b), three-way valve and glass vial for gas sampling from a manual chamber (c) (partly sourced from de Klein and Harvey, 2015).

an expensive option, can be used. The related issues and alternative options were reported by others (Rochette and Bertrand, 2003). Flushing procedures can also be effective (de Klein et al., 2003), where, before sample collection, 25 ml of chamber headspace air is flushed backwards and forwards four times through a 6-ml septum-capped glass vial container. Gas/air sample can be collected using a gas-tight syringe (glass/polypropylene of high quality available in the market) (Fig. 4b and c). The volume (>10 ml) collected will depend on the volume of the sample required for the analyses of the three gases from a single sample. To avoid mixing, it is suggested to use gas-tight valves and needles.

4.3 Sample collection and representative numbers

Measurements of GHGs and data collection should be started preferably after 24 h of collar installation. As stated earlier a syringe attached with a valve plus a needle should be flushed by pumping the plunger for a few times with chamber air to remove any dead volume. This flushed syringe should be used to collect gas/air samples by opening the valve from the chamber headspace via a pre-installed rubber septum (butyl or similar), which should also be replaced occasionally. Immediately the valve must be closed, and the collected gas/air sample should be transferred directly into a pre-evacuated tube/vial, showing an initial spontaneous movement of the syringe plunger. It is suggested to sample about 30–40% more air/gas to keep the vial/tube under pressure to minimize the incursion of ambient air during storage (~0.13% d^{-1}, Rochette and Bertrand, 2003).

The number of chambers for manual measurements depends on land use-associated management practices and area, to minimize spatial variability, as well as manpower availability, so that generally as large a number of chambers as practically possible should be used. Where feasible, a separate sampling for CO_2 emissions and the use of pooling/compositing techniques may be used (Arias-Navarro et al., 2013). For automatic measurements (>5 chambers required), additional manual chambers could be used for supplementary measurements to minimize spatial variability and when the automatic system cannot be operated (e.g. wintertime, intensive field preparations; Pavelka et al., 2018). Where the EC technique is deployed, manual chamber measurements every 2–3 years from at least 25 different positions and twice at the maximum and minimum CO_2 fluxes within the footprint may be required (Pavelka et al., 2018). For either technique, appropriate extensions can be added to accommodate varying plant heights over the growing period if required. For specific soil respiration studies, the chambers must be installed on bare soil.

4.4 Sampling time and frequency

Around noon (12:00 h) gas collections often give a good representation of the GHG fluxes, allowing an increase in the number or size of the gas chambers to minimize any variability (Khalil et al., 2007) although other researchers have suggested a midmorning (9:00–11:00) collection (Darenova et al., 2014). For locations having a distinct climatic variability (except in cold and dry periods), seasonal and diurnal variations should be checked and accommodated to avoid biased estimation of the annual GHG balance. In general, a minimum of four samples should be collected, though this varies from 15 min at an interval of 5 min, to 45 min at an interval of 15 min, depending on the height and overall volume. For a single chamber placed either in between rows or for land uses having plant heights less than the chamber height, a sampling time of 30 min with intervals of 10/15 min should be considered (Khalil et al., 2007, 2009).

It has been recommended that a similar minimum closure time of 5 min is used, to ensure the accurate measurement of very low fluxes (Pavelka et al., 2018). Depending on management-induced and inherent soil characteristics, a shorter interval of 1 min up to 4 or 5 min for the measurement of soil respiration (CO_2 fluxes) against background atmospheric CO_2 may be considered though the first minute reading can be omitted to avoid flux disturbances (e.g. Caprez et al., 2012). Based on the number of chambers and seasons, other workers/studies should be consulted (Darenova et al., 2014; Pavelka et al., 2018) for further details of different approaches. Taking full consideration of temporal variations due to environmental factors or management (e.g. ploughing, fertilization and rainfall events), measurements should start on the day following the 'event', day 1 and every alternate day for a week and at least weekly thereafter, depending on the possibility of modifications in the fluxes. In order to calculate EF, gas measurement should ideally cover 12 months.

4.5 Strategic sampling procedures

In general, the spatial footprint of an individual chamber measurement covers <1 m^2. It is always suggested to quantify spatial variations in the GHG fluxes using manual chambers during various seasons to capture distinct climatic, location and land use-specific uncertainty ranges. For this, the number of chambers needs to be increased or the chambers repeatedly moved to cover a larger area in order to obtain a good representation of soil type, land use and management. For this, a large number of collars and chambers (25 or so depending on the landscape size and characteristics to cover slope, terrain, etc.), along with appropriate manpower, are required. In the case of a shortage of chambers, additional collars should be prepared for installation 24 h before sampling. An automatic chamber system would have limitations for this type of study. In addition, soil heterogeneity and emissions at small spatial scales should be considered (e.g. Stoyan et al., 2000) for national or regional campaigns.

Compared to manual chambers, the automated chamber system allows for measurements at higher temporal resolution (e.g. fertilization, rainfall and seasonal) for extended periods of time though it is still limited by the deployment of a small number of chambers (five to six). In the case of manual sampling, the intensity of gas sampling should be increased following an 'event', either related to seasons or different treatments/management practices.

Both manual and automatic chambers can be used to measure diurnal variability to identify a standard sampling period, particularly for manual sampling using the static chamber method. The number of collars and chambers can be minimized and installed as described earlier, but there is a need to minimize spatial variability. Gas sampling should be done at an hourly resolution (or maximum two hourly) with the intervals stated earlier, commencing early in the morning, for 24 h, and the sample campaign can include a study of the impact of season and land use.

4.6 Storage of air/gas samples

Following the collection of the gas samples, the tube/vial (Exetainer/Vacutainer) must remain gas tight to prevent sample dilution during storage until analysis. Such glass vials have screw-on plastic caps with rubber septa. For N_2O concentrations of 10 $\mu L\ L^{-1}$ with storage periods of 14 and 126 days, between 92% and 98% of the original N_2O concentration could be recovered from containers using butyl rubber and doubled-wadded septa, respectively. This is not required for some pre-evacuated vacutainers/exetainers. For a quality check of tightness, vials may be kept under water for a few hours or until analysis and any vials showing leakage should be removed from the analysis batch.

Vials that can be evacuated can be tested for air-tightness by (i) taking both ambient and standard synthetic GHGs of various concentrations, (ii) keeping them in storage (with or without dipping into water) for several weeks, and (iii) analysing them using a GC to evaluate the dilution effects of the samples of known concentrations. For instrumental efficiency, additional samples can be sent to another laboratory for analysis. Recently, Faust and Liebig (2018) recommended that gas samples in exetainer vials with chlorobutyl septa can be stored no longer than 28 days before analysis for CO_2, CH_4 and N_2O without impacting on the values. Storage can be extended up to or longer than 84 days but decreases in CO_2 concentrations would be expected. A temperature-controlled environment with minimal light is ideal, but a laboratory room with minimal temperature fluctuations is sufficient for GHG sample storage.

4.7 Information and raw data required for flux calculations

It is recommended to regularly record environmental and other data during chamber installation and air/gas samplings: location ID (GPS coordinates), land use type, date and place of collar installation, time measurement commenced, numbering/static ID of collars and chambers and, where feasible, any changed sampling position.

For flux calculations, the raw data required are date and precise time of measurements; CO_2 (ppmv), N_2O and CH_4 concentration (ppbv) dry-air mol fraction; water vapour concentration in the analyser cell (if possible); air temperature inside the chamber (°C); air pressure (Pa); effective system volume (m^3) including the collar minus the volume of any installations inside the chamber (may be neglected if <1%), as well as snow volume, vegetation volume, etc.; area covered by the chamber in m^2; and additional data that identify the flux measurement. In addition, some more considerations for automated systems are measurement frequency and the internal volume of tubing, filters, switching valves analyser and all other parts of the measuring line.

Some auxiliary measurements required are air temperature, soil temperature, soil moisture, soil organic carbon, monitoring the transparency of the chamber and temperature increase during these measurements (if essential) and other related variables such as site and vegetation characteristics, as well as daily meteorological data. The need for and collection of relevant ancillary vegetation in support of GHG measurements has recently been reported (Gielen et al., 2018).

4.8 Gas analysis

The choice of instruments for the measurement of the three GHGs, either separately or together, depends on performance and budget. There are opportunities to use manual sampling or simultaneous auto-sampling/injection in the laboratory and/or automatic sampling for continuous measurement under field conditions. GC has commonly been used for the analysis of the gases, and several companies are available globally for their supply, as well as the supply of GC-MS, FTIR, NDIR spectroscopy, QCLS, CRDS, and Photoacoustic analysers and the respective procedures and protocols should be followed. Given their low concentrations in the atmosphere, instruments for measuring N_2O and CH_4 should have a high (ppb) sensitivity for excellent rate-of-rise quantification, high precision, rapid response time at high temporal resolution and auto water correction. In this section, we will only discuss the GC method for the measurement of GHGs.

All concerned GHGs can be measured by GC. Combining improved injector, gas line and valve-driving modules, a GC can be equipped with a flame ionization detector (FID) for the detection and analysis of CH_4 concentrations and an electron capture detector (ECD, hot >300°C, ^{63}Ni) for N_2O. A methanizer and FID or a thermal conductivity detector (TCD) can be equipped to analyse CO_2 concentrations. In addition to gas sampling, appropriate quality control (QC) should be used for the separation of N_2O and CO_2 as they have similar retention times (and also from other gases in the air sample) by using, for example, porous polymer columns before the ECD. Because of the non-linearity of ECDs at high N_2O concentrations (>1000 ppb), a thorough multipoint calibration is recommended. An adequate (about 20% of analysed samples) number of standards having various concentrations for calibrating the GC system are required to cope with temporal drifts. Further details can be found from the manufacturers and elsewhere (Khalil and Inubushi, 2007; Zheng et al., 2008; Pavelka et al., 2018). In-built computer software is generally available for the measurement of the peak areas of the respective gases.

Gas injection techniques under laboratory conditions: Following the collection of gas samples either directly or from stored samples, they are either injected manually (Fig. 5a) or placed on an auto-sampler (batch for auto-injection) (Fig. 5b). In most cases, gas samples collected from laboratory studies are also injected manually. The volume of gas samples to be injected depends on the GC type but is typically 1 ml. A fixed-volume sample loop is normally employed for gas analysis and equilibrated at the sample's pressure with ambient atmospheric pressure.

For the auto-injection system, a simple, inexpensive, double-ended, upward-facing needle is used to sample gas by penetrating the septa, with the other end inserted into a container filled with water. Alternative sample injection procedure can be used to minimize overpressures in the vial (Hedley et al., 2006). It is recommended to switch off the valve to flush carrier gas through the loop and inject the sample onto the GC column for better precision and accuracy. Commonly, with automated procedures, gas samples are collected in

(a) Manual injection (b) Automatic injection

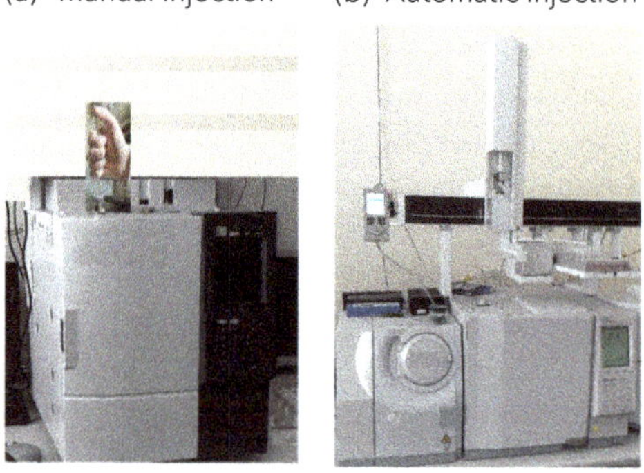

Figure 5 Showing manual (a) and automatic (b) injection of gas samples into the GC system.

the prescribed (capacity varies with manufacture as well as requirements) vials to accommodate them on to sample racks on the platform of an automated liquid handler, and a sampling arm is fitted with a specially constructed double-cavity concentric needle (Stevens et al., 1993).

Needle movement is controlled by the handler's software, synchronized with a valve-switching sequence using a software integral with the GC system. Further details on automated systems can be found elsewhere (e.g. Zheng et al., 2008). It is highly important to use oxygen and vapour-free ultra-high-purity carrier gas. There are many carrier gases (e.g. Ar, He and N_2, and its mixture with other gases such as CH_4) available on the market, and the less expensive one is N_2. GC resolution generally decreases with increasing column temperature: minimal column temperature gives better resolution but at the expense of longer elution times.

These injecting systems are configured according to the needs of the analysis and should be appropriate enough to obtain consistent analytical results. Adjustment systems in the GC method, for example, are pre-cut and backflush that reduces the analysis time and protects columns, heart-cut to improve the separation of trace components and column-switching to improve separation and reduce the burden on analysts. The manufacturer's manual may be consulted for further information.

Automated systems under field conditions: As discussed, there are several instruments such as GC (Pavelka et al., 2018) and QCLAS (Brümmer et al., 2017), either alone or in combination with spectroscopy/gas analyser, which can be used for automatic, simultaneous measurement of GHGs under field/greenhouse conditions (Fig. 5b). However, a suitable range scale, precision and accuracy of a GC, which can record dry through dilution correction instead of water vapour, is essential to ensure the high-quality standard of the chamber measurements. Short-term stability, which is the drift in baseline concentrations over a timescale of a few minutes, should be considered for an analyser to provide accurate outputs. The measurement range must cover all concentrations of monitored GHGs that can appear during the time of measurement. Korkiakoski et al. (2017) demonstrated that very small CH_4 fluxes, varying from uptake to emissions, can be detected with high-precision analysers with a short chamber closure time. Therefore, it is recommended to consider a minimum closure time of 5 min for ensuring the accurate detection and measurement of very low gas concentrations, as well as to install a system enabling continuous measurements of the mixing ratios for N_2O, CH_4 and CO_2 in the field for each chamber at least every 2 h. Further details can be found from the cited literature.

4.9 Calibration of the GC

To accurately measure the concentrations of various GHGs from a gas sample, a GC system must be calibrated. For simultaneous measurements under laboratory and field conditions, an ensemble of at least two instruments: GC with FID, and ECD, or three with TCD is required. Accordingly, calibration protocols should be designed to characterize the response functions of the instruments. In either sampling system, minimum (low mixing ratio) to maximum (high mixing ratio), and in-between concentrations, through the dilution of standard gas species (here N_2O, CH_4 and CO_2), of 7–11 samples of each will be required. Based on their atmospheric concentration (background in either the field and/or the lab) or slightly lower concentration but measurable with high precision and the possible increase during the measurement period, the concentration range may, for example, be 360 to 480 ppm for CO_2, 1600 to 2800 ppb for CH_4 and 280 to 400 ppb for N_2O.

Suggested standards (references), of which the concentrations may vary, can be obtained through consultation with other users (e.g. Hedley et al., 2006; de Klein and Harvey, 2015) and manufacturers for the purpose of calibration, and their mean concentrations should be sampled from pressurized gas cylinders similar to the air samples and transferred into vials. It is suggested that the concentrations of the supplied reference gases be guaranteed at the 95% confidence interval. Both reference cylinders and a third target cylinder may be used to control the quality of the instrument. The F statistic and correlation of determination should be compared to find a better fit. For GC performance, a repeatability standard deviation, SD_r, can be calculated in the usual way from the results obtained using the same method and a set of replicate air samples from the same source, under the same conditions - operator, GC system and laboratory (Ellison et al., 2009).

In addition to separate calibration for QC on a regular basis, a reduced number of reference standards (1–3) should be used after a certain number of gas samples (e.g. 5–10 gas samples) or time period (e.g. 30 min), depending on the time required to complete a sample analysis. The typical measurement sequence should be: reference gas high - samples - reference gas low - samples - reference gas high. Any major deviation may be corrected in calculating the sample fluxes, or the responses of the two references are linearly interpolated to obtain reference values for each measured sample. Inter-laboratory comparisons can be useful and are recommended.

4.10 Data processing and quantification

Unified standards and protocols are suggested to process the data for ensuring QC and comparability. In addition to sampling and measurement errors, scaling of measured data yields a multitude of potential errors and biases that need to be addressed. Despite the availability of software-aided concentration data, the errors associated with measurements and any missing data should separately be adjusted with references. It has been suggested to use both linear and non-linear (exponential) regressions, using the slope of the concentration change inside the chamber headspace (Christiansen et al., 2011; Parkin et al., 2012) to calculate the flux for all gases. The former works best for short chamber closure times, such as for CO_2 (Forbrich et al., 2010), with the lowest detection limits for flux calculations.

In contrast to the exponential model, a significant underestimation of the fluxes was reported (Kutzbach et al., 2007), and the least biased for curves with a convex-upward shape was found and is being widely used (Venterea et al., 2012). Forbrich et al. (2010) suggested the exponential model for calculating fluxes when the closure time is longer with few data points, particularly when using a GC for the measurements. However, measurement-specific calibrations should be derived by testing the best fit, whether linear or non-linear, for the flux calculation.

4.11 Calculation of daily fluxes and cumulative emissions

Chamber method: In general, manufacturer-specific software is being used to record and integrate signals from detectors to produce the data. After the analysis of samples (either single or batch injection), an integrated results file can then be exported to a spreadsheet. There are various methods, although similar in approach, for the calculation of fluxes considering corrections required for temperature and pressure in and outside of a gas chamber (Holland et al., 1999; Parkin et al., 2012). Here, we are providing the approaches of Collier et al. (2014).

Once the concentration of GHGs for each sample is available, the following equation can be used to convert trace gas concentration from volumetric to mass, using the ideal gas law:

$$PV = nRT$$

where P = atmospheric pressure (Pa), V = volume of gas (m^3), n = moles of gas, R = gas constant and T = temperature (°C).

Then, mol trace gas (Gas) as either mol L^{-1} or converted to appropriate units as required for each gas can be calculated using data measured during gas sampling by:

$$\text{Gas}\,(\text{mol } L^{-1}) = \frac{\text{V trace gas L} * 1L^{-1} * \text{P atm}}{(0.08206\text{L atm mol}^{-1}\text{K}^{-1}) * (273 + T)\text{K}}$$

where K = temperature in Kelvin.

For the evaluation of linearity and its goodness of fit (F test and coefficient of determination), a time series concentration of each gas should be used to plot time-by-concentration. As discussed, three to five gas samplings are suggested for calculating the flux (T0, T1, T2, T3 and T5) by establishing a consistent protocol and rejecting any time series that fails to meet that protocol's standards for linearity. Please see Parkin and Venterea (2010) for a thorough discussion of error, bias and variance in flux calculations. Then, the slope of the linear regression can be used to calculate the flux:

$$F = S * V * A^{-1}$$

where F = flux (mol L^{-1} h^{-1}), S = slope of the regression, V = chamber volume including collar and A = chamber area. Thus, the equation becomes:

$$F = \frac{\left(S\,molL^{-1}hr^{-1}\right) * VL}{Am^{-2}}$$

For calculating the slope of the regression to estimate the gaseous flux, a linear approach is generally followed (Fig. 6a and b). However, both linear and non-linear approaches should be considered to avoid errors associated with variable rates/trends of the gaseous build-up in a chamber for the calculation of CO_2 effluxes (Pavelka et al., 2018). For this, data are sorted by time and then extrapolated by two curves - regression line p (red line) and exponential curve K (blue line). The intersection of p and K is the point B where the flux loses its linearity.

A simple equation can also be used to calculate the GHG fluxes:

$$F = \frac{\Delta C}{\Delta t} * \frac{V}{A} * \acute{A} * \frac{273}{273 + T} * cf$$

where F = flux of the concerned gas (mg CO_2; µg CH_4 and µg N_2O; m^{-2} h^{-1}); $\Delta C/\Delta t$ is the concentration change over time (ppm-CO_2, ppb-CH_4 and ppb-N_2O h^{-1}); V = chamber volume (m^3); A = chamber area (footprint; m^2); ρ = gas density (0.717 kg m^{-3} for CH_4 and 1.977 kg m^{-3} for both CO_2 and N_2O at 0°C); T = mean air temperature inside the chamber (°C) and cf = conversion factors for N_2O = N_2O-N, 28/44; CH_4 = CH_4-C, 12/16; and CO_2 = CO_2-C, 12/44.

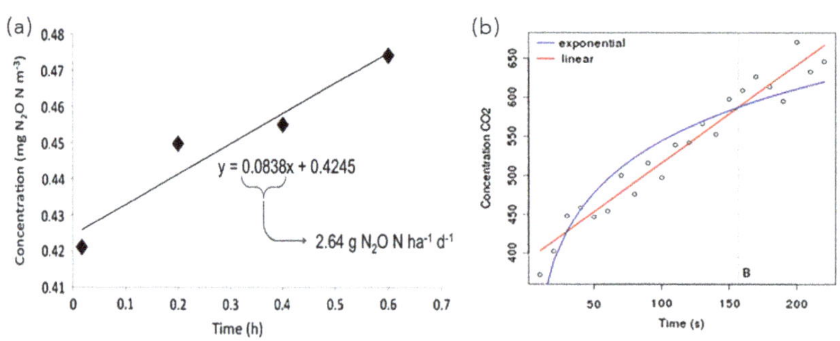

Figure 6 (a) Calculation of N_2O flux from the variation in concentration over time (slope of the linear regression) measured at four-time points over the course of a 36-min sampling period using a typical static chamber. (Collier et al., 2014). (b) Determination of the point B - the time where the CO_2 flux deviated from linearity (Pavelka et al., 2018).

Laboratory method: Following the analysis of gas samples using a GC, the gas concentration can be calculated using the previous equation. In this case, the Gas should be multiplied by the headspace volume of the incubation bottle or jar to obtain the mass of the three gases accumulated during the incubation. Thus, the flux can be calculated using the following equation:

$$GHG_{flux} = \frac{(C_1 - C_0)}{S_m * I * 1000}$$

where GHG_{flux} = flux of the concerned gas (mass of gas produced per unit of time); C_1 and C_0 = the mass of GHG produced at the end and at the beginning of two consecutive samplings, respectively; S_m = mass of air-dried soils; and I = incubation period.

Calculation of cumulative and total flux: Based on the gas sampling/measurement interval commencing at day 1 (or zero), the cumulative production of GHGs can be calculated using the trapezoid rule by linear interpolation between two consecutive samplings, that is the integration approach (Table 1).

A Bayesian approach can also be used to calculate cumulative GHG fluxes. For instance, for N_2O estimates a model is fitted to the data using prior knowledge of microbial N_2O production rates and the observed range of N_2O emission factors (e.g. 0-5%) (Cowan et al., 2020; Stehfest and Bouwman, 2006). This analytical approach does however rely on several assumptions related to the controls on N_2O emissions but provides a more robust method for integrating fluxes and producing uncertainty estimates. Post-fertilization N_2O fluxes typically follow a log-normal distribution with an observed peak and decay temporal pattern, which can be described by the following equation:

$$f(F) = \frac{1}{\sqrt{2\pi}\sigma_{log}F} e^{-(\log(F)-\mu_{log})^2/2\sigma_{log}^2}$$

Where F represents N_2O fluxes at a given time, μ_{log} is the mean and σ_{log} is the standard deviation of the log-transformed variate.

Table 1 Examples of the calculation of cumulative and total gas fluxes

Day of sampling	Gas flux (g N_2O-N ha^{-1} d^{-1})	Equation (integration approach)	Day of sampling	Cumulative flux (g N_2O-N ha^{-1})
0	0.22		D0	
1	2.15	= (D1 − D0)*2.15 + ((D1 − D0)*(2.15-0.22))/2	D1	3.12
3	1.50	= (D3 − D1)*1.50 + ((D3 − D1)*(1.50-2.15)/2	D3	2.35
7	2.80	= (D7 − D3)*2.80+ ((D7 − D3)*(2.80-1.50)/2	D7	13.8
Total flux(week) = ∑(D1 + D3 + D7)				19.27

Applying Bayesian interpolation means that only a single N_2O peak can be fitted to the model. An advantage of this feature is that the log-normal distribution can predict large fluxes, which may not be apparent in the data set. The decay dynamics of the model also allows accurate extrapolation of fluxes beyond the measurement time domain (Levy et al., 2017). One limitation of the model, however, is that it's likely to increase uncertainty where data is randomly distributed and does not display a clear peak and decay pattern. In such cases, the trapezoidal method will integrate the daily sample means irrespective of whether the values reflect valid or noisy data. A more in-depth explanation of the model parameters (Levy et al., 2017; Plummer et al., 2016) and its application in agricultural studies are described in recently published texts (see Cowan et al., 2019, 2020).

5 Micrometeorological techniques for measuring emissions from arable farming: the eddy covariance (EC) method

5.1 Introduction

One of the most popular methods of measuring turbulent gas exchange above vegetation is the eddy covariance (EC) method. With EC, turbulent air flow is sampled at a measuring point for measurements of vertical wind speed, direction and gas concentration. The first EC measurements assessed CO_2 exchange over corn (Desjardins and Lemon, 1974), and following technological advancements in the 1980s, it was possible to carry out CO_2 exchange measurements over crops (Anderson and Verma, 1986; Desjardins, 1985). Subsequent improvements in sensor performance and data acquisition systems enabled the measurement of CO_2 fluxes to be carried out over extended periods (Black et al., 1996; Greco and Baldocchi, 1996; Baldocchi et al., 2018; Wofsy et al., 1993). The following sections discuss the basic principles of EC theory and the measurement specifics and requirements. Aspects of tower establishment, measurement considerations and data processing are also discussed together with alternative techniques.

5.2 Principles of eddy covariance

Four factors underpin the attractiveness of EC over traditional chamber techniques as an investigative tool for both managed and unmanaged landscapes: (i) scale-appropriateness, (ii) non-invasiveness (iii) large flux footprint and (iv) trace gas exchange, which can be integrated across a continuum of timescales ranging from hours to years (Burba and Anderson, 2007). Traditional tower-based micrometeorology examines turbulent fluxes in the surface layer, which occupies the lower 20–50 m of the atmospheric boundary layer (Garratt, 1994). The surface layer consists of two sublayers. The *roughness sublayer* is a region in which vegetation elements dynamically influence the atmospheric turbulence through phenomena such as aerodynamic wakes and thermals (Raupach and Thom, 1981). Typically, the depth of this layer extends from the mean canopy height (*h*) to two to five times *h* (Kaimal and Finnigan, 1994). The layer above the roughness sublayer is known as the *inertial sublayer* or equilibrium boundary layer. Here, logarithmic profiles exist for the wind profile and fluxes with height for near-neutral conditions (Jones, 2013).

The EC method relies on the correlation between instantaneous fluctuations in the vertical wind velocity from its mean value and instantaneous fluctuations in gas or scalar concentrations. The flux of any scalar can be written as:

$$F = \overline{\rho_a w's'}$$

where F is the flux density of gas/scalar (dry mole fraction), ρ_a is the density of air, w′ is the vertical wind speed and s′ is the mixing ratio of the gas/scalar of interest. A positive sign indicates a net CO_2 emission to the atmosphere, whilst a negative sign denotes net CO_2 uptake.

Net exchange is relatively constant over long time periods, but instantaneous contributions to the fluxes vary considerably during the measurements. Reynolds decomposition is used to quantify fluxes into their mean (overbar) and fluctuating parts (prime) over the sampling interval (typically a half-hour time series). The fully expanded equation can be found elsewhere (Aubinet et al., 2012; Burba and Anderson, 2007). Certain terms are omitted based on numerous assumptions that are inferred from the EC method. Sampling must be done over a flat homogenous terrain where: (i) the mean vertical velocity is 0, (ii) density fluctuations are negligible, (iii) fully developed turbulence within the boundary layer of interest. (iv) the upwind area contributes to flux measurements, and (v) this upwind area or 'footprint' is of adequate size.

5.3 Instrumentation

High-frequency CO_2 and H_2O concentrations are measured with a NDIR absorption analyser (otherwise referred to as an IRGA). Fast-response 3D ultrasonic anemometers measure wind speed and direction to calculate the flux and its upwind source. The IRGAs are designed for open-, enclosed or closed-path configurations. For open-path sensors, the ambient air flow is sampled *in situ*, whilst for closed-path analysers, an internal sample cell is flushed with the sample air. The enclosed analyser is based on both a closed and anopen-path configuration, and this reduces data loss during precipitation. This design was selected as the standard design during the establishment phases of ICOS (Rebmann et al., 2018). Each configuration comes with its own advantages and disadvantages (Burba et al., 2010). Besides CO_2, the EC approaches can be used to measure N_2O and CH_4. The LI-7700 is the only open-path sensor currently available for CH_4, but the instrument is susceptible to data loss during precipitation events.

Alternatively, a range of fast-response, closed-path analysers are available for N_2O and CH_4 based on optical absorption techniques. Many of the optical absorption techniques use multi-pass cells and include FTIR, tunable infrared laser differential absorption spectrometers (TILDASs), difference frequency generation (DFG), isotope ratio infrared spectroscopy (IRIS), CRDS and off-axis integrated cavity output spectroscopy (OA-ICOS, i.e. Los Gatos Research, LGR, USA). These techniques may utilize lead-salt lasers, telecom lasers (CH_4) and quantum cascade lasers (CH_4 and N_2O) targeting each gas absorption spectrum in the mid-infrared range. Some instruments based on FTIR can measure all three GHGs (CO_2, N_2O, CH_4) simultaneously.

The concentration detection limits will depend on the instrument's precision as well as the turbulence and site characteristics. Analysers must be able to detect small variations in N_2O (±1 ppb) and maintain short- to long-term repeatability, linearity, low drift and temperature independence, with a correction function for water vapour (Lebegue et al., 2016). Many of the commercial instruments available can correct anomalies associated with interference due to water vapour. The precision of instruments measuring CH_4 must be within a range of 8 to 35 ppb or 10 nmol mol^{-1}, and the value is dependent on the friction velocity (u^*) of turbulent flow. Nemitz et al. (2018) have provided further information about the instrument characteristics and setup.

5.4 Ancillary measurements

Each tower must provide ancillary measurements to complement the continuous EC flux time series. The type of sensors typically required are those for temperature (air - T_{air} and soil - T_{soil}), soil water content (SWC), water table depth (WTD) and soil heat flux (G). The sensors must fulfil the following basic requirements: (i) allow unattended and automated measurements in the field; (ii) be designed for long-term burial in the soil; (iii) require minimal (re)calibration in the factory or the lab once installed, minimizing data gaps and soil disturbance due to sensor removal; (iv) have a fast enough response time to detect changes in the variable measured (within the recorded time interval); (v) have a low susceptibility to drift; (vi) follow a well-validated measurement principle; and (vii) measure the target variable over its entire field range with a given minimal measurement accuracy (specific per variable) (Op de Beeck et al., 2018). Radiation is another key variable that is essential for understanding energy and carbon fluxes at the ecosystem scale. The radiation variables measured include shortwave radiation, long-wave radiation photosynthetically active radiation and diffuse radiation. High-quality net radiometers, pyranometers (shortwave) and pyrgeometers (longwave) are also installed on sites (Carrara et al., 2018).

5.5 Operation and sampling strategy

EC measurements are conducted using a tower (mast) positioned within the ecosystem. Towers should be positioned in such a way as to face the prevailing wind direction and maximize fetch distance. Increases in sensor height also increase the fetch distance (i.e. the upwind source area). A general rule of thumb is 100 m distance for every 1 m increase in height (h). The inertial sublayer depth also increases with the fetch and is a function of canopy architecture (Munro and Oke, 1975). Flux sensors must remain within the limit of the inertial sublayer, whereby its lower boundary is 1.66–2.16 h. Therefore, sensor heights need to be adjusted to accommodate the growing vegetation.

Towers are generally deployed as static structures and situated in one single location in a field. To improve spatial variability, a roving or mobile tower can be used to make replicated measurements of NEE, as performed over a spring barley crop in Ireland (Davis et al., 2010) (Fig. 7).

Figure 7 A mobile eddy covariance system used for making replicated measurements of CO_2 fluxes in an arable field (top panels, Davis et al., 2010), and an eddy covariance tower established over a winter oilseed rape field in Ireland (lower panels) (Forristal, 2019). An open-path IRGA and pole mounted anemometer (left) are orientated into the direction of the prevailing south-westerly wind (left), together with the fetch (right) or upwind distance from which fluxes are detected by the instruments.

Interference with the tower can also affect EC measurements. Experimenters must establish good communication with the landowner to (i) agree on a site location, (ii) coordinate management to safeguard the sensors, (iii) provide access to current and historical site management, (iv) compensate for destructive plant sampling and (v) grant access to electricity, although other options for supplying power, such as batteries or wind, are possible (Moureaux et al., 2012), and these will likely increase in the future as the technology improves. Weekly or bi-weekly visits to sites are necessary to ensure the optimum performance of the sensors with minimum damage or down time. Sensor cleaning will be required from time to time due to the accumulation of dust and debris and also due to birds, which tend to perch (and defecate) on masts located in open areas.

Rapid crop development results in dynamic height changes. Croplands advance from bare soil to maximum development within 1 year, with large variations in sowing and harvest dates. Crop ecosystems also receive numerous specific management practices, with management activities dependent on the crop type, crop rotation, soil properties and climatic conditions (Moureaux et al., 2012).

5.6 Data acquisition

Modern data collection systems use data loggers that collect quasi-parallel signals and create a serial digital signal for transmission to a computer or a storage unit. Many instruments, recording 10 Hz CO_2, N_2O and CH_4 data, can interface with these micro loggers regardless of the sensor's manufacturer. Analogue signals are sent from the instruments and stored on a 2 GB flash memory card. Ancillary data can also be retrieved directly through a number of serial communication interfaces (e.g. via RS232/RS422/RS485 ports or ethernet).

5.7 Raw data processing

Raw data received from the EC, sonic and ancillary sensors is formatted as GHG (.ghg), TOA5 (.txt) and TOB1 (.dat) files. Fluxes are typically processed using dedicated software, although alternative software packages are available (Aubinet et al., 2012). The first step involves a series of QC tests, as high-frequency raw data can produce erroneous readings from time to time. At ICOS ecosystem stations the processing chain consists of 14 steps (Sabbatini et al., 2018). The QC method developed by Mauder and Foken (2004) is a widely applied technique. Here, the best-quality fluxes are assigned a '0', fluxes suitable for annual budgets a '1' and fluxes that should be discarded a '2'. A summary of correctional procedures for open- and closed-path sensors can be viewed in Holl and Kutzbach (2018).

The SmartFlux 2 and 3 systems are compatible with certain Li-COR EC instruments (https://licor.com/env/products/eddy_covariance/smartflux.html). Once the master clock of the microcomputer has been synchronized with GPS, users can view EC fluxes in real time. This innovation offers researchers remote access to their flux data from geographically distant sites. The device also produces fully processed data to facilitate the normal procedures typically conducted by researchers.

5.8 Post-processing

A web-based tool developed by the Max Plank Institute called REddyProc (BGC16) is currently used for gap filling and flux partitioning. Three gap-filling methods are implemented on the half-hourly data: (i) look-up tables, (ii) mean diurnal course (MDC) and (iii) marginal distribution sampling (MDS). Wutzler et al. (2018) provide a comprehensive analysis of how gap filling is parameterized using meteorological variables across each technique.

The CO_2 flux values are expressed in terms of NEE. A principal task of EC studies is to estimate the contribution of GPP and R_e to the vegetation-atmosphere gaseous exchange. Flux partitioning can be performed using the night-time and daytime partitioning methods. Night-time partitioning relies on the temperature response function of night-time NEE fluxes that are representative of R_e (Reichstein et al., 2012). The exponential relationship is extrapolated to daytime, and GPP is estimated by summing NEE and R_e. The Lasslop method estimates GPP using light response curves fitted to daytime NEE and estimates R_e where the curve intercepts the ordinate (Lasslop et al., 2010), typically using a rectangular hyperbolic light response curve (Gilmanov et al., 2007, 2013). The model is modified by including a function to account for the temperature dependency of respiration and the effect of VPD on stomatal closure (i.e. reduced CO_2 uptake).

Both flux partitioning approaches are used extensively by researchers and each method is applied simultaneously in the REddyProc package (Wutzler et al., 2018). Barba et al. (2018) found that the temperature/NEE partitioning approach accurately estimated R_e across contrasting ecosystems (forest to grassland; boreal to semi-arid), indicating the method would be suitable for individual or mixed farms.

6 Other techniques for measuring emissions from arable farming

6.1 Introduction

The EC technique provides a *direct* measurement of turbulent fluxes between vegetative canopies and the atmosphere. *Indirect* techniques also exist for measuring fluxes in the inertial sublayer. The indirect gradient methods can be successfully applied to measure flux entities where the use of expensive fast-response gas analysers are not practical or feasible.

6.2 Flux gradient

The FG method measures the eddy diffusivity of a trace gas with concentration 'x' such that:

$$F_x = -K_x\left(\frac{d_x}{d_z}\right)$$

The exchange coefficient K_x assumes all scalars are transported equally effectively; a single coefficient value is subsequently applied for all scalars. A fast-response anemometer is used to obtain the exchange coefficient for momentum, and the exchange coefficient for the species of interest is empirically determined through a series of steps. Temperature, water vapour and gas entity profiles are determined at a series of heights within the canopy. The dependence on two height points for wind and temperature invokes sensitivities to errors in single instruments or site irregularities. A more detailed description of the theory can be viewed in Monteith and Unsworth (2013).

The FG technique has also been successfully applied to measure atmospheric concentrations of sulphur dioxide and ammonia (Nelson et al., 2018; Nemitz et al., 2000). The low concentrations of ammonia (NH_3) measured at agricultural sites limit the applicability of fast-response 10 Hz detectors. Quantum cascade lasers can be used to quantify NH_3 fluxes, but difficulties arise due to the reactivity of NH_3 with dust and water in the sampling line.

6.3 Relaxed eddy accumulation

A second technique is known as the relaxed eddy accumulation (REA) or conditional sampling (Desjardins, 1977; Hensen et al., 1996). Like EC, measurements are performed at a single point above the surface. REA measurements utilize slow-response, high-precision chemical analysers with fast-response switching valves that separate air according to its vertical wind speed. Eddy accumulation can be expressed as follows:

$$F_x = \beta_o \cdot \sigma_w\left(C_x^+ - C_x^-\right)$$

where β_o is a semi-empirical coefficient (typically 0.56), σ_w the standard deviation of vertical wind speed and C^+_x and C^-_x the concentrations of the gas in the up- and down-wind reservoirs.

An alternative to REA, known as true eddy accumulation (TEA), provides direct flux measurements and may be advantageous over the latter method (reviewed by Siebicke and Emad, 2019).

6.4 Bowen ratio

The Bowen ratio (BR) method is based on the energy conservation law and the equation can be rewritten from the mass balance equation for the energy balance at the surface:

$$H = \frac{(R_n - G)}{(1 + \beta^{-1})}$$

$$LE = \frac{(R_n - G)}{(1 + \beta)}$$

where H is the sensible heat flux, LE is the latent heat flux, G is the soil heat flux and R_n is the net radiation (all in units of W m^{-2}). Sensible and latent heat fluxes are determined via the BR (β), expressed as the ratio of the vertical gradients of temperature and humidity where:

$$\beta = \frac{H}{LE}$$

The BR method does not require stability correction and can be applied in the roughness sublayer as it assumes the similarity of exchange of coefficients between scalars. It is preferred over the flux-gradient method but is generally difficult to apply during stable conditions at night and when R_n values are small at sunrise and sunset. The energy balance approach may also be used to estimate evapotranspiration, the amount of water evaporated from plants (transpiration) and the soil, which is hugely important in agrometeorology.

6.5 The special case of paddy rice

Many of the approaches used for assessing the GHG balance of croplands can be used directly or in a modified way for other related production systems. But special considerations may be needed for other agricultural systems. Rice production systems are one such case. Rice is a staple for almost half the global population, and much of this is produced under flooded conditions. Flooding is created either naturally through precipitation or through deliberate inundation, leading to anaerobic decomposition and enhanced soil emissions or transport of CH_4 through plant tissue to the atmosphere (Carmichael et al., 2014; Ma et al., 2010). Consequently, CH_4 emissions comprise a more significant proportion of the GHG budget than those from arable lands. Methane emissions are concentrated in irrigated areas that make up 60% of the total rice-harvested area but produce 78% of emissions, with an estimated mean of 84–131% globally (Carlson et al., 2017). Paddy rice-based agriculture can also be a source of N_2O depending on the management practice (Khalil and Inubushi, 2007). The IPCC requires local data on CH_4 and N_2O that are measured under field conditions (Tier 2) as the basis for appropriate mitigation measures. As management practices, such as alternate wetting and drying or intermittent flooding, directly dry seeded rice and residues can reduce CH_4 emissions with variable effects on N_2O emissions, which are reported to be either increased (Khalil and Inubushi, 2007; Linquist et al., 2015, 2018; Islam et al., 2020) or decreased (Lagomarsino et al., 2016). Therefore, agricultural management practices (e.g. crop rotations, rice cultivars, irrigation and fertilizer management) and the surrounding soil and environmental conditions could have large impacts on GHG emissions and need to be considered in any measurement protocol.

There are large variations in the measurement approaches that have been used in rice production systems, and a common best practice to measure GHG emissions from paddy rice systems is imperative. Measurement protocols particularly for croplands, with either closed chamber (Fig. 3c) or EC (Fig. 7) as detailed earlier, can be used for paddy rice systems, provided that some additional approaches/protocols and care are taken to measure and calculate their emissions precisely and accurately. Some design modifications will be required to operate the chamber-based approach in a flooded environment, and floating (static) chambers have also been used (Martinez-Eixarch et al., 2018), although the principles behind their operation are still the same. The chamber materials can be same as for other production systems as discussed earlier. However, a large chamber with transparent material used for sampling from wetland paddy rice might have no significant impact in terms of temperature rise (Minamikawa et al., 2015). The consideration for a chamber with plants will be similar as reported earlier. However, the chamber height should be higher than the rice plant (>1 m height or extendable with rice plant height as the season progresses), and the base should be installed up to 10–15 cm soil depth to limit lateral flow of nutrients from outside the chamber and vice versa.

For plants growing in wet or shallow flooded condition, the chamber base should have holes around 2 cm above the soil surface to allow water exchange between inside and outside of the base. If the irrigation water level falls too low and the hole in the base is above the water layer, this hole must be closed before taking samples. Under wet soil conditions, the installation of boardwalks is also important so that the gas exchange characteristics of the sampled area are not modified by personnel taking the gas samples. In case of chamber deployment under wet conditions, construction and use of walking boards may require reducing compaction to avoid variations in soil gas production and vertical transport. Some additional information required for measurements made under flooded conditions is the concentration of GHGs dissolved in water. In addition, Martinez-Eixarch et al. (2018) highlighted the importance of the often-neglected fallow season in the accurate estimation of annual CH_4 emissions and suggested measurement programmes cover the whole crop field cycle.

EC approaches have also been successfully used over paddy fields, and specific structures may be required for their installation and deployment under flooded conditions that may result in an increased disturbance relative to that in terrestrial ecosystems. Towers are typically established on raised platforms with access by walking boards. A large focus has been on CO_2, H_2O and energy fluxes over rice paddies but little attention has been on CH_4, due to the limitations described earlier (Chaichana et al., 2018; Knox et al., 2016; Wassmann et al., 1996). The development of open-path analysers (i.e. LI-7700) has allowed researchers to produce quasi-continuous measurements of CH_4 fluxes over seasonal and annual timescales (see Alberto et al., 2015; Detto et al., 2010; Knox et al., 2016; McDermitt et al., 2011). The LI-7700 analyser is also self-cleaning, which enables the continuous operation of the EC system, reducing manual maintenance. Limitations associated with measurements over rice paddies, ubiquitous to many EC studies, are related to the fetch size and potential spatial heterogeneity in CH_4 sources (Chaichana et al., 2018).

7 Measuring emissions from livestock farms

7.1 Introduction

Though anthropogenic emissions of CO_2 accounts for 76% of global emissions, agricultural emissions are dominated by CH_4 from ruminants and manures and N_2O from fertilizer and animal faeces/urine deposited largely on grassland either as slurry and/or through grazing, contributing 16% and 6%, respectively, to global warming. Due to an increasing demand for dairy and meat (50–80%) there are increasing concerns about the effects of livestock farms on GHG emissions.

Globally, domestic animals contributed around 95 million tonnes of CH_4 emissions in 2010, with an annual growth rate of 0.90% (Patra, 2014), among which ruminant livestock contributes around 17% and 3.3% of global CH_4 and total GHG emissions, respectively (Knapp et al., 2014). Therefore, it is very important to measure GHG emissions from livestock systems, especially N_2O and CH_4 from grassland, and enteric CH_4 emissions from livestock that are either grazing or in housing. A variety of GHG quantification methods from livestock animals are available (Fig. 8), with their own benefits and challenges. Some methods have a higher accuracy, while some are easy to operate, but not all methods are feasible for every situation. There is a lack of perfect

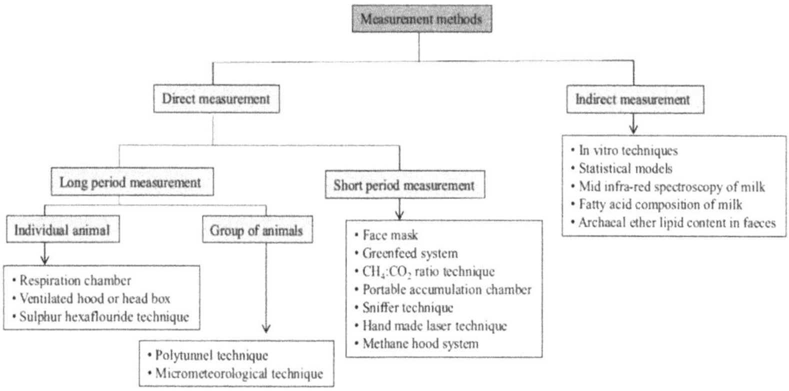

Figure 8 A schematic representation of the different approaches used for quantifying ruminant CH_4 budgets (Source: Patra, 2016).

methods to fit all situations and it is important to be aware of the advantages and disadvantages of each method to ensure proper development of the experimental design and interpretation of the results.

7.2 Greenhouse gas emissions from grassland

Almost all the approaches described for arable land can be directly applied to the assessment of grassland GHG emissions. The key regulatory factors are also similar, and management interventions, such as grazing, and silage cutting would often compromise an assessment of GHG emissions, as similar operations do on arable land. Whilst the closed chamber and EC methods described in detail under croplands can also be used, some additional modifications may be required.

In grassland systems (e.g. grazing and silage), the chamber system should be robust enough to withstand the activities of grazing animals. In some cases, it is recommended that heavy mesh cages are used to cover the chambers to prevent damage by grazing animals. In terms of chamber dimensions, they should be designed in such a way that they cover the area of interest according to the objectives. In general, a cylindrically shaped chamber may be preferred in grazed grassland systems if the objective is to cover urine patches, which are often found to be circular in area. Where feasible, larger chambers should be deployed in patch simulation studies to allow GHG emissions to be captured from the total 'effective area' (Anger et al., 2003; Forrestal et al., 2017; Rochette et al., 2014; Selbie et al., 2015).

For chamber deployment, the following considerations should be undertaken:

- *Grassland (silage):* The chamber diameter can be similar to that given under croplands and the chamber removed prior to each silage application; this could be as often as three times per year, and then re-deployed immediately afterwards.
- *Grazing grassland:* The chamber diameter can be similar to that provided under croplands with the number of replications increased to minimize spatial variability associated with faeces and urine derived from grazed animals. It is also suggested that a fence be included around the chambers to avoid damage caused by grazing animals, but the impact of faeces and urine would then need to be simulated to be representative of that occurring naturally.

Some precautions for the operation and sampling strategy should also be taken. Compared to croplands, flux measurements in grasslands are typically made closer to the ground surface (Fig. 9). Additionally, managed grasslands often have a heterogeneous footprint, including different management practices (e.g. grazing animals, cutting frequency, fertilizer type and quantity) and often complex topography.

7.3 Enteric methane emissions from livestock

To measure GHG emissions from ruminant animals, three methods are often used; these include respiration chambers (RCs) (Grainger et al., 2007), headboxes (Utsumi et al., 2011; Huhtanen et al., 2015) and tracer gas methods (Johnson et al., 1994).

Figure 9 Eddy covariance system installed on a grassland site in south-east Ireland (Teagasc MINE Project, DAFM 15/S/655). The instrumentation includes a yoke-mounted anemometer, an open-path IRGA to measure CO_2 (left) and a closed path QCL (Los Gatos Research, CA, USA) housed in a trailer on the right is used to quantify N_2O emissions.

7.3.1 Respiration chamber technique

Among the techniques that are used to quantify enteric CH_4 emissions, the RC method (Fig. 10) is often considered as the 'gold standard' technique due to high accuracy and repeatability, as well as low animal-to-animal variations (Grainger et al., 2007; Williams et al., 2013). This technique relies on the principle of measuring the concentrations of CH_4 (coming out through all avenues, i.e. mouth, nostrils and rectum from enteric fermentation) in gas samples and the total volume of air removed from the RC (Brouček, 2014). In this technique, an air pump is used, which continuously removes air from chambers through a flow meter in the open-circuit system to calculate the volume of air removed and the outlet gas from the chamber is continuously sampled through a duct system. At least two ventilation fans are used inside the chamber for the proper mixing of expired gases and incoming air. To control the humidity and temperature, air-conditioning systems are often used,

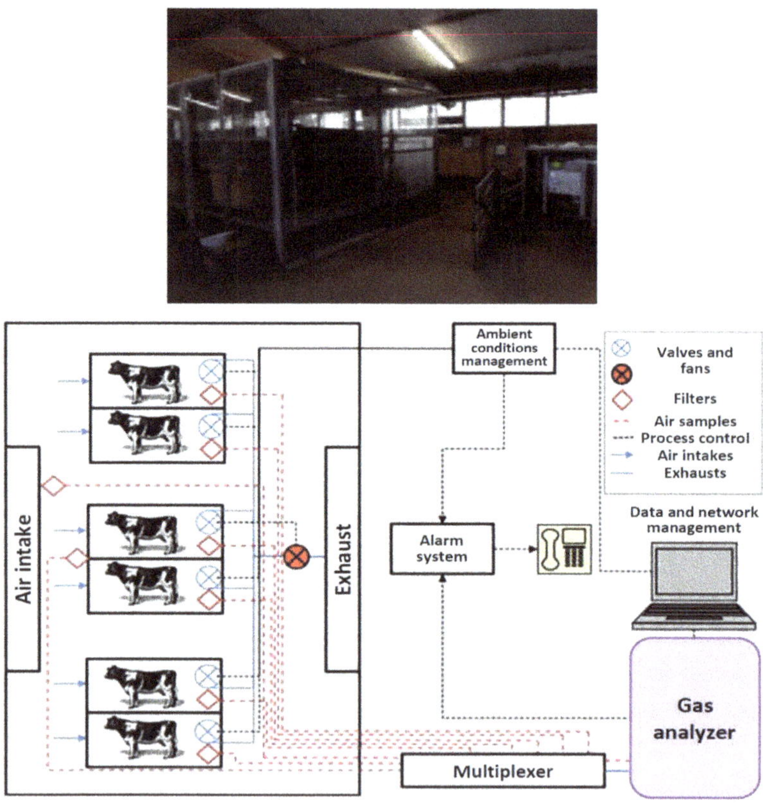

Figure 10 Respiration chamber designs for assessing animal related GHG emissions (Source: INRA, France; Hassouna et al., 2016).

although sometimes as an alternative exhaust fans are also used to exchange fresh air into the RC system to reduce the cost. Temperature, humidity and barometric pressure meters are fitted into the RC systems to regulate gas volume under standard temperature and pressure (STP) conditions (Storm et al., 2012).

Concentration measurements can be made using GC, photoacoustic spectroscopy, OP-FTIR or NDIR spectroscopy. This approach can measure total CH_4 emissions, including losses from anus and rumen fistulas, as well as gas production or consumption of other gases (e.g. CO_2, oxygen, ammonia, hydrogen). Sometimes the chamber can be designed to allow measurements of other nutritional evaluations, such as digestibility, nitrogen balance and energy metabolism.

Despite some limitations, this method can detect relatively small effects of diets and supplements on CH_4 emissions for a small number of animals (Storm et al., 2012). The RC method can be used to characterize diurnal CH_4 emission variations with repeated measurements over the course of daily CH_4 production patterns, which can provide insights into underlying mechanisms of enteric CH_4 formation, as well as relationships with H_2 production when anti-methanogenic compounds are used. In this system, animals are removed from their native environment, which could alter their emission characteristics, although they are reported to adjust well to the chamber environment, despite not exhibiting group behaviour (Chiavegato et al., 2015a).

7.3.2 Headbox technique

Techniques like headboxes include strategies that lock the animal in place or allow the animal to come and go into the headbox (Place et al., 2011; Hristov et al., 2015). With this technique, a box or hood is created to accommodate the head of the animals, and typically these head chambers are large enough to allow the animal to move its head in an unrestricted manner and obtain their feed and water normally. The air samples drawn through the hoods are analysed for CH_4 concentrations (Bhatta et al., 2007). Additionally, this technique can also be used in nutritional and energy metabolism of feed studies, and these ventilated head hoods with different designs and sizes for small and large animals have been installed in different countries (e.g. Place et al., 2011; Patra et al., 2009). As the response times need to be rapid, infrared spectroscopy techniques, such as OP-FTIR or NDIR spectroscopy are best suited for headbox approaches. Flow measures may be made via direct measurement (Maia et al., 2015) or by using a tracer gas approach (Hristov et al., 2015). The flow rate of exhaust air is important for the accuracy of gas analysis and the comfort of the animals in the headbox (Takahashi et al., 1999).

7.3.3 Tracer technique

Marker techniques, like the sulphur hexafluoride (SF_6) tracer technique, have been used extensively during the last two decades to determine enteric CH_4 emissions from ruminants. This tracer technique was developed by Zimmerman (1993). The principle of this technique assumes that if the SF_6 gas production rate from the rumen is known, CH_4 production can be measured (Johnson et al., 1994). For this, SF_6 is filled in small permeation tubes, and then these tubes are placed into the rumen of the test animals (Fig. 11). The gas sampling apparatus is fitted to these test animals, which contains a halter to support capillary tubing whose inlets are placed close to the nose and an evacuated canister to collect gas samples (Brouček, 2014). Gas samples comprising respired and eructated gas are collected usually for 24 h through capillary tubing connected to an

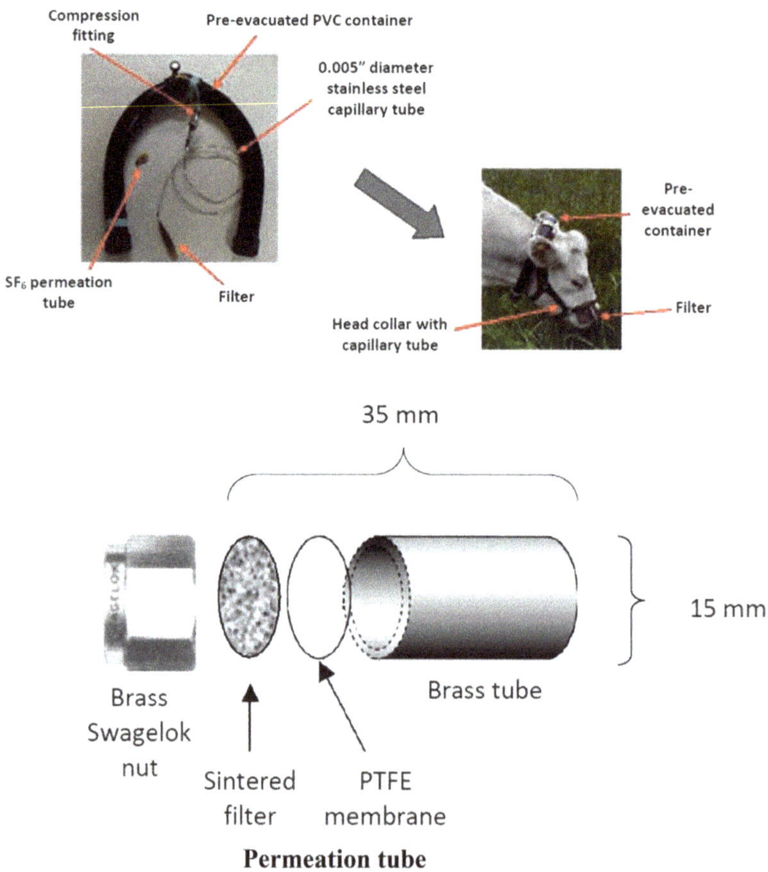

Figure 11 Equipment required to measure CO_2 and enteric CH_4 emissions from ruminants using a tracer gas (SF_6) technique (Source: INRA, France).

evacuated canister, where the tubing regulates the gas sampling rate (Lassey et al., 2001). Finally, a GC is used to analyse the concentration of SF_6 and CH_4 in gas samples collected in the canister. To calculate CH_4 emissions an equation is used (Johnson et al., 1994), which is based on the SF_6 release rate and SF_6 and CH_4 concentrations in the canister gas samples in excess of background concentrations in the air. This technique can quantify enteric CH_4 emissions from large numbers of both grazing and confined animals and has less impact on animal behaviour under normal animal management conditions. The long-term instability of the release rate of the permeation tubes can be a concern for use in longer-term experiments (Deighton et al., 2013; Moate et al., 2015). Furthermore, background atmospheric gas concentrations can impact markedly on the success of this technique (Williams et al., 2011).

7.3.4 GreenFeed technique

This technique was developed by C-Lock Inc. (South Dakota, USA) to measure CH_4 emission from cattle with minimum human exposure or intervention so that the animal behaviour remains unchanged from natural conditions (Zimmerman et al., 2011, 2013a,b). The GreenFeed™ system can be used both inside buildings and in pastures. This system can be used as an alternative to the SF_6 tracer gas technique, which is often difficult to implement and can also measure CH_4 emission from sheep. Each time an animal feeds from the GreenFeed system, both CH_4 and CO_2 are measured on an *ad hoc* basis (Fig. 12). This system relies on the measurement of many short-term CH_4 emission values from an individual animal at different times within a day for a longer period of time, which can be aggregated to calculate average daily CH_4 emissions.

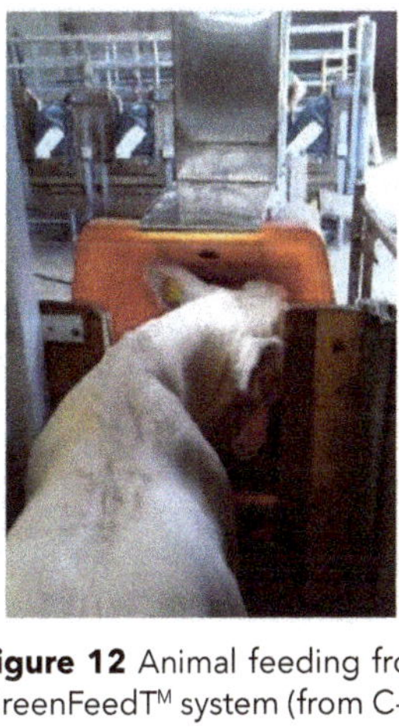

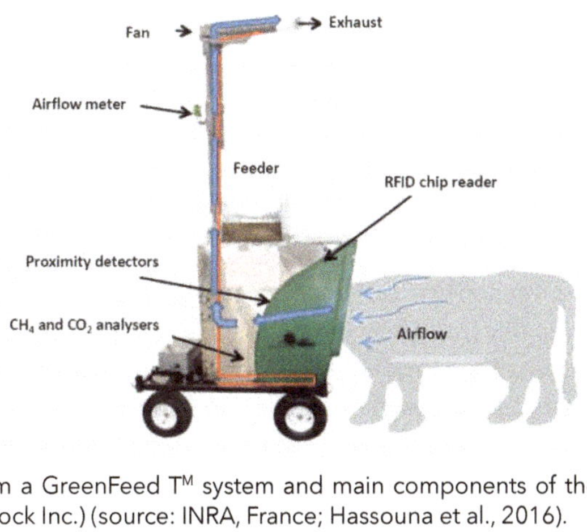

Figure 12 Animal feeding from a GreenFeed ™ system and main components of the GreenFeed™ system (from C-Lock Inc.) (source: INRA, France; Hassouna et al., 2016).

To attract the animal the integral GreenFeed trough is used. Identification of an animal is done by an electronic tag, which triggers the distribution of a portion of feed when the animal puts its head in the trough and records the frequency and number of times the animal comes to feed. During feeding air is extracted at a controlled rate to sample the emissions from the nostrils and transport them to an NDIR analyser, which measures the gas concentrations. Using this system CH_4 and CO_2 can be measured in a frequent *ad hoc* manner for a large number of animals (20 animals per system) over a long period of time. As the system is mobile, it can be deployed in a range of experimental or practical settings (Waghorn et al., 2013).

8 Greenhouse gas measurement techniques at the herd scale

8.1 Polytunnel method for semi-normal grazing conditions

This method involves using a large tunnel made up of polyethylene fitted with end walls and large-diameter ports (Lockyer and Jarvis, 1995) where a measured volume of air is continuously blown into and drawn from the large tunnel containing grazing animals, and the concentration of CH_4 between the incoming and outgoing air is measured (Lockyer and Jarvis, 1995). In addition, temperature and humidity are regularly monitored. Typically, a polytunnel system of 4.3 m wide × 9.9 m long × 2.1 m height with an approximate volume of 66 m^3 is used where different numbers of animals can be enclosed for up to 10 days (Lockyer and Jarvis, 1995; Lockyer, 1997). The recovery percentage of added CH_4 in the tunnel is high using this method. The polytunnel method is suitable for measuring CH_4 from individual or small groups of animals under semi-normal grazing conditions.

8.2 Micrometeorological techniques for large areas

This method measures the fluxes and concentrations of gases in the free atmosphere of a large area containing animals, and gas emissions from animals are calculated from the measured fluxes. This method measures CH_4 concentrations and wind speed through the calculation of emission rates, and the number of points of measurement varies between different methods. While this technique is good for large numbers of animals in large open areas, it is unable to measure emissions from indoor housed animals or individual animals. The technique is also difficult to use for testing mitigation options (McGinn et al., 2011). In recent times, considerable development has taken place in micrometeorological techniques for CH_4 measurement using, for example, inverse dispersion approaches (Flesch et al., 2007). Many feedlot gas emission studies have used inverse dispersion techniques successfully (Flesch et al., 2007; Loh et al., 2008; McGinn et al., 2011). The main advantage of using this method is that the natural behaviour of animals does not interfere with this method and the carbon footprint over large areas can be estimated (van Haarlem et al., 2008). The limitations of this method include wind conditions (high/low) and the need for homogeneity of source (van Haarlem et al., 2008).

Similarly, paddock-scale micrometeorological methods have been used for measuring CH_4 emissions from grazing animals in large fields (Laubach et al., 2008). This method involves analysing the patterns of transportation and dispersion of CH_4 emitted from animals by the wind (Laubach et al., 2008). Here wind direction, wind speed, turbulent airflow characteristics and CH_4 concentrations in and against the direction of the wind are measured to compute the CH_4 emission rates. The measurement accuracy of the paddock-scale technique is dependent upon certain meteorological and landscape conditions, such as wind velocity and direction, topography and the location of animals in the paddock.

8.3 Tracer gas ratio technique for animal housing

The tracer gas ratio technique is the most used method for measuring GHG emissions from animal housing. In addition to measuring enteric CH_4 emissions as discussed earlier, this method has been successfully used to determine CH_4 and other gaseous emissions from animal housing (Kaharabata et al., 2000; Samer et al., 2012). This method requires that a tracer gas be emitted at a known rate near the CH_4 source or other gas of interest and that the concentrations of the gas in question and the tracer gas be measured downwind. The accuracy of this tracer gas ratio technique depends on the assumption that the tracer emissions are representative of the CH_4 emissions and similarly move across the landscape (Kaharabata et al., 2000).

Modelling techniques coupled with downwind concentration measurements have also been widely used to determine GHG emissions from animal housing. For example, feedlot emissions can be measured using an inverse dispersion technique where a single, line-averaged concentration is measured at a downwind distance, and a backward Lagrangian stochastic dispersion model is used to infer the emission rate of the source (Flesch et al., 2007; McGinn et al., 2006b, 2007). The accuracy of the emissions estimate from modelling depends on the accuracy of the model, the data available to develop the model and the accuracy of data input into the model. Sometimes uncertainty with respect to model assumptions can occur due to turbulence across landscapes, wind speed, air stability and lack of a constant emission rate from the source. When the aim of the study is to quantify emissions from large surface areas, this method is recommended.

Alternatively, micrometeorological methods, which usually consider variation in surface characteristics and the resulting influence on turbulence affecting emission measurements, can also be used (Shah et al., 2006). Many previous feedlot or corral studies have used micrometeorological gradient methods that consider measurements at different heights (Hutchinson et al., 1982; Harper et al., 1999; Baek et al., 2006; Todd et al., 2007, 2008). However, these gradient methods are based on the assumption that steady-state conditions exist over a surface area that is uniform in the sources of, and sinks for, emissions (Prueger and Kustas, 2005). When these assumptions cannot be fulfilled in any study, using single-height flux measurements such as those from EC may be the recommended approach.

8.4 Measuring emissions from manure storage areas

The emissions from manure storage and/or production lots can be measured mainly by four methods (i) downwind concentration measures combined with modelling techniques (Koehn et al., 2013), (ii) micrometeorological methods (Park et al., 2006), (iii) static chamber techniques (Hellebrand and Kalk, 2001) and (iv) dynamic flux chamber or flux hood methodology (Mukhtar et al., 2009). The methods for determining emissions and the modelling techniques are the same as those described for animal housing and manure deposition or manure storage areas in a previous section. The primary difference between housing and manure storage area methods is the location for measurement collection.

Both static and dynamic chambers allow the users to obtain instantaneous flux estimates over a short period of time (i.e. 15–30 min). If a GC facility is already available, chamber methods are a relatively low investment option and an easy method to follow. In the static chamber methodology, air samples are collected from chambers using syringes and transferred in vials, which are then analysed for GHG emissions using a GC. However, this method leads to large numbers of samples when at least three-time points per observation period are used to calculate a flux. Moreover, a suitable number of chambers must be deployed to represent the surface area associated with the emissions. Dynamic flux chambers (open, flow-through and active) are used to provide continuous sampling once steady-state conditions are obtained, thereby reducing the number of samples per observation, but requiring additional time to achieve steady-state emission conditions.

Both static and dynamic chamber may include automated samplers, which can provide continuous data and on-site analyses; however, the cost is greater because of the automation and the use of a dedicated field GC. Although both approaches allow for small surface areas to be sampled, the principle of the method applies to liquid storage systems and the logistics of keeping chambers in place at a location that can be sampled multiple times over an observation period makes the dynamic system more feasible for liquid manure storage areas. To sample large surface areas such as loafing areas (place for cows to ruminate and rest following feeding), corrals (Ellis et al., 2001; Misselbrook et al., 2001, 2006; Boadi et al., 2004), barn floors, stacked storage piles and compost piles or windrows (Sommer et al., 2004), both methods are equally suitable. However, it is recommended that a large number of chambers are used to reflect the lack of uniform surface conditions in manure deposition and storage areas.

8.5 Measuring emissions from feed storage areas

Emissions from feed storage areas can be measured by methods similar to those used for manure deposition and storage areas. For example, fluxes from feed containers can be measured using static or dynamic chambers. If the study area is large and not enclosed, micrometeorological techniques and downwind concentration measures, coupled with modelling techniques, can be used. Since typically the feed storage areas are relatively small compared to feedlots and crop fields, the use of chamber techniques is a highly recommended option. The use of chamber techniques will also help to isolate the stored feed emissions from whole farm emissions.

9 Measuring emissions on mixed farms

Mixed farming is a common agricultural system in which a farmer conducts different agricultural practices together. It can be classified based on, for example, land size, crop and animal types, geographical distribution and market orientation, such as (i) on-farm versus between-farm mixing, (ii) mixing within crops and/or animal systems and (iii) diversified versus integrated systems. A number of integrated farming systems that recycle resources efficiently include crop-livestock (agro-pastoral system), crop-forestry (silvo-arable system), forestry-pasturage of livestock (silvo-pastoral system) and crop-forestry-pasturage of livestock (agro-silvo-pastoral system). The practice of using mixed farming systems is often associated with temperate countries but is also found in relatively low-input farms with grain-legume-sheep mixtures. Integration occurs most often in many tropical countries where products or by-products of one component serve as a resource for the other; that is dung goes to the crops and straw to the animals. Diversified systems consist of components such as crop and livestock that co-exist independently from each other, such as pigs, dairy and crops as quite independent units. In this case, the mixing of crops and livestock primarily serves to minimize risk and not to recycle resources.

Globally, mixed farming systems associated with livestock cover about 2.5 billion hectares of land (arable rained: 1.1; irrigated: 0.2; and grassland:

1.2) (de Haan et al., 1997). Environmentally, mixed farming systems are often better at maintaining soil fertility and biodiversity, make the best use of crop residues, minimize soil erosion, help to conserve water and provide suitable habitats for wildlife. It is probably the most benign agricultural production system from an environmental perspective because it is, at least partially, a closed system. However, information on whole farm GHG emissions using mixed farming systems is very limited. Depending on the type of the mixed farm, the measurement approaches described earlier (chamber and/or EC with techniques for herd scale measurements) can be used. The calculations of GHGs should, however, be strictly within the geographical boundary of the farm. To capture spatial variability, the number of replications should be increased, given the likely increased variability within the system.

10 Alternative greenhouse gas measurement techniques

Less common gas measurement techniques also exist that are governed by the same theories and principles indicated earlier but offer variable options. A brief overview of several alternative techniques is provided in this section.

10.1 Fast-box chamber

A fast-box (FB) chamber or 'fast-boxing' is a dynamic approach for measuring GHG emissions (Hensen et al., 2006). A chamber (0.8 m × 0.8 m × 0.5 m) is placed over the emitting surface. Gas is sampled from the chamber at a specified flow rate, and the fans inside the chamber mix the air. The chamber contains an inlet and outlet connected to a QCL or TDLAS, providing continuous sampling of gas. This avoids inducing pressure changes within the headspace. The headspace air volume is monitored for 30-sec to 1-min intervals, allowing the concentration of gas to be determined quickly relative to manual chambers. Emissions are calculated from the change in gas concentration during this time. At least 30 sampling points are required to obtain a reliable field-scale estimate of the flux. Fast-boxing has been applied in several studies (Flechard et al., 2007; Grossel et al., 2014; Hensen et al., 2006; Voglmeier et al., 2019). Voglmeier et al. (2019) applied the FB to quantify N_2O emissions from naturally deposited urine patches in pastures. The larger chamber size allows N_2O to be quantified from the natural patches, which is more representable of actual conditions relative to studies quantifying N_2O using simulated urine patches (see Forrestal et al., 2017).

10.2 Surface renewal

Surface renewal (SR) is a well-known method for estimating rates of evapotranspiration across various ecosystems (Hu et al., 2018). Fine-wire thermocouples are used to measure high-frequency water vapour and temperature data to estimate the energy balance of the system (see section 6.4 "Bowen ratio"). Applying the SR theory to measuring GHGs is still in its early stages, but progress is being made. Suvocarev et al. (2019) shows a good correlation ($r^2 = 0.98$) between CO_2 fluxes made by EC vs SR. This method may have the potential for assessing GHG fluxes on small farms and fields where EC measurements are not possible because of the small footprint.

10.3 Slant-open path

The slant-open path FG technique is a modification of the conventional FG method. As summarized by Flesch et al. (2016), 'A beam of radiation (typically infrared) from an artificial source is sent to a reflector and the spectrum of the returning signal is analysed for concentration information'. A single open-path Fourier transform infrared (OP-FTIR) spectrometer provides the modulated infrared radiation output beam (25 mm diameter), which passes through a ZnSe beamsplitter to a beam expander constructed from a modified 12″ Schmidt- Cassegrain telescope (Meade Instrument Corp., California, USA). The expanded beam is sent along an open path towards a low and high retroreflector (30 cm) and returns via the same measurement path. This enables users to calculate surface emission and deposition rates for a suite of gases (N_2O, CO_2, CH_4 and NH_3, with a precision of 0.4 ppbv typically used for N_2O and NH_3) (Bai et al., 2018, 2019; Flesch et al., 2016). Bai et al. (2019) measured emissions of N_2O over a vegetable farm using a slant-open-path FG method. In their study, the instrument setup consisted of an open-path FTIR spectrometer mounted at the height of 1.45 m above ground aimed at retro-reflectors mounted at 0.8 and 1.8 m above ground. This modified FG technique may also be applied to quantify NH_3 from agricultural fields and may alleviate the limitations associated with sampling lines.

10.4 Satellite measurements and remote sensing

Satellite-based measurements of GHGs have an advantage over many ground-based measurements in their ability to provide estimates ranging from regional to global scales. The atmospheric concentration of GHGs can be estimated using satellite measurements based on the ability of gases to absorb electromagnetic radiation at specific wavelengths (Hardwick and Graven, 2016). Satellite-based spectrometers utilize solar radiation that is reflected off the earth's surface or radiation either emitted by the earth or from laser-induced measurements. To measure CO_2 and CH_4, satellite-based spectrometers use several wavelengths (Fig. 13). For detecting CH_4 the absorption of radiation in the shortwave infrared

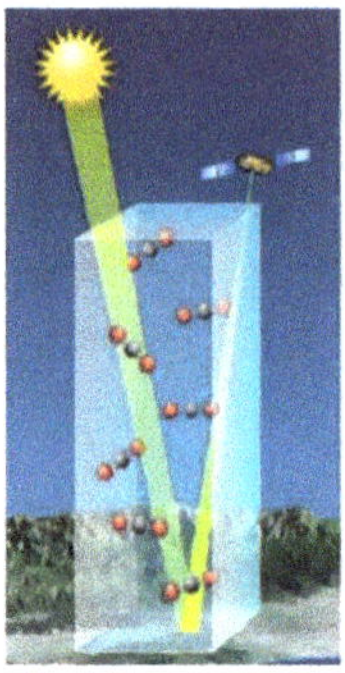
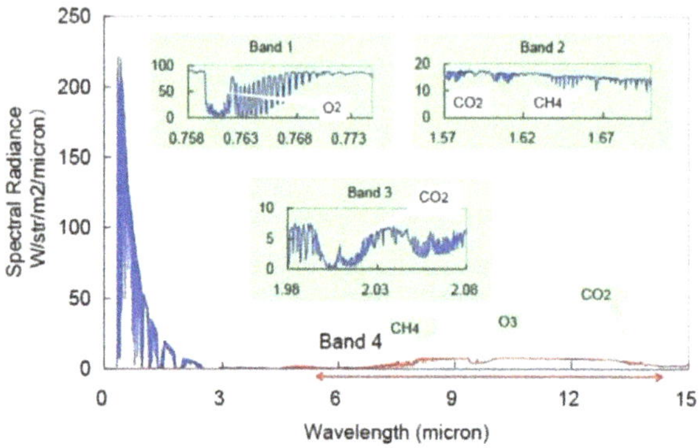

Figure 13 Left: An image of satellite detection of reflected solar radiation by the OCO-2 satellite. Credit: NASA/JPL. Right: Absorption lines to determine column-averaged concentrations of CO_2 and CH_4 observed by GOSAT (The Greenhouse Gases Observing Satellite). The inset panels display spectral bands of GOSAT with expanded wavelength axis; the red arrow indicates the wavelength range of Band 4. Credit: JAXA (Hardwick and Graven, 2016).

(SWIR) at 1.65 and 2.3 μm and the thermal infrared (TIR) in the vicinity of 8 μm (Assareh, 2019) is used. The detection of CO_2 relies on measuring absorption in three near-infrared wavelength bands in the vicinity of 1.58 μm (weak band), 2.06 μm (strong band) and the molecular oxygen (O_2) A-band in the vicinity of 0.76 μm (Hardwick and Graven, 2016). The main limitation of these measurements is that the absorption can be affected by variations in surface pressure due to topography, local weather conditions, clouds or aerosols. To overcome limitations such as variations in surface pressure due to topography and cloud, the absorption spectra in the O_2 A-band at 0.76 μm can be used as atmospheric O_2 concentrations are constant. Generally, it is recommended that only cloud-free observations are used to measure CO_2 and CH_4 concentrations. Most of the satellites are designed to carry out simultaneous measurements of GHG concentrations and cloud cover to allow observations affected by clouds to be flagged and discarded.

All remotely sensed data are linked to calibration laboratories where standard CO_2 and CH_4 concentrations are maintained (WMO/IAEA, 2014). The validation of the observations made by satellites is carried out using TCCON (total column carbon observing network), which is a network of ground-based Fourier transform spectrometers that provide accurate measurements of various GHGs (CO_2, CH_4, etc.). To ensure satellites are providing accurate and consistent measurements, the instruments need to be regularly calibrated after the satellite has been launched. Until now three satellite missions have reported measurements of CO_2 and/or CH_4 concentrations. These are SCIAMACHY (scanning imaging absorption monitoring spectrometer for atmospheric chartography), GOSAT (greenhouse gases observing satellite) and OCO-2 (orbiting carbon observatory 2). Data from satellites are available for free to the public through several access portals. Some of the most useful portals are ESA's Earth Online portal, which gives access to Envisat and Earth Explorer missions (https://earth.esa.int/), NASA's Earth data portal (https://earthdata.nasa.gov), JAXA and NIES's GOSAT User Interface Gateway (http://data. gosat.nies.go.jp) and the EU's Copernicus programme (http://www.copernicus.eu/).

For rice paddies, the combined use of satellite imagery and a GHG emission model can be used. This approach can be used to quantify GHG emissions through the incorporation of information on land use into a GHG emissions model (Fig. 14) (Torbick et al., 2017).

10.5 Unmanned aerial vehicles

Chambers, micrometeorological and remote sensing methods cover fine and large spatial scales. To complement these measurements, *in situ* monitoring with higher accuracy, faster time response and improved spatial resolution is

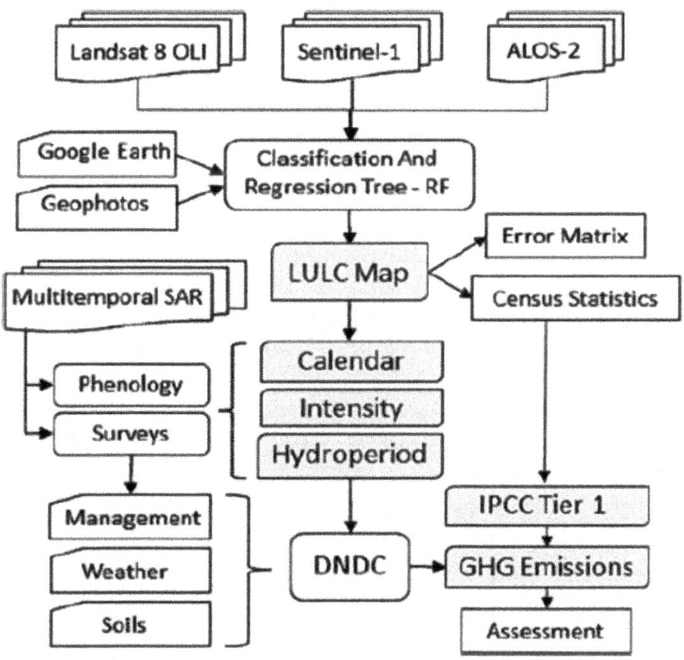

Figure 14 Conceptual flow for mapping and quantifying GHG emissions from rice paddies (Source: Torbick et al., 2017).

required for a more complete understanding of GHG fluxes and the effects of climate change (Berman et al., 2012). Research on using unmanned aerial vehicles (UAVs) or drones is aimed at tackling this gap. A UAV greenhouse gas analyser (GGA) has been developed by Los Gatos Research (CA, USA). It utilizes an OA-ICOS and can simultaneously measure CO_2, CH_4, N_2O and H_2O. The instrument is lightweight (3.5 kg), making it suitable for deployment on UAVs. Already, UAVs have been used to map out the distribution of urine patches on a field. Maire et al. (2018) has estimated the contribution of N_2O emissions from urine patches in a sheep-grazed field, using a combined approach with UAVs (twin spectral camera and mapping) and manual chambers.

10.6 Plume method

In addition, Hensen et al. (2013) present various other approaches (i.e. mass balance, tall tower) that are available for measuring agricultural N_2O emissions. Amongst those low-cost and state-of-the-art methods evaluated is the plume method. Here, the concentration of a gas plume is measured from its point source, using stationary or mobile detectors, at a spatial resolution of 50 m^2 to 1 km^2 (midway between chambers and EC scales). The method requires information on wind speed, wind direction and turbulent parameters to model the source emission values.

Schematic presentations for the various measurement techniques in ruminant, arable and grassland systems are described in Figs. 8 and 15. Approaches are separated based on direct or indirect measurements of GHGs.

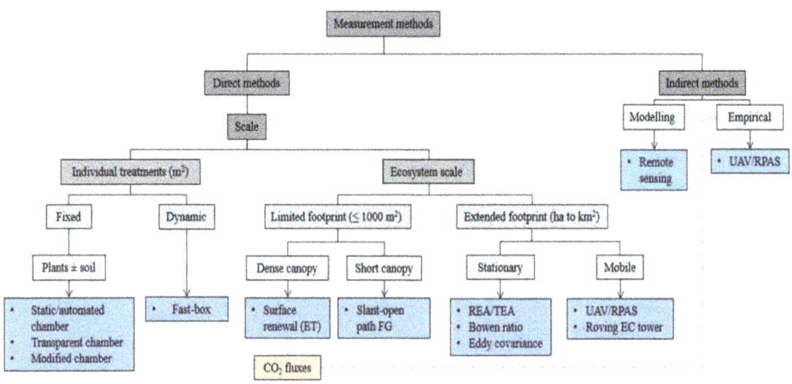

Figure 15 A schematic representation of the chamber, micrometeorology and alternative greenhouse gas measurement techniques available for arable and grassland systems.

11 Comparison of different methods and future challenges

Although a wide variety of techniques have been used for measuring GHGs from various sources, most of the emphasis has been on the use of static chamber or micrometeorological approaches. Many of the challenges associated with the deployment of manual and automated chambers have been covered in Section 2. For arable land, both static and dynamic chambers are invasive, and their proper design and use are important to increase the accuracy of the results. Static chamber measurements have some major shortcomings in their inability to adequately cover temporal and spatial variability (Velthof et al., 1996; Flechard et al., 2007). Also, chamber installation disturbs the environmental conditions and the measured fluxes might not necessarily reflect the actual fluxes. The existence of large variability in chamber design, use and the associated errors in the flux estimations can have a major implication for the reliability and inter-comparability of GHG emission data reported in the literature. For paddy rice, the closed chamber method has been preferred due to its ease of implementation, low-cost and high-logistical feasibility over the EC method. The design of closed chambers for measuring GHGs from paddy rice is different compared to arable lands and pasture. The grassland ecosystems are particularly complex and GHG emissions are often difficult to quantify due to the wide range of management and environmental conditions employed.

In terms of chamber use, the following conditions should be met: (i) include a vent to equalize the pressure between the headspace volume and outside atmosphere, (ii) ensure a proper seal using rubber/neoprene gaskets, (iii) install a fan to mix air inside (particularly where extended chambers are used), (iv) ensure that temperature and humidity show no significant variation in comparison to ambient conditions, (v) maximize spatial coverage at the site, (vi) use the alternative 'HMR method' for calculating fluxes (i.e. Pedersen et al., 2010). In addition, there are a number of published studies that provide recommendations for measuring GHG fluxes from static chambers (Davidson et al., 2002; de Klein and Harvey, 2015; Pedersen et al., 2010) and we have discussed some of the challenges encountered by researchers in this field.

The advantages and disadvantages of chamber and EC methods are described in Table 2. Both techniques for the measurement of agricultural GHGs (N_2O, CH_4 and CO_2) and their quantifications are useful. In comparison to most EC methods, the chamber technique is a much more inexpensive option for assessing the three major GHGs. Static chamber measurements have some major shortcomings in their ability to cover fast temporal and spatial inhomogeneities (Velthof et al., 1996; Flechard et al., 2007) and there are difficulties of making measurements in tall vegetation/wetlands. Issues associated with capturing spatial variations using chamber techniques can

Table 2 Comparison between chamber and micrometeorological methods

Parameter	Chamber method	Micrometeorological method
GHGs	CO_2, N_2O and CH_4	CO_2, N_2O and CH_4
Scale	Fine scale covering soil areas < 1 m^2	Large-scale flat terrain which can span up to regional scales
Suitability for exploring treatment effects in a small area	Possible	Not possible
Interference with the microenvironment	Yes	No
Spatial resolution	Relatively high	Relatively low
Temporal resolution	Low	High
Sampling interval	Low	High
Sensitivity	Relatively low	Relatively high
Instrumental requirement including auxiliary instruments	Low and no auxiliary instruments requirement	Relatively high instrumental requirement with auxiliary instruments, for example, meteorological station and mast
Complexity of instruments	Low	High
Possibility of storing of samples for further analysis	Yes	No
Investment requirement	Relatively lower investment requirement	Very high investment requirement between €54 000 and €72 000
Power supply	No power supply needed in the site	Large power supply needed

be minimized by increasing their numbers, their size and strategic placement. To capture temporal variation, permanently deployed dynamic chambers, coupled with manual sampling using static chambers to cover management interventions, are preferred, and for tall vegetation, chambers with expandable options are recommended (Olfs et al., 2018).

Whilst the EC method is increasingly being used, it also has its disadvantages as certain assumptions, such as a homogenous flat terrain and a predominantly turbulent air flow, may not be fulfilled. A loss of data is incurred when these assumptions are not met. Fluxes are typically integrated over 30 min so any disturbances arising during this time period also result in gaps in the data. As a result, QC measures and gap-filling procedures are conducted to produce annual sums of NEE. EC measurements are also unable to distinguish replicated treatment effects in a small area, although an approach using multiple systems could provide statistically analysable data on treatment effects over areas <2 ha. For example, Davis et al. (2010) deployed a roving or mobile tower to make replicated measurements of NEE over a spring barley crop in Ireland.

To combat the issue of temporal data loss, a technique known as wavelet analysis exists that may limit data loss occurring in half-hourly integrated periods. Wavelet analysis is now applied as a post-processing technique for measuring turbulent GHG fluxes (see Furon et al., 2008; Schaller et al., 2017). One of the strengths of wavelet analysis is that the time averaging can be <30 min (<1 s–60 min). Additionally, wavelet analysis does not require steady-state conditions (i.e. changes in temperature/pressure), where EC measurements, for example, require corrective measures (i.e. the Webb-Pearman-Leuning correction for heat and water vapour effects). However, fluxes calculated using wavelet transformation require higher computational power (e.g. RAM ≥ 8 GB) as signals are processed across different timescales (low- and high-frequency signals).

For N_2O and CH_4, spatial heterogeneity is often the common cause of disparity in flux values (Chaichana et al., 2018; Laville et al., 1999). For N_2O emissions, hotspots that have been characterized using chambers may not be located within the footprint of EC towers. With chambers, an inadequate sampling frequency can underestimate N_2O fluxes by missing emission 'spikes', for example, after fertilization, or episodic rainfall events, whilst linear interpolation of large data gaps may overestimate cumulative N_2O fluxes (Barton et al., 2015). In EC studies, however, the magnitude of these N_2O spikes are reduced, as signals from the hotspots (<1 m^2) are diluted across larger footprints (>1000 m^2) (see Jones et al., 2011). Considerable uncertainty also exists in measuring N_2O using EC at low-flux ranges (see Tallec et al., 2019; Waldo et al., 2019; Wang et al., 2013).

For the measurement of enteric CH_4, the RC method has been preferred due to higher accuracy and repeatability, as well as low animal-to-animal variations. This approach has the capability for measuring total emissions and assessing losses from anus and rumen fistulas and the production or consumption of other gases (e.g. CO_2, oxygen, ammonia, hydrogen), as well as measuring whole animal emissions with or without excreta present. The limitations include the requirement of a large capital investment and operating costs and problems associated with a reduced feed intake when animals are isolated in chambers (McGinn et al., 2004). However, data from grouped animals demonstrate comparable performance, including feed intake, in large group and commercial production settings (Li et al., 2015). In this system, animals are removed from their native environment, although they generally adjust well to the chamber environment (Chiavegato et al., 2015a). The headbox technique is unsuitable where the mimicking of pasture settings is required. The advantages and drawbacks of each method are outlined in more detail in Table 3.

GHG measurement from animal housing, manure deposition and storage areas can be done using methodologies stated earlier but are more advantageous if combined with modelling techniques (Koehn et al., 2013).

Table 3 Summary of methods for measuring greenhouse gas emissions from animal production systems

Method	Application	Principle	Advantages	Limitations	Example citation
In vitro techniques	This technique mimics the rumen environment and used as a screening tool for dietary treatments	Direct measurement of gas produced during an incubation period (hour, day or week)	Quick and inexpensive technique. Requires basic laboratory equipment and a gas chromatograph	As it provides the estimates of rumen gas production, access to fistulated animals is required	Pell and Schofield (1993), Meale et al. (2012)
Chamber techniques (static and passive), including flux hoods	Measures emissions from individual or small groups of animals, with or without manure accumulation. It also measures emissions from surfaces (fields, piles, lots and manure storage surfaces)	Direct measurement of emissions built up in a chamber over time (in case of the static chamber). Measurement of emissions emitted from a chamber with a measured flow rate (in case of passive or flux hood)	Considered the 'gold standard' for measuring animal emissions. Field chambers (static, dynamic or flux hoods) are inexpensive to construct and function with manual or automated sampling systems	Artificial environment for the animal; infrastructure and operating costs are high for animal chambers; chamber placement or large numbers of chambers critical for representative of lot, field, pile or surface; only represent emissions from the surface and not from animals if present (lots, corrals and pastures)	Ellis et al. (2001), Misselbrook et al. (2001), Boadi et al. (2004), Sommer et al. (2004), Grainger et al. (2007), Rochette (2011), Kahmark and Millar (2014), Li et al. (2015)
Headboxes (mobile or stationary, with or without animal restraint)	This technique measures enteric emissions or respired gases (energetics studies)	Gas concentrations are measured in real-time coupled with direct flow measures or flow rate determined using a tracer gas approach	Real-time measures and good application to pastured or – environments where the animals are not restrained	Artificial environment for the animal when restrained, observations may not represent the daily pattern of emissions when animals not restrained and investment cost is high	Place et al. (2011), Utsumi et al. (2011), Hristov et al. (2015), Huhtanen et al. (2015)
Tracer gas methods, including CO_2 mass balance	Estimate ventilation or flow rate (naturally ventilated buildings) and coupled with concentration measurements, can be used to determine emissions from ruminants, lots, corrals or buildings	The mass release of a stable gas is determined and used as an indicator of the mass of gas of interest emitted, similar to the use of marker methods in nutrition studies	Often cheaper than direct methods of measurement and high accuracy of mass measurement	Diurnal or diel effects not discernible and samples are collected for future analyses	Johnson et al. (1994), Kaharabata et al. (2000), McGinn et al. (2006a), Samer et al. (2012), Samer and Abuarab (2014), Chiavegato et al. (2015a)

Emissions measurement	Often ventilated buildings but applicable to any source	Direct measurement of flow or ventilation rate coupled with concentration measures (same principle as chamber methods)	Real-time measurements, multiple gases measured simultaneously, the ability to isolate emission source and continuous measures represent the present conditions over the sampling period	Extensive setup time and calibration expense necessary for continuous measurement	Lin et al. (2012), Joo et al. (2015)
Dispersion modelling	Model emissions across large areas including corrals, lots, pastures and manure storage systems	Measure concentration at a downwind location and use backward dispersion modelling methods to estimate source emission rate	Measurements across large surface areas possible, ease of equipment setup relative to other methods and ability to isolate emission sources	Turbulence due to surface variation creates uncertainty in model assumptions and model output accuracy, dependent on data input and model used	McGinn et al. (2006b), Pacholski et al. (2006), Flesch et al. (2007)
Micrometeorological methods	Most applicable to feedlot, corrals and some pastures where there is a large, relatively uniform surface area	Emissions gradient for any gas of interest is determined at multiple heights, and the flux across the source is determined	Addresses surface variations and the resulting influence of turbulence, well suited for large surface area sources and lack of uniform surfaces can be addressed through the use of eddy covariance adjustments	Assumes steady-state conditions over the emitting surface unless adjustments are made and requires multiple towers and repeated sampling for time period representation of emissions	Hutchinson et al. (1982), Harper et al. (1999), Baek et al. (2006), Shah et al. (2006), Todd et al. (2007), Park et al. (2006)

Source: adapted from Powers and Capelari (2019).

The primary difference between housing and manure storage area methods is location specific, and both chambers allow the users to obtain instantaneous flux estimates over a short period of time. If a GC facility is already available, chamber methods are a relatively low investment option and an easy method to follow. Both allow a small surface area to be sampled with good control, and that also applies to liquid storage systems. For multiple samplings over an observation period, the dynamic system is more feasible for liquid manure storage areas. For feed storage, however, methods similar to those used for manure deposition and storage areas can be used and the chamber techniques are a highly recommended option, including the isolation of the stored feed emission from whole farm emissions.

Depending on the type of mixed farms, the measurement approaches described earlier (chamber and/or EC with techniques for herd scale measurements) can be used. The calculations of GHGs should strictly be within the strict geographical boundary of the farm. To capture spatial variability, the number of replications should be increased by considering the extent of variability within a system. Chambers and EC methods cover fine and large spatial scales. To complement these measurements, *in situ* monitoring with a higher accuracy, faster time response and improved spatial resolution are required for a more complete understanding of GHG fluxes and climate change effects (Berman et al., 2012).

There are alternative methods such as FB requiring large sampling points to obtain a reliable result. This method is constrained by the length of the sampling line, which is typically linked to a closed-path gas analyser (e.g. QCL). The SR method is related to its simplicity in setup and instrumentation (e.g. net radiometer, fine-wire thermocouple, soil heat flux plates, soil thermometers, data logger, anemometer) relative to standard EC tower assemblages. Subsequently, it is less expensive; neither open- nor closed-path IRGAS is required as CO_2 fluxes are measured indirectly. Like the SR method, the slant-open path FG technique offers a viable alternative to the EC methodology where there are small footprints. Thus, UAVs will have a greater role in the future in providing information on GHG emissions from agricultural systems. The plume method is also a low-cost and state-of-the-art one, requiring limited information for its use. Research on using UAVs, which are lightweight and can simultaneously measure CO_2, CH_4, N_2O and H_2O, or drones/satellites (Yang et al., 2019) is aiming to tackle this gap. Most of the techniques discussed here are applicable to field sites. Further information on the advantages and drawbacks of the slant-open path method is described in Table 4.

In addition to measurements, quantification of GHGs from agricultural systems and an understanding of the associated processes involved are extremely complex. Of the future challenges in measurement methods that may be needed to meet all requirements are the following (i) low cost, and

Table 4 Advantages and disadvantages of modified GHG measurement techniques

Modification	Advantages	Drawbacks	Example citation
Chamber design modifications	Generally, only adequate for short plants without major modifications. High spatial resolution Clear chambers allow combined measurements of photosynthesis and respiration (transparent) and respiration only (darkened)	Low spatial coverage Variable chamber design and deployment strategy Potential chamber leakages Less coupled to atmospheric conditions Different gap-filling procedures 'Clouded gas effect' (condensed water on inside walls reduces solar radiation) 'Oasis effect' (less cooling)	Riederer et al. (2014), Wang et al. (2013), Zhao et al. (2019)
Fast-box (chamber, N_2O)	Faster measurements (≤1 min) No collar insertion required Measurements made anywhere on the field Adaptable (i.e. for measurements on urine/dung patches)	Potential issues with poor surface sealing Requires sensitive and expensive instrumentation (QCL, TDLAS) Unsuitable for non-pasture systems (i.e. tall plants)	Flechard et al. (2007), Grossel et al. (2014), Hensen et al. (2006), Vogelmeier et al. (2019)
Surface renewal (FG, Micrometeorology)	Simple, less expensive instrumentation No orientation, levelling, sensor separation adjustments Performs well over dense plant canopies Measures fluxes in both the rough and inertial sub-layers Smaller 'fetch' distance requirements compared to EC	Uncertainties in the FG method calculations (e.g. turbulent Schmidt number, low wind speed) Applicability for CO_2 fluxes limited	Castellvi et al. (2008), McElrone et al. (2013), Suvočarev et al. (2015, 2019)
Slant open-path (FG, Micrometeorology)	Large footprint (> 1000 m^2) Simple field configuration Open-path system (i.e. accounts for 'sticky' gases like NH_3) Free of pumps and tubing	FTIR technique is expensive Uncertainties in the FG method calculations (e.g. turbulent Schmidt number, low wind speed)	Bai et al. (2019), Flesch et al. (2016)
Remote sensing	Large spatial resolution Utilize multiple spectral payloads	High computational power Requires good programming skills Require clear skies Uncertainty with upscaling of fluxes (i.e. with EC) Not always scale-appropriate (i.e. for direct comparisons, upscaling)	Nestola et al. (2016), Ryu et al. (2019)
UAVs/RPAS	TDLAS sensor Identifies point source of CH_4 ≥ 100 m away 5 ppm m^{-1} limit of detection Real-time monitoring Wide operating temperature range (−20–50°C) Autonomous and/or piloted inspections Relatively low-cost technology	Limited information for agricultural systems Operators must hold pilot licence May be restricted flying zones (e.g. aerodrome, airport, military base) Sensor payloads must comply with weight regulations	Berman et al. (2012)

simple to operate and deploy, with rapid implementation options for key land use categories; (ii) ability to use concentration rather than flux; (iii) reduced measurement errors and uncertainties and high reproducibility; (iv) capability of measuring all three GHGs over small areas of land with a greater temporal and spatial resolution; (v) use of inexpensive sensors, and hand-held, automated digital techniques to facilitate their wider deployment and use in large-scale measurement campaigns.

12 Conclusions

Both chamber and micrometeorological techniques can be used for the measurement and quantification of agricultural GHGs (N_2O, CH_4 and CO_2). However, their limitations, particularly in relation to their temporal and spatial coverage, should be recognized, and practical and logistical considerations such as deployment and sampling of an increasing number of chambers recognised.. Whilst dynamic chamber techniques can offer the possibility of improved temporal coverage, cost may limit their ability to cover any spatial variability. Furthermore, permanently deployed dynamic chambers would severely limit the ability to capture the effect of management interventions, such as the effects of fertilization or variations in rainfall, which may require the additional use of either static chambers or EC techniques.

The presence of vegetation limits the use of both types of chambers, requiring options that increase the size of the chamber. However, in comparison to most micrometeorological methods, the chamber technique is an inexpensive option for assessing the three major GHGs. Although it also has its drawbacks, the EC technique does afford the possibility of continuous/near-continuous estimates of GHG exchange with minimum disturbance. This approach is, however, currently limited to relatively large areas of land and is not suitable for comparative assessments of replicated treatment effects. Although the use of movable systems could provide statistically analysable data on treatment effects over areas <2 ha, this is a much larger area than typical experimental field plots and the approach has not received wide attention. Further efforts may therefore be required to enable the use of EC systems with significantly smaller footprints.

For on-farm measurements EC systems are currently too complex for widespread general use and expensive to buy and maintain so that more consideration of alternative micrometeorological methods that could provide the user with real-time information on GHG emissions is required. One such technique, called the SR method, has been proposed as an alternative and inexpensive way to assess evapotranspiration, with the potential for assessing GHG fluxes on small farms and fields where EC measurements are not possible. In all cases, detailed documentation, QC and assurance, as well as the development of appropriate data storage and analytical software, will be important for the precise assessment of GHG fluxes. Although there is a move to a more extensive network of EC systems, their limitations, together with their prohibitive costs, will likely mean that chamber techniques will still be a mainstay of assessments of GHG emissions into the foreseeable future.

Compared to the headbox technique, either RC or tracer technique using SF_6 or GreenFeed Technology, a potentially better alternative to SF_6, can be used to measure enteric CH_4 with high accuracy and repeatability but requires a large capital investment, has high operating cost and needs appropriate technical expertise. There are suitable methods at herd scale, such as the polytunnel method for measuring CH_4 from individual or small groups of animals and the micrometeorological technique can be used for large numbers of animals in large open areas. A more useful approach is to measure GHG fluxes simultaneously with chambers and micrometeorology techniques to overcome individual limitations associated with each method.

Moreover, EC, eddy accumulation, FG methodologies and their derivatives are optimal for larger, ecosystem-scale studies. New developments, such as the slant-open path FG system, may be suitable for use over small areas and has the potential to quantify multiple gases simultaneously. The potential of alternative gas measurement techniques (e.g. FB chamber, SR and slant-open path FG techniques) and UAVs to provide direct and indirect GHG measurements across low-high spatial scales requires further evaluation. GHG measurement from animal housing, manure deposition and storage areas can be assessed using static chamber techniques, dynamic flux chamber measurements or flux hood methodology and micrometeorological methods, in combination with modelling techniques. For mixed farms, a combination of measurement approaches (chamber and/or EC with techniques for herd scale measurements) can be used. Chambers and micrometeorological methods cover fine and large spatial scales, but for a complete understanding of GHG fluxes and assessment of the effects of climate change, *in situ* monitoring with higher accuracy, a faster time response and more spatial resolution is required. There are significant future challenges in any new methodical developments, such as the requirement for low-cost, rapid and precise methods using inexpensive sensors, together with the development of automated systems that can measure all three GHGs.

13 Acknowledgements

The authors gratefully acknowledge the funding by the Department of Agriculture, Food and the Marine (DAFM; Irish Land UsES: 15/S/650), Ireland, and FACCE ERA-GAS (GHG-Manage, ID 47).

14 Where to look for further information

The techniques used to measure GHG emissions have evolved considerably over the last 40 years. Static chambers have been used in the bulk of earlier work. Many measurement techniques are used in generating national stock inventories and subsequently included into the IPCC GHG inventory guidelines (IPCC, 2019). The guideline process requires the most modern scientific approaches with the most up-to-date methodologies (Yona et al., 2020). Continuous measurements can be achieved with automated chambers and eddy covariance towers, however, there exists a shortage of long-term data sets especially over croplands (Baldocchi et al., 2018; Dold et al., 2017).

Emerging technologies will bring new opportunities to refine GHG estimates and provide greater spatial resolution. Remote sensing, coupled with machine learning, offers advantages over chambers and micrometeorology approaches (Reichstein et al., 2019). Empirical upscaling via machine learning can be used to integrate eddy flux, meteorological and satellite measurements for evaluating process-based models (see 'FLUXCOM' below, Jung et al., 2020). This should ensure better predictive modelling of GHGs in the future.

Numerous resources from institutes around the globe are available that provide the latest updates on GHG emissions and the associated measurement approaches/methods. A list of agreed protocols for GHGs and other related measurements for ICOS stations has recently been published in *International Agrophysics* (2018), 32 (4). Some other relevant links are provided below:

- **Intergovernmental Panel on Climate Change (IPCC)** (https://www.ipcc.ch)
- **Global Research Alliance (GRA)** (https://globalresearchalliance.org)
- **European Geosciences Union (EGU)** (https://www.egu.eu)
- **GHG Management Institute** (https://ghginstitute.org)
- **The 4 per 1000 Initiative** (https://www.futurepolicy.org/healthy-ecosystems/the-4-per-1000-initiative-soils-for-food-security-and-climate/)
- **Integrated Carbon Observatory System (ICOS)** (https://www.icos-ri.eu/greenhouse-gases)
- **AmeriFlux** (https://ameriflux.lbl.gov)
- **MODIS** (https://terra.nasa.gov/about/terra-instruments/modis)
- **AsiaFlux** (http://www.asiaflux.net)
- **TERN OzFlux** (http://www.ozflux.org.au)
- **FLUXCOM** (http://www.fluxcom.org)
- **FLUXNET** (https://fluxnet.fluxdata.org)

15 References

Alberto, M. C. R., Wassmann, R., Gummert, M., Buresh, R. J., Quilty, J. R., Correa Jr., T. Q., Centeno, C. A. R. and Oca, G. M. 2015. Straw incorporated after mechanized harvesting of irrigated rice affects net emissions of CH_4 and CO_2 based on eddy covariance measurements. *Field Crops Res.* 184, 162–75. doi:10.1016/j.fcr.2015.10.004.

Anderson, D. E. and Verma, S. B. 1986. Carbon dioxide, water vapor and sensible heat exchanges of a grain sorghum canopy. *Bound-Lay Meteorol.* 34(4), 317–31. doi:10.1007/BF00120986.

Anderson, D. E., Verma, S. B. and Rosenberg, N. J. 1984. Eddy correlation measurements of CO_2, latent heat, and sensible heat fluxes over a crop surface. *Bound-Lay Meteorol.* 29(3), 263–72. doi:10.1007/BF00119792.

Anger, M., Hoffmann, C. and Kühbauch, W. 2003. Nitrous oxide emissions from artificial urine patches applied to different N-fertilized swards and estimated annual N_2O emissions for differently fertilized pastures in an upland location in Germany. *Soil Use Manag.* 19, 104–11.

Arias-Navarro, C., Díaz-Pinés, E., Kiese, R., Rosenstock, T. S., Rufino, M. C., Stern, D., Neufeldt, H., Verchot, L. V. and Butterbach-Bahl, K. 2013. Gas pooling: a sampling technique to overcome spatial heterogeneity of soil carbon dioxide and nitrous oxide fluxes. *Soil Biol. Biochem.* 67, 20–3. doi:10.1016/j.soilbio.2013.08.011.

Assareh, N. 2019. Remote sensing of methane emissions. *Regen Network*. Available at: https://medium.com/regen-network/remote-sensing-of-methane-emissions-2d849cf68fd7.

Aubinet, M., Vesala, T. and Papale, D. (Eds). 2012. *Eddy Covariance: a Practical Guide to Measurement and Data Analysis*. Springer Science and Business Media. https://doi.org/10.1007/978-94-007-2351-1.

Baek, B. H., Todd, R., Cole, N. A. and Koziel, J. A. 2006. Ammonia and hydrogen sulphide flux and dry deposition velocity estimates using vertical gradient method at a commercial beef cattle feedlot. *Int. J. Glob. Environ. Issues* 6(2/3), 189–203. doi:10.1504/IJGENVI.2006.010154.

Bahn, M., Kutsch, W. L., Heinemeyer, A. and Janssens, I. A. 2009. Towards a standardized protocol for the measurement of soil CO_2 efflux. In: Kutsch, W. L., Bahn, M. and Heinemeyer, A. (Eds), *Soil Carbon Dynamics: an Integrated Methodology*. Cambridge University Press, pp. 272–81. https://doi.org/10.1017/CBO9780511711794

Bai, M., Suter, H., Lam, S. K., Davies, R., Flesch, T. K. and Chen, D. 2018. Gaseous emissions from an intensive vegetable farm measured with slant-path FTIR technique. *Agr. Forest Meteorol.* 258, 50–5. doi:10.1016/j.agrformet.2018.03.001.

Bai, M., Suter, H., Lam, S. K., Flesch, T. K. and Chen, D. 2019. Comparison of slant open-path flux gradient and static closed chamber techniques to measure soil N_2O emissions. *Atmos. Meas. Tech.* 12(2), 1095–102. doi:10.5194/amt-12-1095-2019.

Bain, W. G., Hutrya, L., Patterson, D. C., Bright, A. V., Daube, B. C., Munger, J. W. and Wofsy, S. C. 2005. Wind-induced error in the measurement of soil respiration using closed dynamic chambers. *Agr. Forest Meteorol.* 131(3–4), 225–32. doi:10.1016/j.agrformet.2005.06.004.

Baldocchi, D., Chu, H. and Reichstein, M. 2018. Inter-annual variability of net and gross ecosystem carbon fluxes: a review. *Agr. Forest Meteorol.* 249, 520–33. doi:10.1016/j.agrformet.2017.05.015.

Barba, J., Cueva, A., Bahn, M., Barron-Gafford, G. A., Bond-Lamberty, B., Hanson, P. J., Jaimes, A., Kulmala, L., Pumpanen, J., Scott, R. L., Wohlfahrt, G. and Vargas, R. 2018. Comparing ecosystem and soil respiration: review and key challenges of tower-based and soil measurements. *Agr. Forest Meteorol.* 249, 434–43. doi:10.1016/j.agrformet.2017.10.028.

Barton, L., Wolf, B., Rowlings, D., Scheer, C., Kiese, R., Grace, P., Stefanova, K. and Butterbach-Bahl, K. 2015. Sampling frequency affects estimates of annual nitrous oxide fluxes. *Sci. Rep.* 5, 15912. doi:10.1038/srep15912.

Berman, E. S. F., Fladeland, M., Liem, J., Kolyer, R. and Gupta, M. 2012. Greenhouse gas analyzer for measurements of carbon dioxide, methane, and water vapor aboard an unmanned aerial vehicle. *Sensor Actuat. B-Chem.* 169, 128–35. doi:10.1016/j.snb.2012.04.036.

Bhatta, R., Enishi, O. and Kurihara, M. 2007. Measurement of methane production from ruminants. *Asian Australas. J. Anim. Sci.* 20(8), 1305–18. doi:10.5713/ajas.2007.1305.

Black, T. A., Hartog, G., Neumann, H. H., Blanken, P. D., Yang, P. C., Russell, C., Nesic, Z., Lee, X., Chen, S. G., Staebler, R. and Novak, M. D. 1996. Annual cycles of CO_2 and water vapor fluxes above and within a boreal aspen stand. *Glob. Change Biol.* 2(3), 219–29. doi:10.1111/j.1365-2486.1996.tb00074.x.

Boadi, D. A., Wittenberg, K. M., Scott, S. L., Burton, D., Buckley, K., Small, J. A. and Ominski, K. H. 2004. Effect of low and high forage diet on enteric and manure pack greenhouse gas emissions from a feedlot. *Can. J. Anim. Sci.* 84(3), 445–53. doi:10.4141/A03-079.

Bouwman, A. F., Boumans, L. J. M. and Batjes, N. H. 2002. Emissions of N_2O and NO from fertilized fields: summary of available measurement data. *Glob. Biogeochem. Cycles* 16(4), 6.1–6.13. doi:10.1029/2001GB001811.

Brouček, J. 2014. Methods of methane measurement in ruminants. *Slovak J. Anim. Sci.* 47, 51–60.

Brümmer, C., Lyshede, B., Lempio, D., Delorme, J. P., Rüffer, J. J., Fuß, R., Moffat, A. M., Hurkuck, M., Ibrom, A., Ambus, P., Flessa, H. and Kutsch, W. L. 2017. Gas chromatography vs. quantum cascade laser-based N_2O flux measurements using a novel chamber design. *Biogeosciences* 14(6), 1365–81. doi:10.5194/bg-14-1365-2017.

Burba, G. and Anderson, D. 2007. *Introduction to the Eddy Covariance Method: General Guidelines and Conventional Workflow*. Li-Cor Biosciences, p. 141. doi:10.13140/RG.2.1.3723.5683.

Burba, G. G., McDermitt, D. K., Anderson, D. J., Furtaw, M. D. and Eckles, R. D. 2010. Novel design of an enclosed CO_2/H_2O gas analyser for eddy covariance flux measurements. *Tellus B Chem. Phys. Meteorol.* 62(5), 743–8. doi:10.1111/j.1600-0889.2010.00468.x.

Butterbach-Bahl, K., Papen, H. and Rennenberg, H. 1997. Impact of gas transport through rice cultivars on methane emission from rice paddy fields. *Plant Cell Environ.* 20(9), 1175–83. doi:10.1046/j.1365-3040.1997.d01-142.x.

Butterbach-Bahl, K., Baggs, E. M., Dannenmann, M., Kiese, R. and Zechmeister-Boltenstern, S. 2013. Nitrous oxide emissions from soils: how well do we understand the processes and their controls? *Philos. Trans. R. Soc. Lond., B, Biol. Sci.* 368(1621), 20130122. doi:10.1098/rstb.2013.0122.

Butterbach-Bahl, K., Sander, B. O., Pelster, D. and Díaz-Pinés, E. 2016. Quantifying greenhouse gas emissions from managed and natural soils. In: *Methods for Measuring Greenhouse Gas Balances and Evaluating Mitigation Options in Small-Holder Agriculture*. Springer International Publishing, pp. 71–96. https://doi.org/10.1007/978-3-319-29794-1

Caprez, R., Niklaus, P. A. and Körner, C. 2012. Forest soil respiration reflects plant productivity across a temperature gradient in the Alps. *Oecologia* 170(4), 1143-54. doi:10.1007/s00442-012-2371-3.

Carlson, K. M., Gerber, J. S., Mueller, N. D., Herrero, M., MacDonald, G. K., Brauman, K. A., Havlik, P., O'Connell, C. S., Johnson, J. A., Saatchi, S. and West, P. C. 2017. Greenhouse gas emissions intensity of global croplands. *Nat. Clim. Change* 7(1), 63-8. doi:10.1038/nclimate3158.

Carmichael, M. J., Bernhardt, E. S., Bräuer, S. L. and Smith, W. K. 2014. The role of vegetation in methane flux to the atmosphere: should vegetation be included as a distinct category in the global methane budget? *Biogeochemistry* 119(1-3), 1-24. doi:10.1007/s10533-014-9974-1.

Carrara, A., Kolari, P., de Beeck, M. O., Arriga, N., Berveiller, D., Dengel, S., Ibrom, A., Merbold, L., Rebmann, C., Sabbatini, S., Serrano-Ortíz, P. and Biraud, S. C. 2018. Radiation measurements at ICOS ecosystem stations. *Int. Agrophys.* 32(4), 589-605. doi:10.1515/intag-2017-0049.

Castellvi, F., Snyder, R. L. and Baldocchi, D. D. 2008. Surface energy-balance closure over rangeland grass using the eddy covariance method and surface renewal analysis. *Agr. Forest Meteorol.* 148(6-7), 1147-60. doi:10.1016/j.agrformet.2008.02.012.

Chaichana, N., Bellingrath-Kimura, S., Komiya, S., Fujii, Y., Noborio, K., Dietrich, O. and Pakoktom, T. 2018. Comparison of closed chamber and eddy covariance methods to improve the understanding of methane fluxes from rice paddy fields in Japan. *Atmosphere* 9(9), 356. doi:10.3390/atmos9090356.

Chiavegato, M. B., Palumbo, N. and Powers, W. 2015a. Ammonia and greenhouse gas emissions from housed Holstein steers fed different levels of dietary crude protein. *J. Anim. Sci.* 93, 395-404. doi:10.2527/jas.2014-8167.

Christiansen, J. R., Korhonen, J. F. J., Juszczak, R., Giebels, M. and Pihlatie, M. 2011. Assessing the effects of chamber placement, manual sampling and headspace mixing on CH_4 fluxes in a laboratory experiment. *Plant Soil* 343(1-2), 171-85. doi:10.1007/s11104-010-0701-y.

Collier, S. M., Ruark, M. D., Oates, L. G., Jokela, W. E. and Dell, C. J. 2014. Measurement of greenhouse gas flux from agricultural soils using static chambers. *J. Vis. Exp.* 90(90), e52110. doi:10.3791/52110.

Cowan, N., Levy, P., Drewer, J., Carswell, A., Shaw, R., Simmons, I., Bach, C., Marinheiro, J., Brichet, J., Sanchez-Rodriguez, A. R., Cotton, J., Hill, P. W., Chadwick, D. R., Jones, D. L., Misselbrook, T. H. and Skiba, U. 2019. Application of Bayesian statistics to estimate nitrous oxide emission factors of three nitrogen fertilisers on UK grasslands. *Environ. Int.* 128, 362-70. doi:10.1016/j.envint.2019.04.054.

Cowan, N., Carnell, E., Skiba, U., Dragosits, U., Drewer, J. and Levy, P. 2020. Nitrous oxide emission factors of mineral fertilisers in the UK and Ireland: a Bayesian analysis of 20 years of experimental data. *Environ. Int.* 135, 105366. doi:10.1016/j.envint.2019.105366.

Darenova, E., Pavelka, M. and Acosta, M. 2014. Diurnal deviations in the relationship between CO_2 efflux and temperature: a case study. *Catena* 123, 263-9. doi:10.1016/j.catena.2014.08.008.

Davidson, E. A., Savage, K., Verchot, L. V. and Navarro, R. 2002. Minimizing artifacts and biases in chamber-based measurements of soil respiration. *Agr. Forest Meteorol.* 113(1-4), 21-37. doi:10.1016/S0168-1923(02)00100-4.

Davis, P. A., Brown, J. C., Saunders, M., Lanigan, G., Wright, E., Fortune, T., Burke, J., Connolly, J., Jones, M. B. and Osborne, B. 2010. Assessing the effects of agricultural management practices on carbon fluxes: spatial variation and the need for replicated estimates of Net Ecosystem Exchange. *Agr. Forest Meteorol.* 150(4), 564–74. doi:10.1016/j.agrformet.2010.01.021.

de Beeck, M. O., Gielen, B., Merbold, L., Ayres, E., Serrano-Ortiz, P., Acosta, M., Pavelka, M., Montagnani, L., Nilsson, M., Klemedtsson, L., Vincke, C., De Ligne, A., Moureaux, C., Marañon-Jimenez, S., Saunders, M., Mereu, S. and Hörtnagl, L. 2018. Soil-meteorological measurements at ICOS monitoring stations in terrestrial ecosystems. *Int. Agrophys.* 32(4), 619–31. doi:10.1515/intag-2017-0041.

de Haan, C., Steinfeld, H. and Blackburn, H. 1997. Mixed farming systems and the environment. In: *Livestock and the Environment: Finding a Balance*. European Commission Directorate-General for Development, Development Policy Sustainable Development and Natural Resources, 115p. Chapter 3.

Deighton, M. H., O'Loughlin, B. M., Williams, S. R. O., Moate, P. J., Kennedy, E., Boland, T. M. and Eckard, R. J. 2013. Declining sulphur hexafluoride permeability of fluoropolymer membranes causes overestimation of calculated ruminant methane emissions using the tracer technique. *Anim. Feed Sci. Technol.* 183(3–4), 86–95. doi:10.1016/j.anifeedsci.2013.04.021.

de Klein, C. and Harvey, M. 2012. *Nitrous Oxide Chamber Methodology Guidelines*. Ministry for Primary Industries, Wellington, New Zealand.

de Klein, C. and Harvey, M. (Eds). 2015. *Nitrous Oxide Chamber Methodology Guidelines* (version 1.1). Global Research Alliance on Agricultural Greenhouse Gases, 146p.

de Klein, C. A. M., Barton, L., Sherlock, R. R., Li, Z. and Littlejohn, R. P. 2003. Estimating a nitrous oxide emission factor for animal urine from some New Zealand pastoral soils. *Aust. J. Soil Res.* 41, 381–99.

Denmead, O. T. 2008. Approaches to measuring fluxes of methane and nitrous oxide between landscapes and the atmosphere. *Plant Soil* 309(1–2), 5–24. doi:10.1007/s11104-008-9599-z.

Desjardins, R. L. 1974. A technique to measure CO_2 exchange under field conditions. *Int. J. Biometeorol.* 18(1), 76–83. doi:10.1007/BF01450667.

Desjardins, R. L. 1977. Energy budget by an eddy correlation method. *J. Appl. Meteor.* 16(3), 248–50. doi:10.1175/1520-0450(1977)016<0248:EBBAEC>2.0.CO;2.

Desjardins, R. L. 1985. Carbon dioxide budget of maize. *Agr. Forest Meteorol.* 36(1), 29–41. doi:10.1016/0168-1923(85)90063-2.

Desjardins, R. L. and Lemon, E. R. 1974. Limitations of an eddy-correlation technique for the determination of the carbon dioxide and sensible heat fluxes. *Bound-Lay Meteorol.* 5(4), 475–88. doi:10.1007/BF00123493.

Detto, M., Baldocchi, D. and Katul, G. G. 2010. Scaling properties of biologically active scalar concentration fluctuations in the atmospheric surface layer over a managed peatland. *Bound-Lay Meteorol.* 136(3), 407–30. doi:10.1007/s10546-010-9514-z.

Dold, C., Büyükcangaz, H., Rondinelli, W., Prueger, J. H., Sauer, T. J. and Hatfield, J. L. 2017. Long-term carbon uptake of agro-ecosystems in the Midwest. *Agri. For. Meteorol.* 232, 128–40. doi:10.1016/j.agrformet.2016.07.012.

Eggleston, S., Buendia, L., Miwa, K., Ngara, T. and Tanabe, K. (Eds). 2006. *2006 IPCC Guidelines for National Greenhouse Gas Inventories* (vol. 5). Institute for Global Environmental Strategies, Hayama, Japan.

Ellis, S., Webb, J., Misselbrook, T. and Chadwick, D. 2001. Emission of ammonia (NH_3), nitrous oxide (N_2O) and methane (CH_4) from a dairy herd in the UK. *Nutr. Cycl. Agroecosyst.* 60(1/3), 115-22. doi:10.1023/A:1012608100518.

Ellison, S. L. R., Barwick, V. J. and Farrant, T. J. D. 2009. *Practical Statistics for the Analytical Scientist - a Bench Guide* (2nd edn.). Royal Society of Chemistry, Cambridge, UK, p. 268. ISBN:978-1-84755-955-5.

FAO. 2015. *FAO STAT Statistical Database 2013*. Food and Agriculture Organization of the United Nations, Rome, Italy.

Faust, D. R. and Liebig, M. A. 2018. Effects of storage time and temperature on greenhouse gas samples in Exetainer vials with chlorobutyl septa caps. *MethodsX* 5, 857-64. doi:10.1016/j.mex.2018.06.016.

Flesch, T. K., Wilson, J. D., Harper, L. A., Todd, R. W. and. Cole, N. A. 2007. Determining ammonia emissions from a cattle feedlot with an inverse dispersion technique. *Agr. Forest Meteorol.* 144(1-2), 139-55. doi:10.1016/j.agrformet.2007.02.006.

Flesch, T. K., Baron, V. S., Wilson, J. D., Griffith, D. W. T., Basarab, J. A. and Carlson, P. J. 2016. Agricultural gas emissions during the spring thaw: applying a new measurement technique. *Agr. Forest Meteorol.* 221, 111-21. doi:10.1016/j.agrformet.2016.02.010.

Flechard, C. R., Ambus, P., Skiba, U., Rees, R. M., Hensen, A., van Amstel, A., Dasselaar, AvdP., Soussana, J. -F., Jones, M., Clifton-Brown, J., Raschi, A., Horvath, L., Neftel, A., Jocher, M., Ammann, C., Leifeld, J., Fuhrer, J., Calanca, P., Thalman, E., Pilegaard, K., Di Marco, C., Campbell, C., Nemitz, E., Hargreaves, K. J., Levy, P. E., Ball, B. C., Jones, S. K., van de Bulk, W. C. M., Groot, T., Blom, M., Domingues, R., Kasper, G., Allard, V., Ceschia, E., Cellier, P., Laville, P., Henault, C., Bizouard, F., Abdalla, M., Williams, M., Baronti, S., Berretti, F. and Grosz, B. 2007. Effects of climate and management intensity on nitrous oxide emissions in grassland systems across Europe. *Agric. Ecosyst. Environ.* 121(1-2), 135-52. doi:10.1016/j.agee.2006.12.024.

Forbrich, I., Kutzbach, L., Hormann, A. and Wilmking, M. 2010. A comparison of linear and exponential regression for estimating diffusive CH_4 fluxes by closed-chambers in peatlands. *Soil Biol. Biochem.* 42(3), 507-15. doi:10.1016/j.soilbio.2009.12.004.

Forrestal, P. J., Harty, M. A., Carolan, R., Watson, C. J., Lanigan, G. J., Wall, D. P., Hennessy, D. and Richards, K. G. 2017. Can the agronomic performance of urea equal calcium ammonium nitrate across nitrogen rates in temperate grassland? *Soil Use Manag.* 33(2), 243-51. doi:10.1111/sum.12341.

Forristal. P. D. 2019. Alternative Crop Establishment Systems - 6363, Teagasc Technology Update. https://www.teagasc.ie/media/website/publications/2020/6363_Alternative_Crop_Establishment_SystemsCELUP_DForristal.pdf.

Franz, D., Acosta, M., Altimir, N., Arriga, N., Arrouays, D., Aubinet, M., Aurela, M., Ayres, E., López-Ballesteros, A., Barbaste, M., Berveiller, D., Biraud, S., Boukir, H., Brown, T., Brümmer, C., Buchmann, N., Burba, G., Carrara, A., Cescatti, A., Ceschia, E., Clement, R., Cremonese, E., Crill, P., Darenova, E., Dengel, S., D'Odorico, P., Filippa, G., Fleck, S., Fratini, G., Fuß, R., Gielen, B., Gogo, S., Grace, J., Graf, A., Grelle, A., Gross, P., Grünwald, T., Haapanala, S., Hehn, M., Heinesch, B., Heiskanen, J., Herbst, M., Herschlein, C., Hörtnagl, L., Hufkens, K., Ibrom, A., Jolivet, C., Joly, L., Jones, M., Kiese, R., Klemedtsson, L., Kljun, N., Klumpp, K., Kolari, P., Kolle, O., Kowalski, A., Kutsch, W., Laurila, T., de Ligne, A., Linder, S., Lindroth, A., Lohila, A., Longdoz, B., Mammarella, I., Manise, T., Jiménez, S. M., Matteucci, G., Mauder, M., Meier, P., Merbold, L., Mereu, S., Metzger, S., Migliavacca, M., Mölder, M., Montagnani, L., Moureaux, C., Nelson, D.,

Nemitz, E., Nicolini, G., Nilsson, M. B., de Beeck, M. O., Osborne, B., Löfvenius, M. O., Pavelka, M., Peichl, M., Peltola, O., Pihlatie, M., Pitacco, A., Pokorný, R., Pumpanen, J., Ratié, C., Rebmann, C., Roland, M., Sabbatini, S., Saby, N. P. A., Saunders, M., Schmid, H. P., Schrumpf, M., Sedlák, P., Ortiz, P. S., Siebicke, L., Šigut, L., Silvennoinen, H., Simioni, G., Skiba, U., Sonnentag, O., Soudani, K., Soulé, P., Steinbrecher, R., Tallec, T., Thimonier, A., Tuittila, E., Tuovinen, J., Vestin, P., Vincent, G., Vincke, C., Vitale, D., Waldner, P., Weslien, P., Wingate, L., Wohlfahrt, G., Zahniser, M. and Vesala, T. 2018. Towards long-term standardised carbon and greenhouse gas observations for monitoring Europe's terrestrial ecosystems: a review. *Int. Agrophys.* 32(4), 439-55. doi:10.1515/intag-2017-0039.

Furon, A. C., Wagner-Riddle, C., Smith, C. R. and Warland, J. S. 2008. Wavelet analysis of wintertime and spring thaw CO_2 and N_2O fluxes from agricultural fields. *Agr. Forest Meteorol.* 148(8–9), 1305–17. doi:10.1016/j.agrformet.2008.03.006.

Garratt, J. R. 1994. The atmospheric boundary layer. *Earth Sci. Rev.* 37(1-2), 89–134. doi:10.1016/0012-8252(94)90026-4.

Gielen, B., Acosta, M., Altimir, N., Buchmann, N., Cescatti, A., Ceschia, E., Fleck, S., Hörtnagl, L., Klumpp, K., Kolari, P., Lohila, A., Loustau, D., Marañon-Jimenez, S., Manise, T., Matteucci, G., Merbold, L., Metzger, C., Moureaux, C., Montagnani, L., Nilsson, M. B., Osborne, B., Papale, D., Pavelka, M., Saunders, M., Simioni, G., Soudani, K., Sonnentag, O., Tallec, T., Tuittila, E., Peichl, M., Pokorny, R., Vincke, C. and Wohlfahrt, G. 2018. Ancillary vegetation measurements at ICOS ecosystem stations. *Int. Agrophys.* 32(4), 645-64. doi:10.1515/intag-2017-0048.

Gilmanov, T. G., Soussana, J. F., Aires, L., Allard, V., Ammann, C., Balzarolo, M., Barcza, Z., Bernhofer, C., Campbell, C. L., Cernusca, A., Cescatti, A., Clifton-Brown, J., Dirks, B. O. M., Dore, S., Eugster, W., Fuhrer, J., Gimeno, C., Gruenwald, T., Haszpra, L., Hensen, A., Ibrom, A., Jacobs, A. F. G., Jones, M. B., Lanigan, G., Laurila, T., Lohila, A., Manca, G., Marcolla, B., Nagy, Z., Pilegaard, K., Pinter, K., Pio, C., Raschi, A., Rogiers, N., Sanz, M. J., Stefani, P., Sutton, M., Tuba, Z., Valentini, R., Williams, M. L. and Wohlfahrt, G. 2007. Partitioning European grassland net ecosystem CO_2 exchange into gross primary productivity and ecosystem respiration using light response function analysis. *Agric. Ecosyst. Environ.* 121(1–2), 93–120. doi:10.1016/j.agee.2006.12.008.

Gilmanov, T. G., Wylie, B. K., Tieszen, L. L., Meyers, T. P., Baron, V. S., Bernacchi, C. J., Billesbach, D. P., Burba, G. G., Fischer, M. L., Glenn, A. J., Hanan, N. P., Hatfield, J. L., Heuer, M. W., Hollinger, S. E., Howard, D. M., Matamala, R., Prueger, J. H., Tenuta, M. and Young, D. G. 2013. CO_2 uptake and ecophysiological parameters of the grain crops of midcontinent North America: estimates from flux tower measurements. *Agric. Ecosyst. Environ.* 164, 162–75. doi:10.1016/j.agee.2012.09.017.

Grace, P., van der Weerden, T. J., Kelly, K., Rees, R. M. and Skiba, U. M. 2013. Automated greenhouse gas measurement in the field. In: de Klein, C. and Harvey, M. (Eds), *Global Research Alliance on Agricultural Greenhouse Gases: Nitrous Oxide Chamber Methodology Guidelines* (version 1.0). Chapter 5.

Grainger, C., Clarke, T., McGinn, S. M., Auldist, M. J., Beauchemin, K. A., Hannah, M. C., Waghorn, G. C., Clark, H. and Eckard, R. J. 2007. Methane emissions from dairy cows measured using the sulfur hexafluoride (SF6) tracer and chamber techniques. *J. Dairy Sci.* 90(6), 2755–66. doi:10.3168/jds.2006-697.

Greco, S. and Baldocchi, D. D. 1996. Seasonal variations of CO_2 and water vapour exchange rates over a temperate deciduous forest. *Glob. Change Biol.* 2(3), 183-97. doi:10.1111/j.1365-2486.1996.tb00071.x.

Grossel, A., Nicoullaud, B., Bourennane, H., Rochette, P., Guimbaud, C., Chartier, M., Catoire, V. and Hénault, C. 2014. Simulating the spatial variability of nitrous oxide emission from cropped soils at the within-field scale using the NOE model. *Ecol. Mod.* 288, 155-65. doi:10.1016/j.ecolmodel.2014.06.007.

Hardwick, S. and Graven, H. 2016. Satellite observations to support monitoring of greenhouse gas emissions. Briefing Paper No. 16. Grantham Institute, UK.

Harper, L. A., Denmead, O. T., Freney, J. R. and Byers, F. M. 1999. Direct measurements of methane emissions from grazing and feedlot cattle. *J. Anim. Sci.* 77(6), 1392-401. doi:10.2527/1999.7761392x.

Hassouna, M., Eglin, T., Cellier,P., Vincent Colomb, V., Cohan, J-P. et al. 2016. Measuring emissions from livestock farming: greenhouse gases, ammonia and nitrogen oxides. INRA-ADEME, ISBN 2-7380-1392-9. ffhal-01567208f.

Hedley, C. B., Saggar, S. and Tate, K. R. 2006. Procedure for fast simultaneous analysis of the greenhouse gases: methane, carbon dioxide, and nitrous oxide in air samples. *Comm. Soil. Sci. Plan.* 37, 1501-10.

Hellebrand, H. J. and Kalk, W. 2001. Emissions of methane, nitrous oxide, and ammonia from dung windrows. *Nutr. Cycl. Agroecosyst.* 60(1/3), 83-7. doi:10.1023/A:1012662823291.

Hensen, A., Vermeulen, A. T., Wyers, G. P. and Zhang, Y. 1996. Eddy correlation and relaxed eddy accumulation measurements of CO_2 fluxes over grassland. *Phys. Chem. Earth* 21(5-6), 383-8. doi:10.1016/S0079-1946(97)81128-7.

Hensen, A., Groot, T. T., Van den Bulk, W. C. M., Vermeulen, A. T., Olesen, J. E. and Schelde, K. 2006. Dairy farm CH_4 and N_2O emissions, from one square metre to the full farm scale. *Agric. Ecosyst. Environ.* 112(2-3), 146-52. doi:10.1016/j.agee.2005.08.014.

Hensen, A., Skiba, U. and Famulari, D. 2013. Low cost and state of the art methods to measure nitrous oxide emissions. *Environ. Res. Lett.* 8(2), 025022. doi:10.1088/1748-9326/8/2/025022.

Hilker, T., Coops, N. C., Hall, F. G., Black, T. A., Chen, B., Krishnan, P., Wulder, M. A., Sellers, P. J., Middleton, E. M. and Huemmrich, K. F. 2008. A modeling approach for upscaling gross ecosystem production to the landscape scale using remote sensing data. *J. Geophys. Res.* 113(G3). doi:10.1029/2007JG000666.

Hill, J., McSweeney, C., Wright, A. G., Bishop-Hurley, G. and Kalantar-Zadeh, K. 2016. Measuring methane production from ruminants. *Trends Biotechnol.* 34(1), 26-35. doi:10.1016/j.tibtech.2015.10.004.

Holl, D. and Kutzbach, L. 2018. A long-term (2002 to 2017) record of closed-path and open-path eddy covariance CO_2 net ecosystem exchange fluxes from the Siberian Arctic. *Earth Sys. Sci. Data Disc.* (in press).

Holl, D., Wille, C., Sachs, T., Schreiber, P., Runkle, B. R. K., Beckebanze, L., Langer, M., Boike, J., Pfeiffer, E., Fedorova, I., Bolshianov, D. Y., Grigoriev, M. N. and Kutzbach, L. 2019. A long-term (2002 to 2017) record of closed-path and open-path eddy covariance CO2 net ecosystem exchange fluxes from the Siberian Arctic. Earth Syst. Sci. Data 11(1), 221-40. doi:10.5194/essd-11-221-2019.

Holland, E. A., Robertson, G. P., Greenberg, J., Groffman, P. M., Boone, R. D. and Gosz, J. R. 1999. Soil CO_2, N_2O, and CH_4 exchange. In: Robertson, G. P., Coleman, D. C.,

Bledsoe, C. S. and Sollins, P. (Eds), *Standard* Soil Methods *for* Long-*term* Ecological Research. Oxford University Press, Oxford, UK, pp. 185-20.

Hopkins, W. G. and Hüner, N. P. A. 2009. *Introduction to Plant Physiology*. John Wiley and Sons, Inc, USA.

Hristov, A. N., Oh, J., Giallongo, F., Frederick, T., Weeks, H., Zimmerman, P. R., Harper, M. T., Hristova, R. A., Zimmerman, R. S. and Branco, A. F. 2015. The use of an automated system (Greenfeed) to monitor enteric methane and carbon dioxide emissions from ruminant animals. *J. Vis. Exp.* 103(103), e52904. doi:10.3791/52904.

Hu, Y., Buttar, N. A., Tanny, J., Snyder, R. L., Savage, M. J. and Lakhiar, I. A. 2018. Surface renewal application for estimating evapotranspiration: a review. *Adv. Meteorol.* 2018, 1-11. doi:10.1155/2018/1690714.

Huhtanen, P., Cabezas-Garcia, E. H., Utsumi, S. and Zimmerman, S. 2015. Comparison of methods to determine methane emissions from dairy cows in farm conditions. *J. Dairy Sci.* 98(5), 3394-409. doi:10.3168/jds.2014-9118.

Hutchinson, G. L. and Livingston, G. P. 2001. Vents and seals in non-steady-state chambers used for measuring gas exchange between soil and the atmosphere. *Eur. J. Soil Sci.* 52(4), 675-82. doi:10.1046/j.1365-2389.2001.00415.x.

Hutchinson, G. L. and Mosier, A. R. 1981. Improved soil cover method for field measurement of nitrous oxide fluxes. *Soil Sci. Soc. Am. J.* 45(2), 311-6. doi:10.2136/sssaj1981.03615995004500020017x.

Hutchinson, G. L., Mosier, A. R. and. Andre, C. E. 1982. Ammonia and amine emissions from a large cattle feedlot. *J. Environ. Qual.* 11(2), 288-93. doi:10.2134/jeq1982.00472425001100020028x.

IPCC. 2006. *2006 IPCC Guidelines for National Greenhouse Gas Inventories*. prepared by the National Greenhouse Gas Inventories Programme, Eggleston, H. S., Buendia, L., Miwa, K., Ngara, T. and Tanabe, K. (Eds). IGES, Japan.

IPCC. 2007. Summary for policymakers. In: Solomon, S., Qin, D., Manning, M., Chen, Z., Marquis, M., Averyt, K. B., Tignor, M. and Miller, H. L. (Eds), *Climate Change 2007: The Physical Science Basis. Contribution of Working Group I to the Fourth Assessment Report of the Intergovernmental Panel on Climate Change*. Cambridge University Press.

IPCC. 2014. *Climate Change 2014: Impacts, Adaptation and Vulnerability*. Cambridge University Press.

IPCC. 2019. *2019 Refinement to the 2006 IPCC Guidelines for National Greenhouse Gas Inventories*. Calvo Buendia, E., Tanabe, K., Kranjc, A., Baasansuren, J., Fukuda, M., Ngarize S., Osako, A., Pyrozhenko, Y., Shermanau, P. and Federici, S. (Eds). IPCC, Switzerland.

Islam, S. F., van Groenigen, J. W., Jensen, L. S., Sander, B. O. and de Neergaard, A. 2018. The effective mitigation of greenhouse gas emissions from rice paddies without compromising yield by early-season drainage. *Sci. Total Environ.* 612, 1329-39. doi:10.1016/j.scitotenv.2017.09.022.

Islam, S. F. U., de Neergaard, A., Sander, B. O., Jensen, L. S., Wassmann, R. and van Groenigen, J. W. 2020. Reducing greenhouse gas emissions and grain arsenic and lead levels without compromising yield in organically produced rice. *Agric. Ecosyst. Environ.* 295, 106922. doi:10.1016/j.agee.2020.106922.

Johnson, K., Huyler, M., Westberg, H., Lamb, B. and Zimmerman, P. R. 1994. Measurement of methane emissions from ruminant livestock using a SF6 tracer technique. *Environ. Sci. Technol.* 28(2), 359-62. doi:10.1021/es00051a025.

Jones, H. G. 2013. *Plants and Microclimate: a Quantitative Approach to Environmental Plant Physiology*. Cambridge University Press.

Jones, S. K., Famulari, D., Di Marco, C. F., Nemitz, E., Skiba, U. M., Rees, R. M. and Sutton, M. A. 2011. Nitrous oxide emissions from managed grassland: a comparison of eddy covariance and static chamber measurements. *Atmos. Meas. Tech. Discuss.* 4(1), 1079–112. doi:10.5194/amt-4-2179-2011.

Joo, H. S., Ndegwa, P. M., Wang, X., Heber, A. J., Ni, J.-Q., Cortus, E. L., Ramirez-Dorronsoro, J. C., Bogan, B. W. and Chai, L. 2015. Ammonia and hydrogen sulfide concentrations and emissions for naturally ventilated freestall dairy barns. *Trans. ASABE* 58, 1321–31.

Jørgensen, C. J., Struwe, S. and Elberling, B. 2012. Temporal trends in N_2O flux dynamics in a Danish wetland – effects of plant mediated gas transport of N_2O and O_2 following changes in water level and soil mineral-N availability. *Glob. Change Biol.* 18(1), 210–22. doi:10.1111/j.1365-2486.2011.02485.x.

Jung, M., Schwalm, C., Migliavacca, M., Walther, S., Camps-Valls, G., Koirala, S., Anthoni, P., Besnard, S., Bodesheim, P., Carvalhais, N., Chevallier, F., Gans, F., Goll, D., Haverd, V., Köhler, P., Kazuhito, I., Jain, A. K., Liu, J., Lombardozzi, D., Nabel, J. E. M. S., Nelson, J. A., O'Sullivan, M., Pallandt, M., Papale, D., Peters, W., Pongratz, J., Rödenbeck, C., Sitch, S., Trampontana, G., Walker, A., Weber, U. and Reichstein, M. 2020. Scaling carbon fluxes from eddy covariance sites to globe: Synthesis and evaluation of the FLUXCOM approach. *Biogeosci.* 17(5), 1343–65.

Kaharabata, S. K., Schuepp, P. H. and Desjardins, R. L. 2000. Estimating methane emissions from dairy cattle housed in a barn and feedlot using an atmospheric tracer. *Environ. Sci. Technol.* 34(15), 3296–302. doi:10.1021/es990578c.

Kahmark, K. and Millar, N. 2014. Stainless steel chamber construction method. Michigan State University, W.K. Kellogg Biological Station, Hickory Corners, MI.

Kaimal, J. C. and Finnigan, J. J. 1994. *Atmospheric Boundary Layer Flows: Their Structure and Measurement*. Oxford University Press.

Khalil, M. I. and Inubushi, K. 2007. Possibilities to reduce rice-straw-induced global warming potential of a sandy paddy soil by combining hydrological manipulations and urea-N fertilizations. *Soil Biol. Biochem.* 39(10), 2675–81. doi:10.1016/j.soilbio.2007.05.003.

Khalil, M. I. and Osborne, B. 2018. Improving estimates of soil organic carbon (SOC) stocks and their long-term temporal changes in agricultural soils in Ireland. *Geoderma* 322, 172–83. doi:10.1016/j.geoderma.2018.02.038.

Khalil, M. I., Van Cleemput, O., Rosenani, A. B. and Schmidhalter, U. 2007. Daytime, temporal and seasonal variations of N_2O emissions in an upland cropping system of the humid tropics. *Comm. Soil Sci. Plant Anal.* 38(1), 189–204. doi:10.1080/00103620601094122.

Khalil, M. I., Buegger, F., Schraml, M., Gutser, R., Richards, K. G. and Schmidhalter, U. 2009. Gaseous nitrogen emissions from a cambisol cropped to spring wheat with urea granule size and depths of placement. *Soil Sci. Soc. Am. J.* 73(4), 1335–44. doi:10.2136/sssaj2008.0153.

Knapp, J. R., Laur, G. L., Vadas, P. A., Weiss, W. P. and Tricarico, J. M. 2014. Invited review: enteric methane in dairy cattle production: quantifying the opportunities and impact of reducing emissions. *J. Dairy Sci.* 97(6), 3231–61. doi:10.3168/jds.2013-7234.

Knox, S. H., Matthes, J. H., Sturtevant, C., Oikawa, P. Y., Verfaillie, J. and Baldocchi, D. 2016. Biophysical controls on interannual variability in ecosystem-scale CO_2 and CH_4

exchange in a California rice paddy. *J. Geophys. Res. Biogeosci.* 121(3), 978–1001. doi:10.1002/2015JG003247.

Koehn, A. C., Leytem, A. B. and Bjorneberg, D. L. 2013. Comparison of atmospheric stability methods for calculating ammonia and methane emission rates with Windtrax. *Trans. ASABE* 56(2), 763–8. doi:10.13031/2013.42664.

Korkiakoski, M., Tuovinen, J., Aurela, M., Koskinen, M., Minkkinen, K., Ojanen, P., Penttilä, T., Rainne, J., Laurila, T. and Lohila, A. 2017. Methane exchange at the peatland forest floor – automatic chamber system exposes the dynamics of small fluxes. *Biogeosciences* 14(7), 1947–67. doi:10.5194/bg-14-1947-2017.

Kutzbach, L., Schneider, J., Sachs, T., Giebels, M., Nykänen, H., Shurpali, J., Martikainen, P. J., Alm, J. and Wilmking, M. 2007. CO_2 flux determination by closed-chamber methods can be seriously biased by inappropriate application of linear regression. *Biogeosci.* 4, 2279–328. doi:10.5194/bgd-4-2279-2007.

Laan, S. V. D. 2010. Validation of the greenhouse gas balance of the Netherlands: observational constraints on CO_2, CH_4 and N_2O from atmospheric monitoring station Lutjewad. s.n., Groningen.

Lagomarsino, A., Agnelli, A. E., Linquist, B., Adviento-Borbe, M. A., Agnelli, A., Gavina, G., Ravaglia, S. and Ferrara, R. M. 2016. Alternate wetting and drying of rice reduced CH_4 emissions but triggered N_2O peaks in a clayey soil of central Italy. *Pedosphere* 26, 533–48. doi:10.1016/S1002-0160(15)60063-7.

Lassey, K. R., Walker, C. F., McMillan, A. M. S. and Ulyatt, M. J. 2001. On the performance of SF6 permeation tubes used in determining methane emission from grazing livestock. *Chemosphere* 3(4), 367–76. doi:10.1016/S1465-9972(01)00017-4.

Lasslop, G., Reichstein, M., Papale, D., Richardson, A. D., Arneth, A., Barr, A., Stoy, P. and Wohlfahrt, G. 2010. Separation of net ecosystem exchange into assimilation and respiration using a light response curve approach: critical issues and global evaluation. *Glob. Change Biol.* 16(1), 187–208, doi:10.1111/j.1365-2486.2009.02041.x.

Lasslop, G., Migliavacca, M., Bohrer, G., Reichstein, M., Bahn, M., Ibrom, A., Jacobs, C., Kolari, P., Papale, D., Vesala, T., Wohlfahrt, G. and Cescatti, A. 2012. On the choice of the driving temperature for eddy-covariance carbon dioxide flux partitioning. *Biogeosciences* 9(12), 5243–59. doi:10.5194/bg-9-5243-2012.

Laubach, J., Kelliher, F. M., Knight, T. W., Clark, H., Molano, G. and Cavanagh, A. 2008. Methane emissions from beef cattle – a comparison of paddock and animal-scale measurements. *Aust. J. Exp. Agric.* 48(2), 132–7. doi:10.1071/EA07256.

Laville, P., Jambert, C., Cellier, P. and Delmas, R. 1999. Nitrous oxide fluxes from a fertilised maize crop using micrometeorological and chamber methods. *Agr. Forest Meteorol.* 96(1–3), 19–38. doi:10.1016/S0168-1923(99)00054-4.

Lebegue, B., Schmidt, M., Ramonet, M., Wastine, B., Yver Kwok, C., Laurent, O., Belviso, S., Guemri, A., Philippon, C., Smith, J. and Conil, S. 2016. Comparison of nitrous oxide (N_2O) analyzers for high-precision measurements of atmospheric mole fractions. *Atmos. Meas. Tech.* 9(3), 1221–38. doi:10.5194/amt-9-1221-2016.

Lenhart, K., Behrendt, T., Greiner, S., Steinkamp, J., Well, R., Giesemann, A. and Keppler, F. 2019. Nitrous oxide effluxes from plants as a potentially important source to the atmosphere. *New Phytol.* 221(3), 1398–408. doi:10.1111/nph.15455.

Levy, P. E., Cowan, N., Van Oijen, M., Famulari, D., Drewer, J. and Skiba, U. 2017. Estimation of cumulative fluxes of nitrous oxide: uncertainty in temporal upscaling and emission factors. *Eur. J. Soil Sci.* 68(4), 400–11. doi:10.1111/ejss.12432.

Li, Q. F., Trottier, N. and Powers, W. 2015. Feeding reduced crude protein diets with crystalline amino acids supplementation reduce air gas emissions from housing. *J. Anim. Sci.* 93(2), 721-30. doi:10.2527/jas.2014-7746.

Lin, X. J., Cortus, E. L., Zhang, R., Jiang, S. and Heber, A. J. 2012. Air emissions from broiler houses in California. *Trans. ASABE* 55(5), 1895–908. doi:10.13031/2013.42377.

Linquist, B. A., Anders, M. M., Adviento-Borbe, M. A., Chaney, R. L., Nalley, L. L., da Rosa, E. F. and van Kessel, C. 2015. Reducing greenhouse gas emissions, water use, and grain arsenic levels in rice systems. *Glob. Change Biol.* 21, 407–17. doi:10.1111/gcb.12701.

Linquist, B. A., Marcos, M., Adviento-Borbe, M. A., Anders, M., Harrell, D., Linscombe, S., Reba, M. L., Runkle, B. R. K., Tarpley, L. and Thomson, A. 2018. Greenhouse gas emissions and management practices that affect emissions in us rice systems. *J. Environ. Qual.* 47(3), 395–409. doi:10.2134/jeq2017.11.0445.

Liu, G. and Si, B. C. 2009. Multi-layer diffusion model and error analysis applied to chamber-based gas fluxes measurements. *Agr. Forest Meteorol.* 149(1), 169–78. doi:10.1016/j.agrformet.2008.07.012.

Livingston, G. P. and Hutchinson, G. L. 1995. Enclosure-based measurement of trace gas exchange: applications and sources of error. In: Matson, P. A. and Harris, R. C. (Eds), *Biogenic Trace Gases: Measuring Emissions from Soil and Water*. Blackwell Science, Oxford, UK, pp. 14–51.

Lockyer, D. R. 1997. Methane emissions from grazing sheep and calves. *Agric. Ecosyst. Environ.* 66(1), 11–8. doi:10.1016/S0167-8809(97)00080-7.

Lockyer, D. R. and Jarvis, S. C. 1995. The measurement of methane losses from grazing animals. *Environ. Pollut.* 90(3), 383–90. doi:10.1016/0269-7491(95)00009-g.

Loh, Z., Chen, D., Bai, M., Naylor, T., Griffith, D., Hill, J., Denmead, T., McGinn, S. and Edis, R. 2008. Measurement of greenhouse gas emissions from Australian feedlot beef production using open-path spectroscopy and atmospheric dispersion modelling. *Aust. J. Exp. Agric.* 48(2), 244–7. doi:10.1071/EA07244.

Ma, E. D., Zhang, G. B., Ma, J., Xu, H., Cai, Z. C. and Yagi, K. 2010. Effects of rice straw returning methods on N_2O emission during wheat-growing season. *Nutr. Cycl. Agroecosyst.* 88(3), 463–9. doi:10.1007/s10705-010-9369-1.

Maia, G. D. N., Ramirez, B. C., Green, A. R., Rodriguez, L. F., Segers, J. R., Shike, D. W. and Gates, R. S. 2015. A novel ruminant emission measurement system: part I. Design evaluation and description. *Trans. ASABE* 58, 749–62. doi:10.1371/journal.pone.0198081.

Maire, J., Gibson-Poole, S., Cowan, N., Reay, D. S., Richards, K. G., Skiba, U., Rees, R. M. and Lanigan, G. J. 2018. Identifying urine patches on intensively managed grassland using aerial imagery captured from remotely piloted aircraft systems. *Front. Sustain. Food Syst.* 2, 10. doi:10.3389/fsufs.2018.00010.

Martinez-Eixarch, M., Alcaraz, C., Viñas, M., Noguerol, J., Aranda, X., Prenafeta-Boldu, F. X., Saldaña-De la Vega, J. A., Català, M. D. M. and Ibáñez, C. 2018. Neglecting the fallow season can significantly underestimate annual methane emissions in Mediterranean rice fields. *PLoS ONE* 13(5), e0198081. doi:10.1371/journal.pone.0198081.

Mauder, M. and Foken, T. 2004. Documentation and instruction manual of the eddy covariance software package TK2. Arbeitsergebn, Univ Bayreuth, Abt Mikrometeorol, p. 42. ISSN:16148916.

Mauder, M., Cuntz, M., Drüe, C., Graf, A., Rebmann, C., Schmid, H. P., Schmidt, M. and Steinbrecher, R. 2013. A strategy for quality and uncertainty assessment of long-term eddy-covariance measurements. *Agr. Forest Meteorol.* 169, 122–35. doi:10.1016/j.agrformet.2012.09.006.

McDermitt, D., Burba, G., Xu, L., Anderson, T., Komissarov, A., Riensche, B., Schedlbauer, J., Starr, G., Zona, D., Oechel, W., Oberbauer, S. and Hastings, S. 2011. A new low-power, open-path instrument for measuring methane flux by eddy covariance. *Appl. Phys. B* 102(2), 391–405. doi:10.1007/s00340-010-4307-0.

McElrone, A. J., Shapland, T. M., Calderon, A., Fitzmaurice, L., Paw U, K. T. and Snyder, R. L. 2013. Surface renewal: an advanced micrometeorological method for measuring and processing field-scale energy flux density data. *J. Vis. Exp.* 82(82), e50666. doi:10.3791/50666.

McGinn, S. M., Beauchemin, K. A., Coates, T. and Colombatto, D. 2004. Methane emissions from beef cattle: effects of monensin, sunflower oil, enzymes, yeast, and fumaric acid. *J. Anim. Sci.* 82(11), 3346–56. doi:10.2527/2004.82113346x.

McGinn, S. M., Beauchemin, K. A., Iwaasa, A. D. and McAllister, T. A. 2006a. Assessment of the sulfur hexafluoride (SF6) tracer technique for measuring enteric methane emissions from cattle. *J. Environ. Qual.* 35(5), 1686–91. doi:10.2134/jeq2006.0054.

McGinn, S. M., Flesch, T. K., Harper, L. A. and Beauchemin, K. A. 2006b. An approach for measuring methane emissions from whole farms. *J. Environ. Qual.* 35(1), 14–20. doi:10.2134/jeq2005.0250.

McGinn, S. M., Flesch, T. K., Crenna, B. P., Beauchemin, K. A. and Coates, T. 2007. Quantifying ammonia emissions from cattle feedlot using a dispersion model. *J. Environ. Qual.* 36(6), 1585–90. doi:10.2134/jeq2007.0167.

McGinn, S. M., Turner, D., Tomkins, N., Charmley, E., Bishop-Hurley, G. and Chen, D. 2011. Methane emissions from grazing cattle using point-source dispersion. *J. Environ. Qual.* 40(1), 22–7. doi:10.2134/jeq2010.0239.

Meale, S. J., Chaves, A. V., Bash, J. and McAllister, T. A. 2012. Methane production of different forages in *in vitro* ruminal fermentation. *Asian-Australas. J. Anim. Sci.* 25(1), 86–91. doi:10.5713/ajas.2011.11249.

Minamikawa, K., Yagi, K., Tokida, T., Sander, B. O. and Wassmann, R. 2012. Appropriate frequency and time of day to measure methane emissions from an irrigated rice paddy in Japan using the manual closed chamber method. *Greenhouse Gas Meas. Manag.* 2(2–3), 118–28. doi:10.1080/20430779.2012.729988.

Minamikawa, K., Tokida, T., Sudo, S., Padre, A. and Yagi, K. 2015. Guidelines for measuring CH_4 and N_2O emissions from rice paddies by a manually operated closed chamber method. National Institute for Agro-Environmental Sciences, Tsukuba, Japan, 80p.

Misselbrook, T. H., Webb, J., Chadwick, D. R., Ellis, S. and Pain, B. F. 2001. Gaseous emissions from outdoor concrete yards used by livestock. *Atmos. Environ.* 35(31), 5331–8. doi:10.1016/S1352-2310(01)00289-8.

Misselbrook, T. H., Webb, J. and Gilhespy, S. L. 2006. Ammonia emissions from outdoor concrete yards used by livestock - quantification and mitigation. *Atmos. Environ.* 40(35), 6752–63. doi:10.1016/j.atmosenv.2006.05.077.

Moate, P. J., Deighton, M. H., Ribaux, B. E., Hannah, M. C., Wales, W. J. and Williams, S. R. O. 2015. Michaelis-Menten kinetics predict the rate of SF6 release from permeation tubes used to estimate methane emissions from ruminants. *Anim. Feed Sci. Technol.* 200, 47–56. doi:10.1016/j.anifeedsci.2014.12.001.

Moncrieff, J., Valentini, R., Greco, S., Guenther, S. and Ciccioli, P. 1997. Trace gas exchange over terrestrial ecosystems: methods and perspectives in micrometeorology. *J. Exp. Bot.* 48(5), 1133–42. doi:10.1093/jxb/48.5.1133.

Monteith, J. and Unsworth, M. 2013. *Principles of Environmental Physics: Plants, Animals, and the Atmosphere*. Academic Press.

Moureaux, C., Ceschia, E., Arriga, N., Béziat, P., Eugster, W., Kutsch, W. L. and Pattey, E. 2012. Eddy covariance measurements over crops. In: Aubinet, M., Vesala, T. and Papale, D. (Eds), *Eddy Covariance: a Practical Guide to Measurement and Data Analysis*. Springer Science and Business Media, pp. 319–31.

Mukhtar, S., Mutlu, A., Lacey, R. E. and Parnell, C. B. 2009. Seasonal ammonia emissions from a free-stall dairy in central Texas. *J. Air Waste Manag. Assoc.* 59(5), 613–8. doi:10.3155/1047-3289.59.5.513.

Munro, D. S. and Oke, T. R. 1975. Aerodynamic boundary-layer adjustment over a crop in neutral stability. *Bound-Lay Meteorol.* 9(1), 53–61. doi:10.1007/BF00232253.

Nelson, J. A., Carvalhais, N., Migliavacca, M., Reichstein, M. and Jung, M. 2018. Water-stress-induced breakdown of carbon-water relations: indicators from diurnal FLUXNET patterns. *Biogeosciences* 15(8), 2433–47. doi:10.5194/bg-15-2433-2018.

Nemitz, E., Sutton, M. A., Gut, A., San José, R., Husted, S. and Schjoerring, J. K. 2000. Sources and sinks of ammonia within an oilseed rape canopy. *Agr. Forest Meteorol.* 105(4), 385–404. doi:10.1016/S0168-1923(00)00205-7.

Nemitz, E., Mammarella, I., Ibrom, A., Aurela, M., Burba, G. G., Dengel, S., Gielen, B., Grelle, A., Heinesch, B., Herbst, M., Hörtnagl, L., Klemedtsson, L., Lindroth, A., Lohila, A., McDermitt, D. K., Meier, P., Merbold, L., Nelson, D., Nicolini, G., Nilsson, M. B., Peltola, O., Rinne, J. and Zahniser, M. 2018. Standardisation of eddy-covariance flux measurements of methane and nitrous oxide. *Int. Agrophys.* 32(4), 517–49. doi:10.1515/intag-2017-0042.

Nestola, E., Calfapietra, C., Emmerton, C., Wong, C., Thayer, D. and Gamon, J. 2016. Monitoring grassland seasonal carbon dynamics, by integrating MODIS NDVI, proximal optical sampling, and eddy covariance measurements. *Remote Sens.* 8(3), 260. doi:10.3390/rs8030260.

Olfs, H. W., Westerschulte, M., Ruoss, N., Federolf, C. P., Zurheide, T., Vergara Hernandez, M. E., Neddermann, N., Trautz, D., Pralle, H., Fuß, R. and Well, R. 2018. A new chamber design for measuring nitrous oxide emissions in maize crops. *J. Plant Nutr. Soil Sci.* 181(1), 69–77. doi:10.1002/jpln.201700008.

Pacholski, A., Cai, G., Nieder, R., Richeter, J., Fan, X., Zhu, Z. and Roelcke, M. 2006. Calibration of a simple method for determining ammonia volatilization in the field - comparative measurements in Henan Province, China. *Nutr. Cycl. Agroecosyst.* 74(3), 259–73. doi:10.1007/s10705-006-9003-4.

Park, K.-H., Thompson, A. G., Marinier, M., Clark, K. and Wagner- Riddle, C. 2006. Greenhouse gas emissions from stored liquid swine manure in a cold climate. *Atmos. Environ.* 40(4), 618–27. doi:10.1016/j.atmosenv.2005.09.075.

Parkin, T. B. and Venterea, R. T. 2010. Chapter 3. Chamber-based trace gas flux measurements. USDA-ARS GRACEnet Project Protocols. Available at: http://www.ars.usda.gov/SP2UserFiles/Program/212/Chapter%203.%20GRACEnet%20Trace%20Gas%20Sampling%20Protocols.pdf.

Parkin, T. B., Venterea, R. T. and Hargreaves, S. K. 2012. Calculating the detection limits of chamber-based soil greenhouse gas flux measurements. *J. Environ. Qual.* 41(3), 705–15. doi:10.2134/jeq2011.0394.

Patra, A. K. 2014. Trends and projected estimates of GHG emissions from Indian livestock in comparisons with GHG emissions from world and developing countries. *Asian-Australas. J. Anim. Sci.* 27(4), 592–9. doi:10.5713/ajas.2013.13342.

Patra, A. K. 2016. Recent advances in measurement and dietary mitigation of enteric methane emissions in ruminants. *Front. Vet. Sci.* 3, 39. doi:10.3389/fvets.2016.00039.

Patra, A. K., Puchala, R., Animut, G., Gipson, T. A., Sahlu, T. and Goetsch, A. L. 2009. Effects of acclimatization on energy expenditure by meat goats. *Small Rumin. Res.* 81(1), 42–54. doi:10.1016/j.smallrumres.2008.11.002.

Pavelka, M., Acosta, M., Kiese, R., Altimir, N., Brümmer, C., Crill, P., Darenova, E., Fuß, R., Gielen, B., Graf, A., Klemedtsson, L., Lohila, A., Longdoz, B., Lindroth, A., Nilsson, M., Jiménez, S. M., Merbold, L., Montagnani, L., Peichl, M., Pihlatie, M., Pumpanen, J., Ortiz, P. S., Silvennoinen, H., Skiba, U., Vestin, P., Weslien, P., Janous, D. and Kutsch, W. 2018. Standardisation of chamber technique for CO_2, N_2O and CH_4 fluxes measurements from terrestrial ecosystems. *Int. Agrophys.* 32(4), 569–87. doi:10.1515/intag-2017-0045.

Pedersen, A. R., Petersen, S. O. and Schelde, K. 2010. A comprehensive approach to soil-atmosphere trace-gas flux estimation with static chambers. *Eur. J. Soil Sci.* 61(6), 888–902. doi:10.1111/j.1365-2389.2010.01291.x.

Pell, A. N. and Schofield, P. 1993. Computerized monitoring of gas production to measure forage digestion *in vitro*. *J. Dairy Sci.* 76(4), 1063–73. doi:10.3168/jds.S0022-0302(93)77435-4.

Pihlatie, M., Ambus, P., Rinne, J., Pilegaard, K. and Vesala, T. 2005. Plant-mediated nitrous oxide emissions from beech (*Fagus sylvatica*) leaves. *New Phytol.* 168(1), 93–8. doi:10.1111/j.1469-8137.2005.01542.x.

Place, S. E., Pan, Y., Zhao, Y. and Mitloehner, F. M. 2011. Construction and operation of a ventilated hood system for measuring greenhouse gas and volatile organic compound emissions from cattle. *Animals (Basel)* 1(4), 433–46. doi:10.3390/ani1040433.

Plummer, M. 2016. Rjags: Bayesian Graphical Models Using MCMC. R Package Version 4-6 [WWW document], https://CRAN.R-project .org/package=rjags.

Powers, W. and Capelari, M. 2019. Production, Management, and the Environment Symposium titled 'Environmental footprint of livestock production; greenhouse gas emissions and climate change'. *Joint Annual Meeting*, Orlando FL, 12–16 July 2015.

Prueger, J. H. and Kustas, W. P. 2005. Aerodynamic principles of flux-profile relationships. In: Hatfield, J. L. and Baker, J. M. (Eds), *Micrometeorology in Agricultural Systems*. American Society of Agronomy Monograph Series No. 47. ASA-CSSA-SSSA, Madison, WI, pp. 407–36.

Pumpanen, J., Kolari, P., Ilvesniemi, H., Minkkinen, K., Vesala, T., Niinisto, S., Lohila, A., Larmola, T., Morero, M., Pihlatie, M., Janssens, I., Yuste, J. C., Grunzweig, J. M., Reth, S., Subke, J. A., Savage, K., Kutsch, W., Ostreng, G., Ziegler, W., Anthoni, P., Lindroth, A. and Hari, P. 2004. Comparison of different chamber techniques for measuring soil CO_2 efflux. *Agr. For. Meteorol.* 123(3–4), 159–76. doi:10.1016/j.agrformet.2003.12.001.

Pumpanen, J., Longdoz, B. and Kutsch, W. L. 2009. Field measurements of soil respiration: principles and constraints, potentials and limitations of different methods. In: Kutsch, W. L., Bahn, M. and Heinemeyer, A. (Eds), *Soil Carbon Dynamics: an Integrated Methodology*. Cambridge University Press, pp. 16–33.

Qin, Z., Zhuang, Q. and Chen, M. 2012. Impacts of land use change due to biofuel crops on carbon balance, bioenergy production, and agricultural yield, in the conterminous United States. *GCB Bioenergy* 4(3), 277–88. doi:10.1111/j.1757-1707.2011.01129.x.

Raupach, M. R. and Thom, A. S. 1981. Turbulence in and above plant canopies. *Annu. Rev. Fluid Mech.* 13(1), 97–129. doi:10.1146/annurev.fl.13.010181.000525.

Rebmann, C., Aubinet, M., Schmid, H., Arriga, N., Aurela, M., Burba, G., Clement, R., De Ligne, A., Fratini, G., Gielen, B., Grace, J., Graf, A., Gross, P., Haapanala, S., Herbst, M., Hörtnagl, L., Ibrom, A., Joly, L., Kljun, N., Kolle, O., Kowalski, A., Lindroth, A., Loustau, D., Mammarella, I., Mauder, M., Merbold, L., Metzger, S., Mölder, M., Montagnani, L., Papale, D., Pavelka, M., Peichl, M., Roland, M., Serrano-Ortiz, P., Siebicke, L., Steinbrecher, R., Tuovinen, J., Vesala, T., Wohlfahrt, G. and Franz, D. 2018. ICOS eddy covariance flux-station site setup: a review. *Int. Agrophys.* 32(4), 471–94. doi:10.1515/intag-2017-0044.

Reddy, K. R., Parick, W. H. and Lindau, C. W. 1989. Nitrification-denitrification at the plant root-sediment interface in wetlands. *Limnol. Oceanogr.* 34(6), 1004–13. doi:10.4319/lo.1989.34.6.1004.

Reichstein, M., Stoy, P. C., Desai, A. R., Lasslop, G. and Richardson, A. D. 2012. Partitioning of net fluxes. In: Aubinet, M., Vesala, T. and Papale, D. (Eds), *Eddy Covariance*. Springer, Dordrecht, the Netherlands, pp. 263–89.

Reichstein, M., Camps-Valls, G., Stevens, B., Jung, M., Denzler, J., Carvalhais, N., et al. 2019. Deep learning and process understanding for data-driven Earth system science. *Nature* 566, 195–204. doi:10.1038/s41586-019-0912-1

Riederer, M., Serafimovich, A. and Foken, T. 2014. Net ecosystem CO_2 exchange measurements by the closed chamber method and the eddy covariance technique and their dependence on atmospheric conditions. *Atmos. Meas. Tech.* 7(4), 1057–64. doi:10.5194/amt-7-1057-2014.

Rochette, P. 2011. Towards a standard non-steady-state chamber methodology for measuring soil N_2O emissions. *Anim. Feed Sci. Tech.* 166-167, 141–6. doi:10.1016/j.anifeedsci.2011.04.063.

Rochette, P. and Bertrand, N. 2003. Soil air sample storage and handling using polypropylene syringes and glass vials. *Can. J. Soil Sci.* 83(5), 631–7. doi:10.4141/S03-015.

Rochette, P. and Eriksen-Hamel, N. S. 2008. Chamber measurements of soil nitrous oxide flux: are absolute values reliable? *Soil Sci. Soc. Am. J.* 72(2), 331–42. doi:10.2136/sssaj2007.0215.

Rochette, P. and Hutchinson, G. L. 2005. Measurement of soil respiration *in situ*: chamber techniques. In: Hatfield, J. L. (Ed.), *Micrometeorology in Agricultural Systems*. Agronomy Monograph. American Society of Agronomy, Madison, WI, p. 587.

Rochette, P., Chantigny, M. H., Ziadi, N., Angers, D. A., Bélanger, G., Charbonneau, É., Pellerin, D., Liang, C. and Bertrand, N. 2014. Soil nitrous oxide emissions after deposition of dairy cow excreta in eastern Canada. *J. Environ. Qual.* 43(3), 829–41.

Roper, J. D., Burton, D. L., Madani, A. and Stratton, G. W. 2013. A simple method for quantifying dissolved nitrous oxide in tile drainage water. *Canadian J. Soil Sci.* 93, 59–64. doi:10.4141/cjss2012-021.

Running, S. W., Nemani, R. R., Heinsch, F. A., Zhao, M., Reeves, M. and Hashimoto, H. 2004. A continuous satellite-derived measure of global terrestrial primary production.

BioScience 54(6), 547–60. doi:10.1641/0006-3568(2004)054[0547:ACSMOG]2.0.CO;2.

Ryu, Y., Berry, J. A. and Baldocchi, D. D. 2019. What is global photosynthesis? History, uncertainties and opportunities. *Rem. Sens. Env.* 223, 95–114. doi:10.1016/j.rse.2019.01.016.

Sabbatini, S., Mammarella, I., Arriga, N., Fratini, G., Graf, A., Hörtnagl, L., Ibrom, A., Longdoz, B., Mauder, M., Merbold, L., Metzger, S., Montagnani, L., Pitacco, A., Rebmann, C., Sedlák, P., Šigut, L., Vitale, D. and Papale, D. 2018. Eddy covariance raw data processing for CO_2 and energy fluxes calculation at ICOS ecosystem stations. *Int. Agrophys.* 32(4), 495–515. doi:10.1515/intag-2017-0043.

Samer, M. and Abuarab, M. E. 2014. Development of CO_2 mass balance for estimation of ventilation rate in naturally cross-ventilated dairy barns. *Trans. ASABE* 57, 1255–64.

Samer, M., Ammon, C., Loebsin, C., Fiedler, M., Berg, W., Sanftleben, P. and Brunsch, R. 2012. Moisture balance and tracer gas technique for ventilation rates measurement and greenhouse gases and ammonia emissions quantification in naturally ventilated buildings. *Build. Environ.* 50, 10–20. doi:10.1016/j.buildenv.2011.10.008.

Schaller, C., Göckede, M. and Foken, T. 2017. Flux calculation of short turbulent events-comparison of three methods. *Atmos. Meas. Tech.* 10(3), 869–80. doi:10.5194/amt-10-869-2017.

Selbie, D. R., Buckthought, L. E. and Shepherd, M. A. 2015. The challenge of the urine patch for managing nitrogen in grazed pasture systems. *Adv. Agr.* 129, 229–92.

Shah, S. B., Westerman, P. W. and Arogo, J. 2006. Measuring ammonia concentrations and emissions from agricultural land and liquid surfaces: a review. *J. Air Waste Manag. Assoc.* 56(7), 945–60. doi:10.1080/10473289.2006.10464512.

Siebicke, L. and Emad, A. 2019. True eddy accumulation trace gas flux measurements: proof-of-concept. *Atmos. Meas. Tech.* 12(8), 4393–420. doi:10.5194/amt-12-4393-2019.

Sommer, S. G., McGinn, S. M., Hao, X. and Larney, F. J. 2004. Techniques for measuring gas emissions from a composting stockpile of cattle manure. *Atmos. Environ.* 38(28), 4643–52. doi:10.1016/j.atmosenv.2004.05.014.

Smith, K. A. and Dobbie, K. E. 2001. The impact of sampling frequency and sampling times on chamber-based measurements of N_2O emissions from fertilized soils. *Glob. Change Biol.* 7(8), 933–45. doi:10.1046/j.1354-1013.2001.00450.x.

Snyder, C., Davidson, E., Smith, P. and Venterea, R. 2014. Agriculture: sustainable crop and animal production to help mitigate nitrous oxide emissions. *Curr. Opin. Environ. Sustain.* 9–10, 46–54. doi:10.1016/j.cosust.2014.07.005.

Stehfest, E. and Bouwman, L. 2006. N_2O and NO emission from agricultural fields and soils under natural vegetation: summarizing available measurement data and modeling of global annual emissions. Nutrient Cycling in Agroecosystems, 74, 207–228.

Stevens, R. J., Laughlin, R. J., Atkins, G. J. and Prosser, S. J. 1993. Automated determination of nitrogen-15 labelled dinitrogen and nitrous oxide by mass spectroscopy. *Soil Sci. Soc. Am.* 57, 981–8.

Storm, I. M. L. D., Hellwing, A. L. F., Nielsen, N. I. and Madsen, J. 2012. Methods for measuring and estimating methane emission from ruminants. *Animal* 2(2), 160–83. doi:10.3390/ani2020160.

Stoyan, H., De-Polli, H., Böhm, S., Robertson, G. P. and Paul, E. A. 2000. Spatial heterogeneity of soil respiration and related properties at the plant scale. *Plant Soil* 222(1/2), 203–14. doi:10.1023/A:1004757405147.

Suryavanshi, P., Singh, Y. V., Prasanna, R., Bhatia, A. and Shivay, Y. S. 2013. Pattern of methane emission and water productivity under different methods of rice crop establishment. *Paddy Water Environ.* 11(1-4), 321-9. doi:10.1007/s10333-012-0323-5.

Sutton, M. A., Nemitz, E., Milford, C., Fowler, D., Moreno, J., San José, R., Wyers, G. P., Otjes, R. P., Harrison, R., Husted, S. and Schjoerring, J. K. 2000. Micrometeorological measurements of net ammonia fluxes over oilseed rape during two vegetation periods. *Agr. Forest Meteorol.* 105(4), 351-69. doi:10.1016/S0168-1923(00)00203-3.

Suvocarev, K., Reba, M. L. and Runkle, B. 2015. Using the surface renewal technique to estimate CO_2 exchange from a rice field to the atmosphere. *AGU Fall Meeting Abstracts.*

Suvočarev, K., Castellví, F., Reba, M. L. and Runkle, B. R. K. 2019. Surface renewal measurements of H, λE and CO_2 fluxes over two different agricultural systems. *Agr. Forest Meteorol.* 279. doi:10.1016/j.agrformet.2019.107763.

Takahashi, J., Chaudhry, A. S., Beneke, R. G. and Young, B. A. 1999. An open-circuit hood system for gaseous exchange measurements in small ruminants. *Small Rumin. Res.* 32(1), 31-6. doi:10.1016/S0921-4488(98)00163-1.

Tallec, T., Brut, A., Joly, L., Dumelié, N., Serça, D., Mordelet, P., Claverie, N., Legain, D., Barrié, J., Decarpenterie, T., Cousin, J., Zawilski, B., Ceschia, E., Guérin, F. and Le Dantec, V. 2019. N_2O flux measurements over an irrigated maize crop: a comparison of three methods. *Agr. Forest Meteorol.* 264, 56-72. doi:10.1016/j.agrformet.2018.09.017.

Thom, A. S. 1975. *Vegetation and the Atmosphere* (vol. I). Monteith, J. L. (Ed.). Academic Press, London, UK.

Todd, R. W., Cole, N. A., Harper, L. A. and Flesch, T. K. 2007. Flux gradient estimates of ammonia emissions from beef cattle feedyard pens. In: Moody, L. (Ed.), *Proceedings of the International Symposium on Air Quality and Waste Management for Agriculture*, 16-19 September 2007, Broomfield, CO. ASAE Publication No. 701P0907cd. American Society of Agricultural and Biological Engineers, St. Joseph, MI.

Todd, R. W., Cole, N. A., Clark, R. N., Flesch, T. K., Harper, L. A. and Baek, B. H. 2008. Ammonia emissions from a beef cattle feedyard on the southern High Plains. *Atmos. Environ.* 42(28), 6797-805. doi:10.1016/j.atmosenv.2008.05.013.

Torbick, N., Salas, W., Chowdhury, D., Ingraham, P. and Trinh, M. 2017. Mapping rice greenhouse gas emissions in the Red River Delta, Vietnam. *Carbon Manag.* 8(1), 99-108. doi:10.1080/17583004.2016.1275816.

Tubiello, F. N., Salvatore, M., Ferrara, A. F., House, J., Federici, S., Rossi, S., Biancalani, R., Condor Golec, R. D., Jacobs, H., Flammini, A., Prosperi, P., Cardenas-Galindo, P., Schmidhuber, J., Sanz Sanchez, M. J., Srivastava, N. and Smith, P. 2015. The contribution of agriculture, forestry and other land use activities to global warming, 1990_2012. *Glob. Change Biol.* 21(7), 2655-60. doi:10.1111/gcb.12865.

Utsumi, S., Beede, D., Zimmerman, S. and Zimmerman, P. 2011. Whole herd enteric methane emission estimates in three contrasting dairy systems. *J. Dairy Sci.* 94(E-Suppl. 1), 607.

Valentini, R., De Angelis, P., Matteucci, G., Monaco, R., Dore, S. and Mucnozza, G. E. S. 1996. Seasonal net carbon dioxide exchange of a beech forest with the atmosphere. *Glob. Change Biol.* 2(3), 199-207. doi:10.1111/j.1365-2486.1996.tb00072.x.

Van Groenigen, J. W., Velthof, G. L., Oenema, O., Van Groenigen, K. J. and Van Kessel, C. 2010. Towards an agronomic assessment of N_2O emissions: a case study for arable crops. *Eur. J. Soil Sci.* 61(6), 903-13. doi:10.1111/j.1365-2389.2009.01217.x.

van Haarlem, R. P., Desjardins, R. L., Gao, Z., Flesch, T. K. and Li, X. 2008. Methane and ammonia emissions from a beef feedlot in western Canada for a twelve-day period in the fall. *Can. J. Anim. Sci.* 88(4), 641-9. doi:10.4141/CJAS08034.

Velthof, G. L., Jarvis, S. C., Stein, A., Allend, A. G. and Oenema, O. 1996. Spatial variability of nitrous oxide fluxes in mown and grazed grasslands on a poorly drained clay soil. *Soil Biol. Biochem.* 28(9), 1215–25. doi:10.1016/0038-0717(96)00129-0.

Venterea, R. T., Parkin, T. B., Cardenas, L., Petersen, S. O. and Pedersen, A. R. 2012. Data analysis considerations. In: de Klein, C. and Harvey, M. (Eds), *Nitrous Oxide Chamber Methodology Guidelines*. Ministry for Primary Industries, Wellington, New Zealand, pp. 95–121.

Vermeulen, S. J., Campbell, B. M. and Ingram, J. S. I. 2012. Climate change and food systems. *Annu. Rev. Environ. Resour.* 37(1), 195–222. doi:10.1146/annurev-environ-020411-130608.

Vickers, D. and Mahrt, L. 1997. Fetch limited drag coefficients. *Bound-Lay Meteorol.* 85(1), 53–79. doi:10.1023/A:1000472623187.

Voglmeier, K., Six, J., Jocher, M. and Ammann, C. 2019. Grazing-related nitrous oxide emissions: from patch scale to field scale. *Biogeosciences* 16(8), 1685–703. doi:10.5194/bg-16-1685-2019.

Waghorn, G. C., Garnet, E. J., Pinares, C. and Zimmerman, S. 2013. Implementation of GreenFeed for estimating methane in a dairy herd grazing pasture. *5th Greenhouse Gases and Animal Agriculture Conference (GGAA 2013)*, Dublin, Ireland, 23–26 June 2013.

Waldo, S., Russell, E. S., Kostyanovsky, K., Pressley, S. N., O'Keeffe, P. T., Huggins, D. R., Stöckle, C. O., Pan, W. L. and Lamb, B. K. 2019. N_2O emissions from two agroecosystems: high spatial variability and long pulses observed using static chambers and the flux-gradient technique. *J. Geophys. Res., Biogeosci.* 124(7), 1887–904. doi:10.1029/2019JG005032.

Wang, K., Zheng, X., Pihlatie, M., Vesala, T., Liu, C., Haapanala, S., ..., Liu, H. 2013. Comparison between static chamber and tunable diode laser-based eddy covariance techniques for measuring nitrous oxide fluxes from a cotton field. *Agr. For. Meteorol.* 171, 9–19. doi:10.1016/j.agrformet.2012.11.009.

Wassmann, R., Neue, H. U., Alberto, M. C. R., Lantin, R. S., Bueno, C., Llenaresas, D., Arah, J. R., Papen, H., Seiler, W. and Rennenberg, H. 1996. Fluxes and pools of methane in wetland rice soils with varying organic inputs. *Environ. Monit. Assess.* 42(1-2), 163–73. doi:10.1007/BF00394048.

Williams, S. R. O., Moate, P. J., Hannah, M. C., Ribaux, B. E., Wales, W. J. and Eckard, R. J. 2011. Background matters with the SF6 tracer method for estimating enteric methane emissions from dairy cows: a critical evaluation of the SF6 procedure. *Anim. Feed Sci. Technol.* 170(3–4), 265–76. doi:10.1016/j.anifeedsci.2011.08.013.

Williams, I., Kemp, S., Coello, J., Turner, D. A. and Wright, L. A. 2012. A beginner's guide to carbon footprinting. *Carbon Manag.* 3(1), 55–67. doi:10.4155/cmt.11.80.

Williams, S. R. O., Clarke, T., Hannah, M. C., Marrett, L. C., Moate, P. J., Auldist, M. J. and Wales, W. J. 2013. Energy partitioning in herbage-fed dairy cows offered supplementary grain during an extended lactation. *J. Dairy Sci.* 96(1), 484–94. doi:10.3168/jds.2012-5787.

WMO/IAEA. 2014. 17th WMO/IAEA meeting on carbon dioxide, other greenhouse gases and related tracers measurement techniques (GGMT-2013). GAW Report No. 213. World Meteorological Organization, Global Atmospheric Watch.

Wofsy, S. C., Goulden, M. L., Munger, J. W., Fan, S. M., Bakwin, P. S., Daube, B. C., Bassow, S. L. and Bazzaz, F. A. 1993. Net exchange of CO_2 in a mid-latitude forest. *Science* 260(5112), 1314-7. doi:10.1126/science.260.5112.1314.

Wutzler, T., Lucas-Moffat, A., Migliavacca, M., Knauer, J., Sickel, K., Šigut, L., Menzer, O. and Reichstein, M. 2018. Basic and extensible post-processing of eddy covariance flux data with REddyProc. *Biogeosciences* 15(16), 5015–30. doi:10.5194/bg-15-5015-2018.

Xin, H., Li, H., Burns, R. T., Gates, R. S., Overhults, D. G. and Earnest, J. W. 2009. Use of CO_2 concentration difference or CO_2 balance to assess ventilation rate of broiler houses. *Trans. ASABE* 52(4), 1353–61. doi:10.13031/2013.27787.

Xu, L., Furtaw, M. D., Madsen, R. A., Garcia, R. L., Anderson, D. J. and McDermitt, D. K. 2006. On maintaining pressure equilibrium between a soil CO_2 flux chamber and the ambient air. *J. Geophys. Res.* 111(D8).

Yang, S., Lei, L., Zeng, Z., He, Z., and Zhong, H. 2019. An assessment of anthropogenic CO_2 emissions by satellite-based observations in China. Sensors 19, 1118. doi:10.3390/s19051118.

Yona, L., Cashore, B., Jackson, R. B., Ometto, J. and Bradford, M. A. 2020. Refining national greenhouse gas inventories. *Ambio* 49, 1. doi:10.1007/s13280-019-01312-9.

Yu, K. and Chen, G. 2009. Nitrous oxide emissions from terrestrial plants: observations, mechanisms and implications. In: Sheldon, A. I. and Barnbart, E. P. (Eds), *Nitrous Oxide Emissions Research*. Nova Science Publishers, Hauppauge, NY, pp. 85–104.

Zheng, X. H., Mei, B. L., Wang, Y. H., Xie, B. H., Wang, Y. S., Dong, H., Xu, H., Chen, G., Cai, Z., Yue, J., Gu, J., Su, F., Zou, J. and Zhu, J. 2008. Quantification of N_2OFluxes From soil-plant systems may be biased by the applied gas chromatograph methodology. *Plant Soil* 311(1–2), 211–34. doi:10.1007/s11104-008-9673-6.

Zimmerman, P. R. 1993. System for measuring metabolic gas emissions from animals. Patent no. 5 265 618. United States Patent and Trademark Office, Alexandria, VA.

Zimmerman, P., Zimmerman, S., Utsumi, S. and Beede, D. 2011. Development of a user-friendly online system to quantitatively measure metabolic gas fluxes from ruminants. Abstract 856. *2011 ADSA, AMPA, ASAS, CSAS, and WSASAS Joint Annual Meeting*. New Orleans, LA, 10–14 July 2011.

Zimmerman, S., Brito, A., Huhtanen, P., Johnson, K., Michal, J. J., et al. 2013a. Measurement and evaluation of enteric CH4 emissions and variability in production systems. *5th Greenhouse Gases and Animal Agriculture Conference (GGAA 2013)*, Dublin, Ireland, 23–26 June 2013.

Zimmerman, S., Michal, J. J., White, R., Johnson, K. A., Guerouali, A., et al. 2013b. Evaluation of a novel system to measure enteric methane emissions from beef cattle on pasture. *2013 ADSA, AMPA, ASAS, CSAS, and WSASAS Joint Annual Meeting*. Indianapolis, IN, 8–12 July 2013.

Zhao, P., Hammerle, A., Zeeman, M. and Wohlfahrt, G. 2019. On the calculation of daytime CO_2 fluxes measured by automated closed transparent chambers. *Agr. Forest Meteorol.* 263, 267–75. doi:10.1016/j.agrformet.2018.08.022.

Part 3

Adaption and mitigation strategies in agriculture

Chapter 8

Climate-smart crop production: understanding complexity for achieving triple-wins

Katrien Descheemaeker, Pytrik Reidsma and Ken E. Giller, Plant Production Systems, Wageningen University & Research, The Netherlands

1 Introduction

Agriculture worldwide faces the enormous challenge to meet increasing demands for food, feed and fuel, resulting from a growing and increasingly affluent human population (Grafton et al., 2015). This challenge is complicated by climate change, which is expected to negatively affect agriculture in many places through increasing temperature, changing rainfall patterns, increased climate variability and frequency of extreme events (Lipper et al., 2014; Rosenzweig et al., 2014). At the same time, while agriculture emits a large share of global greenhouse gas (GHG) emissions, it holds great potential for reducing these emissions (Smith et al., 2014).

Climate-smart agriculture (CSA) originated from the need to address this three-pronged challenge through a concerted effort to simultaneously improve food security by increasing productivity (pillar 1), strengthen resilience by adapting to climate change (pillar 2) and mitigate GHG emissions (pillar 3; FAO, 2010). The CSA concept aims to reverse the previously existing tendency of climate policies generally treating adaptation and mitigation separately. As such, unintended side effects (both positive and negative) were often overlooked and therefore not addressed (Locatelli et al., 2015).

http://dx.doi.org/10.19103/AS.2020.0064.14

Since its launch in 2010, CSA has become a major focus of agricultural research with a sharp and continuing rise in scientific papers, conferences, tools (e.g. Andrieu et al., 2017; Brandt et al., 2017; Mwongera et al., 2017) and handbooks or guides (e.g. FAO, 2013; CSA guide of CCAFS: https://csa.guide/, accessed 11 February 2019). The concept is increasingly taken up by NGOs, donors, governments and large companies, but also increasingly criticized both in the scientific literature (e.g. Karlsson et al., 2018) and in various media (e.g. http://www.climatesmartagconcerns.info, accessed 11 February 2019). A growing body of literature comprises general reviews on CSA (e.g. Debaeke et al., 2017; Scherer and Verburg, 2017; Steenwerth et al., 2014), reviews of knowledge on specific CSA options (e.g. on agroforestry by Mbow et al. (2014); on conservation agriculture (CA) by Thierfelder et al. (2017); on cover crops by Kaye and Quemada (2017)) and site-specific case studies of CSA implementation (e.g. Long et al. (2016) for European countries and Lee (2017) for Kenya). Based on the existing literature, we provide a concise overview of CSA options with their expected effects on the three pillars and the underlying mechanisms. We aim to add value by reviewing approaches to the categorization of CSA options, and by critically examining the potential of two (groups of) options, namely CA and soil fertility management. From this, we identify gaps in our understanding of CSA and potential pitfalls in the assessment of the CSA potential. Finally, based on two contrasting case studies from the Netherlands and Zimbabwe, we show how farming systems analysis enables the operationalization of the CSA concept through participatory, integrated and cross-scale assessments.

We focus on crop production strategies, recognizing that in many farming systems of the world, cropping activities are intricately linked with livestock keeping, other land uses and a range of rural livelihood activities. Throughout the chapter, we argue that by considering the interactions between crops and the other components of the farming system, important insights can be revealed. Whereas this chapter focusses on options targeting the supply side of agricultural systems, these will have to be combined with solutions from the demand side, for example through changes in consumption behaviour (Scherer and Verburg, 2017; Smith et al., 2013). Yet, the food systems approach needed for such analyses (UNEP, 2016) is beyond the scope of this chapter.

2 Climate-smart agriculture (CSA) cropping options

Several overviews of options for CSA (e.g. Debaeke et al., 2017; Scherer and Verburg, 2017; Steenwerth et al., 2014), adaptation (e.g. Thornton and Herrero, 2014 for mixed crop-livestock systems; Wassmann et al., 2009 for rice) and mitigation (e.g. Smith et al., 2013; Cole et al., 1997) are found in the literature. The information provided in these overviews and the more in-depth papers referred to throughout this chapter is synthesized in the overview of Table 1.

Table 1 CSA options categorized by strategy and level with a qualitative indication of their effects on various CSA pillars and the underlying mechanisms

			Adaptation			Mitigation[a]						
Strategy (Level)	Option	Food security/ Productivity	Risk management	Diversification	(Sustainable) Intensification	N_2O	CH_4	CO_2	C sequestration	REDD+	Constraints to adoption	Comments
Farm production strategies (Field to farm level)	Choice of crop species/ cultivar	++	+	++	++						Limited access, high costs, reluctance to change diet	Effects on mitigation may be both positive and negative, depending on the choice
	Time of sowing	+	+++								Limited availability of labour and machinery	
	Water harvesting	++	+++		+++						Initial cost of investment	
	Conservation agriculture	+/−	++					++			Positive impacts only in long term; problems with weed pressure	CO_2 reduction from reduced fuel use for tillage operations
	Crop diversification	+	+	+++							Limited market opportunities for new crops; more knowledge-intensive and agile management required	Food security impact through diet diversity
	Agroforestry	+/−	+	++					+++	++	Possible negative effect on annual crops; trees take time to mature; young tree seedlings need protection against grazing pressure	Effects on food security depend on scale (+ at farm/ landscape; − at crop level)

(Continued)

Table 1 (*Continued*)

Strategy (Level)	Option	Food security/ Productivity	Adaptation			Mitigation[a]					Constraints to adoption	Comments
			Risk management	Diversification	(Sustainable) Intensification	N_2O	CH_4	CO_2	C sequestration	REDD+		
	Intercropping	+	+	++		+		+			Possible yield penalty on main crop; more laborious crop operations	Positive effect on productivity if LER >1; if legume intercrop, N fixation reduces use of mineral fertilizer, leading to less N_2O and CO_2 (for manufacturing) emissions
	Irrigation	+++			+++						High costs for initial investment and maintenance of irrigation structures; water management governance not always effective; risk of water-borne diseases	
	Integrated pest management	++	+		++			+			Knowledge-intensive	Effects on CO_2 from reduced agrochemical production

Optimize mineral fertilizer use	+++	+/−		+++	+/−	+/−		Access to and costs of mineral fertilizer; financial risk	Effects on risk depend on context, scale and availability of storage facilities; effects on mitigation depend on the context: in low-input systems, fertilizer use will increase, thus increasing emissions - in high-input systems, fertilizer application rates can be reduced, thus reducing emissions
Optimize manure use	++	+/−		+++	+/−−	+/−−	++	Labour and knowledge-intensive; investment costs for proper storage facilities	Effects are similar to mineral fertilizer but more negative in case of increased manure use because of the more difficult synchronization with plant demand, the emissions associated with manure storage and enteric fermentation. The exception is the positive impact on carbon sequestration through the build-up of soil organic matter

(*Continued*)

Table 1 (*Continued*)

Strategy (Level)	Option	Food security/ Productivity	Adaptation: Risk management	Adaptation: Diversification	Adaptation: (Sustainable) Intensification	Mitigation[a]: N_2O	Mitigation[a]: CH_4	Mitigation[a]: CO_2	Mitigation[a]: C sequestration	Mitigation[a]: REDD+	Constraints to adoption	Comments
	Organic amendments (crop residues, compost)	+/–	++			–			++		Competition for crop residues from livestock for use as feed; compost is knowledge- and labour-intensive technology	Positive effect on productivity if improvements in water-use efficiency thanks to better infiltration and/or water-holding capacity; negative effects on productivity if increased C:N ratio is not compensated by higher N application rate. The required N application will lead to a trade-off in N_2O emissions
	Legume crops	+/–		++		+		+			Possible trade-off on farm production may be a disincentive; access to legume varieties and markets may be limited	Effects on food security depend on scale: due to often lower legume productivity compared to cereal, farm-level production may go down, but cereals may benefit in case of no or low mineral fertilizer use

	Cover crops	+/-		++	+	+	+	Possible trade-off on farm production may be a disincentive	Effects on food security depend on type of crop: farm-level food security may decrease if cover crop is not a food crop
	Paddy rice management (straw, drainage, cultivar)	/			+/–	+++		Knowledge-intensive technique	Yield penalties can be avoided. Drainage may increase the emission of N_2O, but this does not offset the huge reductions in CH_4 which can be obtained
	Biochar application	/					+	No clear benefits on crop productivity; access to biomass and processing facilities may be limited	No conclusive evidence on effects on crop yields
	Soil and water conservation measures	+/–	+				+	Possible trade-off on farm production and long time horizon may be a disincentive; labour-intensive	Effects on productivity may be negative at field level due to less space for crops; and positive effects may take time to be realized
Farm financial management (farm level)	Insurance	+	+++					Absence of insurance schemes; costs	Positive effect on crop productivity because of incentive to invest

(Continued)

Table 1 (*Continued*)

Strategy (Level)	Option	Food security/ Productivity	Adaptation			Mitigation[a]					Constraints to adoption	Comments
			Risk management	Diversification	(Sustainable) Intensification	N_2O	CH_4	CO_2	C sequestration	REDD+		
	Income diversification; on/off-farm			+++							Off-farm opportunities may not be available	
	Mixed farming	+	+	++		–	–		+		Knowledge-intensive and agile management required	Effects are considered in comparison with specialized crop farming: negative emissions effects due to manure application ($N_§O$) and enteric fermentation (CH_4). Manure can increase soil organic carbon stocks. In a smallholder context, food security improves due to inclusion of animal-sourced foods in the diet
	Investing in income stabilization	+	++								Options may not be available to all and everywhere	Positive effect on crop productivity because of incentive to invest
Land use planning (landscape level)	Land sharing	–		+++					+	+	Possible trade-off on farm production may be a disincentive; more complex farm management	Positive effect on C sequestration and avoided deforestation if perennial vegetation included

	Agroforestry	+		+++			+	++	Possible negative effect on annual crops; trees take time to mature; young tree seedlings need protection against grazing pressure	Similar effects compared to land sharing, but positive for productivity because tree component also aims at provisioning services, whereas this is not an objective in the land sharing option
	Land sparing	/			+++			++	Unless supported by policies, increased land-use efficiency may lead to more land being cultivated	Productivity on a land basis is increased so that total production can be maintained while setting land aside for nature; (sustainable) intensification is the assumed driver of this option, not the result
Technology (sector level)	Development of crop cultivars	++	+	++	++				Costly; time consuming	
	Weather forecasting	+	++						Limited trust in the forecast	
	Precision agriculture	++			+++	++	+		Costly	Cuts in CO_2 emissions from fuel savings

[a] Mitigation through reduction of CO_2 emissions includes CO_2 from fuel use, manufacturing of agro-chemicals and mineralization of soil organic matter; C sequestration includes above- and below-ground processes; REDD+ refers to the prevention of deforestation and degradation of carbon-rich ecosystems.

Existing reviews typically categorize options with the aim to bring structure in the vast and expanding collection of potential solutions or to explain the mechanisms for adaptation and mitigation. Yet, the approach to the categorization of adaptation options varies considerably. For example, Smit and Skinner (2002) propose a typology of adaptation activities based on the scale of implementation and comprising four categories of farm production practices (e.g. adapting the crop species and/or cultivars, adapting pest and weed control), farm financial management (e.g. diversifying the portfolio of income sources, crop insurance), government programs (e.g. subsidies, land use policies) and technological developments (e.g. breeding of new adapted crop cultivars, improvement in weather forecasting). This approach resembles the one of Scherer and Verburg (2017), who distinguish between technological advancement, adaptive farm management and financial management, and of Hertel and Lobell (2014), who describe three categories of current technologies, new technologies and changes in the institutional environment. Debaeke et al. (2017) follow a different approach by using the time horizon to set apart short-term risk management options, such as shifting cropping patterns or adjusting time and rate of fertilizer application, from long-term adaptations, such as breeding and knowledge-driven optimization of farm management. Finally, options can also be categorized based on the underlying mechanism towards adaptation, such as risk management, diversification and sustainable intensification (SI), as described by Thornton and Herrero (2014) and Descheemaeker et al. (2016). Examples of risk management options include adjusting planting dates and post-harvest storage. Diversification can be achieved with, for example, intercropping and agroforestry, whereas soil fertility management, water harvesting or irrigation can enable SI. In Table 1 we categorize options based on strategy and level of scale and evaluate them based on CSA mechanisms, thus showing that different approaches can be combined.

Studies reviewing knowledge on mitigation options typically employ a categorization approach that focusses on the mitigation mechanisms. The first group of options is geared towards the reduction of GHG emissions. Within this category, the type of GHG determines the mitigation options. For example, N_2O emissions can be reduced by optimizing nutrient-use efficiency with adjusted mineral fertilizer application and inclusion of legume crops; CH_4 emissions in rice systems can be cut back by agronomic management including drainage of irrigated fields, straw management and use of adapted cultivars; CO_2 emissions can be decreased by optimizing energy-use efficiency of farm operations. Likewise, potential mitigation options can be identified by considering the source of the emission, such as enteric fermentation, manure management or agricultural soils. A second category of mitigation options aims at sequestering carbon in the soil and in biomass (e.g. agroforestry). A third category entails the prevention of deforestation and degradation of carbon-rich ecosystems, such as peatlands and grasslands (e.g. the REDD+ strategies; Corbera and Schroeder, 2011). Cropping options only have an indirect influence on the latter strategy, for example through land sparing, which may (or may not) be achieved through SI on existing cropland (Carter et al., 2018). In Table 1, we differentiate effects per GHG, and for C sequestration and REDD+ strategies.

Based on our review of the literature for this chapter, we qualitatively analysed the expected effects of options on the three CSA pillars and the underlying mechanisms (Table 1). Smith and Olesen (2010) claimed that although certain mitigation measures may negatively impact adaptive capacity, most adaptation options have a mitigation co-benefit through their positive effect on nitrogen-use efficiency and soil carbon storage. Yet, Table 1 shows that the options that benefit both adaptation and mitigation are scarce. Some notable exceptions are agroforestry and the incorporation of legume crops. Further, Scherer and Verburg (2017) identified only cover crops, agroforestry and reduced tillage as truly climate-smart cropping options. In contrast, for some options listed in Table 1, clear trade-offs can be expected, such as for manure application. Several studies (see Locatelli et al. (2015) for an overview) pay attention to the possible trade-offs that may exist between adaptation and mitigation when implementing CSA practices that are primarily aimed at one or the other objective.

In the following sections, we focus on CA and soil fertility management to illustrate in more detail how cropping strategies contribute, or not, to the three pillars of CSA. These strategies are chosen because they are believed to influence all of the three CSA pillars, and extensive research has generated a wealth of quantitative information allowing a critical examination of their potential benefits and drawbacks. CA is chosen as a specific intervention, whereas soil fertility management is a more general strategy encompassing various concrete options, including, among others, cover crops, different types of fertilizer, legume crops and rotations.

2.1 Conservation agriculture (CA)

Conservation agriculture combines minimum soil disturbance by reduced or no tillage, crop residue retention on the soil surface and crop diversification through crop rotations or intercropping (FAO, 2015). CA is often claimed to be climate-smart (Kassam et al., 2018). As it is a heavily researched cropping practice both in intensive agriculture (Giller et al., 2015) and in smallholder systems (Thierfelder et al., 2017), a wealth of information is available to substantiate or question this claim.

With respect to the first pillar of CSA, crop yields may benefit from CA because of a general improvement in soil quality (Corbeels et al., 2014), but the effect varies depending on the agro-ecological context and the type of season.

A meta-analysis of CA evidence suggests that yield benefits are only obtained in dry climates (Pittelkow et al., 2015), and positive effects of CA are stronger in dry seasons (Kirkegaard et al., 1994). In addition, almost invariably, if yields increase, they do so only after some time (Rusinamhodzi et al., 2011), which may discourage farmers from adopting the practice.

The adaptation potential of CA primarily relates to its ability to improve soil water availability, thus rendering the cropping system more resilient to less and more erratic rainfall in future climates (Corbeels et al., 2014). This effect is explained both by increased infiltration rates (Thierfelder and Wall, 2009) and by reduced evaporation in CA systems compared to conventional cropping. Reduced or no-till combined with residue retention and deep-rooting legume crops in rotation or intercropping favours macro-fauna and root activity in the soil, resulting in a continuous pore system that facilitates water flows (Thierfelder et al., 2017). The higher organic carbon content in the surface soil layer improves soil structure, benefitting infiltration. By covering the soil, mulch not only reduces temperature and evaporation (Klocke et al., 2009), thus conserving water, but also tempers the direct impact of raindrops on the soil, thereby reducing erosion (Thierfelder et al., 2017). Whereas residue retention has the capacity to buffer temperature fluctuations and therefore could protect the cropping system against rising temperatures and more frequent heat stress with climate change, quantitative evidence on this aspect is hard to find. Besides the soil-related effects discussed earlier, the reduction or elimination of tillage operations also improves the adaptive capacity of the system by reducing labour requirements for land preparation. This allows more agile farm management, including early sowing or sowing with the first rains, typically leading to better and more reliable yields through a better use of in-season rainfall (Masvaya et al., 2018).

With respect to the third pillar of CSA claims of GHG emission mitigation (e.g. Corsi et al., 2012) are not generally justified. Studies on the effects of CA on carbon sequestration found that CA generally results in larger soil organic carbon content near the soil surface (Powlson et al., 2016). However, as this is due mainly to C redistribution, it goes hand in hand with a reduced carbon content in deeper layers, and thus no net increase in the carbon stock. In addition, the soil carbon sequestration is usually not maintained in the long run, as the carbon is lost when soils are conventionally tilled, which, for agronomic reasons, typically happens occasionally even in no-till systems. Also, seemingly fast initial C sequestration rates quickly drop over time as the soil becomes saturated with carbon (Baveye et al., 2018). With respect to GHG emissions, there is contrasting evidence in the literature on N_2O and CH_4 emissions, as some studies report increases and others decreases of the emissions under CA systems (Corbeels et al., 2018; Powlson et al., 2016). Also, as most studies did not investigate N_2O and CH_4 simultaneously, it is risky to draw conclusions on the overall potential of CA to mitigate GHG emissions. Another aspect on which the jury is still out relates to erosion. Although there is consensus in the literature that CA is effective against erosion, contrasting opinions exist on whether reduced erosion lowers C emissions or not, as some see erosion as a sink, others as a source of C, whereas it could also be neither a source nor a sink (Van Oost et al., 2007). In the end, the only mechanism through which CA undoubtedly contributes to the mitigation of GHG emissions is through the fuel savings from the elimination of tillage operations, resulting in less CO_2 emissions compared to conventional tillage systems.

Increased weed pressure is a common drawback of CA (Giller et al., 2009). Its associated rise in labour and/or herbicide requirements, the latter incurring costs and potential environmental trade-offs, may be a barrier to adoption in a variety of contexts. Due to the intensive use of herbicides in some CA systems, herbicide resistance in weeds is building up and becoming a headache for farmers (Kirkegaard et al., 2014). Other drawbacks are more context-specific. For example, the infiltration and soil water enhancing features of CA are a clear advantage in drier and more variable climates, whereas they present a drawback in wetter climates, in case of delayed sowing or waterlogging. Especially in smallholder agriculture, the lack of adequate planting machinery is often a barrier for adoption, as is the absence of immediate income benefits after the implementation of CA. Biomass constraints and the competing use of crop residues as animal feed is a major barrier to CA adoption in smallholder systems (Giller et al., 2009).

CA is one of the most heavily researched cropping systems worldwide. The resulting evidence from empirical and modelling studies indicates that CA is not convincingly climate-smart. This is because although the adaptation potential is relatively clear, positive effects on crop productivity and farmer livelihoods are noticed only in some contexts and after some years. Also, the mitigation potential of CA is doubtful. Moreover, important trade-offs and barriers to adoption clearly indicate that it is a worthwhile option only in some contexts and farming systems.

2.2 Soil fertility management

Soil fertility management encompasses various practices and technologies, such as the use of mineral and organic fertilizer (animal manure, compost, green manure), crop residue management, the incorporation of legume crops for N fixation, rotations and intercropping. Soil fertility management is an important part of agronomic management and through its effects on crop yield and nutrient (particularly N) dynamics in the soil, it directly influences the productivity and mitigation pillars of CSA. Its effects on the adaptive capacity and resilience of cropping systems are more complex, sometimes counteracting and often indirect (see Section 2.2.4). As soil fertility management has been part and parcel of agronomic sciences (e.g. Vanlauwe et al., 2015), it goes beyond the scope of this chapter to review the existing evidence on the effects of different soil fertility management options on crop productivity (CSA pillar 1). This section first delves into the evidence on mitigation effects (CSA pillar 3), with attention to concurrent effects on productivity (CSA pillar 1). After that, the effects on adaptive capacity and resilience (CSA pillar 2) are described.

With respect to the third CSA pillar, soil fertility management influences mitigation in four ways. First, it affects emissions, particularly of N_2O, through its influence on soil C and N dynamics. In addition, emissions of CH_4 are influenced through manure management, whereas organic C mineralization in soils, manure and organic fertility amendments influence CO_2 emissions. Secondly, soil fertility management influences C sequestration through the addition of organic materials to soils. Thirdly, as soil fertility management is required to (sustainably) intensify crop production, it may also serve to avoid the expansion of cultivated land (land sparing), thus avoiding deforestation and degradation of carbon-rich ecosystems (Carter et al., 2018; van Loon et al., 2019). Fourthly, as soil fertility management aimed at mitigation can lead to a reduced demand for mineral (N) fertilizers, it may lower the 'upstream' fossil energy consumption and associated CO_2 emissions for their manufacturing (Zhang et al., 2013). In the following sections, we will look into the first three mitigation mechanisms.

2.2.1 Mitigation through cutting emissions

Cutting the nitrous oxide (N_2O) emissions is a high-potential mitigation pathway that can be influenced with soil fertility management (Griscom et al., 2017; Venterea et al., 2012; Van Groenigen et al., 2010). N_2O is a long-lived GHG with a large warming potential that is produced during microbially governed denitrification and nitrification reactions in soils and manure, or from downstream denitrification of nitrate that is leached from agricultural soils (Bos et al., 2017; Kaye and Quemada, 2017; Smith, 2017). Many soil conditions (e.g. temperature, water content and aeration, pH) control N_2O production, but as it is largely dependent on nitrate and ammonium concentrations, the most direct and effective strategy to reduce the N_2O emissions is to cut the N input to the system by reducing the application of organic and inorganic fertilizer (Debaeke et al., 2017). Obviously, because of the direct trade-off with crop yield, a balance must be found between farmers' profit-making and/or food-production objectives, mitigation objectives and other environmental concerns such as nitrate leaching. Nevertheless, especially in high-input agriculture where only a maximum of about 60% of the applied N is taken up by the crops, there is scope to reduce the inputs without compromising yield (Bos et al., 2017; Silva et al., 2017; Zhang et al., 2015). Promising strategies for increasing the nutrient-use efficiency of the cropping system include (1) matching the fertilizer rate, time of (split) application and fertilizer composition to crop requirements, (2) precise localization and incorporation of the fertilizer to enhance root uptake and avoid losses through volatilization, (3) complete N balance calculations including the manure and crop residue sources of N that are often underestimated (Debaeke et al., 2017). Mineral N fertilization rates can also be reduced by the inclusion of nitrogen-fixing legume crops as a main crop, intercrop or cover crop, because the legume crop itself does not require N fertilization and it supplies N to the following crop. Although N fixation by legumes itself does not result in N_2O emissions (Rochette and Janzen, 2005), the inclusion of legume crops may. This depends on the fate of the legume crop biomass and the resulting time and amount of N released from its decomposition, subsequently influencing denitrification rates and N_2O losses. Another promising strategy is to include cover crops (leguminous or other), which take up any leftover nitrogen after the main crop, thus reducing the risk of nitrate leaching and its downstream denitrification. Finally, N_2O emissions can also be reduced by applying nitrification inhibitors, which are however costly (Lam et al., 2017).

Manure is an important source of nutrients and organic matter used in soil fertility management and cropping strategies. As manure is a by-product of the livestock industry, GHG emissions from manure are often considered in analyses of the livestock sector. Manure management encompasses a continuum from its production by livestock, over storage, treatment and spreading on crop or grassland. Whereas Chadwick et al. (2011) give an overview of the emissions and mitigation options in each part of this continuum, here we highlight only areas where the evidence is convincingly pointing to mitigation gains. N_2O is produced from nitrification and denitrification processes in stored manure, while CH_4 is the result of anaerobic decomposition of the organic matter in manure. Both processes are affected by environmental factors such as temperature and aeration, biomass composition and management of the manure (Chadwick et al., 2011). By affecting the N content of the diet, the livestock feeding strategy influences the N concentration in the manure and hence its N_2O production potential, while it also affects the CH_4 emissions through enteric fermentation during feed digestion in the animal. The potential to reduce the emissions of N_2O and CH_4 during storage depends on the type of manure (e.g. liquid or solid, with the latter having lower emissions). While for N_2O, it is important to keep the solid manure heaps in anaerobic conditions (e.g. compacting and covering the heap), CH_4 the emissions can be curbed by two opposing strategies aimed at promoting aeration (e.g. through composting) or promoting anaerobic conditions, so that slower decomposition of manure leads to less heating and less stimulation of CH_4 emissions in the anaerobic micro-environments. During storage, manure can also be converted to biogas through a process of anaerobic digestion, which significantly lowers GHG emissions (Cuéllar and Webber, 2008). Moreover, the process creates a methane-rich biogas which can be used as a fuel for cooking, heating or light, as well as sludge that can be used as a fertilizer. With respect to manure application, timing, rate and method influence the C and N dynamics in the plant-soil continuum. To avoid nutrient losses, timing and rate should be fine-tuned with crop nutrient demand, while taking into account that the microbial processes responsible for manure breakdown take time and are influenced by soil temperature and water. In relation to crop nutrient demand, manure is a more effective supplier of P than of N, because excess P is stabilized in the soil, while excess N is lost (e.g. Conijn and van Dijk, 2018). In this respect, processing of manure to alter its composition can offer a solution (Velthof, 2015). Effects of application methods, roughly ranging from incorporation to broadcasting, have been investigated in several studies, but findings on emissions do not all point in the same direction and trade-offs between N_2O and NH_3 emissions require further research (Chadwick et al., 2011).

2.2.2 Mitigation through carbon (C) sequestration

The soil carbon stock is the result of carbon inputs (from roots, crop residues and external organic amendments) and carbon losses through mineralization, making these two mechanisms the main levers for C sequestration (Debaeke et al., 2017). Soil fertility management strategies that benefit C sequestration (for overviews see e.g. Ogle et al., 2005; Powlson et al., 2011; Smith, 2008) include the application of organic amendments and the inclusion of cover crops or green manures in rotations. With respect to the application of organic amendments, mineralization of the organic material after incorporation in the soil results in much of the carbon being lost again to the atmosphere as CO_2. Yet, in the long run, these amendments are observed to positively affect the build-up of the soil carbon stock (Autret et al., 2016; Johnston et al., 2009).

The '4 per 1000' initiative (http://4p1000.org/understand), launched in 2015 at the COP21 in Paris, has attracted a lot of political and scientific attention to the potential of C sequestration in agricultural soils. The initiative aims at increasing global soil organic carbon stocks by 0.4% per year to halt the increase of CO_2 in the atmosphere. However, whereas recent studies (e.g. a multi-country overview by Minasny et al. (2017) and a global gridded modelling study by Zomer et al. (2017)) paint optimistic pictures of the offsetting potential, the reality is likely less rosy due to a number of biophysical caveats (Baveye et al., 2018; Poulton et al., 2018) and socio-economic realities (Amundson and Biardeau, 2018). Firstly, when C sequestration measures are implemented, the carbon sequestration rate may be fast at first, but quickly drops down as the soil saturates with C and reaches a new equilibrium at which point the soil will no longer function as a C sink (Johnston et al., 2009). Secondly, to maintain the C stock, the addition of organic material needs to be continued over time, otherwise the sequestration will be reversed (Smith, 2012). This organic material needs to be produced, thus requiring nutrients and potentially leading to more emissions. Thirdly, with global warming, increasing temperatures will fuel microbial activity leading to a faster decomposition of soil organic matter and release of CO_2 in the atmosphere (Crowther et al., 2016). Fourthly, some of the management practices aimed at increasing the soil C stocks (e.g. manure use) may increase the emissions of other, more potent, GHG like CH_4 and N_2O, as explained in the previous section. Also, increased N input (with associated N_2O emissions) is needed to counteract the lower N availability associated with suboptimal soil stoichiometry of C and nutrients resulting from the process of increasing SOC (van Groenigen et al., 2017; Kirkby et al., 2013). Finally, leakage or displacement issues have been described where the organic amendments applied to a certain area for C sequestration come from another area where losses in carbon are incurred due to the displacement. This is especially problematic in areas of low biomass availability (e.g. in Burkina Faso, as described by Félix et al. (2018)). The above caveats caution against overly optimistic claims, resulting from ignoring scientific evidence. These claims may decrease the sense of urgency, which is badly needed for the implementation of climate-smart solutions (Baveye et al., 2018). In addition, the stipulations point to the importance of holistic and multi-scale analyses to understand the true potential of proposed solutions.

2.2.3 Mitigation through land sparing – a result of sustainable intensification?

Land use change and forestry are responsible for the largest share of the annual GHG emissions from the Agriculture, Forestry and Other Land Use (AFOLU) sector (Smith et al., 2014). In their review of a range of natural climate solutions, Griscom et al. (2017) point to avoided forest conversion as a pathway with an enormous mitigation potential, second only after the reforestation pathway. However, worldwide and throughout human history, the opposite trend has been the most dominant: the conversion of land for agricultural practices has allowed production increases in response to increasing demands for food, feed and fuel (Pretty and Bharucha, 2014; Mandemaker et al., 2011). But, as suitable land is becoming scarce and the conversion leads to ecosystem degradation and GHG emissions, intensification of agricultural production on existing agricultural land is claimed as the way forward (van Ittersum et al., 2016; Tilman et al., 2011). Yet, as agricultural intensification has led to environmental, health and social problems (Kerr, 2012; Stoate et al., 2001; Matson et al., 1997), calls for SI have come to the fore. Even though the SI concept and its meaning are debated (Struik and Kuyper, 2017; Garnett et al., 2013), the principle of producing more output from the same (or less) area of land while reducing negative environmental externalities (Pretty et al., 2011) is appealing in its simplicity. Not surprisingly, next to genetic improvements, soil fertility measures that manage the trade-offs between maximizing productivity and minimizing environmental externalities are high on the SI agenda (for an overview, see Weltin et al., 2018). Extensive information on what constitutes SI in different contexts is available. In high-input systems, there is often scope to maintain good yields while reducing input levels, thus improving efficiency and reducing environmental impact (see e.g. Silva et al. (2017) for a study on Dutch arable farming systems). In low-input systems, emphasis is placed on improving yields by environmentally sound increases in (nutrient) input use, for example through integrated soil fertility management (Vanlauwe et al., 2014, 2015). This will require a so-called 'tunnelling through' of the environmental Kuznets curve through management and technological practices that enable high nutrient-use efficiency and avoid surpluses that may degrade the environment (Zhang et al., 2015).

In summary, by improving agricultural productivity and efficiency, SI may not only reduce the emission intensity of agricultural production, it may also contribute to reforestation and avoiding forest degradation through land sparing (Rounsevell et al., 2005; Schröter et al., 2005). This premise relies on the assumption that with improved resource-use efficiency (e.g. land productivity or yield), the use of that resource (i.e. the land) will decline. However, as described in 'Jevons Paradox', often the opposite is observed due to an increased demand for that resource. This paradox points to the important role of policies and land use legislation to achieve the land sparing effect (Ceddia et al., 2013; Mandemaker et al., 2011; Rudel et al., 2009).

2.2.4 Adaptation

From the previous sections, it is clear that soil fertility management has direct impacts on the first productivity pillar of CSA and on many mechanisms related to the third pillar of mitigation. For adaptation, however, effects are not always straightforward and may even be counteracting. An indirect positive effect is that as a result of soil fertility management, soil organic carbon content and general soil health may improve, leading to better soil functioning (Bos et al., 2017). Indeed, associated increased water-holding capacity and infiltration rates (Franzluebbers, 2002) may help crops to better bridge dry spells in the growing season, which may become more frequent in the future. Reduced soil erodibility is another effect of soil management that helps cropping systems to be more resilient against more intense and larger storm events. Mulching as a form of crop residue management protects the soil not only against raindrop effects, but also creates a micro-climate by buffering temperature amplitudes and heat, which is a useful feature in a warming climate.

With respect to nutrient input, the evidence points in different directions depending on the context and scale of analysis. In general, crops that are not nutrient-limited can take better advantage of CO_2 fertilization effects (Masikati et al., 2019) and cope with pests and diseases (Dordas, 2008). Also, higher fertilizer use generally reduces yield variability (Reidsma et al., 2009). Yet, increasing the nutrient input may also turn out to be risky, especially in low-input systems, typical of smallholder agriculture in Africa. In variable climates, these systems are usually managed to minimize downside risk and characterized by no or very little fertilizer input (Tittonell and Giller, 2013). Resulting crop yields are low on average, mainly due to nutrient limitations, but relatively constant from year to year and relatively insensitive to climate change (Masikati et al., 2019). With increased nutrient input, average yields would improve, but the year-to-year variability would also increase (Keating et al., 2010), as crop production would now be less constrained by nutrients and more determined by seasonal variations in the climate. With investments in inputs not sure to pay off, this increased risk may be a serious barrier preventing the uptake of improved soil fertility management by smallholders. Similarly, the larger sensitivity to climate change (Traoré et al., 2017; Rurinda et al., 2015) would reduce rather than enhance resilience of the crops. At the larger farm level, however, increased nutrient input may improve farmer incomes on average and in good years more substantially. With good financial management and storage facilities (Milgroom and Giller, 2013), this may help to buffer losses in poor years and create possibilities for further re-investment in farm management, thus benefiting the farmer's adaptive capacity.

2.2.5 Trade-offs and constraints

Soil fertility management may entail important trade-offs between the three pillars of CSA. These can be assessed either from the level of the mechanisms or with a specific practice as a starting point. As an example of the first type, N_2O emissions can be reduced by cutting N input to cropping systems, with potential repercussions on crop yield (Debaeke et al., 2017). A good example of the second type is cover crops, which have the potential to increase C sequestration and reduce N_2O emissions, but consume water, which may lead to poorer adaptation to drying climates (Kaye and Quemada, 2017).

As soil fertility management influences the C and N dynamics in the soil, which in turn influence the soil carbon stock and the emissions of GHG from the soil, it is not surprising that trade-offs between the various mitigation mechanisms exist. For example, applying manure may increase carbon stocks, but at the same time increase N_2O emissions from the soil or from downstream denitrification of leached nitrate, thus partially offsetting the benefits of C sequestration (Bos et al., 2017). In the same vein, it is important to be aware of pollution swapping. For example, the use of (imported) manure may lead to savings in mineral fertilizer use, but this is counteracted by the emissions during manure storage and due to enteric fermentation. Another example is that the decrease in ammonia emissions through incorporating slurry may stimulate N_2O losses in some cases (Chadwick et al., 2011).

Whether to apply manure and/or mineral fertilizer is a question that strongly depends on the scale of the analysis, the context and the diversity of CSA objectives and trade-offs taken into account. At field and farm level, using manure may lead to a higher emission intensity compared to using mineral fertilizer, because the slow decomposition of manure complicates the exercise of matching crop demand with N supply, leading to losses in various forms. The fact that the increase in N_2O emissions outweighs the gains in C sequestration from manure application (Bos et al., 2017) is another argument for using mineral fertilizers. In addition, contrary to common beliefs, a meta-analysis by Hijbeek et al. (2017) showed that in Europe, increased soil organic matter does not on average lead to higher yields, if sufficient nutrients are applied by mineral fertilizers. However, at a larger regional level, a different conclusion could be reached. For example, in areas with large livestock densities such as the Netherlands, it makes perfect sense to deal with manure surplus problems by using the available manure on arable fields. Also in a different context the evidence could be in favour of using manure. For example, in low-input systems characterized by poor soil fertility, manure not only helps in building soil organic matter, but also provides a range of other (micro-)nutrients, the lack of which limits crop yield and response to mineral fertilizer (Rusinamhodzi et al., 2013).

Soil fertility management influences (both positively and negatively) a range of agricultural and environmental issues and broader sustainability dimensions beyond the three pillars of CSA that may strongly affect its relevance and attractiveness for farmers. For example, increased nitrogen-use efficiency in agricultural systems may also benefit water and air quality through reduced nitrate leaching and ammonia volatilization, as well as below- and above-ground biodiversity (Debaeke et al., 2017; Zhang et al., 2015). In the social dimension, equity is an important sustainability perspective, particularly in the smallholder context. Better-endowed farmers are often those who own cattle producing manure and have capital to buy fertilizers. Hence, these farmers are more likely to adopt soil fertility management practices that enhance their productivity and resilience (Zingore et al., 2007), thus enlarging the gap with the less-endowed farmers in a community and increasing inequity. On the positive side, co-benefits of including leguminous crops can be found in the economic domain, as the legumes diversify the farmers' income portfolio and hence increase resilience. Another synergy is that legumes diversify the human diet, thus improving the food security of subsistence farmers.

Incremental and transformational change

Anticipated drastic changes in the future climate and resulting changes in ecosystems may exceed thresholds beyond which human actors and natural systems cannot adapt by incremental adjustments (Klein et al, 2014). Such small adaptations in technologies and farm management, including changing crop cultivars, shifting planting dates or water harvesting, have, in the past, allowed farmers to adapt continuously to gradual changes in their environment. Growing evidence for substantial changes in future climate, for which these incremental adaptations will not suffice, has underscored the need for more drastic adaptation (Panda, 2018; Rickards and Howden, 2012). In other words, when the current system is not able to adapt within its biophysical, economic and technological limits, a transformation of the system is needed. For example, climate change is likely to exclude the cultivation of particular crops in certain places due to a shortening of the growing season, or exceeding various suitability thresholds related to temperature and rainfall (e.g. Rippke et al., 2016; Jones and Thornton, 2009). In such cases, transformational change comprises shifts in the production location of crops or shifts to new crops or production systems for a particular location (Rippke et al., 2016).

Transformational adaptation is a relatively new concept in the literature, which still suffers from ambiguity in its definition and the numerous perspectives on its dimensions and required degree of change (Panda, 2018; Klein et al., 2014). Perhaps as useful as a clear definition is the realization that boundaries between types of adaptation are fluid and that adaptation is a continuum from coping through incremental changes, over systemic adaptations (e.g. major changes in the interaction between crops and livestock, the introduction of new technologies such as precision agriculture, the inclusion of shade trees), to transformational change (Rickards and Howden, 2012). Panda (2018) distinguished five types of transformational adaptation, including (a) adaptation actions adopted at a larger scale, (b) shifting crops and changing agricultural systems, (c) changing business scale, structure and location, (d) creating new croplands/irrigation and (e) forced farm abandonment and migration.

Furthermore, in many agricultural systems globally, it is not climate change but other drivers, for example in demography, environmental degradation, or markets and trade, that put most pressure on the current system and coerce it to transform (Reidsma et al., 2015). Hence, the transformation towards a climate-smart agricultural system will require changes in interdependent socio-economic, institutional and political aspects of society (Wise et al., 2014), as well as in the behaviour of consumers. As such, the change in the agricultural sector towards better productivity, resilience and reduced GHG emissions needs to be considered within a larger food systems context (UNEP, 2016).

So far, most research efforts on CSA have been directed at assessing effects of incremental options through empirical and modelling studies (Challinor et al., 2014). The resulting lack of information on the effects of more radical changes is a challenge for decision makers. Visioning and scenario analyses conducted with stakeholders and linked to integrated agricultural models can help fill that knowledge gap (Antle et al., 2018; Meuwissen et al., 2019; and Section 4.2).

3 Gaps and problems in our current understanding of the climate-smart agriculture (CSA) concept and its potential

The number of scientific studies on CSA has expanded rapidly since the concept was launched in 2010. The number of journal papers rose from less than 10 per year in 2012–2014, to over 50 in 2017 and 82 in 2018 (based on papers indexed by Web of Science, accessed 7th of January 2019). Journal papers dealing with both adaptation and mitigation in relation to agriculture, but without necessarily mentioning CSA specifically, are more numerous, with numbers starting to rise quickly from over 20 in 2010 to more than 100 in 2017 and over 140 in 2018. However, from a review of journal articles, Locatelli et al. (2015) concluded that empirical and quantitative information is lacking for many options and contexts. They concluded that CSA assessments are often qualitative and based on generalized statements. Systematic reviews and meta-analyses of effects of specific CSA options on the different CSA pillars are indeed not common. A good exception is the study by Kaye and Quemada (2017) that deals with effects of cover crops on seven mitigation-related mechanisms and adaptation to climate change aspects of drought, extreme rainfall and increased temperatures. For CA, Thierfelder et al. (2017) summarize information for southern Africa. For other CSA options, the quantitative evidence is biased towards one CSA pillar, such as C sequestration for the case of agroforestry (e.g. Mbow et al., 2014; Verchot et al., 2007). On yet other options that claim to be climate-smart (e.g. soil fertility management), empirical information on the effects on the CSA pillars is scattered in papers targeted to a specific pillar or to a specific technology or practice, without systematic gathering of the current knowledge. This knowledge gap complicates decision-making for agricultural practitioners and for policy makers in terms of which CSA options to prioritize in which context, and how to underpin the uptake of the CSA options with adequate policies. Related to this, the CSA concept is often interpreted so broadly that almost all good agricultural practices fit under it (see e.g. FAO,

2013). As such, CSA runs the risk of becoming an empty signifier, amenable to any kind of political intervention. In addition, it is a common problem that many studies refer to the same few sources, where values may have been reported based on rough assumptions. Particularly in relation to the C sequestration potential of agricultural soils, this has led to overly optimistic claims when the uncertainties associated with original values were ignored (Powlson et al., 2011). Related to this, standard IPCC Tier 1 values are often used as if they are true everywhere, while much scientific evidence already indicates that these values vary and may differ from earlier suggestions.

The CSA literature often mentions trade-offs and the three-pillar concept lends itself very well to considering trade-offs between productivity, adaptation and mitigation objectives. Yet, CSA assessments often mention trade-offs in qualitative or general terms without quantification (Bos et al., 2017), and there could be a bias in reporting win-wins and ignoring trade-offs (Locatelli et al., 2015). For example, with respect to mitigation, studies on soil fertility management focus mostly on either carbon sequestration or N losses, whereas options that increase soil C stocks often lead to larger N_2O emissions. With a modelling study, Bos et al. (2017) quantified these trade-offs for specific soil fertility management options in the Netherlands, but were the first to do so for arable land. They also studied the trade-offs in profits associated with reducing emissions and found that with current carbon prices, C credits could not compensate the expected financial losses. Whether a practice leads to a win-win or a trade-off may also depend on the unit of analysis, as illustrated next for the tension between food security and mitigation objectives. As discussed before, meeting future global food requirements will have to rely on increased productivity per area of land, in particular in currently low-yielding areas. This will require increased input use. Hence, whereas the GHG emissions on a product basis may go down, the GHG emissions on a land basis may increase. Contrary to this, organic agriculture and agricultural extensification, with overall lower yields, may lead to lower emissions per hectare, but larger emissions per unit of produce (Carlson et al., 2016; Bos et al., 2007). Globally, with more food being produced it will be very hard to bring down the absolute amounts of GHG emissions (Smith, 2017).

Beyond the three CSA pillars, trade-offs associated with the implementation of CSA options may occur in other environmental domains and in the social and human well-being domains of sustainability. Taking these trade-offs into account in ex-ante impact assessments is important for addressing unintended side effects (Klapwijk et al., 2014). With respect to the environmental domain, a narrow focus on mitigation may lead to so-called pollution swapping, as described previously for the specific case of manure use. With respect to the social domain, side effects of CSA options on the gender balance within households may occur if the implementation leads to shifts in the crop mix or other management practices, resulting in disproportionally more labour or more cash for a specific gender group (Jerneck, 2018). Social equity in communities may be adversely affected if the promotion of certain CSA practices mostly benefit the better-endowed farmers who can afford the investment or are able to implement the change in management. However, the social dimension in the impact of CSA options remains largely understudied (Karlsson et al., 2018).

The potential bias towards win-win and risk of overlooking or underestimating trade-offs may be aggravated when constraints and barriers to adoption are not carefully considered, resulting in an overestimation of the CSA potential and thus an underestimation of the future impacts of climate change. The potential effect of CSA options on mitigation and adaptation is usually derived from experimental or modelling studies. Very often, the CSA options tested also have the potential to increase the productivity in the current climate and system configuration. However, the contrast between the theoretical benefits and the fact that farmers are not currently using the options points to barriers in the current system that prevent adoption. An example from sub-Saharan Africa is the often advocated shift from maize to so-called climate-robust crops such as millet and sorghum. Given the fact that we currently see the exact opposite trend in many countries and that maize yields more than sorghum and millet under virtually all but the very driest climates (Rurinda et al., 2014; Traoré et al., 2014), this seems very unlikely to happen. Taste, ease of processing and markets are further factors for what seems an inexplicable trend if assessed from only a climate perspective. Another example from the Netherlands illustrates the importance of prices and income. A shift to wheat is often advocated because it is more robust in climate extremes than potato and it better stimulates soil organic matter build-up. However, compared to potato, wheat is a lower value crop, which discourages farmers to grow it (Mandryk et al., 2017). It is naive to think that the constraints (or drivers) may simply disappear in the future, and a thorough understanding is needed as a basis for their alleviation (or bending). For this, the well-developed literature on adoption of agricultural technologies and practices (for overview papers, see e.g. Kassie et al., 2015; Knowler and Bradshaw, 2007; Doss, 2006; Feder and Umali, 1993) can shed light on possible constraints and barriers limiting the adoption of CSA options. A common conclusion from these studies is that the adoption potential of a certain option depends on its fit with the farmer's context, which is determined by combined agro-ecological, sociocultural, economic and institutional dimensions at scales varying from the farm to landscape, regional and national level (Descheemaeker et al., 2019; Ojiem et al., 2006). At the farm level, poor access to natural, human and financial capital restricts the incentives and possibilities for farmers to invest in technologies or alter their farm management. In particular, small farm sizes prevent investments in improved technologies or practices to be economically viable (Harris and

Orr, 2014). Farm diversity in resource endowment may thus explain why CSA options could be interesting for some and unfeasible for other farmers in the same community or agro-ecological zone, which is why tailoring of options and providing farmers with baskets of options is advocated for improved adoption (Descheemaeker et al., 2019). Factors at larger spatial and organizational scales beyond the farm, such as community organization or extension services, may influence information flows, knowledge and skills, which have been observed to influence adoption (Kassie et al., 2015). Cultural norms and beliefs are also important determinants of adoption factors such as motivation, perception and attitude towards risk (Meijer et al., 2015; Peterson et al., 2012). Finally, common institutional determinants include market infrastructure and organization, price settings (levels, uncertainty, decision-making, information) and the presence/absence of payment (e.g. for environmental services) and insurance schemes (Mullins et al., 2018; Mandryk et al., 2015; Garbach et al., 2012).

Besides the bias from ignoring adoption constraints, other factors related to modelling capabilities may cause an overestimation of the CSA potential. These include the limitations of models to include factors interacting with climate such as weeds, pests and diseases, future changes in water resource availability and effects of climate extremes and increased variability (Challinor et al., 2014; Schaap et al., 2013). However, also the opposite trend of underestimating the true adaptation potential may occur as crop models can only simulate a limited set of CSA options (e.g. planting dates, cultivars, certain soil fertility management options) and mostly exclude systemic and transformational adaptations by focussing on current system configurations (Mandryk et al., 2012; Reidsma et al., 2010). Additionally, there are still huge uncertainties with respect to the climate predictions themselves, and more work is required to understand the interactions and feedbacks between climate, crop growth and soil biogeochemical processes under elevated CO_2 concentrations, especially in tropical soils and climates. Besides uncertainty in the climate predictions, there are various sources of crop model uncertainty which may cause either over- or underestimates of climate change impacts on crops (e.g. Asseng et al., 2013; Müller et al., 2011).

4 Operationalizing climate-smart agriculture (CSA)

Moving towards CSA requires sound decision and policy making from farm to landscape and from national to regional and global levels. CSA prioritization tools (e.g. Andrieu et al., 2017; Mwongera et al., 2017) are designed to support the decision-making process. Yet, to be effective, these tools need quantitative information on the likely effects of a range of options on the different CSA pillars; and this information needs to be specific for the context where the decisions are made. Besides that, quantitative information is also needed on the associated effects in other sustainability dimensions and on the potential constraints to adoption by diverse farmers. Furthermore, as agricultural systems are constantly changing, information is needed on likely effects in future system settings.

This section mainly focuses on two case studies from the Netherlands and Zimbabwe, where researchers in collaboration with stakeholders aimed to fill the above-mentioned knowledge gaps. The cases represent contrasting farming systems and environments in terms of the likely impact of climate change and the institutional setting determining the adaptive capacity of the agricultural sector. Despite the contrasting setting, a similar modelling approach was used, integrating climate, farm component (crop and/or livestock) and whole farm models. The modelling exercises were embedded in a broader farming systems analysis and stakeholder engagements. The case studies illustrate how farming systems analysis can assist in operationalizing CSA tools by generating contextualized and quantitative evidence on the effects of options on the CSA pillars and by integrating stakeholders' views to carve out climate-smart pathways to a sustainable future.

4.1 Climate-smart agriculture (CSA) in Flevoland, The Netherlands

The Netherlands exhibits a mild maritime climate with mean temperatures varying between 3°C in January to 18°C in July. Rainfall averages 832 mm per year. Warming is expected more so in winter than in summer, but summer variability is expected to increase. Precipitation has increased by 14% annually since the 1950s; all seasons have become wetter except summer. Winter rain will continue to increase at similar rates, summers will likely become drier.

The province Flevoland is an important agricultural area in the Netherlands, as its marine clay soils are very fertile (Janssen, 2017). Arable farming is the dominant land use (75% of the area) and is one of the major drivers of the economy (Schaap et al., 2013). Most of the farms are characterized by a substantial share of seed and ware potatoes in the rotation that are both high-value crops. Other high-value crops are seed onions and sugar beet. Winter wheat is a relatively low-value crop, used mainly for feed, and it is merely grown to improve soil quality and to keep disease pressures down.

Three related integrated assessments were performed for arable farming in Flevoland, to assess impacts of climate change towards 2050 (Mandryk et al., 2017; Reidsma et al., 2015; Wolf et al., 2015). These studies all combined a crop model, bio-economic farm model and participatory approaches. In addition, Bos et al. (2017) specifically focussed on mitigation in relation to soil fertility management.

Reidsma et al. (2015) illustrated that (1) crop models cannot account for all relevant climate change impacts and adaptation options, and (2) changes in technology, policy and prices have had and are likely to have larger impacts on farms than climate change. While crop modelling indicated positive impacts of climate change on yields of major crops in 2050, a semi-quantitative and participatory method assessing impacts of extreme events (Schaap et al., 2013) showed that there are nevertheless several climate risks. A range of adaptation measures were, however, available to reduce possible negative effects at crop level.

Potential yield improvement due to climate change simulated by WOFOST ranged between 2% (ware potato) and 29% (sugar beet) in the different scenarios (Reidsma et al., 2015). For ware potato, the main crop, a change in cultivar (higher temperature sum and longer life span) and an earlier sowing date, further improved the yield increase to 11%. In the driest scenario, water-limited yield reduced from 77% to 67% of the potential, so more irrigation was also an adaptation option. Five main climate risks were identified, which had the largest economic losses. For heat waves, causing second-growth in seed and ware potato, drip irrigation was identified as the best adaptation measure in most scenarios. To reduce the impact of warm winters inducing early sprouting in seed and ware potato, air-conditioning seemed the best adaptation measure, and to protect seed onions against fungi infections during warm and wet conditions, chemical protection was identified as the best measure. However, a portfolio of adaptation measures was available, and costs and benefits depended on the scenario (Schaap et al., 2013).

In addition to crop-level measures, at farm level, farmers can change cropping patterns, and adjust inputs and outputs. Optimal adaptation depended on objectives and constraints of farmers. If farmers can cope with extreme events, increasing the share of high-value crops such as potato improved their gross margins (Mandryk et al., 2017; Wolf et al., 2015). However, within the current rotation constraints this was only possible if arable farms cooperated with dairy farms, and rented land to rotate their high-value crops with grass or maize. As climate change had less impact on wheat, increasing the share of wheat was a more robust option, wheat prices allowing. Wheat also helped to improve the organic matter balance, and was therefore more climate-smart from a mitigation perspective (Bos et al., 2017; Mandryk et al., 2017).

In terms of mitigation, doubling the winter wheat area combined with the cultivation of cover crops to increase soil organic carbon accumulation resulted in a net GHG emission benefit, but it was associated with a financial trade-off (Bos et al., 2017). With regard to the use of organic fertilizers, only the use of compost resulted in net mitigation benefits, with larger relative soil organic carbon increases compared to N_2O emissions. When using cattle or pig slurry, the increase in N_2O emissions was higher than the emission offset through soil organic carbon accumulation. Wolf et al. (2015) projected increases in

N_2O emissions under future adaptation scenarios, but these may be reduced by increasing nitrogen-use efficiency, as nitrogen application can be reduced while maintaining yields (Silva et al., 2017).

Finally, it should be noted that farm structural change will influence productivity, adaptation and mitigation. Larger farms have more potential to adopt adaptation and mitigation options, but whether they do so depends on their objectives. The Dutch government envisions a more circular agriculture, which should lead to more cooperation among farms, increased resource-use efficiency at the regional level, and an agricultural sector, which is more climate-smart.

4.2 Transforming mixed cereal-livestock farming systems in semi-arid Zimbabwe

Semi-arid Zimbabwe is characterized by variable rainfall and poor soil fertility. Mixed crop-livestock systems predominate with crop residues used as dry-season feed, and livestock providing draft power and manure to crop production (Homann-Kee Tui et al., 2015). Current crop yields are low, with maize yielding on average 0.7 t ha^{-1} and other crops such as sorghum, millet and groundnut even less. Also, livestock productivity is poor and constrained by high mortality rates. With more than 76% of the rural population below the poverty line (ZimVAC, 2013) and food self-sufficiency achieved for only 3–10 months per year, rural households are extremely vulnerable to the adverse effects of climate change.

To assess the likely impacts of climate change and design more resilient systems, the Agricultural Model Intercomparison and Improvement Project (AgMIP) Regional Integrated Assessment (RIA) approach (Antle et al., 2018) was applied, embedded in an iterative process of stakeholder engagements. The RIA links climate, crop, livestock and economic data and models for impact assessment in heterogeneous farm populations (AgMIP, 2015). During stakeholder consultation workshops, scenarios, called 'Representative Agricultural Pathways' (RAPs; Valdivia et al., 2015) describing future socio-economic and biophysical settings, and adaptation options were conceived and consecutively refined based on modelling results. Mitigation options and effects were not considered.

Results from 29 climate models, run for the mid-century period for two representative concentration pathways, showed consistent increases in temperature, and more variable rainfall projections. Here, we highlight results contrasting a relatively 'hot-dry' with a 'hot-wet' future. A total of 160 households were surveyed and stratified into three types (extremely poor, poor and non-poor) based on resource endowments (Homann-Kee Tui et al., 2013). The modelling framework was run with specific farm structure and management settings per household. The crop growth model APSIM (Holzworth et al., 2014) and the livestock model LIVSIM (Rufino et al., 2009) simulated field- and herd-level productivity and fed into the TOA-MD model, which estimated economic performance at the farm level (Antle et al., 2014).

In the current agricultural systems, crop sensitivity to climate change was relatively small, due mainly to poor soil fertility and low fertilizer application rates, causing severe nutrient stress. The observed small decline was attributed to increased temperature accelerating phenological development, and exacerbated by drought stress in the hot-dry climate. As non-poor farms cultivated maize on better soils and with more fertilizer input, their maize was more sensitive to yield loss compared with the less-endowed farms. Livestock productivity was affected through altered production of crop residue and rangeland biomass, influencing feed intake, ranging from a positive to a negative effect in the hot-wet and the hot-dry climate, respectively. Non-poor farms, typically with larger stocking density, were more sensitive to feed gaps, and more strongly impacted by climate change than poor farms (Descheemaeker et al., 2018). At the farm level, the simulated change in farm net returns depended on the climate and the farm type. Ninety-five per cent of the extremely poor households was currently below the poverty line. Their low-input production system was not sensitive to climate change, and only 40–50% of these farmers would lose (a little). In the group of best-endowed farmers, who cultivate better soils and keep more cattle, about 75% faced reduced returns, which increased the poverty rate from 60% to 70% of the households.

During a first round of stakeholder meetings, these expected impacts were discussed and a package of adaptation options was proposed for testing. This consisted of a drought-tolerant maize variety, rotation with the forage legume mucuna (*Mucuna pruriens* var. *utilis*) and fertilizer application on maize (Masikati et al., 2015). The effects of the adaptation package were evaluated in the context of a likely future system of the mid-century, conceived with stakeholders in a first RAP. In this future system, the performance of the crop and livestock components improved due to increased nutrient input for maize and high-quality feed for the cattle. However, at the farm and community levels, improvements were disappointing with no reduction in poverty rates for the extremely poor and poor. Only the better-endowed households considerably increased their farm net returns and the poverty rate in this group dropped from 60% to 40%.

Researchers and stakeholders discussed these results in a second round of workshops and decided to design RAPs and adaptation packages that would transform the agricultural system and generate truly positive impacts on farmer livelihoods. Two contrasting RAPs were developed: a 'sustainability' pathway (RAP4) and a 'fast-economic growth' pathway (RAP5). In RAP4, improved access to technologies, markets and services enabled intensification, diversification and crop-livestock integration. Farms would increase in size by up to 40%.

RAP5 supported the better-off farms, who would increase their size by 80% and rely more heavily on external inputs. The extremely poor farms would shrink and rely more on off-farm income. Within both RAPs, the adaptation options consisted of heat-tolerant cereal varieties able to retain life cycle, and drought-tolerant legume varieties.

Crop yields in the future agricultural systems of RAP4 and RAP5 were better but also more sensitive to climate change as compared to current agricultural systems, because increased fertilizer use alleviated nutrient limitations. Contrastingly, livestock production was less sensitive, because feeding concentrates mitigated the feed gaps. At the farm level, the net returns were much higher and poverty rates much lower in the future than today, with RAP4 resulting in more profitable and less vulnerable farms compared to RAP5. As in the current system, the least endowed farm group was least sensitive to climate change. This was explained by their stronger reliance on the more robust groundnut in RAP4 and on off-farm income in RAP5. Better-adapted crop varieties effectively improved yield under climate change, particularly in RAP4 where manure use enabled more stable production compared to RAP5 where crop production relied solely on inorganic fertilizers.

However, even with the drastically better agricultural productivity in the RAPs and the reduced climate vulnerability with the adaptations, poverty rates remained high, at about 40% overall, and particularly in the less-endowed groups. This highlights the limited capability of agricultural development to lift people out of poverty and the farm size as a binding constraint for generating a decent income from agriculture. Hence, job opportunities outside agriculture will be needed to provide a living for the poorest groups and allow farms to grow bigger in area and/or economic size. By upscaling findings from field and herd to farm level, seemingly positive effects due to incremental adaptations evaporated, with virtually no impact on poverty rates. With this insight, the approach of co-designing transformative change by relaxing current binding constraints (e.g. farm size, input access) showed potential ways forward and underscored the required institutional and policy change to enable sustainable development. The role of local stakeholders was essential in contextualizing the (modelling) work and incorporating local realities in future pathways.

4.3 Lessons learnt from farming systems analysis in the case studies

Despite the large contrasts, the case studies illustrate the importance of (1) stakeholder involvement, (2) integrated, multi-criteria and cross-scale assessment and (3) consideration of farm diversity for a holistic understanding of the CSA potential. The participatory nature of the studies allowed to reveal risks and options that were not captured by the model and to co-design transformative pathways for future development. By taking on board multiple criteria, beyond the CSA pillars, trade-offs were revealed that need to be addressed for improving the adoption potential of options. By scaling from field and herd level to farm and larger levels, important limits and constraints, such as farm size, came to the fore. Likewise, this revealed why choices that seemed obvious at the field level, worked out differently at the farm level. Interestingly, in both studies, larger farms showed a larger potential to adopt and benefit from options. This illustrates the importance of taking farm diversity into account when trying to understand the true impact and potential of CSA options.

Global studies on climate change impact and CSA options (e.g. van Meij et al., 2018; Rosenzweig et al., 2014) need to be complemented with contextualized information at smaller scales to account for locally varying factors that determine crop productivity (e.g. soil fertility, crop management) and limit adaptation and mitigation (Challinor et al., 2014). The two case studies in this chapter show the strength of such contextualized research for unravelling the complexity in expected impacts of climate change and for illustrating what matters and what is feasible. In the Netherlands' example of highly productive agriculture, the pillars of adaptation and mitigation are of high priority. Nevertheless, the economic consequences of CSA options determine their uptake by farmers. In this context, input levels and related GHG emissions can be cut without a penalty on crop productivity. In contrast, in smallholder systems, represented here by the Zimbabwean case, only food security and adaptation are on the minds of farmers. One can argue whether it is even ethical to promote climate change mitigation (Doelman et al., 2018; Page, 2008) in smallholder agriculture, where amounts of input need to rise from the currently low levels in order to meet future food requirements.

5 Conclusion

The agricultural sector is both a culprit and a victim of climate change. Hence, there is a strong need for adapting to climate change and mitigating agriculture's contribution to GHG emissions. At the same time, agriculture will have to meet demands for food that are expected to sharply increase in the coming decades. The CSA concept offers scope to analyse this three-pronged challenge of food security, resilience and mitigation in an integrated way.

Numerous cropping options are claimed to be climate-smart, with agroforestry and the incorporation of legume crops as powerful examples. Yet, many claims of triple-wins do not withstand detailed scrutiny, as benefits for one CSA pillar often go hand in hand with drawbacks for another pillar or compromises in terms of social or economic sustainability. In addition, constraints to the adoption of CSA options limit the adaptation and mitigation capacity of the current system. Policies to underpin pathways towards CSA must take account of the possible trade-offs and synergies between the CSA pillars. This chapter shows that these interactions depend on the context and the scale of the analysis. To understand this complexity and find holistic solutions, integrated, cross-scale analyses that look beyond the binding constraints of the current system are necessary. We therefore argue that the quest for climate-smart cropping options is best conducted within a farming and food systems perspective.

The '4 per 1000' initiative illustrates the risk of hyping a strategy without a sound scientific basis. Unrealistic expectations of the capacity of the agricultural sector to curb emissions may reduce the sense of urgency of policy makers to invest and promote effective solutions that may be more costly. That same risk must be avoided by the science community when it comes to promoting CSA. There is an urgent need to strengthen the qualitative and generalized statements on CSA with quantitative and contextualized information on the effects of options and their feasibility for farmers. Besides decision and policy makers looking for evidence-based information, also science would benefit from systematic reviews and meta-analyses allowing the identification of knowledge gaps and new areas of research.

Finally, our contrasting case studies illustrate that context matters. Developing countries require an intensification of agricultural production to close yield gaps and meet sharply rising food demands. In this context, there are less possibilities to reduce GHG emissions, and it makes perfect sense to target efforts to food security and resilience. In contrast, for developed countries with intensive agriculture, it is not a priority to increase production, but to reduce emissions, while also adapting to climate change. Climate-smart cropping options tailored to the context can help achieving these targets.

6 Acknowledgements

The studies on climate change impacts in the Netherlands were supported by the programme 'Climate changes spatial planning' funded by the Dutch government, the EU Network of Excellence 'LIAISE' and the strategic programme of WUR 'Scaling and governance'. The study in Zimbabwe was funded by the UK Department for International Development (UKAID) as part of the AgMIP Regional Integrated Assessment project in southern Africa. Ken Giller is grateful to NWO-WOTRO for funding though the Netherlands-CGIAR Scientific Expert Programme.

7 References

Agricultural Model Inter-comparison and Improvement Project (AgMIP) (2015). *Guide for Regional Integrated Assessments: Handbook of Methods and Procedures*. Version 6.0. http://www.agmip.org/wp-content/uploads/2015/09/AgMIP-RIA-Protocols-V6 sm.pdf (accessed 30 October 2016).

Amundson, R., and Biardeau, L. (2018). Opinion: soil carbon sequestration is an elusive climate mitigation tool. *Proceedings of the National Academy of Sciences* 115(46), 11652-6.

Andrieu, N., Sogoba, B., Zougmore, R., Howland, F., Samaké, O., Bonilla-Findji, O., Lizarazo, M., Nowak, A., Dembele, C., and Corner-Dolloff, C. (2017). Prioritizing investments for climate-smart agriculture: lessons learned from Mali. *Agricultural Systems* 154(C), 13-24.

Antle, J. M., Stoorvogel, J., and Valdivia, R. (2014). New parsimonious simulation methods and tools to assess future food and environmental security of farm populations. *Philosophical Transactions of the Royal Society B* 369, 20120280.

Antle, J. M., Homann-Kee Tui, S., Descheemaeker, K., Masikati, P., and Valdivia, R. O. (2018). Using AgMIP regional integrated assessment methods to evaluate vulnerability, resilience and adaptive capacity for climate smart agricultural systems. In: *Climate Smart Agriculture: Building Resilience to Climate Change*. Natural Resource Management and Policy, 52. Springer, Rome, pp. 307-33.

Asseng, S., Ewert, F., Rosenzweig, C., Jones, J. W., Hatfield, J. L., Ruane, A. C., Boote, K. J., Thorburn, P. J., Rötter, R. P., Cammarano, D., Brisson, N., Basso, B., Martre, P., Aggarwal, P. K., Angulo, C., Bertuzzi, P., Biernath, C., Challinor, A. J., Doltra, J., Gayler, S., Goldberg, R., Grant, R., Heng, L., Hooker, J., Hunt, L. A., Ingwersen, J., Izaurralde, R. C., Kersebaum, K. C., Müller, C., Naresh Kumar, S., Nendel, C., O'Leary, G., Olesen, J. E., Osborne, T. M., Palosuo, T., Priesack, E., Ripoche, D., Semenov, M. A., Shcherbak, I., Steduto, P., Stöckle, C., Stratonovitch, P., Streck, T., Supit, I., Tao, F., Travasso, M., Waha, K., Wallach, D., White, J. W., Williams, J. E., and Wolf, J. (2013). Uncertainty in simulating wheat yields under climate change. *Nature Climate Change* 3, 827-32.

Autret, B., Mary, B., Chenu, C., Balabane, M., Girardin, C., Bertrand, M., Grandeau, G., and Beaudoin, N. (2016). Alternative arable cropping systems: a key to increase soil organic carbon storage? Results from a 16 year field experiment. *Agriculture, Ecosystems and Environment* 232, 150-64.

Baveye, P. C., Berthelin, J., Tessier, D., and Lemaire, G. (2018). The '4 per 1000' initiative: a credibility issue for the soil science community? *Geoderma* 309, 118-23.

Bos, J., de Haan, J., and Sukkel, W. (2007). *Energieverbruik, broeikasgasemissies en koolstofopslag: de biologische en gangbare landbouw vergeleken*. Wageningen.

Bos, J. F. F. P., ten Berge, H. F. M., Verhagen, J., and Van Ittersum, M. K. (2017). Trade-offs in soil fertility management on arable farms. *Agricultural Systems* 157, 292-302.

Brandt, P., Kvakić, M., Butterbach-Bahl, K., and Rufino, M. C. (2017). How to target climate-smart agriculture? Concept and application of the consensus-driven decision support framework 'targetCSA'. *Agricultural Systems* 151, 234-45.

Carlson, K. M., Gerber, J. S., Mueller, N. D., Herrero, M., Macdonald, G. K., Brauman, K. A., Havlik, P., O'Connell, C. S., Johnson, J. A., Saatchi, S., and West, P. C. (2016). Greenhouse gas emissions intensity of global croplands. *Nature Climate Change* 7(1), 63.

Carter, S., Arts, B., Giller, K. E., Soto Golcher, C., Kok, K., de Koning, J., van Noordwijk, M., Reidsma, P., Rufino, M. C., Salvini, G., Verchot, L., Wollenberg, E., and Herold, M. (2018). Climate-smart land use requires local solutions, transdisciplinary research, policy coherence and transparency. *Carbon Management* 9, 291-301.

Ceddia, M. G., Sedlacek, S., Bardsley, N. O., and Gomez-y-Paloma, S. (2013). Sustainable agricultural intensification or Jevons paradox? The role of public governance in tropical South America. *Global Environmental Change* 23, 1052-63.

Chadwick, D., Sommer, S., Thorman, R., Fangueiro, D., Cardenas, L., Amon, B., and Misselbrook, T. (2011). Manure management: implications for greenhouse gas emissions. *Animal Feed Science and Technology* 166-167, 514-31.

Challinor, A. J., Watson, J., Lobell, D. B., Howden, S. M., Smith, D. R., and Chhetri, N. (2014). A meta-analysis of crop yield under climate change and adaptation. *Nature Climate Change* 4(4), 287-91.

Cole, C., Duxbury, J., Freney, J., Heinemeyer, O., Minami, K., Mosier, A., Paustian, K., Rosenberg, N., Sampson, N., Sauerbeck, D., and Zhao, Q. (1997). Global estimates of potential mitigation of greenhouse gas emissions by agriculture. *Nutrient Cycling in Agroecosystems* 49, 221-28.

Conijn, J. G., and van Dijk, W. (2018). Closing the life cycle of phosphorus in an urban food system: the case Almere (NL). Wageningen Plant Research, Business Unit Applied Arable and Vegetable Research, Lelystad.

Corbeels, M., de Graaff, J., Ndah, T. H., Penot, E., Baudron, F., Naudin, K., Andrieu, N., Chirat, G., Schuler, J., Nyagumbo, I., Rusinamhodzi, L., Traore, K., Mzoba, H. D., and Adolwa, I. S. (2014). Understanding the impact and adoption of conservation agriculture in Africa: a multi-scale analysis. *Agriculture, Ecosystems and Environment* 187, 155-70.

Corbeels, M., Cardinael, R., Naudin, K., Guibert, H., and Torquebiau, E. (2018). The 4 per 1000 goal and soil carbon storage under agroforestry and conservation agriculture systems in sub-Saharan Africa. *Soil and Tillage* 188, 16-26.

Corbera, E., and Schroeder, H. (2011). Governing and implementing REDD+. *Environmental Science and Policy* 14, 89-99.

Corsi, S., Friedrich, T., Kassam, A., Pisante, M., and de Moraes Sa, J. (2012). *Soil Organic Carbon Accumulation and Greenhouse Gas Emission Reductions from Conservation Agriculture: a Literature Review*. FAO, Rome, pp. 103.

Crowther, T. W., Todd-Brown, K. E. O., Rowe, C. W., Wieder, W. R., Carey, J. C., Machmuller, M. B., Snoek, B. L., Fang, S., Zhou, G., Allison, S. D., Blair, J. M., Bridgham, S. D., Burton, A. J., Carrillo, Y., Reich, P. B., Clark, J. S., Classen, A. T., Dijkstra, F. A., Elberling, B., Emmett, B. A., Estiarte, M., Frey, S. D., Guo, J., Harte, J., Jiang, L., Johnson, B. R., Kröel-Dulay, G., Larsen, K. S., Laudon, H., Lavallee, J. M., Luo, Y., Lupascu, M., Ma, L. N., Marhan, S., Michelsen, A., Mohan, J., Niu, S., Pendall, E., Peñuelas, J., Pfeifer-Meister, L., Poll, C., Reinsch, S., Reynolds, L. L., Schmidt, I. K., Sistla, S., Sokol, N. W., Templer, P. H., Treseder, K. K., Welker, J. M., and Bradford, M. A. (2016). Quantifying global soil carbon losses in response to warming. *Nature* 540, 104-8.

Cuéllar, A. D., and Webber, M. E. (2008). Cow power: the energy and emissions benefits of converting manure to biogas. *Environmental Research Letters* 3(3).

Debaeke, P., Pellerin, S., and Scopel, E. (2017). Climate-smart cropping systems for temperate and tropical agriculture: mitigation, adaptation and trade-offs. *Cahiers Agricultures* 26(3), 34002-312.

Descheemaeker, K., Oosting, S. J., Homann-Kee-Tui, S., Masikati, P., Falconnier, G. N., and Giller, K. E. (2016). Climate change adaptation and mitigation in smallholder crop-livestock systems in sub-Saharan Africa: a call for integrated impact assessments. *Regional Environmental Change* 16, 2331-43.

Descheemaeker, K., Ronner, E., Ollenburger, M., Franke, A. C., Klapwijk, C. J., Falconnier, G. N., Wichern, J., and Giller, K. E. (2019). Which options fit best? Operationalizing the socio-ecological niche concept. *Experimental Agriculture*, 55 (S1), 169–190. https://doi.org/10.1017/S001447971600048X.

Descheemaeker, K., Zijlstra, M., Masikati, P., Crespo, O., and Homann-Kee Tui, S. (2018). Effects of climate change and adaptation on the livestock component of mixed farming systems: a modelling study from semi-arid Zimbabwe. *Agricultural Systems* 159, 282–95.

Doelman, J. C., Stehfest, E., Tabeau, A., van Meijl, H., Lassaletta, L., Gernaat, D. E. H. J., Hermans, K., Harmsen, M., Daioglou, V., Biemans, H., van der Sluis, S., and van Vuuren, D. P. (2018). Exploring SSP land-use dynamics using the IMAGE model: regional and gridded scenarios of land-use change and land-based climate change mitigation. *Global Environmental Change* 48, 119–35.

Dordas, C. (2008). Role of nutrients in controlling plant diseases in sustainable agriculture. A review. *Agronomy for Sustainable Development* 28, 33–46.

Doss, C. R. (2006). Analyzing technology adoption using microstudies: limitations, challenges, and opportunities for improvement. *Agricultural Economics* 34, 207–19.

FAO (2010). *Climate-Smart Agriculture: Policies, Practices and Financing for Food Security, Adaptation and Mitigation*. Food and Agriculture Organization of the United Nations, Rome. http://www.fao.org/docrep/013/i1881e/i1881e00.htm.

FAO (2013). *Climate-smart Agriculture. Sourcebook*. FAO, Rome, 570p. http://www.fao.org/docrep/018/i3325e/i3325e.pdf (accessed 29 October 2019).

FAO (2015). Conservation agriculture. http://www.fao.org/conservation-agriculture/en/ (accessed 10 February 2019).

Feder, G., and Umali, D. L. (1993). The adoption of agricultural innovations: a review. *Technological Forecasting and Social Change* 43, 215–39.

Félix, G. F., Clermont-Dauphin, C., Hien, E., Groot, J. C. J., Penche, A., Barthès, B. G., Manlay, R. J., Tittonell, P., Cournac, L. (2018). Ramial wood amendments (*Piliostigma reticulatum*) mitigate degradation of tropical soils but do not replenish nutrient exports. *Land Degradation and Development* 29, 2694–706.

Franzluebbers, J. A. (2002). Water infiltration and soil structure related to organic matter and its stratification with depth. *Soil and Tillage Research* 66, 197–205.

Garbach, K., Lubell, M., and DeClerck, F. A. J. (2012). Payment for ecosystem services: the roles of positive incentives and information sharing in stimulating adoption of silvopastoral conservation practices. *Agriculture, Ecosystems and Environment* 156, 27–36.

Garnett, T., Appleby, M., Balmford, A., Bateman, I., Benton, T., Bloomer, P., Burlingame, B., Dawkins, M., Dolan, L., Fraser, D., Herrero, M., Hoffmann, I., Smith, P., Thornton, P., Toulmin, C., Vermeulen, S., and Godfray, H. (2013). Sustainable intensification in agriculture: premises and policies. *Science* 341(6141), 33–4.

Giller, K. E., Witter, E., Corbeels, M., and Tittonell, P. (2009). Conservation agriculture and smallholder farming in Africa: the heretics' view. *Field Crops Research* 114, 23–34.

Giller, K. E., Andersson, J. A., Corbeels, M., Kirkegaard, J., Mortensen, D., Erenstein, O., and Vanlauwe, B. (2015). Beyond conservation agriculture. *Frontiers in Plant Science* 6, Article 870.

Grafton, R. Q., Daugbjerg, C., and Qureshi, M. E. (2015). Towards food security by 2050. *Food Security* 7(2), 179–83.

Griscom, B. W., Adams, J., Ellis, P. W., Houghton, R. A., Lomax, G., Miteva, D. A., Schlesinger, W. H., Shoch, D., Siikamäki, J. V., Smith, P., Woodbury, P., Zganjar, C., Blackman, A.,

Campari, J., Conant, R. T., Delgado, C., Elias, P., Gopalakrishna, T., Hamsik, M. R., Herrero, M., Kiesecker, J., Landis, E., Laestadius, L., Leavitt, S. M., Minnemeyer, S., Polasky, S., Potapov, P., Putz, F. E., Sanderman, J., Silvius, M., Wollenberg, E., and Fargione, J. (2017). Natural climate solutions. *Proceedings of the National Academy of Sciences of the United States of America* 114(44), 11645-50.

Harris, D., and Orr, A. (2014). Is rainfed agriculture really a pathway from poverty? *Agricultural Systems* 123(C), 84-96.

Hertel, T. W., and Lobell, D. B. (2014). Agricultural adaptation to climate change in rich and poor countries: current modeling practice and potential for empirical contributions. *Energy Economics* 46(C), 562-75.

Hijbeek, R., van Ittersum, M., ten Berge, H. Gort, G., Spiegel, H., and Whitmore, A. P. (2017). Do organic inputs matter - a meta-analysis of additional yield effects for arable crops in Europe. *Plant Soil* 411, 293.

Holzworth, D. P., Huth, N. I., deVoil, P. G., Zurcher, E. J., Herrmann, N. I., McLean, G., Chenu, K., van Oosterom, E. J., Snow, V., Murphy, C., Moore, A. D., Brown, H., Whish, J. P. M., Verrall, S., Fainges, J., Bell, L. W., Peake, A. S., Poulton, P. L., Hochman, Z., Thorburn, P. J., Gaydon, D. S., Dalgliesh, N. P., Rodriguez, D., Cox, H., Chapman, S., Doherty, A., Teixeira, E., Sharp, J., Cichota, R., Vogeler, I., Li, F. Y., Wang, E., Hammer, G. L., Robertson, M. J., Dimes, J. P., Whitbread, A. M., Hunt, J., van Rees, H., McClelland, T., Carberry, P. S., Hargreaves, J. N. G., MacLeod, N., McDonald, C., Harsdorf, J., Wedgwood, S., and Keating, B. A. (2014). APSIM - evolution towards a new generation of agricultural systems simulation. *Environmental Modelling and Software* 62, 327-50.

Homann-Kee Tui, S., Bandason, E., Maute, F., Nkomboni, D., Mpofu, N., Tanganyika, J., van Rooyen, A., Gondwe, T., Dias, P., Ncube, S., Moyo, S., Hendricks, S., and Nisrane, F. (2013). Optimizing livelihood and environmental benefits from crop residues in smallholder crop-livestock systems in Southern Africa. ICRISAT, Socioeconomics Discussion Paper Series, vol. 11. ICRISAT, Patancheru.

Homann-Kee Tui, S., Valbuena, D., Masikati, P., Descheemaeker, K., Nyamangara, J., Claessens, L., Erenstein, O., van Rooyen, A., and Nkomboni, D. (2015). Economic trade-offs of biomass use in crop-livestock systems: exploring more sustainable options in semi-arid Zimbabwe. *Agricultural Systems* 134, 48-60.

Janssen, B. H. (2017). Crop yields and NPK use efficiency of a long-term experiment on a former sea bottom in the Netherlands. In *Plant Production Systems*. Wageningen University. doi: 10.18174/409554.

Jerneck, A. (2018). What about gender in climate change? Twelve feminist lessons from development. *Sustainability* 10, 627.

Johnston, A. E., Poulton, P. R., and Coleman, K. (2009). Soil organic matter: its importance in sustainable agriculture and carbon dioxide fluxes. *Advances in Agronomy* 101, 1-57.

Jones, P. G., and Thornton, P. K. (2009). Croppers to livestock keepers: livelihood transitions to 2050 in Africa due to climate change. *Environmental Science and Policy* 12, 427-37.

Karlsson, L., Naess, L. O., Nightingale, A., and Thompson, J. (2018). 'Triple wins' or 'triple faults'? Analysing the equity implications of policy discourses on climate-smart agriculture (CSA). *The Journal of Peasant Studies* 45(1), 150-74.

Kassam, A., Friedrich, T., and Derpsch, R. (2018). Global spread of conservation agriculture. *International Journal of Environmental Studies* 76, 29-51.

Kassie, M., Teklewold, H., Jaleta, M., Marenya, P., and Erenstein, O. (2015). Understanding the adoption of a portfolio of sustainable intensification practices in eastern and southern Africa. *Land Use Policy* 42, 400–11.

Kaye, J. P., and Quemada, M. (2017). Using cover crops to mitigate and adapt to climate change. A review. *Agronomy for Sustainable Development* 37, 4.

Keating, B. A., Carberry, P. S., Bindraban, P. S., Asseng, S., Meinke, H., and Dixon, J. (2010). Eco-efficient agriculture: concepts, challenges, and opportunities. *Crop Science* 50(Suppl. 1), S109–19.

Kerr, R. (2012). Lessons from the old Green Revolution for the new: social, environmental and nutritional issues for agricultural change in Africa. *Progress in Development Studies* 12(2–3), 213–29.

Kirkby, C. A., Richardson, A. E., Wade, L. J., Batten, G. D., Blanchard, C., and Kirkegaard, J. A. (2013). Carbon-nutrient stoichiometry to increase soil carbon sequestration. *Soil Biology and Biochemistry* 60, 77–86.

Kirkegaard, J. A., Angus, J., Gardner, P. A., and Müller, W. (1994). Reduced growth and yield of wheat with Conservation Cropping. I. Field studies in the first year of the cropping phase. *Australian Journal of Agricultural Research* 45, 511–28.

Kirkegaard, J. A., Conyers, M. K., Hunt, J. R., Kirkby, C. A., Watt, M., and Rebetzke, G. J. (2014). Sense and nonsense in conservation agriculture: principles, pragmatism and productivity in Australian mixed farming systems. *Agriculture, Ecosystems and Environment* 187, 133–45.

Klapwijk, C. J., van Wijk, M. T., Rosenstock, T. S., van Asten, P., Thornton, P. K., and Giller, K. E. (2014). Analysis of trade-offs in agricultural systems: current status and way forward. *Current Opinion in Environmental Sustainability* 6, 110–15.

Klein, R. J. T., Midgley, G. F., Preston, B. L., Alam, M., Berkhout, F. G. H., Dow, K., and Shaw, M. R. 2014. Adaptation opportunities, constraints, and limits. In: Field, C. B., Barros, V. R., Dokken, D. J., Mach, K. J., Mastrandrea, M. D., Bilir, T. E., Chatterjee, M., Ebi, K. L., Estrada, Y. O., Genova, R. C., Girma, B., Kissel, E. S., Levy, A. N., MacCracken, S., Mastrandrea, P. R. and White, L. L. (Eds), *Climate Change 2014: Impacts, Adaptation, and Vulnerability. Part A: Global and Sectoral Aspects. Contribution of Working Group II to the Fifth Assessment Report of the Intergovernmental Panel on Climate Change*. Cambridge University Press, Cambridge, United Kingdom and New York, NY, USA, pp. 899–943.

Klocke, N., Currie, R., and Aiken, R. (2009). Soil water evaporation and crop residues. *Transactions of the ASABE* 52(1), 103–10.

Knowler, D., and Bradshaw, B. (2007). Farmers' adoption of conservation agriculture: a review and synthesis of recent research. *Food Policy* 32, 25–48.

Lam, S. K., Suter, H., Mosier, A. R., and Chen, D. (2017). Using nitrification inhibitors to mitigate agricultural N_2O emission: a double-edged sword? *Global Change Biology* 23, 485–9.

Lee, J. (2017). Farmer participation in a climate-smart future: evidence from the Kenya Agricultural Carbon Project. *Land Use Policy* 68, 72–9.

Lipper, L., Thornton, P., Campbell, B. M., Baedeker, T., Braimoh, A., Bwalya, M., Caron, P., Cattaneo, A., Garrity, D., Henry, K., Hottle, R., Jackson, L., Jarvis, A., Kossam, F., Mann, W., McCarthy, N., Meybeck, A., Neufeldt, H., Remington, T., Sen, P. T., Sessa, R., Shula, R., Tibu, A., and Torquebiau, E. F. (2014). Climate-smart agriculture for food security. *Nature Climate Change* 4(12), 1068–72.

Locatelli, B., Pavageau, C., Pramova, E., and Di Gregorio, M. (2015). Integrating climate change mitigation and adaptation in agriculture and forestry: opportunities and trade-offs. *Wiley Interdisciplinary Reviews: Climate Change* 6(6), 585–98.
Long, T. B., Blok, V., and Coninx, I. (2016). Barriers to the adoption and diffusion of technological innovations for climate-smart agriculture in Europe: evidence from the Netherlands, France, Switzerland and Italy. *Journal of Cleaner Production* 112, 9–21.
Mandemaker, M., Bakker, M., and Stoorvogel, J. (2011). The role of governance in agricultural expansion and intensification: a global study of arable agriculture. *Ecology and Society* 16(2), 8.
Mandryk, M., Reidsma, P., and van Ittersum, M. K. (2012). Scenarios of long-term farm structural change for application in climate change impact assessment. *Landscape Ecology* 27(4), 509–27.
Mandryk, M., Reidsma, P., Kartikasari, K., van Ittersum, M. K., and Arts, B. (2015). Institutional constraints for adaptive capacity to climate change in Flevoland's agriculture. *Environmental Science and Policy* 48, 147–62.
Mandryk, M., Reidsma, P., and van Ittersum, M. K. (2017). Crop and farm level adaptation under future climate challenges: an exploratory study for Flevoland, the Netherlands. *Agricultural Systems* 152, 154–64.
Masikati, P., Homann-KeeTui, S., Descheemaeker, K., Crespo, O., Walker, S., Lennard, C. J., Claessens, L., Gama, A. C., Famba, S., and van Rooyen, A. F. (2015). Crop-livestock intensification in the face of climate change: exploring opportunities to reduce risk and increase resilience in Southern Africa using an integrated multi-modeling approach. In: Rosenzweig, C. and Hillel, D. (Eds), *Handbook of Climate Change and Agroecosystems: The Agricultural Model Intercomparison and Improvement Project (AgMIP) Integrated Crop and Economic Assessments*. ICP Series on Climate Change Impacts, Adaptation, and Mitigation, vol 3. Imperial College Press, London, pp. 90–112.
Masikati, P., Descheemaeker, K., and Crespo, O. (2019). Understanding the role of soils and management on crops in the face of climate uncertainty in Zimbabwe: a sensitivity analysis. In: Rosenstock T., Nowak A., Girvetz E. (Eds), *The Climate-Smart Agriculture Papers*. Springer, Cham, pp. 49–64.
Masvaya, E. N., Nyamangara, J., Giller, K. E., and Descheemaeker, K. (2018). Risk management options in maize cropping systems in semi-arid areas of Southern Africa. *Field Crops Research* 228, 110–21.
Matson, P., Parton, W., Power, A., and Swift, M. (1997). Agricultural intensification and ecosystem properties. *Science* 277(5325), 504–9.
Mbow, C., Smith, P., Skole, D., Duguma, L., and Bustamante, M. (2014). Achieving mitigation and adaptation to climate change through sustainable agroforestry practices in Africa. *Current Opinion in Environmental Sustainability* 6, 8–14.
Meijer, S. S., Catacutan, D., Ajayi, O. C., Sileshi, G. W., and Nieuwenhuis, M. (2015). The role of knowledge, attitudes and perceptions in the uptake of agricultural and agroforestry innovations among smallholder farmers in Sub-Saharan Africa. *International Journal of Agricultural Sustainability* 13(1), 40–54.
Meuwissen, M., Feindt, P., Spiegel, A., Mathijs, E., de Mey, Y., Finger, R., Balmann, A., Wauters, E., Urquhart, J., Vigani, M., Zawalińska, K., Herrera, H., Nicholas-Davies, P., Hansson, H., Paas. W., Slijper, T., Coopmans, I., Vroege, W., Ciechomska, A., Accatino, F., Kopainsky, B., Poortvliet, M., Candel, J., Maye, D., Severini, S., Senni, S., Soriano,

B., Lagrkvist, C-J., Peneva, M., Gavrilescu, C. and Reidsma, P. (2019). A framework to assess the resilience of farming systems. *Agricultural Systems* 176, 102656. Available online: https://doi.org/10.1016/j.agsy.2019.102656.

Milgroom, J., and Giller, K. E. (2013). Courting the rain: rethinking seasonality and adaptation to recurrent drought in semi-arid southern Africa. *Agricultural Systems* 118(C), 91–104.

Minasny, D., Malone, B. P., McBratney, A. B., Angers, D. A., Arrouays, D., Chambers, A., Chaplot, V., Chen, Z.-S., Cheng, K., Das, B. S., Field, D. J., Gimona, A., Hedley, C. B., Hong, S. Y., Mandal, B., Marchant, B. P., Martin, M., McConkey, B. G., Mulder, V. L., O'Rourke, S., Richer-de-Forges, A. C., Odeh, I., Padarian, J., Paustian, K., Pan, G., Poggio, L., Savin, I., Stolbovoy, V., Stockmann, U., Sulaeman, Y., Tsui, C.-C., Vågen, T. G., van Wesemael, B., and Winowiecki, L. (2017). Soil carbon 4 per mille. *Geoderma* 292, 59–86.

Müller, C., Cramer, W., Hare, W. L., and Lotze-Campen, H. (2011). Climate change risks for African agriculture. *Proceedings of the National Academy of Sciences* 108(11), 4313–15.

Mullins, J., Zivin, J. G., Cattaneo, A., Paolantonio, A., and Cavatassi, R. (2018). The adoption of climate smart agriculture: the role of information and insurance under climate change. In: Lipper, L., McCarthy, N., Zilberman, D., Asfaw, S. and Branca, G. (Eds), *Climate Smart Agriculture*. Natural Resource Management and Policy, Vol. 52. Springer, Cham.

Mwongera, C., Shikuku, K. M., Twyman, J., Läderach, P., Ampaire, E., Van Asten, P., Twomlow, S., and Winowiecki, L. A. (2017). Climate smart agriculture rapid appraisal (CSA-RA): a tool for prioritizing context-specific climate smart agriculture technologies. *Agricultural Systems* 151, 192–203.

Ogle, S. M., Breidt, F. J., and Paustian, K. (2005). Agricultural management impacts on soil organic carbon storage under moist and dry climatic conditions of temperate and tropical regions. *Biogeochemistry* 72(1): 87–121.

Ojiem, J. O., de Ridder, N., Vanlauwe, B., and Giller, K. E. (2006). Socio-ecological niche: a conceptual framework for integration of legumes in smallholder farming systems. *International Journal of Agricultural Sustainability* 4(1), 79–93.

Page, E. A. (2008). Distributing the burdens of climate change. *Environmental Politics* 17, 556–75.

Panda, A. (2018). Transformational adaptation of agricultural systems to climate change. *Wiley Interdisciplinary Reviews: Climate Change* 9(4), e520–15.

Peterson, H. H., Barkley, A., Chacón-Cascante, A., and Kastens, T. L. (2012). The motivation for organic grain farming in the United States: profits, lifestyle, or the environment? *Journal of Agricultural and Applied Economics* 44(02), 137–55.

Pittelkow, C. M., Liang, X., Linquist, B. A., van Groenigen, K. J., Lee, J., Lundy, M. E., van Gestel, N., Six, J., Venterea, R. T., and van Kessel, C. (2015). Productivity limits and potentials of the principles of conservation agriculture. *Nature* 517, 365–8.

Poulton, P., Johnston, J., Macdonald, A., White, R., and Powlson, D. (2018). Major limitations to achieving '4 per 1000' increases in soil organic carbon stock in temperate regions: evidence from long-term experiments at Rothamsted Research, United Kingdom. *Global Change Biology* 24, 2563–84.

Powlson, D. S., Whitmore, A. P., and Goulding, K. W. T. (2011). Soil carbon sequestration to mitigate climate change: a critical re-examination to identify the true and the false. *European Journal of Soil Science* 62(1), 42–55.

Powlson, D. S., Stirling, C. M., Thierfelder, C., White, R. P., and Jat, M. L. (2016). Does conservation agriculture deliver climate change mitigation through soil carbon

sequestration in tropical agro-ecosystems? *Agriculture, Ecosystems and Environment* 220, 164–74.
Pretty, J., and Bharucha, Z. P. (2014). Sustainable intensification in agricultural systems. *Annals of Botany* 114(8), 1571–96.
Pretty, J. N., Toulmin, C., and Williams, S. (2011). Sustainable intensification in African agriculture. *International Journal of Agricultural Sustainability* 9(1), 5–24.
Reidsma, P., Oude Lansink, A., and Ewert, F. (2009). Economic impacts of climatic variability and subsidies on European agriculture and observed adaptation strategies. *Mitigation and Adaptation Strategies for Global Change* 14, 35.
Reidsma, P., Ewert, F., Oude Lansink, A., and Leemans, R. (2010). Adaptation to climate change and climate variability in European agriculture: the importance of farm level responses. *European Journal of Agronomy* 32, 91–102.
Reidsma, P., Wolf, J., Kanellopoulos, A., Schaap, B. F., Mandryk, M., Verhagen, J., and van Ittersum, M. K. (2015). Climate change impact and adaptation research requires integrated assessment and farming systems analysis: a case study in the Netherlands. *Environmental Research Letters* 10, 045004.
Rickards, L., and Howden, S. M. (2012). Transformational adaptation: agriculture and climate change. *Crop and Pasture Science* 63(3), 240.
Rippke, U., Ramirez-Villegas, J., Jarvis, A., Vermeulen, S. J., Parker, L., Mer, F., Diekkrüger, B., Challinor, A. J., and Howden, M. (2016). Timescales of transformational climate change adaptation in sub-Saharan African agriculture. *Nature Climate Change* 6, 605–9.
Rochette, P., and Janzen, H. H. (2005). Towards a revised coefficient for estimating N_2O emissions from legumes. *Nutrient Cycling in Agroecosystems* 73, 171.
Rosenzweig, C., Elliott, J., Deryng, D., Ruane, A. C., Müller, C., Arneth, A., Boote, K. J., Folberth, C., Glotter, M., Khabarov, N., Neumann, K., Piontek, F., Pugh, T. A. M., Schmid, E., Stehfest, E., Yang, H., and Jones, J. W. (2014). Assessing agricultural risks of climate change in the 21st century in a global gridded crop model intercomparison. *Proceedings of the National Academy of Sciences* 111, 3268–73.
Rounsevell, M., Ewert, F., Reginster, I., Leemans, R., and Carter, T. (2005). Future scenarios of European agricultural land use II. Projecting changes in cropland and grassland. *Agriculture Ecosystems and Environment* 107, 117–35.
Rudel, T. K., Schneider, L., Uriarte, M., Turner, B. L., DeFries, R., Lawrence, D., Geoghegan, J., Hecht, S., Ickowitz, A., Lambin, E. F., Birkenholtz, T., Baptista, S., and Grau, R. (2009). Agricultural intensification and changes in cultivated areas, 1970-2005. *Proceedings of the National Academy of Sciences of the United States of America* 106(49), 20675–80.
Rufino, M. C., Herrero, M., van Wijk, M. T., Hemerik, L., de Ridder, N., and Giller, K. E., 2009. Lifetime productivity of dairy cows in smallholder farming systems of the highlands of Central Kenya. Animal 3, 1044–1056.
Rurinda, J., Mapfumo, P., van Wijk, M. T., Mtambanengwe, F., Rufino, M. C., Chikowo, R., and Giller, K. E. (2014). Comparative assessment of maize, finger millet and sorghum for household food security in the face of increasing climatic risk. *European Journal of Agronomy* 55, 29–41.
Rurinda, J., van Wijk, M. T., Mapfumo, P., Descheemaeker, K., Supit, I., and Giller, K. E. (2015). Climate change and maize yield in southern Africa: what can farm management do? *Global Change Biology* 21, 4588–601.
Rusinamhodzi, L., Corbeels, M., van Wijk, M., Rufino, M. C., Nyamangara, J., and Giller, K. E. (2011). A meta-analysis of long-term effects of conservation agriculture on maize

grain yield under rain-fed conditions. *Agronomy for Sustainable Development* 31, 657-763.

Rusinamhodzi, L., Corbeels, M., Zingore, S., Nyamangara, J., and Giller, K. E. (2013). Pushing the envelope? Maize production intensification and the role of cattle manure in recovery of degraded soils in smallholder farming areas of Zimbabwe. *Field Crops Research* 147, 40-53.

Schaap, B. F., Reidsma, P., Verhagen, J., Wolf, J., and van Ittersum, M. K. (2013). Participatory design of farm level adaptation to climate risks in an arable region in The Netherlands. *European Journal Agronomy* 48, 30-42.

Scherer, L., and Verburg, P. H. (2017). Mapping and linking supply- and demand-side measures in climate-smart agriculture. A review. *Agronomy for Sustainable Development* 37, 66.

Schröter, D., Cramer, W., Leemans, R., Prentice, I. C., Araújo, M. B., Arnell, N. W., Bondeau, A., Bugmann, H., Carter, T. R., Gracia, C. A., de la Vega-Leinert, A. C., Erhard, M., Ewert, F., Glendining, M., House, J. I., Kankaanpää, S., Klein, R. J., Lavorel, S., Lindner, M., Metzger, M. J., Meyer, J., Mitchell, T. D., Reginster, I., Rounsevell, M., Sabaté, S., Sitch, S., Smith, B., Smith, J., Smith, P., Sykes, M. T., Thonicke, K., Thuiller, W., Tuck, G., Zaehle, S., and Zierl, B. (2005). Ecosystem service supply and vulnerability to global change in Europe. *Science* 310, 1333-37.

Silva, J. V., Reidsma, P., and van Ittersum, M. K. (2017). Yield gaps in Dutch arable farming systems: analysis at crop and crop rotation level. *Agricultural Systems* 158, 78-92.

Smit, B., and Skinner, M. W. (2002). Adaptation options in agriculture to climate change: a typology. *Mitigation and Adaptation Strategies for Global Change* 7(1), 85-114.

Smith, P. (2008). Land use change and soil organic carbon dynamics. *Nutrient Cycling in Agroecosystems* 81, 169-78.

Smith, P. (2012). Soils and climate change. *Current Opinion in Environmental Sustainability* 4, 539-44.

Smith, K. A. (2017). Changing views of nitrous oxide emissions from agricultural soil: key controlling processes and assessment at different spatial scales. *European Journal of Agronomy* 68, 137-55.

Smith, P., and Olesen, J. E. (2010). Synergies between the mitigation of, and adaptation to, climate change in agriculture. *The Journal of Agricultural Science* 148(05), 543-52.

Smith, P., Haberl, H., Popp, A., Erb, K., Lauk, C., Harper, R., Tubiello, F. N., Siqueira Pinto, A., Jafari, M., Sohi, S., Masera, O., Böttcher, H., Berndes, G., Bustamante, M., Ahammad, H., Clark, H., Dong, H., Elsiddig, E. A., Mbow, C., Ravindranath, N. H., Rice, C. W., Robledo Abad, C., Romanovskaya, A., Sperling, F., Herrero, M., House, J. I., and Rose, S. (2013). How much land-based greenhouse gas mitigation can be achieved without compromising food security and environmental goals? *Global Change Biology* 19, 2285-302.

Smith, P., Bustamante, M., Ahammad, H., Clark, H., Dong, H., Elsiddig, E. A., Haberl, H., Harper, R., House, J., Jafari, M., Masera, O., Mbow, C., Ravindranath, N. H., Rice, C. W., Robledo Abad, C., Romanovskaya, A., Sperling, F., and Tubiello, F. (2014). Agriculture, Forestry and Other Land Use (AFOLU). In: Edenhofer, O., Pichs-Madruga, R., Sokona, Y., Farahani, E., Kadner, S., Seyboth, K., Adler, A., Baum, I., Brunner, S., Eickemeier, P., Kriemann, B., Savolainen, J., Schlömer, S., von Stechow, C., Zwickel, T., and Minx, J. C. (Eds), *Climate Change 2014: Mitigation of Climate Change. Contribution of Working Group III to the Fifth Assessment Report of the Intergovernmental Panel on Climate*

Change. Cambridge University Press, Cambridge, United Kingdom and New York, NY, USA.

Steenwerth, K. L., Hodson, A. K., Bloom, A. J., Carter, M. R., Cattaneo, A., Chartres, C. J., Hatfield, J. L., Henry, K., Hopmans, J. W., Horwath, W. R., Jenkins, B. M., Kebreab, E., Leemans, R., Lipper, L., Lubell, M. N., Msangi, S., Prabhu, R., Reynolds, M. P., Solis, S. S., Sischo, W. M., Springborn, M., Tittonell, P., Wheeler, S. M., Vermeulen, S. J., Wollenberg, E. K., Jarvis, L. S., and Jackson, L. E. (2014). Climate-smart agriculture global research agenda: scientific basis for action. *Agriculture and Food Security* 3(1), 11.

Stoate, C., Boatman, N., Borralho, R., Carvalho, C., Snoo, G., and Eden, P. (2001). Ecological impacts of arable intensification in Europe. *Journal of Environmental Management* 63(4), 337–65.

Struik, P. C., and Kuyper, T. W. (2017). Sustainable intensification in agriculture: the richer shade of green. A review. *Agronomy for Sustainable Development* 37, 39.

Thierfelder, C., and Wall, P. C. (2009). Effects of conservation agriculture techniques on infiltration and soil water content in Zambia and Zimbabwe. *Soil and Tillage Research* 105(2), 217–27.

Thierfelder, C., Chivenge, P., Mupangwa, W., Rosenstock, T. S., Lamanna, C., and Eyre, J. X. (2017). How climate-smart is conservation agriculture (CA)? – Its potential to deliver on adaptation, mitigation and productivity on smallholder farms in southern Africa. *Food Security* 9, 537–60.

Thornton, P. K., and Herrero, M. (2014). Climate change adaptation in mixed crop–livestock systems in developing countries. *Global Food Security* 3(2), 99–107.

Tilman, D., Balzer, C., Hill, J., and Befort, B. L. (2011). Global food demand and the sustainable intensification of agriculture. *Proceedings of the National Academy of Sciences* 108(50), 20260–64.

Tittonell, P., and Giller, K. E. (2013). When yield gaps are poverty traps: the paradigm of ecological intensification in African smallholder agriculture. *Field Crops Research* 143, 76–90.

Traore, B., van Wijk, M. T., Descheemaeker, K., Corbeels, M., Rufino, M. C., and Giller, K. E. (2014). Evaluation of climate adaptation options for Sudano-Sahelian cropping systems. *Field Crops Research* 156, 63–75.

Traore, B., Descheemaeker, K., van Wijk, M. T., Corbeels, M., Supit, I., and Giller, K. E. (2017). Modelling cereal crops to assess future climate risk for family food self-sufficiency in southern Mali. *Field Crops Research* 201, 133–45.

UNEP (2016). *Food Systems and Natural Resources. A Report of the Working Group on Food Systems of the International Resource Panel*. Westhoek, H., Ingram J., Van Berkum, S., Özay, L., and Hajer M.

Valdivia, R. O., Antle, J. M., Rosenzweig, C., Ruane, A. C., Vervoort, J., Ashfaq, M., Hathie, I., Homann-Kee Tui, S., Mulwa, S., Nhemachena, C., Ponnusamy, P., Rasnayaka H., and Singh, H. (2015). Representative agricultural pathways and scenarios for regional integrated assessment of climate change impact, vulnerability and adaptation. In: Rosenzweig, C. and Hillel, D. (Eds), *Handbook of Climate Change and Agroecosystems: The Agricultural Model Intercomparison and Improvement Project Integrated Crop and Economic Assessments, Part 1*. Imperial College Press, London.

Van Groenigen, J. W., Velthof, G. L., Oenema, O., Van Groenigen, K. J., and Van Kessel, C. (2010), Towards an agronomic assessment of N_2O emissions: a case study for arable crops. *European Journal of Soil Science* 61, 903–13.

Van Groenigen, J. W., van Kessel, C., Hungate, B. A., Oenema, O., Powlson, D. S., and van Groenigen, K. J. (2017). Sequestering soil organic carbon: a nitrogen dilemma. *Environmental Science and Technology* 51(9), 4738-9.

Van Ittersum, M. K., van Bussel, L. G., Wolf, J., Grassini, P., Van Wart, J., Guilpart, N., Claessens, L., de Groot, H., Wiebe, K., Mason-D'Croz, D., Yang, H., Boogaard, H., van Oort, P. A. J., van Loon, M. P., Saito, K., Adimo, O., Adjei-Nsiah, S., Agali, A., Bala, A., Chikowo, R., Kaizzi, K., Kouressy, M., Makoi, J. H. J. R., Ouattara, K., Tesfaye, K. and Cassman, K. G. (2016). Can sub-Saharan Africa feed itself? *Proceedings of the National Academy of Sciences of the United States of America* 113(52), 14964-9.

Vanlauwe, B., Coyne, D., Gockowski, J., Hauser, S., Huising, J., Masso, C., Nziguheba, G., Schut, M., and Van Asten, P. (2014). Sustainable intensification and the African smallholder farmer. *Current Opinion in Environmental Sustainability* 8, 15-22.

Vanlauwe, B., Descheemaeker, K., Giller, K. E., Huising, J., Merckx, R., Nziguheba, G., Wendt, J., and Zingore, S. (2015). Integrated soil fertility management in sub-Saharan Africa: unravelling local adaptation. *Soil* 1(1), 491-508.

van Loon, M. P., Hijbeek, R., ten Berge, H. F. M., De Sy, V., ten Broeke, G. A., Solomon, D., and van Ittersum, M. K., 2019. Impacts of intensifying or expanding cereal cropping in sub-Saharan Africa on greenhouse gas emissions and food security. *Global Change Biology* 25, 3720-30. https://doi.org/10.1111/gcb.14783

Van Meij, H., Havlik, P., Lotze-Campen, H., Stehfest, E., Witzke, P., Pérez Domínguez, I., Bodirsky, B. L., van Dijk, M., Doelman, J., Fellmann, T., Humpenöder, F., Koopman, J. F. L., Müller, C., Popp, A., Tabeau, A., Valin, H., and van Zeist, W.–J. (2018). Comparing impacts of climate change and mitigation on global agriculture by 2050. *Environmental Research Letters* 13, 064021.

Van Oost, K., Quine, T., Govers, G., De Gryze, S., Six, J., Harden, J. W., Ritchie, J. C., McCarty, G. W., Heckrath, G., Kosmas, C., Giraldez, J. V., Marques da Silva, J. R., and Merckx, R. (2007). The impact of agricultural soil erosion on the global carbon cycle. *Science* 318(5850), 626-9.

Velthof, G. L. (2015). Mineral concentrate from processed manure as fertiliser. Alterra Report 2650. Alterra Wageningen UR (University and Research Centre), Wageningen.

Venterea, R. T., Halvorson, A. D., Kitchen, N., Liebig, M. A., Cavigelli, M. A., Grosso, S. J., Motavalli, P. P., Nelson, K. A., Spokas, K. A., Singh, B. P., Stewart, C. E., Ranaivoson, A., Strock, J., and Collins, H. (2012). Challenges and opportunities for mitigating nitrous oxide emissions from fertilized cropping systems. *Frontiers in Ecology and the Environment* 10(10), 562-70.

Verchot, L. V., Van Noordwijk, M., Kandji, S., Tomich, T., Ong, C., Albrecht, A., Mackensen, J., Bantilan, C., Anupama, K. V., and Palm, C. (2007). Climate change: linking adaptation and mitigation through agroforestry. *Mitigation and Adaptation Strategies for Global Change* 12(5), 901-18.

Wassmann, R., Jagadish, S. V. K., Heuer, S., Ismail, A., Redona, E., Serraj, R., Singh, R. K., Howell, G., Pathak, H., and Sumfleth, K. (2009). Climate change affecting rice production: the physiological and agronomic basis for possible adaptation strategies. *Advances in Agronomy* 101, 59-122.

Weltin, M., Zasada, I., Piorr, A., Debolini, M., Geniaux, G., Moreno Perez, O., Scherer, L., Tudela Marco, L., and Schulp, C. J. E. (2018). Conceptualising fields of action for sustainable intensification - A systematic literature review and application to regional case studies. *Agriculture, Ecosystems and Environment* 257, 68-80.

Wise, R. M., Fazey, I., Stafford Smith, M., Park, S. E., Eakin, H. C., Archer Van Garderen, E. R. M., and Campbell, B. (2014). Reconceptualising adaptation to climate change as part of pathways of change and response. *Global Environmental Change* 28, 325-36.

Wolf, J., Kanellopoulos, A., Kros, J., Webber, H., Zhao, G., Britz, W., Reinds, G. J., Ewert, F., and de Vries, W. (2015). Combined analysis of climate, technological and price changes on future arable farming systems in Europe. *Agricultural Systems* 140, 56-73.

Zhang, W. F., Dou, Z. X., He, P., Ju, X. T., Powlson, D., Chadwick, D., Norse, D., Lu, Y. L., Zhang, Y., Wu, L., Chen, X. P., Cassman, K. G., and Zhang, F. S. (2013). New technologies reduce greenhouse gas emissions from nitrogenous fertilizer in China. *Proceedings of the National Academy of Sciences of the United States of America* 110(21), 8375-80.

Zhang, X., Davidson, E. A., Mauzerall, D. L., Searchinger, T. D., Dumas, P., and Shen, Y. (2015). Managing nitrogen for sustainable development. *Nature* 528(7580), 51-9.

ZimVAC (2013). *Rural Livelihoods Assessment*. Harare.

Zingore, S., Murwira, H. K., Delve, R. J., and Giller, K. E. (2007). Variable grain legume yields, responses to phosphorus and rotational effects on maize across soil fertility gradients on African smallholder farms. *Nutrient Cycling in Agroecosystems* 80(1), 1-18.

Zomer, R. J., Bossio, S. A., Sommer, R., and Verchot, L. V. (2017). Global sequestration potential of increased organic carbon in cropland soils. *Scientific Reports* 7(1), 15554.

Chapter 9

The contribution of integrated crop-livestock systems in combatting climate change and improving resilience in agricultural production to achieve food security

Mark van Wijk and James Hammond, International Livestock Research Institute, Kenya; Simon Fraval, International Livestock Research Institute, Kenya and Wageningen University, The Netherlands; Jannike Wichern, Wageningen University, The Netherlands; Randall Ritzema, Olivet Nazarene University, USA; and Ben Henderson, Natural Resources Policy, Organisation for Economic Co-operation and Development (OECD), France

1 Introduction

Smallholder farming systems in sub-Saharan Africa (SSA) are often characterized by diverse production activities, spreading food production and income generation across a variety of different crops and types of livestock (Waha et al., 2018). Such mixed farming systems are the most common form of agriculture in SSA. This diversified approach is often interpreted as a response to reduce risk, for example to market or climatic shocks. Many African farmers own livestock as an insurance during periods of drought (Ayantunde et al., 2018). Combined crop and livestock systems (i.e. mixed systems) in low- and middle-income

http://dx.doi.org/10.19103/AS.2020.0064.15

countries produce significant quantities of livestock and crop food products both for consumption by farmers and also for sale to rural and urban markets (Herrero et al., 2010). There are also synergies in well-integrated mixed systems: for example, livestock is used to provide draft power to cultivate the land and manure to fertilize the soil, while crop residues are often a key feed resource for livestock.

Most of these mixed farming systems rely on rainfall for water supply and are, therefore, at risk from climate change. Analyses show that, even if rainfall remains constant, average crop yields are expected to decline about 15% across SSA (with, of course, large regional variations) due to rising temperatures. Expected temperature rises and increases in extreme rainfall events will damage crop production, increase market volatility and increase food insecurity and poverty (Ongoma et al., 2018). How climate change will affect livestock and its production is more uncertain. Heat stress affecting animals is likely to increase, thereby adversely affecting animal health and productivity, with negative effects on fodder production (grass and crop residues), but quantitative assessments are largely lacking (Malik et al., 2015). Climate change is already affecting households who are applying diverse coping strategies in response to these risks while also trying to adapt to rapid socio-economic change (e.g. Below et al., 2012; Helgeson et al., 2013).

In this chapter, we first look at current patterns in crop-livestock systems. Our focus is on the role of crop and livestock production in achieving food security in smallholder systems across low- and middle-income countries. We have brought together existing data on farm households, developed indicators of food security and applied them to interpret this existing data. We have developed our own harmonized farm household survey tool - the Rural Household Multiple Indicator Survey (RHOMIS; www.rhomis.org; Hammond et al., 2017a). The RHOMIS tool has experienced a very rapid uptake by a wide range of research and development partners since its conception in 2015. We have now built up a harmonized database of more than 29 000 farm households in 32 countries. Data from both approaches (i.e. data re-use and RHOMIS) can be combined easily as they focus on a consistent set of variables and welfare indicators, resulting in a database containing information on more than 50 000 households. Such a large database provides an immensely rich resource to link different indicators relating, for example, to food security, poverty, dietary diversity, gender equity and land use.

This chapter is organized in five sections as follows:

1 A short overview of the range of existing studies focussing on climate change and mixed crop-livestock systems in the recent scientific literature.
2 Analyses that quantify how food security and dietary diversity are being shaped in crop-livestock systems.

3 Analyses of coping strategies that farmers currently apply in mixed crop-livestock systems to cope with short-term climate variability.
4 Analyses of adaptation options to climate change in mixed farming systems. We use a modelling tool to explore (a) possible farming futures, and quantify how climate change impacts farm-level performance and (b) what adaptation options (that exploit the synergies between crop and livestock production) can do to alleviate these effects.
5 Applying a framework that integrates the findings of the previous sections in a coherent way to think about the resilience of mixed crop-livestock systems in the face of existing climate variability and possible future climate change.

2 A short overview of analyses and approaches focussing on climate change and mixed crop-livestock systems

Despite the importance of mixed crop-livestock systems in low- and middle-income countries, and their vulnerability to climate change, there is limited work that assesses the effects of climate change on these farming systems. Recent work on climate change and smallholder farming systems has focussed mainly on crops and evaluated single commodities. Many impact assessments are available at global, regional and local level for individual crops such as maize, wheat, sorghum, beans, rice and cassava (e.g. Rosenzweig et al., 2014; Zinyengere et al., 2013; Roudier et al., 2011). Assessments of climate change effects on livestock production are more scarce (Thornton et al., 2009; Descheemaeker et al., 2018). Assessing climate change effects on livestock production is complex because, besides direct effects (less or more drinking water available, heat stress), indirect climate change effects also play a role in livestock production. This includes the availability of fodder (either through crop residues, fodder grasses and/or grazing land), which is likely to change and will affect livestock production and the size of herds that can be maintained by farmers.

There has been substantial work at the regional level looking at drivers and patterns of change in farming systems. For example, Thornton et al. (2002) mapped likely transitions from rangeland (LG) to mixed (MR) systems in response to increasing population densities to 2050. A similar study showed that considerable shifts in cropping and livestock activities are likely to occur in parts of East and Southern Africa (Jones and Thornton, 2009).

Despite some gaps in our quantitative knowledge of the interactions between crop and livestock production, a wide range of models exist to assess climate change impacts on mixed crop-livestock systems (see Nicholson (2007) and van Wijk et al. (2014)). Individual case studies exist for specific farming systems (e.g. Henderson et al., 2018; Descheemaeker et al., 2018; Rigolot et al., 2017). However, there is a lack of studies that look at larger-scale patterns of change in mixed systems and take into account crop-livestock interactions. In a case study focussing on southern Zimbabwe, Descheemaeker et al. (2018) showed that relative changes in livestock net revenue compared to the baseline climate varied from a 6% increase to a 43% decrease, and the proportion of farmers negatively affected varied from 20% to 100%, depending on the climate scenario. Farms with larger stocking density faced more severe feed gaps and were more sensitive to climate change than less densely stocked farms. Inclusion of grain and forage legumes in the fodder basket could increase milk productivity and net revenues by 30% if they were used to alleviate dry-season feed gaps. Similarly, in an analysis for Burkina Faso, Henderson et al. (2018) estimated annual profit losses resulting in 15% less income on average by 2045 under a 'middle of the road' climate change scenario (RCP6). In a study of farming in Burkina Faso, Rigolot et al. (2017) also estimated an average decline of 27% in production under a moderate climate change scenario, with similar severe consequences for net income.

These case studies provide useful insights and additional regional case studies are urgently needed to (e.g. Thornton et al., 2009):

1 obtain more reliable estimates of the effect of climate change on livestock in mixed systems;
2 derive general conclusions from individual studies;
3 get a better insight into the likely future vulnerability of mixed systems and their associated livelihoods to climate (and other) changes; and
4 identify with more certainty the appropriate options for adaptation to minimize these risks.

This is not straightforward as mixed farming systems are highly complex, consisting of a mixture of different crops and livestock species. Interactions between these crops and livestock (through, e.g. supply of draught power for land preparation, crop residues for feed, manure for soil improvement) show that assessment has to be done at the level of the system. This is discussed in the following sections of this chapter.

3 Agricultural production, consumption, sale and food security

Agricultural production contributes to food security in two ways: through generating food that can be consumed directly, or that can be sold, thereby generating cash with which food can be bought. However, whether increased agricultural production always leads to better food security and better nutrition is much less straightforward. Research has shown that technologies solely focussing on productivity can have negative consequences for dietary diversity and nutrition (e.g. Ruel et al., 2018). In order to better understand the relationship between agricultural production and food security in smallholder systems, we explore different pathways towards food security and diet diversification. We evaluate the pros and cons of different technologies before suggesting the most appropriate approaches.

An important approach we have adopted is to develop simple pathway models that describe how different drivers of agricultural production and food security interact to shape food security outcomes. Figure 1 shows the results from the first of these exercises, in which household survey data for Burkina Faso were analysed to develop a conceptual pathway (Fraval, 2019). Farm resources and productivity provide the basis for food production. Production can then follow subsistence and/or market avenues to ensure food security. Market participation, off-farm income and other forms of support all contribute to net income. Mediated through expenditure and consumption decisions, these two avenues of income and subsistence as sources of food then have a bearing on food security and dietary diversity. These two ways of achieving food security and diverse diets can be called the 'farm-based' route (i.e. producing food on the farm and consuming it) and the 'purchased' route (i.e. selling farm produce and/or generating off-farm income, and then buying food). The relative importance of these two avenues will differ between households in a given region, and between farm households in contrasting agro-ecological and socio-economic conditions.

Two sets of results for contrasting systems are presented here as an illustration of these pathway models (Table 1). We show the results of the two avenues/pathways of achieving diverse diets for two regions in Burkina Faso, that is the 'farm-based' and 'purchased' avenues, and their variation over time. To assess dietary diversity, we used a modified household dietary diversity score (HDDS). The food groups in our HDDS are based on the food categories used in the Minimum Diet Diversity for Women (MDD-W) indicator, which is based on a count of ten food categories defined in Table 1 (FAO and

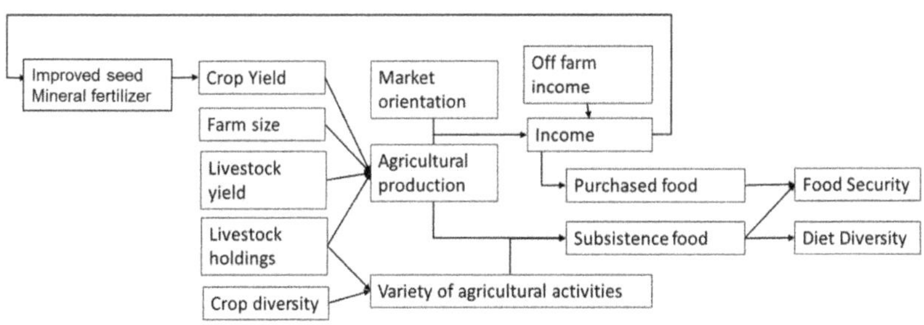

Figure 1 Simple pathway model of food security and dietary diversity.

Table 1 Site summary of household diet diversity score (HDDS; median and IQR)

	Seno	Yatenga
HDDS - good period	8 (4)	9 (1)
HDDS - lean period	3.5 (3)	6 (2)
HDDS from farm-based - lean period	1 (1)	2 (2)
HDDS purchased - lean period	3 (3)	6 (2)
HDDS from farm-based - good period	2.5 (2)	2 (1.5)
HDDS purchased - good period	7 (5)	7 (2)

The HDSS food categories are: 1. Grains, roots and tubers and plantains; 2. Pulses; 3. Nuts and seeds; 4. Dairy; 5. Meat, poultry and fish; 6. Eggs; 7. Dark green-leafy vegetables; 8. Other vitamin A-rich fruits and vegetables; 9. Other vegetables; and 10. Other fruits.

FHI 360, 89, 2016). Besides this modification to the HDDS indicator to capture food categories that are better linked to nutrition, we also asked households to recall the frequency of access to each food category in the best and worst months (referred to as 'good' and 'lean' periods). The HDDS score is the sum of the food categories that are at least consumed a couple of times per week. As an extension of diet diversity indicators, we also asked households to recall the channel of access for each food category - farm, purchased or free/traded (for a detailed analysis see Fraval et al., 2019). The importance of these routes varies within a year. In this exercise conducted in Burkina Faso, HDDS from farm production was mainly determined by the number of different crops grown, while the purchased items contributing to the HDDS depended on the income generated, which was underpinned by off-farm income, cash crops grown and livestock holdings (Fraval, 2019).

Contrasting results comparing the 'farm-based' versus 'purchased' avenue to diverse diets were also obtained for two sites in South-East Asia: Cambodia and Laos (Table 2; Ritzema et al., 2019). Table 1 shows that household diet diversity differs as a result of different factors, while Table 2 lists the most important drivers to achieve high household diet diversity. A key result of Table 2 is the identification of which pathways increase/decrease HDDS. For example, the positive effect of female control, off-farm income and value of crops on overall HDDS in Cambodia occurs through the 'purchased' avenue. In Laos, overall HDDS is negatively affected by market orientation (i.e. how much of the farm produce is being sold) and positively by the value of crops, both working through the 'farm-based' avenue.

The value of crops is the only variable that was statistically significant for total HDDS in all three sites, while crop diversity, livestock holdings, value of livestock production and crop productivity were not significant for total HDDS across all the sites. A positive association between off-farm income and HDDS in the Laos and Cambodia sites shows the importance of off-farm income in

Table 2 Significant livelihood strategies and farm performance regressor variables for the household dietary diversity score using data from South-East Asia ($p < 0.05$)

	Cambodia			Laos			Vietnam		
Variable	HDDS total	HDDS purch	HDDS farm-b	HDDS total	HDDS purch	HDDS farm-b	HDDS total	HDDS purch	HDDS farm-b
Female decision control	+	+							
Off-farm income	+	+		+	+				
Market orientation	–	–	–	–		–			
Value of crops	+	+		+		+	+	+	
Value of livestock produce			+			+			
Land owned	+			+	+				
Crop diversity			+		–				+
N fertilizer use				+	+				
Land irrigated			+	+	+	+			
Crop productivity						–			
Livestock holdings		–	+						
Livestock diversity				+	+		+		+
Livestock orientation	+	+							
Livestock productivity				+		+			+

'purch' = purchased; 'farm_b' = farm-based.
Empty cells are non-significant regressor variables.

purchasing diverse foods from the market. A strong negative association between market orientation for food availability versus HDDS in the Laos and Cambodia sites indicates that increased use of crop and livestock sales to access food energy is matched with a loss of dietary diversity.

A key component of the analysis (presented in much more detail in Ritzema et al., 2019) was linking HDDS to a simple but comprehensive set of farm, livelihood and household characteristics. Forging these links requires an understanding of two factors for effective interpretation of results within and across sites:

1 the levels of each variable at each site; and
2 the variation of those variables within each site and the significance of the association of that variation with HDDS and the different components of HDDS.

For example, the value of crop production in the Vietnam site shows significant variation that is correlated to HDDS. However, a lack of correlated variation in nitrogen (N) fertilizer-use rates to HDDS in the Vietnam site does not mean it is not important as a 'driver': in fact, high crop production levels in the Vietnam site depend upon fertilizer inputs. There is little variation in N fertilizer input rates correlating to HDDS because fertilizer use is high across the spectrum of farm households in the Vietnam site. Interpretation of the data within its site context is, therefore, a key component of accurate analysis.

The results we obtained analysing the HDDS information have an important set of consequences. First, only asking the farmer about dietary diversity in the last 24 h (or of the last week) will not capture the seasonal variation in overall HDDS (e.g. Table 1), and the responses would mask a lot of underlying variation in how these dietary scores are achieved. A widely held hypothesis is that diversifying production leads to better nutrition of food insecure smallholders. However, evidence for this hypothesis in the scientific literature is limited. Our results analysing HDDS data show why. The relationship between production diversity and dietary diversity depends significantly on the relative importance of the purchased avenue versus the farm-based avenue. In almost all cases, the farm-based avenue showed a strong relationship between production diversity and farm-based dietary diversity. However, if the purchased-based avenue is the most important avenue in achieving diverse diets, it is unlikely that production diversity will be related to overall dietary diversity. This, however, does not mean that production diversification is not a good solution to improve diets of smallholders; it will actually diminish the reliance of smallholders on food market availability for achieving diverse diets.

A second consequence of these results is that this information can help design more appropriate interventions. In analyses using the overall RHOMIS database of 8000 households in SSA, we analysed in detail the food groups underpinning the overall dietary diversity scores, as well as the purchased and farm-based routes for different farm types. In Fig. 2, we show results for two contrasting farm types, based on analyses of RHOMIS data in eight countries in SSA. We again show the results of the two pathways of achieving diverse diets for each farm type (i.e. the 'farm-based' and 'purchased' routes). It is important to note that these pathways do not have to be mutually exclusive. The food group breakdown (Fig. 2) shows that, in contrasting farm types, the pathways towards diverse diets are different. For example, in the farm types with livestock present, milk consumption occurred in almost all families, even when overall diets are not diverse. In the crop-based farm types, milk has to be purchased, but this only happens in families with higher dietary diversity scores, which form a minority in the population. This shows that, in smallholder systems, the production system does influence what people eat, and that following a purely cash-oriented pathway to increased dietary diversity may not be as beneficial as encouraging diverse production for consumption, with some resulting in cash income. It can also help us to identify the key food groups that are

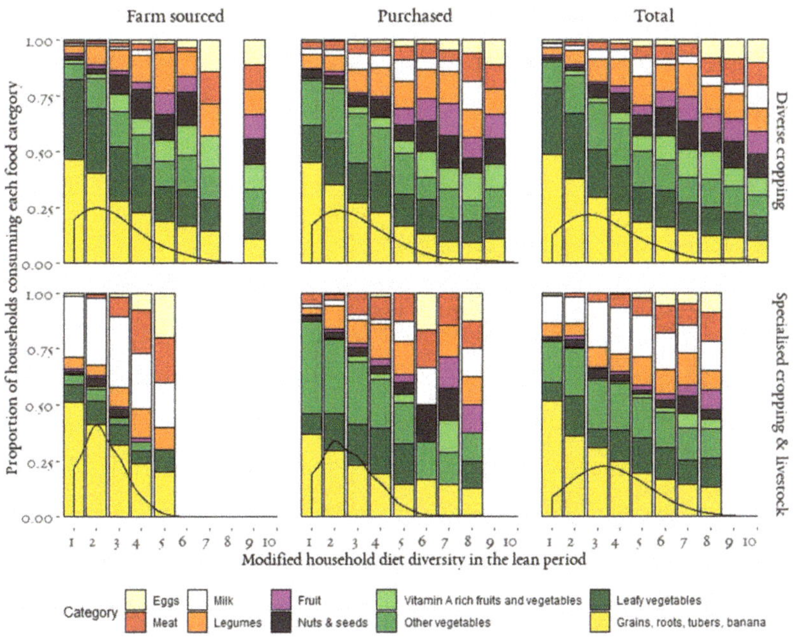

Figure 2 Dietary diversity scores (black lines) and food group composition for two contrasting farm household types, diverse cropping and specialized cropping and livestock. Results based on Fraval et al. (2019). Values on the x-axis are the HDDS scores for farm sourced, purchased and total scores.

important for specific micro-nutrients that specific interventions might target to improve dietary diversity of these households. This can for example be a specific group of vegetables that can improve the intake of vitamin A or iron (Fraval et al., 2019).

Importantly, the results of Fig. 2 also stress the importance of mixed production systems in supplying a balanced diet. The dietary diversity scores of households with mixed farming systems (with both crops and livestock) are substantially higher than for households that just rely on crop production. Mixed production, therefore, plays a key role in the welfare of smallholder households. Although there is increasing consumption of animal-sourced food in the human diet in many countries in SSA (such as Burkina Faso, Ghana, Ethiopia, Kenya), the diet is still largely dominated by the intake of basic cereal-based staple foods which are usually deficient in the protein and micro-nutrients necessary for healthy human development. Consumption of animal protein is particularly important for children under 5 years and women of reproductive age. The importance of consumption of animal-sourced food for cognitive development of children is well documented (de Pee et al., 2017).

4 Current methods used by farmers in mixed crop-livestock systems to deal with climate variability

An understanding of how farmers are currently dealing with adverse climate events can guide assessments of future climate change effects on smallholder farming systems. Responses to extreme climate events such as droughts and flooding can be monitored, while existing climate gradients can be studied to better assess how the vulnerability of smallholder farmers might decrease in the future and what suitable mitigation options might be available. Here, we give an example of how we have studied coping strategies for climate and price variability (Fig. 3). The results are based on a household survey carried out in Rakai, Uganda, in 2017 (see Wichern, 2019 for more details). The household survey collected information on household characteristics, on- and off-farm

Figure 3 Number of households applying different ex-ante coping strategies when they expect a (a) good season, (b) bad season, (c) price increase for one of their major crops. A very limited number of coping strategies were identified for price decreases.

sources of income, 'good', 'bad' and 'normal' production of major food and cash crops, as well as coping strategies that households apply to their major food and cash crops. The shocks studied were if they:

1 expect a good/bad season in terms of productivity; and
2 expect a high/low price.

Coping strategies were focussed on how households dealt with the need to get food or income:

1 after their major crops were damaged or destroyed; and
2 after the price of their major cash crop declined at the end of the season.

Despite the importance of both annual crops and perennial crops such as coffee and banana, livestock is key for many households in Rakai. As a result, livestock manure management is mentioned as the second most important option to make best use of an expected good season. Overall, 53% of the households would apply coping strategies when expecting a good season and only 28% when expecting a bad season. A social differentiation was noticeable. Households with a gross income or a cash value of activities below $1.25 per capita per day were more likely to apply coping strategies in a good season. When dealing with a bad season, the three most important coping strategies were using off-farm income, selling livestock and eating less. The choice between these strategies was determined by household access to (productive) resources, families could only use off-farm income or sell livestock if they had these resources readily available while, for the poorest households, only eating less was left as an option. The selling of livestock to deal with a hunger situation is the second most important coping strategy, highlighting again that livestock is often used as a buffer in these systems.

5 Possible methods of assessing climate change adaptation options

These findings can be used to explore what the future might look like for smallholder farmers, and how they can exploit synergies between crop and livestock production to better combat and deal with climate change. We do this by quantifying the possible pros and cons of intensification options for different farmers in different locations. We evaluate packages of intervention options rather than single interventions to take into account the benefits of synergies between options. An example is shown in Fig. 4 (Henderson et al., 2018). In this study, we quantified which intensification options could improve

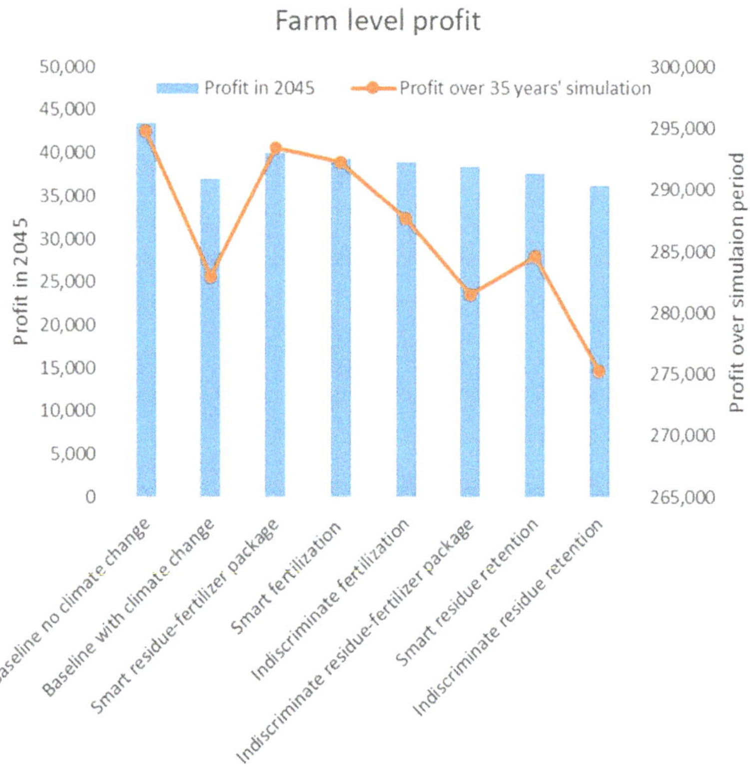

Figure 4 Total profit and final-year profit for different adaptation packages, tested for a mixed crop-livestock system in central Burkina Faso. Based on Henderson et al. (2018).

the current and future profitability of smallholder farms in Burkina Faso (the simulation period was 2010-2045, i.e. 35 years). To do this, we used a positive mathematical programming (PMP) farm household model in combination with data from our farm household characterization database. Each individual household was included in the analysis, thereby giving us a critical insight into the different responses that households can show. We also incorporated the likely production and price effects of the climate change predicted for the region, giving us an insight into the consequences of the interactions between production resources, climate and socio-economic developments that farmers are likely to face in the coming decades.

Two baseline scenarios were defined, one with and one without climate change in order to calculate the costs of climate change in the absence of new practices to mitigate the effects of climate change. The other scenarios involve the use of residue retention and N fertilization, applied either singularly or as a package, to assess how much they can offset the simulated economic losses from climate change. For each scenario, an indiscriminate and a 'smart' version is simulated. The indiscriminate version is applied to all farms irrespective of whether the practice is profitable or not. In contrast, the smart version is applied only to farms for which the measures are profitable. This assumes that farmers will have perfect foresight about the profitability of the intervention on each farm. In each intervention, the synergies between crop and livestock production are explored. An example is fertilization of cereals recommended by experts in the study area as an option to increase average cereal yields in the context of climate change, but which also leads to increased availability of crop residues for livestock feeding.

The results of Fig. 4 show that predicted climate change for Burkina Faso has a strong negative effect on the economic performance of mixed smallholder farming systems. None of the intervention options can completely compensate for that loss when climate change impacts are significant. Up to 2030, the intervention options scenario still outperforms the 'no climate change' scenario but, after 2030, negative climate change effects cannot be compensated for by better management. Both the indiscriminate and smart versions of the fertilizer and crop residue measures outperformed the climate change baseline over the whole simulation period. For the package of the two measures, there is quite a large contrast between the smart and indiscriminate versions, with the smart version of the package being the most profitable of all the scenarios.

The contrasting results of the smart residue retention scenario (relative good average performance over the 35 years of simulation, while not performing very well in 2045) show that intervention packages that perform well in the first part of the 35 years simulations are not necessarily the best performing ones at the end of the simulation period. These results show that best options for farmers to intensify current production are not optimum in the longer term. Diverse packages of intensification options are needed to best serve smallholder production in the near future.

6 Conclusion

Throughout this chapter, we have used the concept of resilience to better understand current mixed smallholder systems, and have then used that knowledge to gauge what might happen under future climate change conditions. In Hammond et al. (2019), we performed a systematic household survey exercise in which more than 600 households in four contrasting sites in East Africa were revisited over 4 years in a panel survey. We identified variables in agricultural production, consumption, sales, and off-farm income, and related these variables to a series of welfare indicators for hunger, food security, poverty and nutrition. This allowed us to assess the socio-economic sustainability of smallholder farming in the region and to identify the key drivers of the rapid short-term changes going on in smallholder households in East Africa. We suggest that a more nuanced understanding of the heterogeneity of the smallholder sector is essential to target development interventions efficiently.

Figure 5 provides a conceptual diagram summarizing our key findings. Building on the notion of resilience (Müller et al., 2014), we conceptualize three poverty 'plateaux' for smallholder farmers. The results of Hammond et al. (2019) showed that, during an approximately 4-year period, more than half of the surveyed households shifted between these plateaux, with almost as many falling as rising. Assuming that households wish to get richer and not poorer, and given the development efforts in each site, this shows how difficult it is to rise up and how relatively easy it is to slide down. Development efforts may be made more effective by focussing on making it harder to fall down and easier to move up. Factors associated with falling households can be identified and then applied to other households so that they do not fall down. Similarly, factors identified from rising households can be applied to static households, making it easier for them to move up. For example, falling households tended to have increasing family sizes, and to have lost sources of off-farm income and produce sales. Furthermore, some households are more resilient when moving between plateaux, which is good if they are wealthier but bad if they are poorer. Households from the poorest plateau who appeared to be 'stuck' are constrained by very small farm sizes, and would, therefore, need a different development intervention compared to 'steady' households on the medium

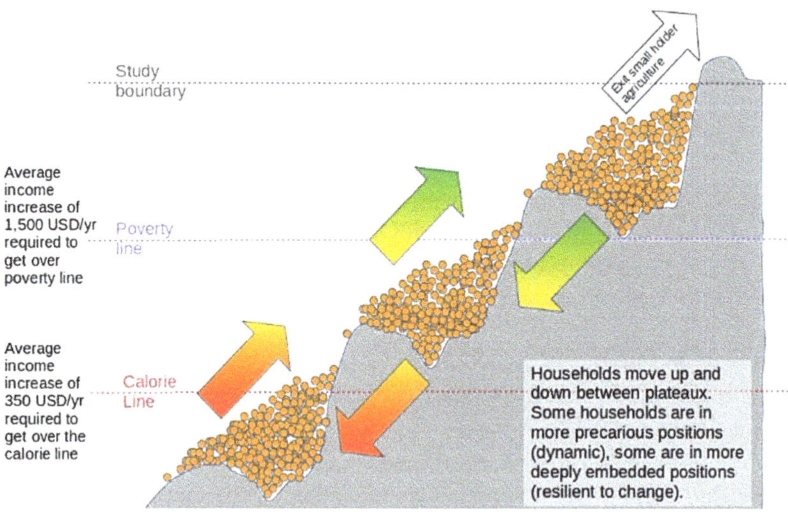

Figure 5 A schematic sketch of the concept of resilience in smallholder farming systems. Values based on Hammond et al. (2019).

plateaux. Interventions aimed to increase off-farm incomes for those with the smallest farms would, therefore, be especially well suited to helping the most vulnerable households escape from their poverty trap. Households looking to move from medium to upper plateaux (or beyond) could benefit more from agricultural intensification.

When overlaying the possible effects of climate change - especially the likelihood of increased climate variability and occurrence of extreme weather events - it is likely that the probability of sliding down from a higher plateau to a lower plateau will increase. In the study by Hammond et al. (2019), this sliding down was mainly driven by the loss of off-farm income. However, in the future, climate change might become a key driver. A (partial) loss of maize yield of say 1000 kg due to drought or flooding translates into a monetary loss of roughly US$250. This is sufficient to shift a household from one plateau down to the one below. A loss of cattle (with a conservative estimated price of US$500 per head, or several goats and sheep) can also mean a shift in plateau. The study by Hammond et al. (2019) shows the more complex dynamics of smallholder welfare with off-farm income and on-farm intensification going hand-in-hand. However, it is clear climate change can put a downward pressure on the ability of farmers to move up the poverty ladder and make agricultural intensification less easy to achieve in a smallholder setting.

In this chapter, we illustrate through a contrasting set of examples how integrated crop-livestock systems currently contribute to food security and diverse diets of smallholder livelihoods and how farmers use these systems to cope with a variable climate. We use this knowledge to better understand what might happen to these systems under different climate change scenarios and what this all means for the resilience of these systems. Despite the importance of mixed crop-livestock systems in low- and middle-income countries, and the vulnerability of their production to climate change, there is limited integrated work that assesses the effects of climate change on these farming systems. Recent work on climate change and smallholder farming systems has focussed particularly on crops and single commodities. While many impact assessments are available at global, regional and local level for individual crops, assessments of climate change effects on livestock production are scarce. More studies are urgently needed to:

1 obtain more reliable assessments of the effect of climate change on livestock in mixed systems;
2 derive general conclusions from these individual studies;
3 get a better insight into the likely future vulnerability of mixed systems to climate (and other) changes; and
4 identify with more certainty appropriate adaptation options.

In identifying adaptation options, it is essential that the diversity of farm households in different populations and regions is captured and that likely 'winners' and 'losers' of global change are identified (van Wijk, 2014; Rodriguez et al., 2017; Hammond et al., 2017b). These analyses help us to target diverse packages of interventions appropriate for the diversity of agricultural livelihoods present at local and regional levels, and better match interventions to farm and household characteristics. To really address the issue of 'the contribution of integrated crop-livestock systems in combatting climate change and improving resilience in agricultural production', more work building on the research presented in this chapter is needed.

7 Where to look for further information

The lead authors of this chapter are scientists of the International Livestock Research Institute (ILRI; www.ilri.org). ILRI's research is directed to improving food and nutrition security through increased production and access to animal-source foods; stimulating economic development and poverty reduction through enhanced livestock value chains and increased productivity; improving human health through improved access to animal-source foods and a reduction in the burden of zoonotic and food-borne diseases; and managing the adaptation of livestock systems to climate change and mitigating the impact of livestock on the environment. They also collaborate intensively with CSIRO's Global Food and Nutrition Security group, a multi-disciplinary team of systems specialists that partner to deliver agricultural outcomes for improved global food and nutrition security. These teams and our collaborators are producing a set of interesting blogs where key new developments in agricultural research and food science can be found at:

https://medium.com/the-nature-of-food
https://www.rhomis.org/blog
https://research.csiro.au/foodglobalsecurity/
http://www.environmentreports.com/foodmatters/

8 References

Ayantunde, A. A., Duncan, A. J., van Wijk, M. T. and Thorne, P. 2018. Review: role of herbivores in sustainable agriculture in sub-Saharan Africa. *Animal: an International Journal of Animal Bioscience* 12(s2), s199–209. doi:10.1017/S175173111800174X.

Below, T. B., Mutabazi, K. D., Kirschke, D., Franke, C., Sieber, S., Siebert, R. and Tscherning, K. 2012. Can farmers' adaptation to climate change be explained by socio-economic household-level variables? *Global Environmental Change* 22(1), 223–35. doi:10.1016/j.gloenvcha.2011.11.012.

de Pee, S., Taren, D. and Bloem, M. W. (Eds). 2017. *Nutrition and Health in Developing World* (3rd edn.). Humana Press, New York.

Descheemaeker, K., Zijlstra, M., Masikati, P., Crespo, O. and Homann-Kee, S. 2018. Effects of climate change and adaptation on the livestock component of mixed farming systems: a modelling study from semi-arid Zimbabwe. *Agricultural Systems* 159, 282-95. doi:10.1016/j.agsy.2017.05.004.

Fraval, S. 2019. Food security in rural sub-Saharan Africa. A household level assessment of crop-livestock systems. Dissertation, Wageningen University.

Fraval, S., Hammond, J., Ayantunde, A., de Boer, I. J. M., Bogard, J., Bosire, C., Chesterman, S., Chilingulo, K., Chinyophiro, A., Dontsop, P., van Etten, J., Herrero, M., Kihoro, E., Lamanna, C., Lannerstad, M., Ng'endo, M., Njingulula, P., Okafor, C., Oosting, S. J., Rao, J., Rosenstock, T., Steinke, J., Stirling, C. M., Teufel, N., Vanlauwe, B., Waha, K., Wichern, J., Yameogo, V. and van Wijk, M. T. 2019. Food access deficiencies in sub-Saharan Africa: prevalence and implications for agricultural interventions. *Front. Sustain. Food Syst.* 3, 104. doi:10.3389/fsufs.2019.00104.

Hammond, J., Fraval, S., van Etten, J., Suchini, J. G., Mercado, L., Pagella, T., Frelat, R., Lannerstad, M., Douxchamps, S., Teufel, N., Valbuena, D. and van Wijk, M. T. 2017a. The Rural Household Multi-Indicator Survey (RHoMIS) for rapid characterisation of households to inform climate smart agriculture interventions: description and applications in East Africa and Central America. *Agricultural Systems* 151, 225-33. doi:10.1016/j.agsy.2016.05.003.

Hammond, J., Pagella, T., Smajgl, A., Yi, Z., van Wijk, M., Xu, J., Ward, J., Su, Y. and Harrison, R. 2017b. Farm types and farmer motivations to adapt: implications for design of sustainable agricultural interventions using the example of rubber plantations in South West China. *Agricultural Systems* 154, 1-12.

Hammond, J., Pagella, T., Fraval, S., Wichern, J., Teufel, N., Kihoro, E., Herrero, M., Rosenstock, T. S. and van Wijk, M. T. 2019. Rapid pace of change for rural smallholders in East Africa: prosperity is driven by off farm income, in tandem with agricultural intensification. *Proceedings of the National Academy of Sciences of the United States of America* (submitted).

Helgeson, J. F., Dietz, S. and Hochrainer, S. 2013. Vulnerability to weather disasters: the choice of coping strategies in rural Uganda. *Ecology and Society* 18(2). doi:10.5751/ES-05390-180202.

Henderson, B., Cacho, O., Thornton, P. K., van Wijk, M. and Herrero, M. 2018. Climate change impacts and options for mixed smallholder farmers in Burkina Faso. *Agricultural Systems* 167, 195-205. doi:10.1016/j.agsy.2018.09.012.

Herrero, M., Thornton, P. K., Notenbaert, A. M., Wood, S., Msangi, S., Freeman, H. A., Bossio, D., Dixon, J., Peters, M., van de Steeg, J., Lynam, J., Parthasarathy Rao, P., Macmillan, S., Gerard, B., McDermott, J., Seré, C. and Rosegrant, M. 2010. Smart investments in sustainable food production: revisiting mixed crop-livestock systems. *Science* 327(5967), 822-5. doi:10.1126/science.1183725.

Jones, P. G. and Thornton, P. K. 2009. Croppers to livestock keepers: livelihood transitions to 2050 in Africa due to climate change. *Environmental Science and Policy* 12(4), 427-37. doi:10.1016/j.envsci.2008.08.006.

Malik, P., Bhatta, R. and Takahashi, J. 2015. *Livestock Production and Climate Change*. CABI Publishing International, Wallingford, Oxford.

Müller, D., Sun, Z., Vongvisouk, T., Pflugmacher, D., Xu, J. and Mertz, O. 2014. Regime shifts limit the predictability of land-system change. *Global Environmental Change* 28, 75–83. doi:10.1016/j.gloenvcha.2014.06.003.

Nicholson, C. F. 2007. Review of methods for modelling systems evolution. ILRI Targeting and Innovation Discussion Paper No. 3. Nairobi, Kenya, p. 123.

Ongoma, V., Chen, H. S., Gao, C. J., Nyongesa, A. M. and Polong, F. 2018. Future changes in climate extremes over Equatorial East Africa based on CMIP5 multimodel ensemble. *Natural Hazards* 90(2), 901–20. doi:10.1007/s11069-017-3079-9.

Rigolot, C., De Voil, P., Douxchamps, S., Prestwidge, D., van Wijk, M., Thornton, P. K., Rodriguez, D., Henderson, B., Medina, D. and Herrero, M. 2017. Interactions between intervention packages, climatic risk, climate change and food security in mixed crop-livestock systems in Burkina Faso. *Agricultural Systems* 151, 217–24. doi:10.1016/j.agsy.2015.12.017.

Ritzema, R. S., Douxchamps, S., Fraval, S., Bolliger, A., Hok, L., Phengsavanh, P., Long, C. T., Hammond, J. and van Wijk, M. T. 2019. Household-level drivers of dietary diversity in transitioning agricultural systems: Evidence from the Greater Mekong Subregion. *Agricultural Systems* 176, 102657.

Rodriguez, D., de Voil, P., Rufino, M. C., Odendos, M. and van Wijk, M. T. 2017. To mulch or to munch? Big modeling of big data. *Agricultural Systems* 153, 32–42. doi:10.1016/j.agsy.2017.01.010.

Rosenzweig, C., Elliott, J., Deryng, D., Ruane, A. C., Müller, C., Arneth, A., Boote, K. J., Folberth, C., Glotter, M., Khabarov, N., Neumann, K., Piontek, F., Pugh, T. A. M., Schmid, E., Stehfest, E., Yang, H. and Jones, J. W. 2014. Assessing agricultural risks of climate change in the 21st century in a global gridded crop model intercomparison. *Proceedings of the National Academy of Sciences of the United States of America* 111(9), 3268–73. doi:10.1073/pnas.1222463110.

Roudier, P., Sultan, B., Quirion, P. and Berg, A. 2011. The impact of future climate change on West African crop yields: what does the recent literature say? *Global Environmental Change* 21(3), 1073–83. doi:10.1016/j.gloenvcha.2011.04.007.

Ruel, M. T., Quisumbing, A. R. and Balagamwala, M. 2018. Nutrition-sensitive agriculture: what have we learned so far? *Global Food Security* 17, 128–53 doi:10.1016/j.gfs.2018.01.002.

Thornton, P. K., Kruska, R. L., Henninger, N., Kristjanson, P. M., Reid, R. S., Atieno, F., Odero, A. and Ndegwa, T. 2002. *Mapping Poverty and Livestock in the Developing World*. International Livestock Research Institute, Nairobi, Kenya, 124pp.

Thornton, P. K., van de Steeg, J., Notenbaert, A. and Herrero, M. 2009. The impacts of climate change on livestock and livestock systems in developing countries: a review of what we know and what we need to know. *Agricultural Systems* 101(3), 113–27. doi:10.1016/j.agsy.2009.05.002.

van Wijk, M. T. 2014. From global economic modelling to household level analyses of food security and sustainability: how big is the gap and can we bridge it? *Food Policy* 49, 378–88. doi:10.1016/j.foodpol.2014.10.003.

van Wijk, M. T., Rufino, M. C., Enahoro, D., Parsons, D., Silvestri, S., Valdivia, R. O. and Herrero, M. 2014. Farm household modelling and its role in designing climate-resilient agricultural systems. *Global Food Security* 3(2), 77–84. doi:10.1016/j.gfs.2014.05.001.

Waha, K., van Wijk, M. T., Fritz, S., See, L., Thornton, P. K., Wichern, J. and Herrero, M. 2018. Agricultural diversification as an important strategy for achieving food security in Africa. *Global Change Biology* 24(8), 3390-400. doi:10.1111/gcb.14158.

Wichern, J. 2019. Food security in a changing world: disentangling the diversity of rural livelihood strategies across Uganda. Wageningen, 204pp..

Zinyengere, N., Crespo, O. and Hachigont, S. 2013. Crop response to climate change in southern Africa: a comprehensive review. *Global and Planetary Change* 111, 118-26. doi:10.1016/j.gloplacha.2013.08.010.

Chapter 10

Agroforestry as a solution for multiple climate change challenges in Africa

C. Mbow, Future Africa at University of Pretoria, South Africa and Michigan State University, USA; E. Toensmeier, Perennial Agriculture Institute, USA; M. Brandt, University of Copenhagen, Denmark; D. Skole, Michigan State University, USA; M. Dieng, Senegalese Institute of Agricultural Research (ISRA), Senegal; D. Garrity, World Agroforestry Centre, Kenya; and B. Poulter, NASA Goddard Space Flight Center, USA

1 Introduction

Agroforestry is not just the inclusion of woody perennials within farming and pastoral systems; it encompasses tree-based livelihoods that are both a traditional land use approach and an economic enterprise developed by farmers in Africa to address many of their land management challenges. Agroforestry practices survived the widely adopted monocropping farming imposed by colonial systems, and now maintained by subsequent so-called 'modern' agricultural policies, more often associated with an export-oriented economic model. It is good news that the monocropping production systems are losing ground across the globe, with strong progress toward the development of more diverse, multifunctional, agroecological farming systems and landscapes (Minang et al., 2015; Simelton and Ostwald, 2019).

In Africa, farming practices inherited from colonial times were typically based on centralized, heavily subsidized approaches (Mbow et al., 2008) that proved costly and financially unsustainable (AGRA, 2017). Many small landholder farmers lack the minimum asset endowments (mainly financial) to

http://dx.doi.org/10.19103/AS.2020.0064.17

enable them to efficiently adopt all aspects of mechanization due to its relatively high cost. Given the context of small holder farming systems, using improved technologies along with agroforestry practices is quite compatible with profitability and better natural resources management, particularly when such practices are using natural solutions based on trees and perennials to secure food and nutrition, advance social protection, environmental safeguards, land rights security, build resilience, and support vulnerable groups.

Sustainable agriculture in Africa entails a set of practices that are suitable for the diversified set of enterprises where small landholder farmers spread their sources of living across a range of mixed farming system based on agroforestry and on the maintenance of biodiversity and other important ecosystem services. Livelihood options promoted by farmers themselves show that the various agroforestry practices are part of successful responses to land degradation and climate change, and are therefore, expanding in various landscapes (Garrity et al., 2010). Agroforestry is an accepted approach for farming in Africa, with or without international development efforts. This fusion between agriculture and culture–local knowledge, perceptions, and attitudes–is an important relationship that holds the basis for transformation, resilience of production, and improved land management systems in Africa (Meijer et al., 2015).

Using trees to induce healthy ecosystems in Africa has become a genuine and ubiquitous land use option, considering the many environmental challenges, such as climate change impacts, biodiversity loss, deforestation, and habitat degradation. Improving tree cover is intimately connected to the reduction of land degradation, particularly in non-forested landscapes (Lewis et al., 2019), to enhancing conservation (corridors, reduced fragmentation) and in addressing the global warming targets of the UNFCCC. The benefits of agroforestry for adaptation to, and mitigation of, climate change are linked to many direct and indirect impacts on ecosystems and land uses, such as food production, water regulation, better health through balanced diets and medicinal plants, for shelter, and for land quality and cash returns (Mbow et al., 2014a; Takimoto et al., 2008a).

It is obvious that the challenges of maintaining biodiversity arise particularly in areas with low resources and high-resource extraction rates, and where external drivers such as climate change modulate the trend toward reduced ecosystems and species diversity. While the drivers of deforestation include modern agriculture, it has been reported that in Africa, some traditional farming systems (e.g. slash and burn agriculture, using fires for land management) can also lead to rapid land degradation (Scherr and Mcneely, 2008; MacFadyen et al., 2012; Barral et al., 2015; Mbow et al., 2000).

Conserving biodiversity within human-influenced landscapes is a challenging task in the face of high land pressure and diverging stakeholders' interests. The role of agroforestry cannot be limited to expanding the numbers of trees per unit area. The diversity of tree varieties, with associated diversity of wild and weedy species, is an important part of the beneficial role that agroforestry contributes to communities. There are many places and practices where the diversification of agroecosystems enhances agrobiodiversity, ecological processes, biodiversity conservation, and the delivery of ecosystem services (Tscharntke et al., 2012).

Despite a number of claims that agroforestry practices contribute to climate change responses through various services, there is very limited substantive supporting materials that systematically show what role agroforestry systems play for adaptation to climate change (Muthee et al., 2017). In this chapter, we assembled the action-oriented knowledge that builds a more consolidated narrative, and a set of feasible actions to advance climate change responses based on agroforestry systems.

2 Typology and dynamics of agroforestry in Africa

2.1 Agroforestry in practice

African agricultural landscapes are traditionally known for the importance of trees for ecosystem functions, but also as a source of food, fiber, and energy. Many of the farming practices are tree-based systems combined with various sets of land management practices for food and animal production. These practices may have little more in common with each other than that they include the coincidence of woody perennials with agricultural crops and/or livestock. Each practice has a set of objectives that can differ from similar ones in different agroecological contexts. Across Africa, agroforestry has proven suitable for small landholder farmers and less-developed economies, because of the range of practices that offer a source of livelihoods and a basis for local economies (Nair, 1993).

There are various forms of agroforestry in Africa leading to complex typologies as outlined by Nair (1993), Torquebiau (2000), and Mbow et al. (2014a). Here, we suggest a simplified approach for agroforestry classification that determines which type of agroforestry systems exist in various agroecologies but highlighting the land use objective. We avoided concepts such as parkland or mosaic as they seem to overlap in various landscapes. Similarly, practices such as household woodlots and home gardens can be mixed with other agroforestry practices. The characterization we suggest for differentiating agroforestry systems is more functional (the type of use) than structural (composition and appearance). For example, multistoried systems can be applied to family plots as home gardens, as well as in tree crops produced on a much bigger scale. Rather, we suggest a typology based on a first entry that is the agro-system where the agroforestry is taking place, followed by the type of practices; and it ends with the objectives of the various practices (Fig. 1).

2.2 Changes in Africa's woody cover in relation to demography and livelihood

Deforestation linked to human population growth in Africa's humid regions is an ongoing issue; however, the sparsely populated African drylands have been experiencing a pronounced greening trend during the past 30 years (Brandt et al., 2017). While herbaceous vegetation has a high interannual variation, the greening trend seen in drylands is mainly caused by the woody layer, and it is largely driven by climatic factors, for example, by an increase in the frequency of heavy rainfall events and extra-seasonal rainfall (Brandt et al., 2017, 2019; Zhang et al., 2019). Recent studies challenge the narrative of an exclusively negative correlation between human population growth and changes in woody cover (Brandt et al., 2018a). Population growth cannot exclusively be linked with deforestation. In particular, farmers in arid and semiarid Sahelian lands promote a medium-to-high woody cover in their fields, often even exceeding the woody cover of the surrounding savannas (Brandt et al., 2017, 2018a).

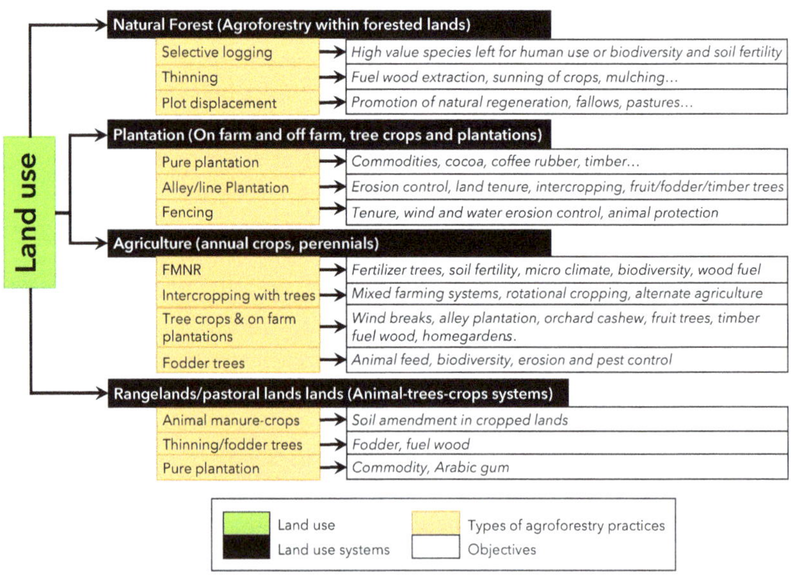

Figure 1 Agroforestry typology based on land use objectives.

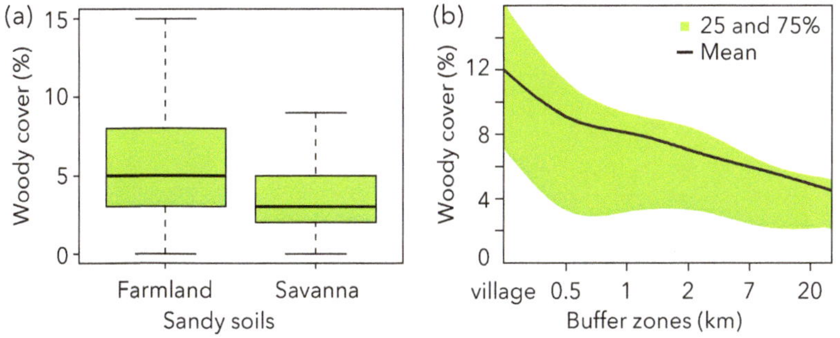

Figure 2 Woody cover in the Sahel: (a) Farmlands and savanna with sandy soils. (b) By distance to settlements. Source: adapted from Brandt et al. (2018a).

Farmers and pastoral communities depend on a variety of ecosystem services provided by trees: trees improve soil fertility, the woody foliage is used as fodder for livestock, the wood is used for household energy and for construction, and tree fruits are an important source of nutrition for humans and animals (Kalaba et al., 2013). In the more humid regions, the transformation of savanna woodlands to farmlands implies an initial reduction of woody cover; however, the overall tree density in farmlands remains high (Fig. 2). This is good news, as the Bonn Challenge is committed to restoring 350 Mha of forest by 2030, which can only be achieved if tree cover in agricultural lands is increased (Lewis et al., 2019), while at the same time local livelihoods are enhanced. Moreover, while large parts of the dryland savanna woody vegetation consist of small trees and shrubs with a relatively low carbon stock, farmland trees take advantage of the reduced competition to grow larger than in the rangeland savannas, and are thus a considerable sink for carbon stocks. Figure 3 shows the distribution patterns of trees in relation with the distance from settlements.

At the continental scale, most drylands have seen an increase in woody cover. Deforestation is mainly occurring in the densely and sparely populated zones of the more humid areas (Fig. 3).

3 Agroforestry and climate change mitigation

3.1 Mitigation potential

Atmospheric carbon removal, combined with reducing emissions from fossil fuel use, is critical to limit global warming and to ensure a stable climate. Rogelj et al. (2018; in IPCC-SR 1.5, 2019) report indicates that all pathways that limit global warming to 1.5–2.0°C with limited or no overshoot will require the use of carbon dioxide removal (CDR) on the order of 100–1000 $GtCO_2$ during the twenty-first century. CDR will be needed to compensate for residual emissions and, in most cases, to achieve enough net-negative emissions to return global

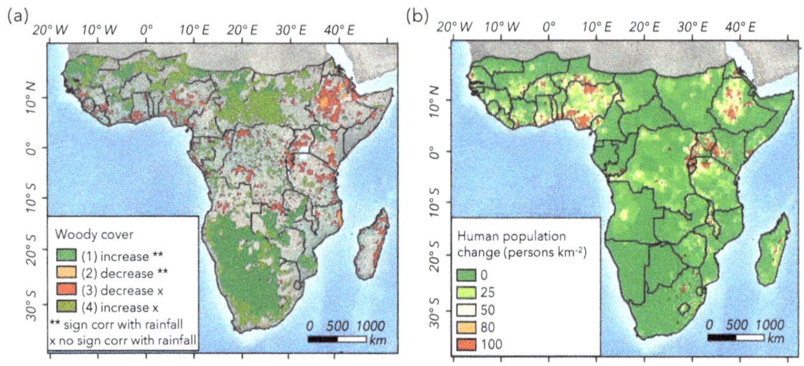

Figure 3 Woody cover and human population growth. (a) Woody cover changes, 1992-2011. (b) Human population change, 1990-2010. Source: adapted from Brandt et al. (2017).

warming to 1.5-2.0°C following an overshoot peak. The IPCC concludes that biological approaches to carbon capture appear to be by far the most promising ones. This is an ambitious mitigation goal that nature-based solutions can help achieve (Griscom et al., 2017a). These measures include protecting current forestlands, restoring degraded forest lands, increasing tree cover on agricultural lands through agroforestry, and increasing the biomass production of pasturelands.

In Africa, the obvious starting point is to use nature. Trees have been storing carbon in their trunks and roots for millions of years. By restoring degraded forests and degraded agricultural lands, and by using smarter farming practices, we can capture much more carbon on the land. There is overwhelming evidence that these natural solutions can pay for themselves quickly, improving food security, creating jobs, reducing poverty, and extracting billions of tons of carbon (Griscom et al., 2017b).

It is not only timber and agroforestry on croplands that sequester carbon. Tree crops provide the same service, while yielding food and other products. Figure 4 shows the sequestration rates in Mg/ha/yr of native African tree crops, along with figures from the IPCC 2006 Guidelines for woody crops in various tropical rainfall regimes, compared with conservation agriculture. Rates of carbon storage for tropical tree crops production range from 1.8 Mg/ha/yr to 10 Mg/ha/yr, compared with 0.6 Mg/ha/yr for conservation agriculture.

The global sequestration potential of increased adoption of tropical staple tree crops (trees producing protein, carbohydrates, and oils) has been estimated on degraded grassland (Hawken, 2017). Their model calculated a global sequestration impact of 20.2-46.7 Pg of CO_2 between 2020 and 2050. Figure 4 gives various rates of carbon sequestration for different crops and practices.

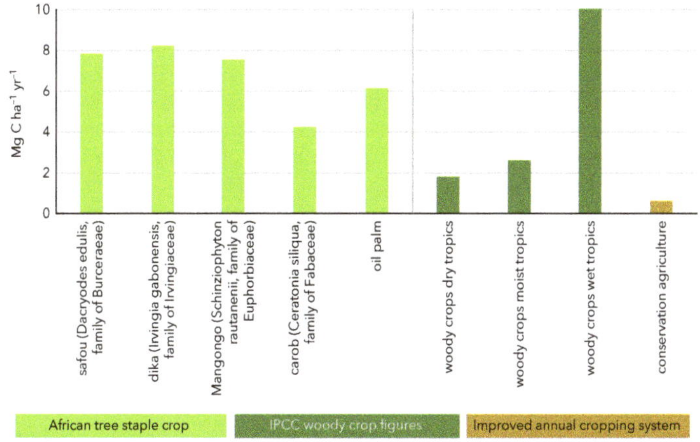

Figure 4 Carbon sequestration rates of African tree crops in Mg/ha/yr. Source: data from Egbe and Tabot (2011), Geraldo et al. (2010), IPCC (2006) and Srinivasarao et al. (2015).

Box 1 Long-term trend in carbon uptake in agroforestry landscapes of the Sahel

The very high mitigation response based on agroforestry, along with co-benefits to increased food security and animal production, has been widely accepted (Smith et al., 2014; Carsan et al., 2014). Agroforestry is the second-best land-based mitigation option after natural-forest restoration. It is seven times more effective than timber tree plantations (Lewis et al., 2019) and does not compromise, but rather in most instances reinforces the achievement of other adaptation goals, such as food and nutrition, livestock productivity, and soil fertility; while it reduces heat stress and excessive insolation inducing a more balanced microclimate in cropping and livestock systems.

The role of semiarid agroforestry systems in the carbon budget has been underestimated because of many uncertainties in the drivers affecting carbon dynamics in these highly variable spatial and seasonal biomes. The strong positive and persistent long-term changes in tree cover observed from ground and space-borne measurements have a clear and favorable influence on the global carbon budget. For example, the contribution of the Sahel to global terrestrial carbon uptake has reversed from source to sink since 1981. The carbon sink in the Sahel has become strongly positive since 1981, according to the 10 DGVM models used in the TRENDY ensemble (Sitch et al., 2015). The vegetation productivity in the Sahel has increased and it has led to net carbon uptake, with their results supported by remote-sensing and data-driven models of carbon uptake (Brandt et al., 2015). The increase in the strength of the carbon sink is largely driven by an increase in gross primary productivity during the growing season (JASO).

An increase in precipitation in the July-August-September-October (JASO) season has led to reduced drought (based on Standardized Precipitation Evapotranspiration Index-SPEI) and thus, to better growing conditions. At the same time, a long dry season in the Sahel does not mean there is no productivity during that period. The very resilient agroforestry trees (e.g. Acacia, Balanites, Shea, Ziziphus, etc.) are tapping into the water table during that season to sustain an accumulation of biomass over time–despite large effluxes during the dry season and transition periods–explaining an improvement in the net sink over the last 30 years (Dardel et al., 2014).

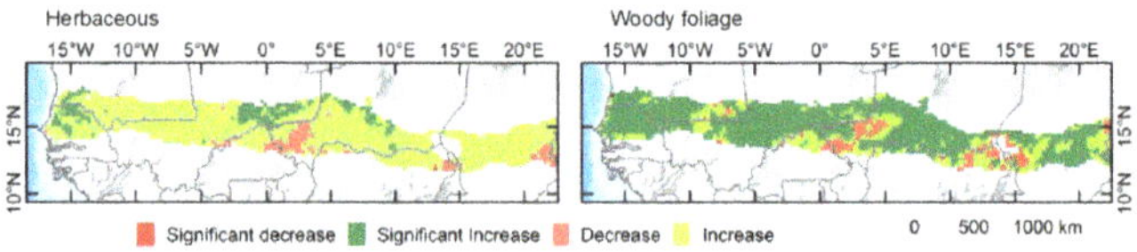

Figure Box 1.1 The Sahel: Changes in (a) herbaceous vegetation and (b) woody foliage in the Sahel (1992-2012). Source: adapted from Brandt et al. (2019).

The recent large variation of the land dynamics is associated with several forms of agroforestry ecosystems and tree composition, leading to differences in land use changes, cropland expansion, importance of livestock and the area of grazing lands. These management processes are not captured by the TRENDY model ensemble, and so the Sahel sink may in fact be larger than what the models suggest. Land degradation or tree recovery leads to differential flux exchanges. These are modulating factors of carbon seasonal change.

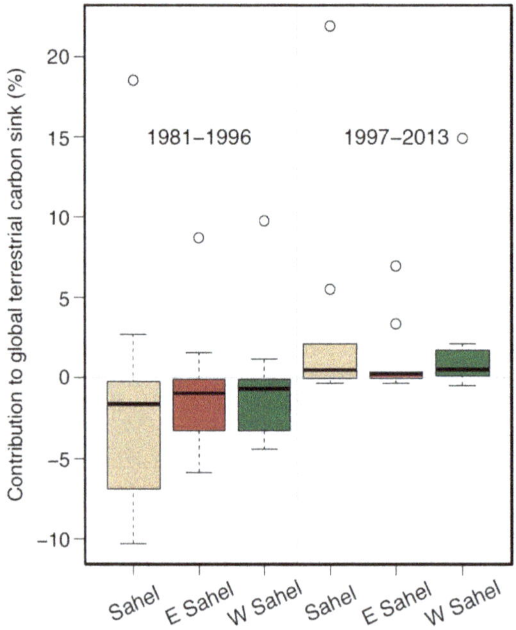

Figure Box 1.2 The Sahel: From a net source to a net sink of carbon.

3.2 Agroforestry potential for global emissions neutrality

Currently, more than 43% of the world's agricultural land has greater than 10% tree cover (Zomer et al., 2016). These trees do not appear in any forest classification, but they are an important carbon stock. The FAO would classify these landscapes as systems of trees outside of forests (TOF). This is an extremely large area of tree cover that is outside of the land that is classified as 'forest' (Zomer et al., 2016). Between 2000 and 2010, there was a 2% increase in tree cover on agricultural lands globally, and 36% of Africa's land area had an increase in woody cover between 1992 and 2011 (Brandt et al., 2017).

Agricultural land is accumulating carbon through this increased tree cover. The annual expansion of tree cover on agricultural land is storing an additional 740 Mg CO_2 per year on a global basis (Zomer et al., 2016); thus, agroforestry is offsetting greenhouse gas (GHG) emissions by over 0.7 billion tons CO_2 eq/yr (Lewis et al., 2019).

Tree cover on agricultural land is now 10–30% in subhumid Africa. Most agricultural land and crop production lands are suited to growing trees. Traditionally, farmers have managed the natural regeneration of trees on their farmlands. This practice is being adopted at an accelerating pace, and it is being supplemented by the active planting of useful trees on farm and field boundaries and in crop fields.

The mitigation benefits from agroforestry are now offsetting a large portion of the total global emissions from the agriculture sector (viz. livestock and rice crop methane emissions, and soil carbon losses, including embedded emissions coming from C soil oxidation, biomass export, and nitrogen emissions). The total balance shows that agricultural land has only a net GHG emissions of 1.4 billion tons (Garrity, 2017). It will be possible to improve these numbers, since the future expansion of agroforestry can increase carbon stocks at a low cost, while generating a high income and improving livelihoods, particularly in the tropics. Garrity (2017) has called for a global commitment to doubling the annual rate of increase in useful tree cover on agricultural lands during the next 15 years, and to tripling the current annual rate by 2050; at which time agroforestry could offset the emissions on current agricultural land entirely–and thus make it possible for the agricultural sector to become emissions neutral by 2050.

There are many ways to further increase carbon stocks in agricultural lands:

1 Establish trees on all field and farm boundaries and in alleyways and on contours;
2 Expand farmer-managed natural regeneration of trees on farmlands, particularly in the drylands;
3 Support a major increase in perennial crops for food production;

4 Expand farm woodlot expansion; and
5 Establish trees and shrubs for fertilizer, food, and fodder throughout crop fields (discussed in more detail in the next section).

The Sahel is a promising example, where the combination of positive rainfall anomalies and active tree cover promotion by communities has led to a positive carbon balance during recent decades (Brandt et al., 2019; Fig. 5). Carbon fluxes in African drylands are highly variable in space and time, and they are related to weather conditions, but a consistent trend in carbon uptake is related to the increase in tree cover. Because of the complex relationships between climate and human processes, the future fate of the net uptake of CO_2 in semiarid Africa is uncertain, but there is good cause to be optimistic if the current trend toward accelerating the restoration of degraded farmlands can be sustained and further improved. This is a major goal of the African Restoration Initiative (AFR100), which is coordinated by the Africa Union. Twenty-eight countries have now set national land restoration targets that amount to an aggregate commitment of over 112 million ha of restored land by 2030.

Droughts have caused considerable carbon losses in the drylands. At 5–10°N, large C gains can be observed which are unrelated to climate. It is very likely that a large portion of this gain is related to human management of natural regeneration as a response to food, climate, and soil fertility objectives.

A brief review of a range of cases suggests that the carbon stock potential of agroforestry systems, as measured in the field, is highly variable, but it consistently contributes an important addition to the whole carbon balance of these systems (Table 1). The contribution to soil C is significant. Aboveground carbon is also variable, depending on the land use system, the particular species that are present, and the rotation period. Table 1 shows the carbon stock in various agroforestry systems in Africa (Fig. 6).

For comparison, Table 2 shows that various commodity trees have an important potential for mitigation when cultivated as plantation tree crops for land restoration (Lewis et al., 2019).

3.3 African perennial staple crops and climate change mitigation

Perennial crops show impressive potential to sequester carbon above and belowground. Perennial fruits are widely known, and some perennial vegetables such as *Moringa oleifera* are gaining acceptance, but the great majority of cropland produces staple crops. Staple crop production in Africa occupies 235.1 Mha, 87% of African cropland (FAO, 2019). Sequestration in annual staple crop systems can be achieved by incorporating agroforestry trees in the crop fields and/or on field boundaries; but some staple crops are themselves perennial.

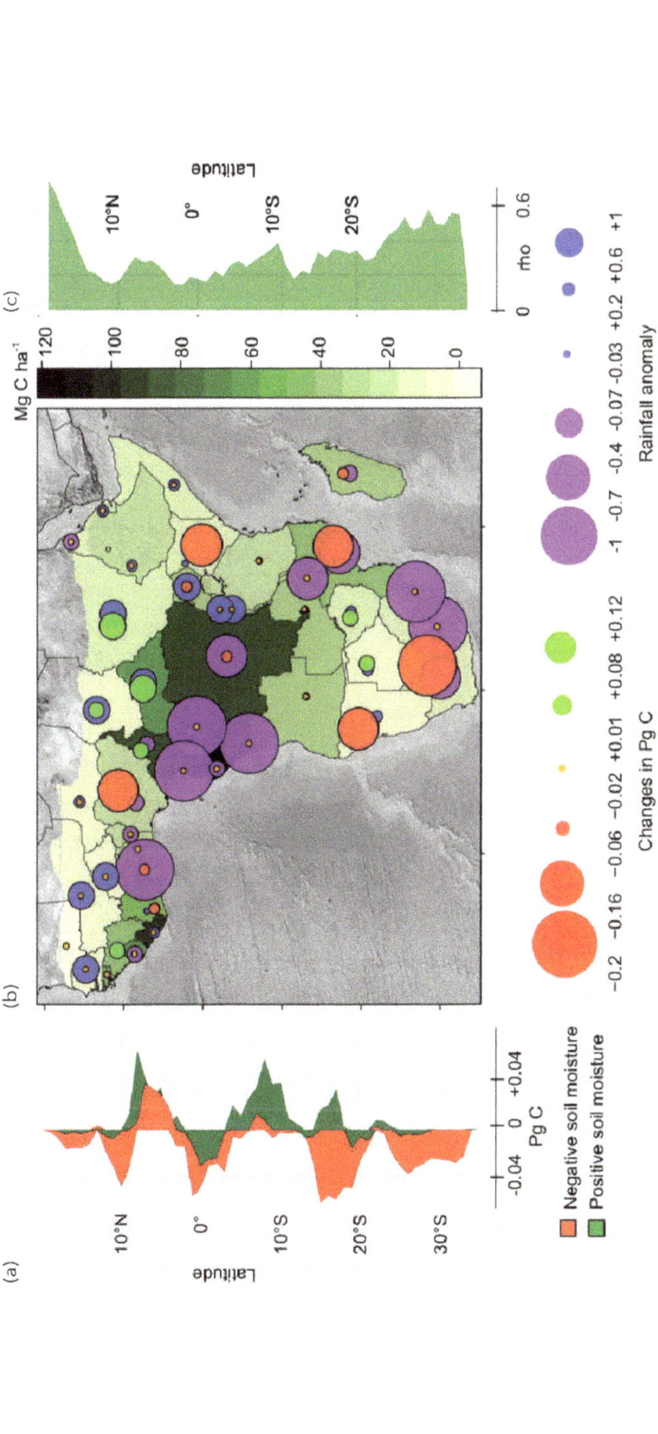

Figure 5 Recent carbon fluxes across Africa in relation to rainfall and soil moisture (2010–2016). Climate as driver of carbon stock dynamics. (a) Direction and magnitude of net carbon change 2010–2016 (summed per latitude) are shown for areas with positive (green) and negative (red) linear trends in soil moisture (2010–2016). (b) Average carbon density (in Mg C ha^{-1}) and net carbon change 2010–2016 summed for each country (in Pg C). Anomalies in annual rainfall (2010–2016 compared to 1981–2016) averaged for each country are shown as purple (negative trend) and blue (positive trends) circles. (c) Correlations between annual carbon stocks and annual soil moisture (Spearman's rho, n = 26 711) averaged along the latitudes. Source: adapted from Brandt et al. (2018b).

Table 1 Carbon in soil and biomass of selected agroforestry systems

Agroforestry systems	Mg C/ha	Source
Soil carbon estimates		
Soil C in parklands' agroforestry–Senegal	84.85	Tschakert et al. (2004)
Soil organic carbon under tree fallows–Tanzania Morogoro	13-25.6	Liu et al. (2008)
Alley cropping Leucaena–Nigeria	14	Lal (2005)
Shaded coffee, *Coffea canephora var. robusta + Albizia adianthifolia*-South West Togo	97	Dossa et al. (2008)
Parkland, *Faidherbia albida*–Mali	33	Takimoto et al. (2008b)
Live fence (*Acacia nilotica, A. senegal, Bauhinia rufescens, Lawsonia inermis,* and *Ziziphus mauritiana*)–Mali	24	Takimoto et al. (2008b)
Fodder bank (*Gliricidia sepium, Pterocarpus lucens, and P. erinaceus*)–Mali	33	Takimoto et al. (2008b)
Gliricidia sepium + maize (Zea mays), Zomba–Malawi	123	Makumba et al. (2007)
Parklands–Senegal	30	Nair et al. (2009)
Parklands cultivated fields–Senegal	18-20	Tschakert et al. (2004)
Parklands fallow–Senegal	14	Tschakert et al. (2004)
Parkland *F. albida*–Mali	30	Takimoto et al. (2008b)
Parkland *V. paradoxa*–Mali	25	Takimoto et al. (2008b)
Live fence–Mali	25	Takimoto et al. (2008b)
Fodder bank–Mali	35	Takimoto et al. (2008b)
Agroforest soil (semiarid tropics)	3-60	Williams et al. (2008)
Biomass estimates		
AF systems in Segou, Mali-Sahel/parklands of *Faidherbia albida*	70.8	Kumar and Nair (2012)
Agriculture and parkland–Woody	39.51	Woomer et al. (2004a)
Tree parklands–woody biomass	32	Woomer et al. (2004a)
Agro-silvo-pastoral (humid tropics)	30-230	Glenday (2008)
Cropland, parklands, and fallows with trees	31.8-52.1	Albrecht and Kandji (2003)
Old peanut basin in Senegal–*Faidherbia albida*	28	Kimaro et al. (2011)
AF systems around the Arabuke Sokoke forest on the Kenyan	19	Takimoto et al. (2008b)
Anacardium occidentale–India	19.5	Sundarapandian et al. (2013)
Long-term fallow–Mozambique	15.7	Walker and Desanker (2004)
Agroforestry and integrated land use (different ecoregions)	12-228	Woomer et al. (2004a)

Agroforestry systems	Mg C/ha	Source
Rotational woodlots–Tanzania	11.6-25.5	Dixon et al. (1994)
Acacia senegal improved fallow	9.60-18.11	Temgoua et al. (2018)
Silvopastoral zone of Senegal–woody biomass	9	Kimaro et al. (2011)
Cordyla pinnata–Senegal	8.81	Diatta et al. (2016)
Tamarindus indica–Madagascar	6.4	Ranaivoson et al. (2015)
Acacia nilotica parks	6.15	Haokip (2016)
Young agroforestry parks–West Africa	5.7-6.3	Kursten and Burschel (1993)
Cacao agroforest–Cameroon	165	Duguma et al. (2001)
Planted tree fallows–Kenya	17-41	Stahl et al. (2002)
Parkland *F. albida*–Mali	55	Takimoto et al. (2008a)
Parkland *V. paradoxa*–Mali	20	Takimoto et al. (2008a)
Live fence–Mali	5	Takimoto et al. (2008a)
Fodder bank–Mali	5	Takimoto et al. (2008a)
Tree gardens high density–Tropics	240	Roshetko et al. (2007)
Rubber intercropping–Tropics	200	Roshetko et al. (2007)
Coffee intercropping–Tropics	160	Roshetko et al. (2007)
Traditional cocoa–Ghana	39-131	Somarriba et al. (2013)
Tropical agroforestry–Cameroon	70-100	Mutuo et al. (2005)

These perennial staple crops produce protein, carbohydrates, and oils. They include trees, shrubs, vines, succulents, grasses, and herbaceous perennials. The edible parts include fruits, legume pods, nuts, seeds, pulses, egusi (melon seeds), aerial tubers, trunk starch, sap, and milk products. Though these perennial staples are already consumed by people, it may be that replacing annual grains that are fed to livestock with fodder produced by perennial species is the easiest initial market for increased production (Toensmeier, 2016).

Perennial staple crops are grown in orchard or plantation systems, and in agroforestry systems, including silvo-arable, silvo-pasture, and multistrata models (Toensmeier, 2017). As of 2016, 9% of Africa's staple crop production area was in perennial crops, notably bananas and plantains (5.9 Mha), oil palm (4.5 Mha), olives (3.4 Mha), cashews (3.3 Mha), and coconuts (1.2 Mha; FAO, 2019).

Many perennial staple crops are native to the African continent. Fifty-three species indigenous to Africa are listed in Toensmeier (2016), representing 28% of all perennial staple crops. Figure 7 shows the form of these species. It is notable that 50% of these species are trees.

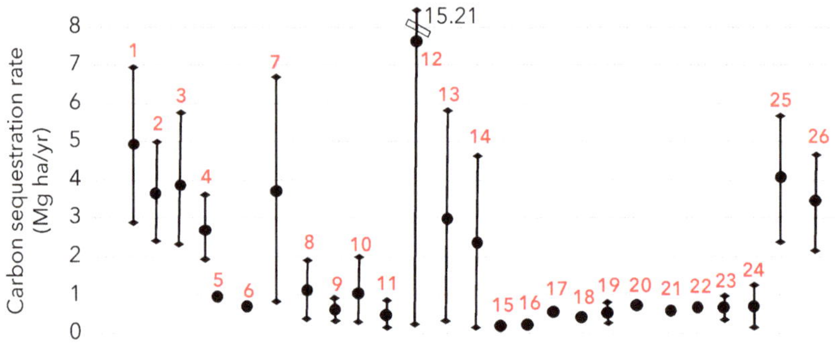

1	Pandey (2002)	Carbon sequestration rate in tropical agroforestry systems
2	Kimaro et al. (2011)	Rotational woodlots in Tanzania
3	Luedeling (2012)	Rotational woodlots in Africa
4	Kuersten and Burschel (1993)	Shade trees in coffee and cocoa
5	Dixon (1994)	Fuel wood system in DRC
6	Walker and Desanker (2004)	Long term Fallow-Mozambique/regrowth
7	Jarecki (2003)	Aboveground biomass in agroforestry systems
8	Kuersten and Burschel (1993)	Agroforestry for fuel wood
9	Luedeling (2012)	AF systems in East Africa (tree planting, intensification of windrows, conversion of cropland to home-gardens)
10	Kuersten and Burschel (1995)	Fuel wood plantation
11	Luedeling (2012)	AF systems in East Africa (tree planting, intensification of windrows, conversion of cropland to home-gardens)
12	Kumar and Nair (2011)	Carbon uptake for various agroforestry systems in the tropics
13	Luedeling (2012)	AF systems in Southern Africa (regrowth of woodland in abandoned farms, coppiced Miombo, Faidherbia plantation, rotational woodlots)
14	Dixon (1995)	Agroforestry and integrated land use (different ecoregions)
15	Tschakert (2004)	*Faidherbia albida* plantation in Senegal
16	Tschakert (2004)	*Faidherbia albida* plantation in Senegal
17	Woomer et al. (2004b)	Restoring degraded grassland to woody grassland in Senegal
18	Takimoto et al. (2008b); Tschakert et al. (2004); Woomer et al. (2004b)	Establishment of new parklands in the Sahel
19	Batjes (2004)	Tree planting to restore highly degraded land in East Africa
20	Henry et al. (2009)	Intensification of windrows and tree biomass
21	Henry et al. (2009)	Conversion of cropland to home-gardens in Southern Africa
22	Walker and Desanker (2004)	Regrowth of woodland on abandoned farms in Mozambique
23	Stromgaard (1985); Chidumayo (1997)	Coppiced Miombo woodland in Zambia
24	Okorio and Maghembe (1994)	Faidherbia albida plantation in Tanzania
25	Nyadzi et al. (2003); Kimaro (2009)	Rotational woodlots in Tanzania
26	Kaonga and Bayliss-Smith (2009)	Rotational woodlots in Zambia

Figure 6 Carbon sequestration rate of various agroforestry systems in Africa (Table appendix gives the sources).

The domestication level of native African perennial staple crops is variable (see Fig. 8). 5% percent are global crops with over $1 billion in annual trade. Another 11% are minor global crops, which are grown across the tropics but with

Table 2 Tree commodities sequestration rate

Tree commodity	Mg C/ha/yr
Timber (Inc. Acacia and Eucalyptus spp.)	5.0
Avocado (*Persea americana*)	2.6
Cashew (*Anacardium occidentale*)	3.6
Cocoa (*Theobroma cacao*)	1.5
Mango (*Mangifera indica*)	1.6
Rubber (*Hevea brasiliensis*)	3.7

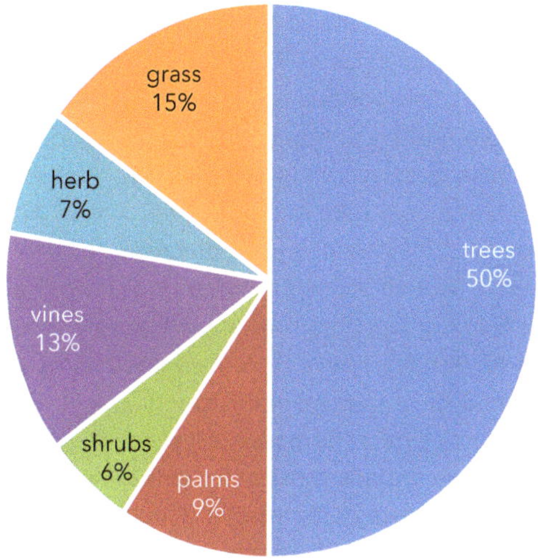

Figure 7 The form of African perennial staple crop species.

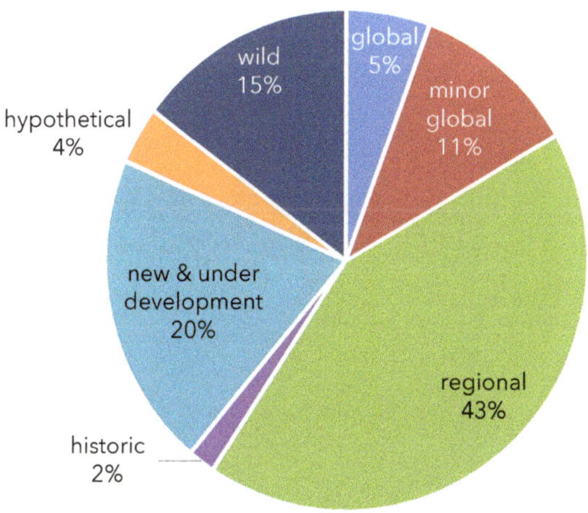

Figure 8 Domestication status of African perennial staple crop species.

lower global production levels. 42% are regional crops, which are cultivated in their region of origin but not grown elsewhere. Eleven percent are new crops, or are under development (e.g. perennial sorghum is under development and close to release (Toensmeier, 2016)). 2% are historic (cultivation has been largely abandoned), and 4% are hypothetical (annual crops which can be crossed with perennial forms, but where this effort is not underway at present). The remaining 15% are wild staple foods identified as being worthy of cultivation. Most perennial staple crops, including the African species, are neglected and underutilized. Expanded production of these species could make a notable contribution to crop biodiversity and could be an important basis for the future perennialization of agriculture. Note that many perennial staple crops native to other regions could be suitable as well.

Toensmeier (2017) categorizes staple crop species in five categories, based on their protein and fat content. Basic starch crops contain 5% or less of protein and oil (like cassava), balanced carbohydrates feature oil and/or protein between 6% and 15% (like maize), protein crops have protein at 16% or higher but oil under 15% (like cowpea), oil crops have oil at 16% or higher but protein under 15% (like oil palm), and protein-oil crops have protein and oil content both of 16% or higher (like groundnut). Figure 9 shows the breakdown of nutrient classes for Africa's native perennial staple crops. Protein-oil is the largest category, with 35% of species in this category.

Perennial staple crops must demonstrate yields that are competitive with annual staples in order to replace them without the need for land use change. Several challenges make these comparisons difficult. Comparison must be on a dry-weight, edible portion basis, as many perennial staples have a higher water content and more inedible parts than staple grains (making it difficult,

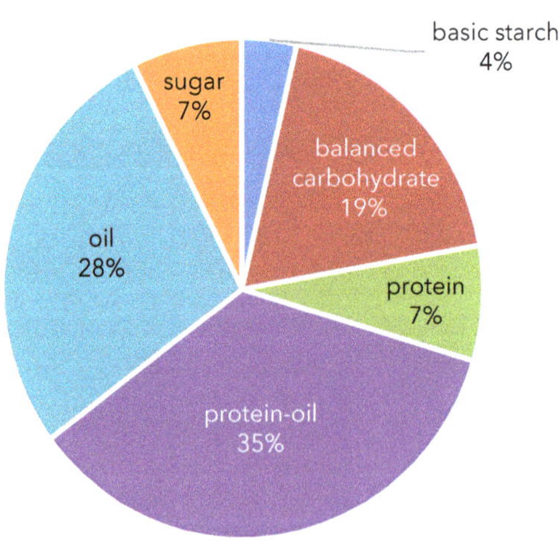

Figure 9 Nutrient classes of African perennial staple crop species.

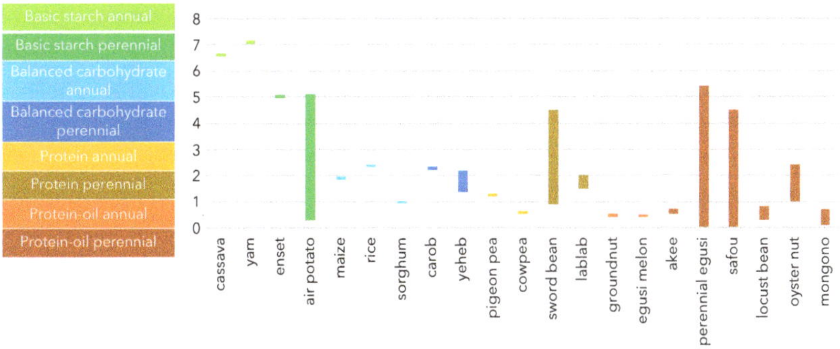

Figure 10 Average African annual staple crop yields in dry-weight, edible portion Mg/ha compared with African perennial staple crop yield ranges in dry-weight, edible portion Mg/ha. Source: data from FAO (2019) and Toensmeier (2016).

for example, to compare banana yields with rice). The next challenge is that FAO data on average yields is not available for most perennial staple crops. Yields for perennial staples are typically reported as wide ranges (Toensmeier, 2016). Figure 10 compares average African yields for annual staples in Mg/ha, as reported by FAO (2019), with the reported yield ranges in Mg/ha reported in Toensmeier (2017).

These results are clustered by nutrient class for more relevant comparison. In the basic starch category, the average yields of the annual staples, that is, yam and cassava (6.6–7.1 Mg/ha African average), exceed the reported yields for the perennial staple crops and air potato (0.3–5.1 Mg/ha), though the higher end of the range for the perennials is respectable. For balanced carbohydrate crops, yields of annual and perennial staples fall within the same range (1.0–2.4 Mg/ha for annuals African average, and 1.4–2.3 for perennials). In the case of protein crops, annual crops (0.6–1.3 Mg/ha African average) yielded lower than the reported ranges of perennials (0.9–4.5 Mg/ha). Protein-oil crops show an even stronger pattern, with relatively low-yielding annual crops (0.4–0.5 Mg/ha African average) versus a number of high-yielding perennials (0.3–5.4 Mg/ha; see Fig. 10). Thus, it can be seen that there is evidence that some African perennial staple crops may show competitive yields with annual staples in their nutrition class.

4 Agroforestry, biodiversity, and ecosystem-based adaptation

4.1 Ecosystem benefits and adaptation

Land-based adaptation measures to reduce the impact of climate change and soil degradation on food production will be central to any strategy for poor communities, particularly in a highly variable climate (Mbow et al., 2014b;

Smith et al., 2014). In order to fill the food gap and build resilient ecosystems, African communities rely on pathways for sustainable yield improvement through improved soil management and mitigation of the adverse effects of climate change (Garnett and Godfray, 2012). Agroforestry systems counter many environmental stressors afflicting poor communities, including unusually high winds, heavy precipitation (flooding), heat stress, drought, pest outbreaks, salinity, and nutrient deficiencies (Iskandar, 2016; Brandt et al, 2014). Trees on farms generate biomass that provides soil organic matter (leaf litter, root decomposition) while at the same time they produce many other benefits, such as firewood and environmental services (Lipper et al., 2014).

In addition, biodiversity provides various environmental, economic, and social benefits that show the important contribution that agroforestry can play in achieving the climate goals. Another challenge with agroforestry-based / ecosystem-based adaptation (EBA) is slower turnover. Agroforestry like most perennials will take a few years to unleash its benefits, while other sectors may be able to act in a shorter time window. Long-term planning for adaptation in most African ecoregions means securing the natural assets through preservation of productive capacities, almost all related to sustainable land management.

All levels of biodiversity (ecosystem, species, and genetic diversity) are exhibited by various agroforestry systems. Conserving of biodiversity within an agro-ecosystem is crucial to ensure the consistent supply of goods and services, such as species evolution, plant improvement through breeding, organisms that create the biological variation of agro-ecosystems, and also a wide range of ecological functions (Atta-Krah et al., 2004). To achieve this goal, multifunctional landscapes (ML) are seen as a possible alternative to habitat exclusion in human-dominated landscapes, as they allow the coexistence of multiple objectives, such as nature conservation and resource use (Ross-Cuéllar et al., 2019).

Conservation of natural resources to reduce biodiversity loss is still a big challenge. For example, Kleijn et al. (2011) claimed that during four decades, even though conservation efforts increased rapidly and significantly, biodiversity threats also increased. This indicated that despite the awareness among conservation scientists and decision makers, efforts targeted to minimize biodiversity loss have been largely unsuccessful. Agroforestry has a role to play in conservation and sustainable agriculture, along with its importance through the services it provides to achieve the climate goals. It is a compromise solution to the apparent land use conflict between forest conservation, reforestation, and agricultural land use.

4.2 Human benefits and adaptation

EBA has gained a large space in national adaptation plans in Africa (Muthee et al., 2017). But investments in the forest and agroforestry sector are still low, and it comes after agriculture, water, and energy in importance. The profitability of agroforestry, as compared to a pure commodity business, is seen to be lower, although this is not the case due to its higher land equivalent ratio. Sustaining agroforestry is a condition to support agriculture, water, energy, and biodiversity.

Rural livelihoods depend on ecosystem services to adapt to climate change impacts (Muthee et al., 2017). Ecosystems provide products such as food, fiber, energy, building materials, and medicine (to name just a few) and services such as water and biodiversity. The latter itself provides services such as pollination and pest control.

Additionally, agroforestry could also contribute to restoring ecosystem services on land that has been transformed from natural ecosystems to intensive agricultural systems, a process which is among the underlying causes of the high biodiversity loss throughout the world (Vallejo-Ramos et al., 2016).

Appropriate resource conservation and production ecology has focused on providing the knowledge base and management practices for biodiversity conservation and enhancing sustainable production. Developments in tree-based ecology are giving more attention to the design of agroecosystems to achieve biodiversity conservation and other ecologically and environmentally crucial ecosystem services. Advanced policies, institutions, and production systems can play a crucial role in reconciling biodiversity conservation and ecological services with agricultural production systems (Brussaard et al., 2010).

Agroforestry has great potential for marketable products such as fruits, vegetables, oil, timber and fuel wood, and so forth. Most of these agroforestry products are now attracting advanced industrial processing and marketing. This includes the seed oils from Baobab, Shea, Balanites, Allanblackia, Azadirachta, and gums and resins such as gum Arabic, which is used in numerous food products, cosmetics, and organic pesticides. The economic importance of agroforestry can be partly understood by examining data on the export value of major tree products. Drawing upon data from FAOSTAT (2011), conservative estimates of the international trade in selected tree products were valued at a whopping US$140 billion in 2009.

The actual valuation levels are much higher, considering that the FAO list includes only well-known and common tree products, and that many tree products in developing countries are not marketed internationally (e.g. fruits, firewood, fodder, medicinal uses). For products such as fruit, as much as 90% of production is consumed domestically. In addition, the positive ecosystem services represented by trees (e.g. carbon sequestration, nutrient cycling, provision of shade, etc.) are not counted. The correlation established between the increase in plant diversity and increased incomes in agroforestry systems (Salazar-Díaz and Tixier, 2017) is a positive indicator of the real benefits of agroforestry. Harnessing the full potential of marketable products will be a great further source of revenue generation, and a basis for developing new enterprises and job creation for improved adaptation to climate change.

5 Agroforestry and the bioenergy sector in Africa

5.1 Advancing the clean energy agenda

Most adaptation measures require enough quality energy to produce, transform, and consume or trade food products. Africa has a lot of natural resources and productive lands, but it requires much greater efforts to develop and scale-up clean energy to boost the processing of natural products to help create jobs and reduce the vulnerability of communities. Bioenergy will help address basic needs, but most importantly it will increase income from the products of conservation and transformation.

The energy gap in Africa is much higher in rural areas (less than 25% of rural households) as compared to urban (as high as 70% or urban households), as revealed in the Africa Sustainable Development Report of 2018 (African Union, 2018). Throughout rural Africa, the demand for domestic energy is placing increasing pressure on standing forests and woodlands for the harvest of fuelwood and charcoal. Much of the forest degradation from fuelwood culling is in sparse woodlands and other low biomass dryland systems, as well as in the forest blocks remaining in protected areas. Although community-based forest management has had some success in reducing the pressure on protected forests and woodlands in some areas, efforts to enhance conservation by increasing the stocks of woody biomass on agricultural lands, for the purpose of producing sustainable energy sources, have not been widespread.

There are many practical explanations for the lack of widespread adoption of agroforestry in community projects outside of protected or managed forest areas. These include a lack of labor, skills, and available land. The delay in producing bioenergy from new woody perennial projects diminishes the incentive for harvesting wood from the existing stocks in the forest estate. Furthermore, the delayed benefits of woody perennial planting for energy on cultivated land have resulted from a lack of awareness and attention to the unique solutions that are now available for the adoption of energy-focused agroforestry that actually increases food crop yields (see Section 5.2).

The challenge is to develop small-scale biomass production for domestic energy production, while simultaneously keeping it compatible with other land and resource requirements of rural households. New initiatives in many countries for Forest Landscape Restoration (FLR), for example Malawi's National Strategy (Government of Malawi, 2017), have helped refocus strategic options for increasing TOF, but have not been specifically focused on energy *per se*. The issue is how to allocate local land use to bioenergy production and conservation, while also increasing the quantity and value of agricultural outputs that can increase financial returns to households from food markets, all of which are key aspects of resilience.

Most policies and project-oriented energy interventions focus on the demand side, aimed to reduce the pressure on intact forests and woodlands. These include improved and more efficient cookstove technologies and similar efforts. But options do exist that emphasize development of the wood supply side through forest management and agroforestry. There are three related focal points for increasing the development of supply-side landscape-scale energy-based forest management and agroforestry: (1) increasing biomass from trees as solid fuels for energy: investigating new models for sustainable harvest of forest woodlands, perhaps on prescribed rotations, coupled with increased development of community woodlots outside the forest areas on farms or communal land; (2) increasing the utilization of natural products from trees as liquid fuels for energy, from agroforestry energy systems with oil-producing nuts and fruits; and (3) the development of evergreen energy biomass production by intercropping leguminous shrubs in croplands for wood fuel production while dramatically increasing crop productivity in the same fields.

In the first case, the strategy is to use both improved sustainable forest management coupled with agroforestry on agricultural land for the harvesting of biomass. Harvested wood comes from intact forests and woodlots outside of forests. For example, in the case of charcoal energy, the focus is on sustainable charcoal production through innovations in sustainable forest harvest and management. In the second case, the strategy is to use products, such as oil-bearing nuts and fruits, from agroforestry on agricultural land to produce liquid fuels at the community scale.

5.2 Feasibility and benefits of biomass energy from agroforestry

With investments in the supply of tree-based feedstocks (e.g., biomass and oil-bearing fruits or nuts), complementary investments in community-based energy production systems are needed. On the one hand, this means innovations in fuelwood and charcoal production systems, such as improved kiln efficiencies (the World Agroforestry Center estimates a doubling of efficiency is possible; Neufeldt et al., 2015a,b). On the other hand, this means more innovation in small-scale systems for bio-based liquid fuel and for power generation from biomass. While much of the development focus is currently being given to investments on the demand side (e.g. improved cook stoves), the supply side innovations and investments have been neglected.

Supply-side measures call for new models of community-based biomass production and processing for electrical energy. The outputs can be electricity, liquid biofuels, and biochar. Biomass power generation must meet three requirements to be sustainable: it must be economically viable, environmentally sound, and beneficial to society. Because agroforestry integrates directly into the farming system in most parts of Africa, it is possible to develop technologies that both sequester carbon in soils and landscapes from these systems and produce biomass for electrical power.

The IPCC model pathways to limit global warming are projecting the conversion of up to 500 M ha of agricultural land (currently producing food and feed grains), and up to 800 M ha of pastureland, into bioenergy crops, while also increasing the global forest area by up to 1 billion ha. Given the large land area requirement for bioenergy crop production, there are concerns about how food security and other ecosystem services will be maintained (Fuss et al., 2018; Minx et al., 2018). The IPCC 1.5 C Report does take note that such an enormous conversion of land use, particularly from annual crop production to bioenergy production with perennials, poses profound challenges for food production. In fact, the conversion of vast amounts of agricultural land to the production of annual or perennial bioenergy crops is not feasible. This would endanger the whole ability of the food production system to keep up with increases in population growth. And, as noted previously, food production is already under serious threat of decline in the face of current climate change.

However, agroforestry offers potential solutions by which we can vastly increase carbon capture that would not require the conversion of agricultural land and would not compete with food crops for land. As noted previously, nearly half of the agricultural land globally has already greater than 10% tree cover, and it is accumulating an additional 0.74 billion tons of CO_2 annually (Lewis et al., 2019). The system that best provides smallholder farmers with a portfolio approach to income and increased crop and livestock production involves the integrated cultivation of food crops with shrubs and trees.

There are also ways of deploying carbon capture and storage on farmlands by incorporating leguminous shrubs into crop fields to increase crop productivity while simultaneously producing several tons of woody biomass per hectare per year.

Millions of farmers are establishing leguminous shrubs and trees in their crop fields in many countries throughout the tropics. They harvest the foliage to fertilize their crops, provide fodder for their livestock, and they harvest the fuelwood for household energy or for sale. These shrubs and trees improve vegetative soil cover year-round, and they increase soil moisture, soil fertility, and crop yields. Trees also buffer crop production from the increased drought stress and higher temperatures that are now occurring due to climate change. And they store vast additional quantities of carbon in the soil, while enhancing biodiversity.

These systems of tree regeneration in farming lands are currently being massively scaled-up in many countries in semiarid Africa, particularly in eastern and southern Africa.

Leguminous shrubs are planted in maize fields by small-scale farmers in Zambia and Malawi. They are cut down to the ground before the crops are sown, providing rich biofertilizer to the crops and high-protein livestock fodder. The shrubs gradually begin to regrow by the time the maize is harvested (as seen here). They continue to grow during the dry season, and they produce their woody biomass while there is no crop growing in the field. Photo: D. Garrity.

Farmers can establish leguminous shrubs at densities of 1000–5000 plants per ha throughout their crop fields and grazing lands (Garrity et al., 2010). The N-rich foliage increases crop yields and livestock production (Sileshi et al., 2008; Akinnifesi et al., 2010). These systems accumulate carbon above and belowground in the range of 2–4 tC/ha/yr (Makumba et al., 2007; Kaonga and Bayliss-Smith, 2009). These techniques have been pioneered by World Agroforestry and its partners for the last three decades.

The shrub wood can be harvested at the end of the dry season for sale to local producers of electrical power, thus providing carbon emission reductions with major co-benefits for poor rural populations of increased incomes and job creation. These agroforestry techniques are already supplying vast quantities of wood fuel from crop fields to power electricity generation facilities in Sri Lanka (EverGreen Agriculture Partnership, 2017). Such systems are being developed in Africa, where there is potential for application on hundreds of millions of hectares (IRENA, 2019). These investments will also assist smallholder food crop agriculture to become more resilient to adverse climate change by reducing yield losses due to drought (Syampungani et al., 2010; Neufeldt et al., 2009).

Leguminous shrubs are increasingly the basis for sustainably boosting maize production in countries such as Malawi, Zambia, and Kenya. Their cultivation in farm fields dramatically increases crop production, providing smallholders with greater food security, income, and better nutrition. In Eastern Zambia, for instance, through a major extension program implemented by COMACO, over 80 000 smallholder farm families have recently planted a total of 41 million Gliricidia shrubs, on over 27 000 ha of land (IRENA, 2019).

These shrub-based food-energy systems are already providing substantial electrical power generation in Sri Lanka (see Box 2).

Box 2 The example of Gliricidia shrubs

Across Africa and South Asia, millions of farm families are already familiar with woody shrubs such as Gliricidia, and they have already incorporated them into their farming systems to increase their food production. The potential for selling the surplus wood to local producers of electrical power would provide them with an incentive to expand the cultivation of these valued agroforestry systems. This is exactly what is being done in Sri Lanka.

Gliricidia shrubs are the fourth most important tree crop in Sri Lanka. This source of bioenergy is already providing substantial electrical power generation in the country. There are now eight tree-based power plants built and operated by private sector companies in operation in the country, and ten or more plants are under construction or in the feasibility stage. These electrical power generation systems are based on the cultivation of these nitrogen-fixing shrubs. The shrubs capture nitrogen for use as fertilizer by smallholder farmers in their crop fields, while they sell the wood produced to the power plants. The scale of power generation ranges from small-scale electrical production for local use (producing 4–9 KW), to large plants capable of producing 5–10 MW of electricity. Gliricidia is an excellent fuelwood, producing 19.8 MJ/kg and it burns slowly with little smoke or sparks (IRENA, 2019).

The production of biomass energy involves and engages the local community. Instead of importing expensive diesel fuel from urban centers, or high-tech solar or wind equipment from foreign countries, the bioenergy needed to fuel a power plant comes from the smallholder farmers in the community themselves. The wood is produced by smallholder farmers on their farm fields, by the shrubs growing alongside their maize and other crops.

The offsetting gains in agroforestry carbon stocks can be further accelerated, because they are low cost, and they have high income, livelihoods, and resilience benefits for the farming populations, particularly in the tropics. They can ensure a continuous supply of power that can be increased or decreased when needed, avoiding involuntary power fluctuations and peaks in demand. And beyond addressing the challenge of electrification for rural and remote cellular base stations, these systems also significantly contribute to local communities in youth education, health, food conservation, communications and so forth, that are not evident with traditional and uncertain gridded systems.

It is therefore important to promote assertive goals for increasing agroforestry in the development agenda for Africa, and increase the pace of overall shrub and tree cover on agricultural land.

On the 200 million ha of croplands in Africa alone, the promotion of this solution on a much more expansive scale could also address the global community's goal of increasing carbon capture on a massive scale, in order to mitigate against rising global temperatures in the fight against climate change, while also achieving the United Nations Sustainable Development Goals.

Conventional conservation farming systems tend to sequester a maximum of 0.1-0.2 t C $ha^{-1}yr^{-1}$. Evergreen Agriculture systems with trees incorporated in croplands accumulate carbon both above and belowground in the range of 2-4 t C $ha^{-1}yr^{-1}$, roughly an order of magnitude higher than with conservation farming alone. This is particularly true for systems incorporating fertilizer trees such as *Faidherbia albida*, or shrubs such as *Gliricidia sepium* (Makumba et al., 2007; Kaonga and Bayliss-Smith, 2009).

There are 1.4 billion ha of cropland in the world (FAO). If evergreen agriculture with leguminous shrub systems were implemented on 30% of global croplands (420 million ha), then the annual carbon capture in the soil per year would be about 3.1 billion tons CO_2 stored per year. This figure does not account for the amount of emissions saved if the wood were used to generate electrical power that would have been provided by the burning of fossil fuels. Added to that is the major boost in crop production that would occur by incorporating these biological fertilizer factories in the fields, and the reduced emissions by substituting this organic source of nitrogen for chemical fertilizers that are produced with large quantities of fossil fuel. The food security, land restoration, income generation, and job creation benefit for the rural poor would be enormous.

The deployment of this agroforestry-based energy solution, known as evergreen energy, can be done in phases.

The first phase is the massive scaling-up of fertilizer shrubs and tree systems in croplands across many countries in the tropics, and using the N-rich foliage for crop and livestock production, while using the wood for household fuelwood, local sale, charcoal-making, biochar, and briquette-making. The carbon benefits here are in the buildup of soil organic carbon stocks, and the deflection of local deforestation because wood fuel and timber needs are generated on the farm.

The second phase is the development of electrical power generation with rural power stations to electrify rural communities. This will induce the expansion of shrub cultivation for wood fuel at a bigger scale, which is the phase in which Sri Lanka is now.

Bioenergy is the most common renewable energy source used today and the one with a large future potential. But we need to consider the various strategic objectives for promoting bioenergy and their trade-offs with other ecosystem services: (1) Agroforestry-based bioenergy is a low-carbon development strategy, where the main objective is climate change mitigation with associated costs and benefits; (2) it is a catalytic co-benefit while increasing food production and food security and enhancing climate-smart agriculture strategies; and (3) it is used to produce, process, and distribute food more effectively as a key adaptation strategy in developing countries.

6 Conclusion

Degradation of productive lands, especially the loss of soil fertility, in combination with climate-driven shifts of agricultural zones, leads to a substantial decrease of agricultural area and it reduces the capacity of smallholders to cope with their changing environments. Policy efforts to regenerate degraded areas tend to support simple economically profitable land use systems (e.g. tree plantations), which often increase the depletion of natural resources (including biodiversity) instead of conserving or enhancing them. Nevertheless, the adoption of more sustainable tree-based agricultural systems is often not effectively promoted by public policies. There has been considerable emphasis paid to input subsidy programs and campaigns under the umbrella of climate-smart agriculture. These subsidy programs for inorganic fertilizers have been focused on increasing resilience in annual cropping systems. They have often, intentionally or inadvertently, de-emphasized options that include the establishment of agroforestry systems, even those that provide food products or increase crop production. This neglect needs to be vigorously addressed as soon as possible.

Trees help fight climate change by storing carbon (mitigation). Carbon sequestered by trees and stored in aboveground biomass and in the soil contributes to reducing GHG concentrations in the atmosphere. There is an urgent need to more comprehensively estimate the carbon sequestration potential of agroforestry systems in various conditions and land management systems.

Trees buffer against weather-related production losses, enhancing resilience against climate impacts and build more fertile soils. They provide additional income and diversity of food sources through tree-based products, while supporting the growing bioenergy needs of poor households. The multipurpose agroforestry concept should be revisited in order to better understand the variety of benefits these systems can afford to vulnerable populations.

Trees on farms enhance adaptation to climate change by reducing vulnerability to climate impacts. A key strategy used by climate-vulnerable farmers and herders to increase food security is the integration of trees in their farms and rangelands. This reduces their dependency on a single staple crop, and it improves fodder availability for their animals, particularly in the dry season. Trees can buffer the effects of weather extremes such as droughts or heavy rain and can minimize their impacts. Trees can stabilize the soil against landslides and erosion, and they improve water infiltration rates by improving soil properties. Recent studies have showed that trees can improve the crop microclimate and improve rainfall via rainbow water. And they can increase groundwater recharge (particularly important for drylands). Annual crops have improved productivity in agroforestry systems, due to greater availability of water and soil nutrients.

The agroforestry bioenergy approach discussed previously shows how to combine climate change goals with food security goals with major benefits to farm families, to local communities, and to the environment. There is a synergy between these two entry points. Unfortunately, the dominant rhetoric in bioenergy has been more focused on the trade-offs side. When it comes to land-based mitigation, agroforestry systems and agroforestry-based bioenergy should be predominantly seen as a trade-in between mitigation and food security, not a trade-off: A trade-in that can be achieved without land competition.

The development of agroforestry is one of the most cost-effective and technically sound options to realize climate mitigation and adaptation with major benefits for local livelihoods and food security. Consequently, climate change mitigation and adaptation plans based on meeting the Paris Agreement temperature targets, and implemented through the NDCs, could be achieved by relying very highly on agroforestry solutions.

7 Where to look for further information

Important information that shows how many smallholder farmers use agroforestry practices to face their lack of the minimum asset endowments (physical, human and financial) while sustainably producing food under severe climate change challenges can be found at:

- https://www.sciencedirect.com/journal/current-opinion-in-environmental-sustainability/vol/6.
- https://link.springer.com/article/10.1007/s10457-009-9229-7.

The carbon potential of agroforestry has been extensively studied by Prof P. K. Nair and colleagues. Trees combined with other land use systems holds immense potential for various land-based mitigation options (Nair), bioenergy supply (Garrity) and carbon sequestration and contribute both to the global carbon budget and buffering rising temperatures and droughts.

- https://www.springer.com/gp/book/9789400746756
- http://library.uniteddiversity.coop/Permaculture/Agroforestry/Carbon_Sequestration_Potential_of_Agroforestry_Systems-Opportunities_and_Challenges.pdf

Evidences of land recovery in Africa and the utility of perennial crop yield opens new avenues to explore ways accelerate the deep and rapid transformation needed to scale up and support agroforestry to face climate change (By Dr Eric Toensmeir).

- https://www.chelseagreen.com/product/the-carbon-farming-solution/

Various environmental challenges and people's wellbeing including food production and quality nutrition are met with agroforestry.

- https://link.springer.com/article/10.1007/s10457-009-9229-7

In the emerging shift towards agroecology (Sinclair, GCA), there are establish systems that are highly resilient. One such model is the 'collection' of approach that covers agro-biodiversity, livelihoods, food and nutrition.

- https://cdn.gca.org/assets/2019-09/TheContributionsOfAgroecologicalApproaches.pdf

8 References

African Union (2018). *Africa Sustainable Development Report: Towards a Transformed and Resilient Continent*. Addis Ababa: Economic Commission for Africa and African Development Bank and United Nations Development Programme.

AGRA (2017). *Africa Agriculture Status Report: The Business of Smallholder Agriculture in sub-Saharan Africa*. Issue No. 5. Nairobi, Kenya: Alliance for a Green Revolution in Africa (AGRA).

Akinnifesi, F. K., Ajayi, O. C., Sileshi, G., Chirwa, P. W. and Chianu J. (2010). Fertilizer tree systems for sustainable food security in the maize-based production systems of East and Southern Africa Region: A review. *Journal of Agronomy for Sustainable Development*. DOI:10.1051/agron/2009058.

Albrecht, A. and Kandji, S. T. (2003). Carbon sequestration in tropical agroforestry systems. *Agriculture, Ecosystems and Environment* 99(1): 15-27.

Atta-Krah, K., Kindt, R., Skilton, J. N. and Amaral, W. (2004). Managing biological and genetic diversity in tropical agroforestry. *Agroforestry Systems* 61-62: 183-94.

Barral, M. P., Benayas, J. M. R., Meli, P. and Maceira, N. O. (2015). Quantifying the impacts of ecological restoration on biodiversity and ecosystem services in agroecosystems: A global meta-analysis. *Agriculture, Ecosystems and Environment* 202: 223-31.

Brandt, R., Lachmuth, S., Landschulz, C., Haß, F. and Hensen, I. (2014). Species-specific responses to environmental stress on germination and juvenile growth of two Bolivian Andean agroforestry species. *New Forests* 45(6), 777-795.

Brandt, M., Mbow, C., Diouf, A. A., Verger, A., Samimi, C. and Fensholt, R. (2015). Ground- and satellite-based evidence of the biophysical mechanisms behind the greening Sahel. *Global Change Biology* 21(4). DOI:10.1111/gcb.12807.

Brandt, M., Rasmussen, K., Peñuelas, J., Tian, F., Schurgers, G., Verger, A., Mertz, O., Palmer, J. R. B. and Fensholt, R. (2017). Human population growth offsets climate-driven increase in woody vegetation in sub-Saharan Africa. *Nature Ecology and Evolution* 1, 81. DOI:10.1038/s41559-017-0081.

Brandt, M., Rasmussen, K., Hiernaux, P., Herrmann, S., Tucker, C. J., Tong, X. and Fensholt, R. (2018a). Reduction of tree cover in West African woodlands and promotion in semi-arid farmlands. *Nature Geoscience*. DOI:10.1038/s41561-018-0092-x.

Brandt, M., Wigneron, J.-P., Chave, J., Tagesson, T., Penuelas, J., Ciais, P. and Fensholt, R. (2018b). Satellite passive microwaves reveal recent climate-induced carbon losses in African drylands. *Nature Ecology and Evolution*. DOI:10.1038/s41559-018-0530-6.

Brandt, M., Hiernaux, P., Rasmussen, K., Tucker, C. J., Wigneron, J.-P., Diouf, A. A. and Fensholt, R. (2019). Changes in rainfall distribution promote woody foliage production in the Sahel. *Communications Biology* 2(1): 133. DOI:10.1038/s42003-019-0383-9.

Brussaard, L., Caron, P., Campbell, B., Lipper, L., Mainka, S., Rabbinge, R., Babin, D. and Pulleman, M. (2010). Reconciling biodiversity conservation and food security: Scientific challenges for a new agriculture. *Current Opinion in Environmental Sustainability* 2, 34-42.

Carsan, S., Stroebel, A., Dawson, I., Kindt, R., Mbow, C., Mowo, J. and Jamnadass, R. (2014). Can agroforestry option values improve the functioning of drivers of agricultural intensification in Africa? *Current Opinion in Environmental Sustainability* 6, 35-40. DOI:10.1016/j.cosust.2013.10.007.

Dardel, C., Kergoat, L., Hiernaux, P., Mougin, E., Grippa, M. and Tucker, C. J. (2014). Re-greening Sahel: 30 years of remote sensing data and field observations (Mali, Niger). *Remote Sensing of Environment* 140: 350-64. DOI:10.1016/j.rse.2013.09.011.

Diatta, A. A., Ndour, N., Manga, A., Sambou, B., Faye, C. S., Diatta, L., Mbow, C. and Dieng, S. D. (2016). Services ecosystémiques du parc agroforestier à Cordyla pinnata (Lepr.

Ex A. Rich.). Milne-Redh. dans le Sud du Bassin Arachidier (Sénégal). *International Journal of Biological and Chemical Sciences* 10(6): 2511–25.

Dixon, R. K., Winjum, J. K., Andrasko, K. J., Lee, J. J. and Schroeder, P. E. (1994). Integrated land-use systems: Assessment of promising agroforest and alternative land-use practices to enhance carbon conservation and sequestration. *Climatic Change* 27: 71. DOI:10.1007/BF01098474.

Dossa, E. L. Fernandes, E. C. M., Reid, W. and Guillaume Ezui, K. S. G. (2008). Above- and belowground biomass, nutrient and carbon stocks contrasting an open-grown and a shaded coffee plantation. *Agroforestry Systems* 72(2): 103–15.

Duguma, B., Gockowski, J. and Bakala, J. (2001). Smallholder Cacao (*Theobroma cacao* Linn.). cultivation in agroforestry systems of West and Central Africa: Challenges and opportunities. *Agroforestry Systems* 51: 177. https://doi.org/10.1023/A:1010747224249.

Egbe, E. and Tabot, P. (2011). Carbon sequestration in eight woody non-timber forest species and their economic potentials in southwestern Cameroon. *Applied Ecology and Environmental Research* 9(4): 369–85.

EverGreen Agriculture Partnership (2017). *Creating Evergreen Food-energy Systems for Rural Electrification in Africa*. Nairobi, Kenya: EverGreen Agriculture Partnership, 4p,

FAO (2019). FAOSTAT Database. Rome, Italy: FAO. Available at: http://faostat3.fao.org/home/E (accessed on 3 June 2019).

FAOSTAT (2011). http://faostat.fao.org.

Fuss, S., Lamb, W. F., Callaghan, M. W., Hilaire, J., Creutzig, F., Amann, T., Beringer, T., de Oliveira Garcia, W., Hartmann, J., Khanna, T., Luderer, G., Nemet, G. F., Rogelj, J., Smith, P., Vicente, J. L. V., Wilcox, J., Dominguez, M. del M. Z. and Minx, J. C. (2018). Negative emissions – Part 2: Costs, potentials and side effects. *Environmental Research Letters* 13(6): 63002. DOI:10.1088/1748-9326/aabf9f.

Garnett, T. and Godfray, C. (2012). Sustainable intensification in agriculture. Navigating a course through competing food system priorities. In: *Food Climate Research Network and the Oxford Martin Programme on the Future of Food*. UK: University of Oxford.

Garrity, D. (2017). Agriculture could achieve emissions neutrality by 2050 – through agroforestry. Nairobi, Kenya: EverGreen Agriculture Partnership, World Agroforestry Centre, 4p.

Garrity, D., Akinifesi, F., Ajayo, O., Sileshi, G. W., Mowo, J. G., Kalinganire, A., Larwanou, M. and Bayala, J. (2010). Evergreen Agriculture: A robust approach to sustainable food security in Africa. *Food Security* 2(3): 197–214.

Geraldo, D., Pedro J. C. and Luis M. N. (2010). Carob-tree as a CO_2 sink in the carbon market. In: *Advances in Climate Changes, Global Warming, Biological Problems and Natural Hazards*. Faro, Portugal: WSEAS Press, pp. 119–23.

Glenday J. (2008). Carbon storage and emissions offset potential in an African dry forest, the Arabuko-Sokoke Forest, Kenya. *Environmental Monitoring and Assessment* 142: 85–95.

Government of Malawi (2017). *National Forest Landscape Restoration Strategy*. Lilongwe: Ministry of Natural Resources, Energy and Mining, p. 44.

Griscom, B. W., Goodman, R. C., Burivalova, Z. and Putz, F. E. (2017a). Carbon and biodiversity impacts of intensive versus extensive tropical forestry. *Conservation Letters* 10, 160–5.

Griscom, B. W., Adams, J., Ellis, P. W., Houghton, R. A., Lomax, G., Miteva, D. A., Schlesinger, W. H., Shoch, D., Siikamäki, J. V., Smith, P., Woodbury, P., Zganjar, C., Blackman, A., Campari, J., Conant, R. T., Delgado, C., Elias, P., Gopalakrishna, T., Hamsik, M. R., Herrero, M., Kiesecker, J., Landis, E., Laestadius, L., Leavitt, S. M., Minnemeyer, S., Polasky, S., Potapov, P., Putz, F. E., Sanderman, J., Silvius, M., Wollenberg, E., Fargione, J. (2017b). Natural climate solutions. *Proceedings of the National Academy of Sciences* 114(44): 11645–50. DOI:10.1073/pnas.1710465114.

Haokip, V. (2016). Biomass, carbon stock and carbon sequestration potential in Acacia nilotica-paddy based traditional agroforestry systems. M.Sc. (Forestry) Thesis. Department of Forestry, College of Agriculture, Indira Gandhi Krishi Vishwavidyalaya Raipur.

Hawken, P. (2017). *Drawdown: The Most Comprehensive Plan Ever Proposed to Reverse Global Warming*. New York: Penguin Random House.

IPCC (2006). *IPCC Guidelines for National Greenhouse Gas Inventories*. Japan: National Greenhouse Gas Inventories Programme, IGES.

Iskandar, J. (2016). Responses to environmental and socio-economic changes in the Karangwangi traditional agroforestry system, South Cianjur, West Java system. *Biodiversitas* 17(1): 332–41. DOI: 10.13057/biodiv/d170145.

Kalaba, F. K., Quinn, C. H. and Dougill, A. J. (2013). The role of forest provisioning ecosystem services in coping with household stresses and shocks in Miombo woodlands, Zambia. *Ecosystem Services* 5: 143–8.

Kaonga, M. and Bayliss-Smith, T. P. (2009). Carbon pools in tree biomass and the soil in improved fallows in eastern Zambia. *Agroforest Systems* 76: 37–51.

Kimaro, A. A., Isaac, M. E. and Chamshama, S. A. O. (2011). Carbon pools in tree biomass and soils under rotational woodlot systems in Eastern Tanzania. In: B. M. Kumar and P. K. R. Nair (Eds), *Carbon Sequestration Potential of Agroforestry Systems: Opportunities and Challenges*. Advances in Agroforestry (vol. 8). Dordrecht: Springer. DOI:10.1007/978-94-007-1630-8_7.

Kleijn, D., Rundlöf, M., Scheper, J., Smith, H. G. and Tscharntke, T. (2011). Does conservation on farmland contribute to halting the biodiversity decline? *Trends in Ecology and Evolution* 26(9): 474–81. DOI:10.1016/j.tree.2011.05.009.

Kumar, B. M. and Nair, P. K. R. (2012). *Carbon Sequestration Potential of Agroforestry Systems: Opportunities and Challenges*. Springer.

Kursten, E. and Burschel, P. (1993). CO_2-mitigation by agroforestry. *Water, Air and Soil Pollution* 70(1): 533–44. https://doi.org/10.1007/BF01105020.

Lal, R. (2005). Soil carbon sequestration in natural and managed tropical ecosystems. *Journal of Sustainable Forestry* 2: 1–30.

Lewis, S. L., Wheeler, C. E., Mitchard, E. T. A. and Koch, A. (2019). Regenerate natural forests to store carbon. *Nature* 568: 25–8.

Lipper, L., Thornton, P., Campbell, B. M., Baedeker, T., Braimoh, A., Bwalya, M., Caron,P., Cattaneo, A., Garrity, D., Henry, K., Hottle, R., Jackson, L., Jarvis, A., Kossam, F., Mann, W., McCarthy, N., Meybeck, A., Neufeldt, H., Remington, T., Thi Sen, P., Sessa, R., Shula, R., Tibu, A. and Torquebiau, E. F. (2014). Climate-smart agriculture for food security. *Nature Climate Change* 4(12), 1068–72. DOI:10.1038/nclimate2437.

Liu, J., Fritz, S., Wesenbeeck, C. F. A. V., Fuchs, M., You, L., Obersteiner, M. and Yang, H. (2008). A spatially explicit assessment of current and future hotspots of hunger in Sub-Saharan Africa in the context of global change. *Global and Planetary Change* 64: 222–35.

MacFadyen, S., Cunningham, S. A., Costamagna, A. C. and Schellhorn, N. A. (2012). Managing ecosystem services and biodiversity conservation in agricultural landscapes: Are the solutions the same? *Journal of Applied Ecology* 49: 690–4.

Makumba, W., Akinnifesi, F., Janssen, B. and Oenema, O. (2007). Long-term impact of a Gliricidia-maize intercropping system on carbon sequestration in southern Malawi. *Agriculture Ecosystem and Environment* 118: 237–43.

Mbow, C., Nielsen, T. T. and Rasmussen, K. (2000). Savanna fires in East-Central Senegal: Distribution patterns, resource management and perceptions. *Human Ecology* 28(4): 561–83, https://doi.org/10.1023/A:1026487730947.

Mbow, C, Mertz, O. Diouf, A., Rasmussen, K., Reenberg, A. (2008). The history of environmental change and adaptation in eastern Saloum-Senegal – driving forces and perceptions. *Global and Planetary Change-Elsevier* 64: 210–221.

Mbow, C., Noordwijk, M. V., Luedeling, E., Neufeldt, H., Minang, P. A. and Kowero, G. (2014a). Agroforestry solutions to address food security and climate change challenges in Africa. *Current Opinion in Environmental Sustainability* 6: 61–7.

Mbow, C., Smith, P., Skole, D., Duguma, L. and Bustamante, M. (2014b). Achieving mitigation and adaptation to climate change through sustainable agroforestry practices in Africa. *Current Opinion in Environmental Sustainability* 6: 8–14. DOI:10.1016/j.cosust.2013.09.002.

Meijer, S. S., Catacutan, D., Ajayi, O. C., Sileshi, G. W. and Nieuwenhuis, M. (2015). The role of knowledge, attitudes and perceptions in the uptake of agricultural and agroforestry innovations among smallholder farmers in sub-Saharan Africa. *International Journal of Agricultural Sustainability* 13(1): 40–54, DOI:10.1080/1473 5903.2014.912493.

Minang, P. A., van Noordwijk, M., Freeman, O. E., Mbow, C., de Leeuw, J. and Catacutan, D. (Eds). (2015). *Climate-Smart Landscapes: Multifunctionality in Practice*. Nairobi: World Agroforestry Centre (ICRAF), pp. 21–35.

Minx, J. C., Lamb, W. F., Callaghan, M. W., Fuss, S., Hilaire, J., Creutzig, F., Amann, T., Beringer, T., Garcia, W. D. O., Hartmann, J., Khanna, T., Lenzi, D., Luderer, G., Nemet, G. F., Rogelj, J., Smith, P., Vicente, J. L. V., Wilcox, J. and del Mar Zamora Dominguez, M. (2018). Negative emissions - Part 1: Research landscape and synthesis. *Environmental Research Letters* 13(6): 63001. DOI:10.1088/1748-9326/aabf9b.

Muthee, K., Mbow, C., Macharia, G. and Leal Filho, W. (2017). Ecosystem-Based Adaptation (EbA) as an adaptation strategy in Burkina Faso and Mali. In: Leal Filho, W., Belay, S., Kalangu, J., Menas, W., Munishi, P. and Musiyiwa, K. (Eds), *Climate Change Adaptation in Africa: Climate Change Management*. Cham: Springer, pp. 205–15. DOI: 10.1007/978-3-319-49520-0_13.

Mutuo, P. K., Cadisch, G. Albrecht, A., Palm, C. A. and Verchot, L. (2005). Potential of agroforestry for carbon sequestration and mitigation of greenhouse gas emissions from soils in the tropics. *Nutrient Cycling in Agroecosystems* 71: 43–54.

Nair, P. K. R. (1993). *An Introduction to Agroforestry*. Kluwer Academic Publishers, 489p.

Nair, P. K. R., Nair, V. D., Gama-Rodrigues, E. F., Garciad, R., Haile, S. G., Howlett, D. S., Kumar, B. M., Mosquera-Losada, M. R., Saha, S. K., Takimoto, A. N. G. and Tonucci, R. G. (2009). Soil carbon in agroforestry systems: An unexplored treasure? *Nature Proceedings*. hdl:10101/npre.2009.4061.1 (Posted 9 December 2009).

Neufeldt, H., Wilkes, A., Zomer, R. J., Xu, J., Nangole, E., Munster, C. and Place, F. (2009). Trees on farms: Tackling the triple challenges of mitigation, adaptation and food security. World Agroforestry Centre Policy Brief 07. Nairobi, Kenya, World Agroforestry Centre.

Neufeldt, H., Dobie, P., Iiyama, M., Njenga, M., Mohan, S. and Neely, C. (2015a). Developing sustainable tree-based bioenergy systems in sub-Saharan Africa. ICRAF Policy Brief No. 28. Nairobi, Kenya: ICRAF.

Neufeldt, H., Langford, K., Fuller, J., Iiyama, M. and Dobie, P. (2015b). From transition fuel to viable energy source: Improving sustainability in the sub-Saharan charcoal sector. ICRAF Working Paper No. 196. Nairobi: World Agroforestry Centre. DOI:10.5716/WP15011.PDF.

Ranaivoson, T., Brinkmann, K., Rakouth, B., Buerkert, A. (2015). Distribution, biomass and local importance of tamarind trees in south-western Madagascar. *Global Ecology and Conservation* 4: 14–25.

Rogelj, J. D., Shindell, K., Jiang, S., Fifita, P., Forster, V., Ginzburg, C., Handa, C., Kheshgi, H., Kobayashi, S., Kriegler, E., Mundaca, L., Séférian, R., Virginia Vilariño, M., Calvin, K., Edelenbosch, O., Emmerling, J., Fuss, S., Gasser, T., Gillet, N., He, C., Hertwich, E., Höglund Isaksson, L., Huppmann, D., Luderer, G., Markandya, A., McCollum, D., Millar, R., Meinshausen, M., Popp, A., Pereira, J., Purohit, P., Riahi, K., Ribes, A., Saunders, H., Schadel, C., Smith, C., Smith, P., Trutnevyte, E., Xiu, Y., Zickfeld, K. and Zhou, W. (2018). Mitigation pathways compatible with 1.5°C in the context of sustainable development. In: Masson-Delmotte et al. (Eds), *Global Warming of 1.5°C. An IPCC Special Report on the Impacts of Global Warming of 1.5°C Above Pre-industrial Levels and Related Global Greenhouse Gas Emission Pathways, in the Context of Strengthening the Global Response to the Threat of Climate Change, Sustainable Development, and Efforts to Eradicate Poverty*. Intergovernmental Panel on Climate Change.

Roshetko, J. M., Lasco, R. D. and Angeles, M. S. D. (2007). Smallholder agroforestry systems for carbon storage. *Mitigation and Adaptation Strategies for Global Change* 12(2): 219–42. https://doi.org/10.1007/s11027-005-9010-9.

Ross-Cuéllar, J., Bolland, L. P. and Moheno, M. B. (2019). Can multifunctional landscapes become effective conservation strategies? Challenges and opportunities from a Mexican case study. *Land* 8: 6. doi:10.3390/land8010006.

Salazar-Díaz, R. and Tixier, P. (2017). Effect of plant diversity on income generated by agroforestry systems in Talamanca, Costa Rica. *Agroforestry Systems* 93(2): 571–80.

Scherr, S. and Mcneely, J. (2008). Biodiversity conservation and agricultural sustainability: Towards a new paradigm of 'ecoagriculture' landscapes. *Philosophical Transactions of the Royal Society B Biological Sciences* 363: 477–94.

Sileshi, G., Akinnifesi, F. K., Ajayi, O. C. and Place, F, (2008). Meta-analysis of maize yield response to woody and herbaceous legumes in the sub-Saharan Africa. *Plant Soil* 307: 1–19.

Simelton, E. and Ostwald, M. (2019). *Multifunctional Land Uses in Africa (Open Access): Sustainable Food Security Solutions*. Routledge: Earthscan Food and Agriculture, 176p.

Sitch, S., Friedlingstein, P., Gruber, N., Jones, S. D., Murray-Tortarolo, G., Ahlström, A., Doney, S. C., Graven, H., Heinze, C., Huntingford, C., Levis, S., Levy, P. E., Lomas, M., Poulter, B., Viovy, N., Zaehle, S., Zeng, N., Arneth, A., Bonan, G., Bopp, L., Canadell, J. G., Chevallier, F., Ciais, P., Ellis, R., Gloor, M., Peylin, P., Piao, S. L., Le Quéré, C., Smith, B., Zhu, Z. and Myneni, R. (2015). Recent trends and drivers of regional sources and sinks of carbon dioxide. *Biogeosciences* 12: 653–79, https://doi.org/10.5194/bg-12-653-2015.

Smith P., Bustamante, M., Ahammad, H., Clark, H., Dong, H., Elsiddig, E. A., Haberl, H., Harper, R., House, J., Jafari, M., Masera, O., Mbow, C., Ravindranath, N. H., Rice, C. W., et al. (2014). Agriculture, Forestry and Other Land Use (AFOLU). In O. Edenhofer, R. Pichs-Madruga, Y. Sokona, E. Farahani, S. Kadner, K. Seyboth, A. Adler, I. Baum, S. Brunner, P. Eickemeier, B. Kriemann, J. Savolainen, S. Schlömer, C. von Stechow, T. Zwickel and J. C. Minx (Eds), *Climate Change 2014: Mitigation of Climate Change. Contribution of Working Group III to the Fifth Assessment Report of the Intergovernmental Panel on Climate Change*. Cambridge, UK and New York, NY: Cambridge University Press.

Somarriba, E., Cerda, R., Orozco, L., Cifuentes, M., Dávila, H., Espin, T., Mavisoy, H., Ávila, G., Alvarado, E., Poveda, V., Astorga, C., Say, E. and Deheuvels, O. (2013). Carbon stocks and cocoa yields in agroforestry systems of Central America. *Agriculture, Ecosystems and Environment* 173: 46–57.

Srinivasarao, C., Lal, R., Kundu, S. and Thakur, B. P. (2015). Conservation agriculture and soil carbon sequestration. In M. Farooq and M. K. H. Siddique (Eds), *Conservation Agriculture*. Cham: Springer International Publishing, pp. 479-524.

Stahl, L., Nyberg, G., Hogberg, P. and Buresh, R. J. (2002). Effects of planted tree fallows on soil nitrogen dynamics, above-ground and root biomass, N_2-fixation and subsequent maize crop productivity in Kenya. *Plant and Soil* 243(1): 103–17.

Sundarapandian, S. M., Amritha, S., Gowsalya, L., Kayathri, P., Thamizharasi, M., Dar, J. A., Srinivas, K. and Gandhi, S. D. (2013). Estimation of biomass and carbon stock of woody plants in different land-uses. *Forest Research* 3(1): 1-5.

Syampungani, S., Chirwa, P. W., Akinnifesi, F. K. and Ajayi, O. C. (2010). The potential of using agroforestry as a win-win solution to climate change mitigation and adaptation and meeting food security challenges in Southern Africa. *Agricultural Journal* 5: 80–8.

Takimoto, A., Nair, P. K. R. and Alavalapati, J. R. R. (2008a). Socioeconomic potential of carbon sequestration through agroforestry in the West African Sahel. *Mitigation and Adaptation Strategies for Global Change* 13(7): 745–61.

Takimoto, A., Nair, P. K. R. and Nair, V. D. (2008b). Carbon stock and sequestration potential of traditional and improved agroforestry systems in the West African Sahel. *Agriculture Ecosystems and Environment* (125): 159–66. https://doi.org/10.1016/j.agee.2007.12.010.

Temgoua, L. F., Momo S. M. C., Palou, M. O. and Gonne, T. B. (2018). Floristic structure and carbon sequestration potential of Acacia Senegal (L.). wild. (Fabaceae). improved fallows in far north region of Cameroon. *European Journal of Agriculture and Forestry Research* 6(2): 29–41.

Toensmeier, E. (2016). *The Carbon Farming solution: A Global Toolkit of Perennial Crops and Regenerative Agriculture Practices for Climate Change Mitigation and Food Security*. White River Junction, VT: Chelsea Green Publishing.

Toensmeier, E. (2017). Perennial staple crops and agroforestry for climate change mitigation. In: F. Montagnini (Ed.), *Integrating Landscapes: Agroforestry for Biodiversity Conservation and Food Sovereignty*. Advances in Agroforestry (vol. 12), pp. 439–51.

Torquebiau, E. (2000). A renewed perspective on agroforestry concepts and classification. *Comptes rendus de l'Académie des Sciences/Life Sciences* 323: 1009–17.

Tschakert, P., Khouma, M. and Sene, M. (2004). Biophysical potential for soil carbon sequestration in agricultural systems of the Old Peanut Basin of Senegal. *Journal of Arid Environments* 59(3): 511–33.

Tscharntke, T., Tylianakis, J. M., Rand, T. A., Didham, R. K., Fahrig, L., Batáry, P., Bengtsson, J., Clough, Y., Crist, T. O., Dormann, C. F., Ewers, R. M., Fründ, J., Holt, R. D., Holzschuh, A., Klein, A. M., Kleijn, D., Kremen, C., Landis, D. A., Laurance, W., Lindenmayer, D., Scherber, C., Sodhi, N., Steffan-Dewenter, I., Thies, C., van der Putten, W. H. and Westphal, C. (2012). Landscape moderation of biodiversity patterns and processes: Eight hypotheses. *Biological Reviews* 87: 661–85.

Vallejo-Ramos, M., Moreno-Calles, A. I. and Casas, A. (2016). TEK and biodiversity management in agroforestry systems of different socioecological contexts of the Tehuacán Valley. *Journal of Ethnobiology and Ethnomedicine* 12: 31. DOI:10.1186/s13002-016-0102-2.

Walker, S. M. and Desanker, P. V. (2004). The impact of land use on soil carbon in Miombo Woodlands of Malawi. *Forest Ecology and Management* 203: 91–30, 345–60.

Williams, M., Ryan, C. M., Rees, R. M., Sambane, E., Fernando, J. and Grace, J. (2008). Carbon sequestration and biodiversity of re-growing Miombo woodlands in Mozambique. *Forest Ecology Management* 254: 145–55.

Woomer, P. L., Tieszen, L. L., Tappan, G., Touré, A. and Sall, M. (2004a). Land use change and terrestrial carbon stocks in Senegal. *Journal of Arid Environments* 59(3): 625–42.

Zhang, L., Qiao, N., Huang, C. and Wang, S. (2019). Monitoring drought effects on vegetation productivity using satellite solar-induced chlorophyll fluorescence. *Remote Sensing* 11(4): 378. DOI:10.3390/rs11040378.

Zomer, R. J., Neufeldt, H., Xu, J., Ahrends, A., Bossio, D., Trabucco, A. and Wang, M. V. (2016). Global tree cover and biomass carbon on agricultural land: The contribution of agroforestry to global and national carbon budgets. *Science Reports* 6.

Index

www.ingramcontent.com/pod-product-compliance
Ingram Content Group UK Ltd.
Pitfield, Milton Keynes, MK11 3LW, UK
UKHW021825100926
8556IPUK00003B/167

9 781786 763204